W9-CDD-338

SECOND EDITION

Algorithms

Robert Sedgewick

Princeton University

ADDISON-WESLEY PUBLISHING COMPANY

Reading, Massachusetts • Menlo Park, California • New York
Don Mills, Ontario • Wokingham, England • Amsterdam • Bonn • Sydney • Singapore
Tokyo • Madrid • San Juan

Keith Wollman: Sponsoring Editor
Karen Guardino: Managing Editor
Karen Myer: Production Supervisor
Hugh Crawford: Manufacturing Supervisor
Wendy Lewis: Production Coordinator

Linda Sedgewick: Cover Art

This book is in the Addison-Wesley Series in **Computer Science**
Michael A. Harrison: Consulting Editor

The programs and applications presented in this book have been included for their instructional value. They have been tested with care, but are not guaranteed for any particular purpose. The publisher does not offer any warranties or representations, nor does it accept any liabilities with respect to the programs or applications.

Library of Congress Cataloging-in-Publication Data

Sedgewick, Robert, 1946 –
Algorithms / by Robert Sedgewick. — 2d ed.
650 p. 24 cm.
Includes bibliographies and index
ISBN 0-201-06673-4
1. Algorithms. I. Title.
QA76.6.S435 1988
519.4 — dc19 87-36435
CIP

Reproduced by Addison-Wesley from camera-ready copy supplied by the author.
Reprinted with corrections November, 1988

CDEFGHIJ-DO-898

To Adam, Andrew, Brett, Robbie
and especially Linda

Preface

This book is intended to survey the most important computer algorithms in use today and to teach fundamental techniques to the growing number of people in need of knowing them. It can be used as a textbook for a second, third or fourth course in computer science, after students have acquired some programming skills and familiarity with computer systems, but before they have taken specialized courses in advanced areas of computer science or computer applications. Additionally, the book may be useful for self-study or as a reference for those engaged in the development of computer systems or applications programs, since it contains a number of implementations of useful algorithms and detailed information on their performance characteristics. The broad perspective taken in the book makes it an appropriate introduction to the field.

Scope

The book contains forty-five chapters grouped into eight major parts: fundamentals, sorting, searching, string processing, geometric algorithms, graph algorithms, mathematical algorithms and advanced topics. A major goal in developing this book has been to bring together the fundamental methods from these diverse areas, in order to provide access to the best methods known for solving problems by computer. Some of the chapters give introductory treatments of advanced material. It is hoped that the descriptions here can give readers some understanding of the basic properties of fundamental algorithms ranging from priority queues and hashing to simplex and the fast Fourier transform.

One or two previous courses in computer science or equivalent programming experience are recommended for a reader to be able to appreciate the material in this book: one course in programming in a high-level language such as Pascal, and perhaps another course which teaches fundamental concepts of programming systems. This book is thus intended for anyone conversant with a modern programming language and with the basic features of modern computer systems.

Most of the mathematical material supporting the analytic results is self-contained (or labeled as "beyond the scope" of this book), so little specific preparation in mathematics is required for the bulk of the book, though a certain amount of mathematical maturity is definitely helpful. A number of the later chapters deal with algorithms related to more advanced mathematical material—these are intended to place the algorithms in context with other methods throughout the book, not to teach the mathematical material. Thus the discussion of advanced mathematical concepts is brief, general, and descriptive.

Use in the Curriculum

There is a great deal of flexibility in how the material here can be taught. To a large extent, the individual chapters in the book can be read independently of the others, though in some cases algorithms in one chapter make use of methods from a previous chapter. The material can be adapted for use for various courses by selecting perhaps twenty-five or thirty of the forty-five chapters, according to the taste of the instructor and the preparation of the students.

An elementary course on "data structures and algorithms" might omit some of the mathematical algorithms and some of the advanced topics, then emphasize how various data structures are used in the implementations. An intermediate course on "design and analysis of algorithms" might omit some of the more practically oriented sections, then emphasize the identification and study of the ways in which algorithms achieve good asymptotic performance. A course on "software tools" might omit the mathematical and advanced algorithmic material, then emphasize how to integrate the implementations given here into large programs or systems. A course on "algorithms" might take a survey approach and introduce concepts from all these areas.

Some instructors may wish to add supplementary material to the courses described above to reflect their particular orientation. For "data structures and algorithms," extra material on basic data structures could be taught; for "design and analysis of algorithms," more mathematical analysis could be added; and for "software tools," software engineering techniques could be covered in more depth. In this book, attention is paid to all these areas, but the emphasis is on the algorithms themselves.

This book (and preliminary versions) have been used in recent years at Princeton University and Brown University for the second or third course in computer science, which is prerequisite to all later courses. Typically, about half of the students taking the course are computer science majors. Our experience has been that the breadth of coverage of material in this book provides our majors with an introduction to computer science that can later be expanded upon in later courses on analysis of algorithms, systems programming and theoretical computer science, while at the same time providing all the students with a large set of techniques that they can immediately put to good use. Furthermore, we have found that students are coming to college with increasingly better preparation in programming fundamentals, ready to tackle the variety of algorithms in this book.

There are 450 exercises, ten following each chapter, that generally divide into one of two types. Most are intended to test students' understanding of material in the text, and ask students to work through an example or apply concepts described in the text. A few of them, however, involve implementing and putting together some of the algorithms, perhaps running empirical studies to compare algorithms and to learn their properties.

Algorithms of Practical Use

The orientation of the book is toward algorithms likely to be of practical use. The emphasis is on teaching students the tools of their trade to the point that they can confidently implement, run and debug useful algorithms. Full implementations of the methods discussed (in an actual programming language) are included in the text, along with descriptions of the operations of these programs on a consistent set of examples.

Properties of the algorithms and situations in which they might be useful are discussed in detail. Though not emphasized, connections to the analysis of algorithms and theoretical computer science are not ignored. When appropriate, empirical and analytic results are discussed to illustrate why certain algorithms are preferred. When interesting, the relationship of the practical algorithms being discussed to purely theoretical results is described.

The programming language used throughout the book is Pascal. The advantage of Pascal is that it is widely available and widely known; the disadvantage is that it lacks many features needed by sophisticated algorithms. The programs can easily be translated to other modern programming languages, since relatively few Pascal constructs are used. Some of the programs can be simplified by using more advanced language features (some not available in Pascal), but this is true less often than one might think.

A goal of this book is to present the algorithms in as simple and direct a form as possible. The programs are intended to be read not by themselves, but as part of the surrounding text. This style was chosen as an alternative, for example, to having inline comments. The style is consistent whenever possible, so that programs that are similar look similar.

Changes in the Second Edition

This second edition incorporates corrections, updated information, and improvements throughout, including three major changes with significant impact overall.

First, a new section covering introductory material on data structures and the design and analysis of algorithms has been added. This sets the tone for the rest of the book and provides a framework within which more advanced algorithms are treated. Some readers may skip or skim this section; others may learn the basics there.

Second, the analysis of the algorithms is developed more fully and more rigorously than in the first edition. Specific information on performance characteristics of algorithms is encapsulated throughout in "properties," important facts about the algorithms that deserve further study.

Third (and this change is most evident), hundreds of figures have been added. Many algorithms are brought to light on an intuitive level through the visual dimension provided by these figures.

Acknowledgments

Many people gave me helpful feedback on earlier drafts of this book. In particular, students at Princeton and Brown have suffered through preliminary versions of the material in this book over the past several years. Special thanks are due to Trina Avery, Tom Freeman and Janet Incerpi. Janet provided extensive detailed comments and suggestions which helped me fix innumerable technical errors and omissions; Tom checked many of the programs; and Trina's copy-editing (of both editions) helped me make the text clearer and more nearly correct. I would also like to thank the many readers who provided me with detailed comments about the second edition, including Guy Almes, Jay Gischer, Kennedy Lemke, Udi Manber, Dana Richards, John Reif, M. Rosenfeld, Stephen Seidman, and Michael Quinn.

Special thanks are due to Janet Incerpi who initially converted the book into TEX format, added the thousands of changes I made after the "last draft" of the first edition, guided the files through various systems to produce printed pages and even wrote the scan-conversion routine for TEX used to produce draft manuscripts, among many other things. Only after performing many of these tasks myself for the second edition do I truly appreciate Janet's contribution.

Many of the designs in the figures of the second edition are based on joint work with Marc Brown in the "electronic classroom" project at Brown University in 1983. Marc's support and assistance in creating the designs (not to mention the system with which we worked) are gratefully acknowledged.

This second edition owes its existence largely to the support of Keith Wollman at Addison-Wesley, who initially convinced me to proceed. Among many other things, Keith set an impossible schedule which I was somehow able to meet (nearly).

Much of what I've written here I've learned from the teaching and writings of Don Knuth, my advisor at Stanford. Though Don had no direct influence on this work, his presence may be felt in the book, for it was he who put the study of algorithms on a scientific footing that makes a work such as this possible.

I am very thankful for the support of Brown University and INRIA where I did most of the work on the book, and the Institute for Defense Analyses and the Xerox Palo Alto Research Center, where I did some work on the book while visiting. Many parts of the book are dependent on research that has been generously supported by the National Science Foundation and the Office of Naval Research. Finally, I would like to thank Bill Bowen, Aaron Lemonick, and Neil Rudenstine at Princeton University for their support in building an academic environment in which I was able to prepare this second edition, despite numerous other responsibilities.

Robert Sedgewick
Marly-le-Roi, France, February, 1983
Princeton, New Jersey, January, 1988

Contents

Fundamentals

1. Introduction **3**
Algorithms. Outline of Topics.

2. Pascal **7**
Example: Euclid's Algorithm. Types of Data. Input/Output. Concluding Remarks.

3. Elementary Data Structures **15**
Arrays. Linked Lists. Storage Allocation. Pushdown Stacks. Queues. Abstract Data Types.

4. Trees **35**
Glossary. Properties. Representing Binary Trees. Representing Forests. Traversing Trees.

5. Recursion **51**
Recurrences. Divide-and-Conquer. Recursive Tree Traversal. Removing Recursion. Perspective.

6. Analysis of Algorithms **67**
Framework. Classification of Algorithms. Computational Complexity. Average-Case Analysis. Approximate and Asymptotic Results. Basic Recurrences. Perspective.

7. Implementation of Algorithms **81**
Selecting an Algorithm. Empirical Analysis. Program Optimization. Algorithms and Systems.

Sorting Algorithms

8. Elementary Sorting Methods **93**
Rules of the Game. Selection Sort. Insertion Sort. Digression: Bubble Sort. Performance Characteristics of Elementary Sorts. Sorting Files with Large Records. Shellsort. Distribution Counting.

9. Quicksort **115**
The Basic Algorithm. Performance Characteristics of Quicksort. Removing Recursion. Small Subfiles. Median-of-Three Partitioning. Selection.

10. Radix Sorting **133**
Bits. Radix Exchange Sort. Straight Radix Sort. Performance Characteristics of Radix Sorts. A Linear Sort.

11. Priority Queues **145**
Elementary Implementations. Heap Data Structure. Algorithms on Heaps. Heapsort. Indirect Heaps. Advanced Implementations.

12. Mergesort **163**
Merging. Mergesort. List Mergesort. Bottom-Up Mergesort. Performance Characteristics. Optimized Implementations. Recursion Revisited.

13. External Sorting **177**
Sort-Merge. Balanced Multiway Merging. Replacement Selection. Practical Considerations. Polyphase Merging. An Easier Way.

Searching Algorithms

14. Elementary Searching Methods **193**
Sequential Searching. Binary Search. Binary Tree Search. Deletion. Indirect Binary Search Trees.

15. Balanced Trees **215**
Top-Down 2-3-4 Trees. Red-Black Trees. Other Algorithms.

16. Hashing **231**
Hash Functions. Separate Chaining. Linear Probing. Double Hashing. Perspective.

17. Radix Searching **245**
Digital Search Trees. Radix Search Tries. Multiway Radix Searching. Patricia.

18. External Searching **259**
Indexed Sequential Access. B-Trees. Extendible Hashing. Virtual Memory.

String Processing

19. String Searching **277**
A Short History. Brute-Force Algorithm. Knuth-Morris-Pratt Algorithm. Boyer-Moore Algorithm. Rabin-Karp Algorithm. Multiple Searches.

20. Pattern Matching **293**
Describing Patterns. Pattern Matching Machines. Representing the Machine. Simulating the Machine.

21. Parsing **305**
Context-Free Grammars. Top-Down Parsing. Bottom-Up Parsing. Compilers. Compiler-Compilers.

22. File Compression **319**
Run-Length Encoding. Variable-Length Encoding. Building the Huffman Code. Implementation.

23. Cryptology **333**
Rules of the Game. Simple Methods. Encryption/Decryption Machines. Public-Key Cryptosystems.

Geometric Algorithms

24. Elementary Geometric Methods **347**
Points, Lines, and Polygons. Line Segment Intersection. Simple Closed Path. Inclusion in a Polygon. Perspective.

25. Finding the Convex Hull **359**
Rules of the Game. Package-Wrapping. The Graham Scan. Interior Elimination. Performance Issues.

26. Range Searching **373**
Elementary Methods. Grid Method. Two-Dimensional Trees. Multidimensional Range Searching.

27. Geometric Intersection **389**
Horizontal and Vertical Lines. Implementation. General Line Intersection.

28. Closest-Point Problems **401**
Closest-Pair Problem. Voronoi Diagrams.

Graph Algorithms

29. Elementary Graph Algorithms **415**
Glossary. Representation. Depth-First Search. Nonrecursive Depth-First Search. Breadth-First Search. Mazes. Perspective.

30. Connectivity **437**
Connected Components. Biconnectivity. Union-Find Algorithms.

31. Weighted Graphs **451**
Minimum Spanning Tree. Priority-First Search. Kruskal's Method. Shortest Path. Minimum Spanning Tree and Shortest Paths in Dense Graphs. Geometric Problems.

32. Directed Graphs **471**
Depth-First Search. Transitive Closure. All Shortest Paths. Topological Sorting. Strongly Connected Components.

33. Network Flow **485**
The Network Flow Problem. Ford-Fulkerson Method. Network Searching.

34. Matching **495**
Bipartite Graphs. Stable Marriage Problem. Advanced Algorithms.

Mathematical Algorithms

35. Random Numbers **509**
Applications. Linear Congruential Method. Additive Congruential Method. Testing Randomness. Implementation Notes.

36. Arithmetic **521**
Polynomial Arithmetic. Polynomial Evaluation and Interpolation. Polynomial Multiplication. Arithmetic Operations with Large Integers. Matrix Arithmetic.

37. Gaussian Elimination **535**
A Simple Example. Outline of the Method. Variations and Extensions.

38. Curve Fitting **545**
Polynomial Interpolation. Spline Interpolation. Method of Least Squares.

39. Integration **555**
Symbolic Integration. Simple Quadrature Methods. Compound Methods. Adaptive Quadrature.

Advanced Topics

40. Parallel Algorithms **569**
General Approaches. Perfect Shuffles. Systolic Arrays. Perspective.

41. The Fast Fourier Transform **583**
Evaluate, Multiply, Interpolate. Complex Roots of Unity. Evaluation at the Roots of Unity. Interpolation at the Roots of Unity. Implementation.

42. Dynamic Programming **595**
Knapsack Problem. Matrix Chain Product. Optimal Binary Search Trees. Time and Space Requirements.

43. Linear Programming **607**
Linear Programs. Geometric Interpretation. The Simplex Method. Implementation.

44. Exhaustive Search **621**
Exhaustive Search in Graphs. Backtracking. Digression: Permutation Generation. Approximation Algorithms.

45. NP-Complete Problems **633**
Deterministic and Nondeterministic Polynomial-Time Algorithms. NP-Completeness. Cook's Theorem. Some NP-Complete Problems.

Index **643**

Fundamentals

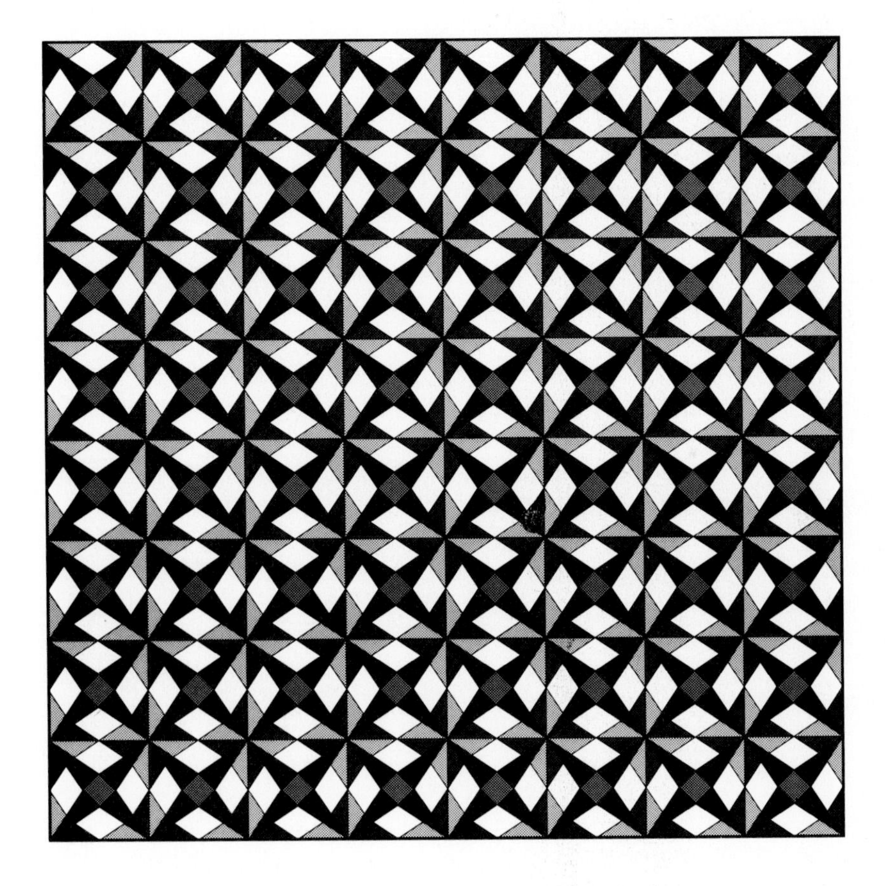

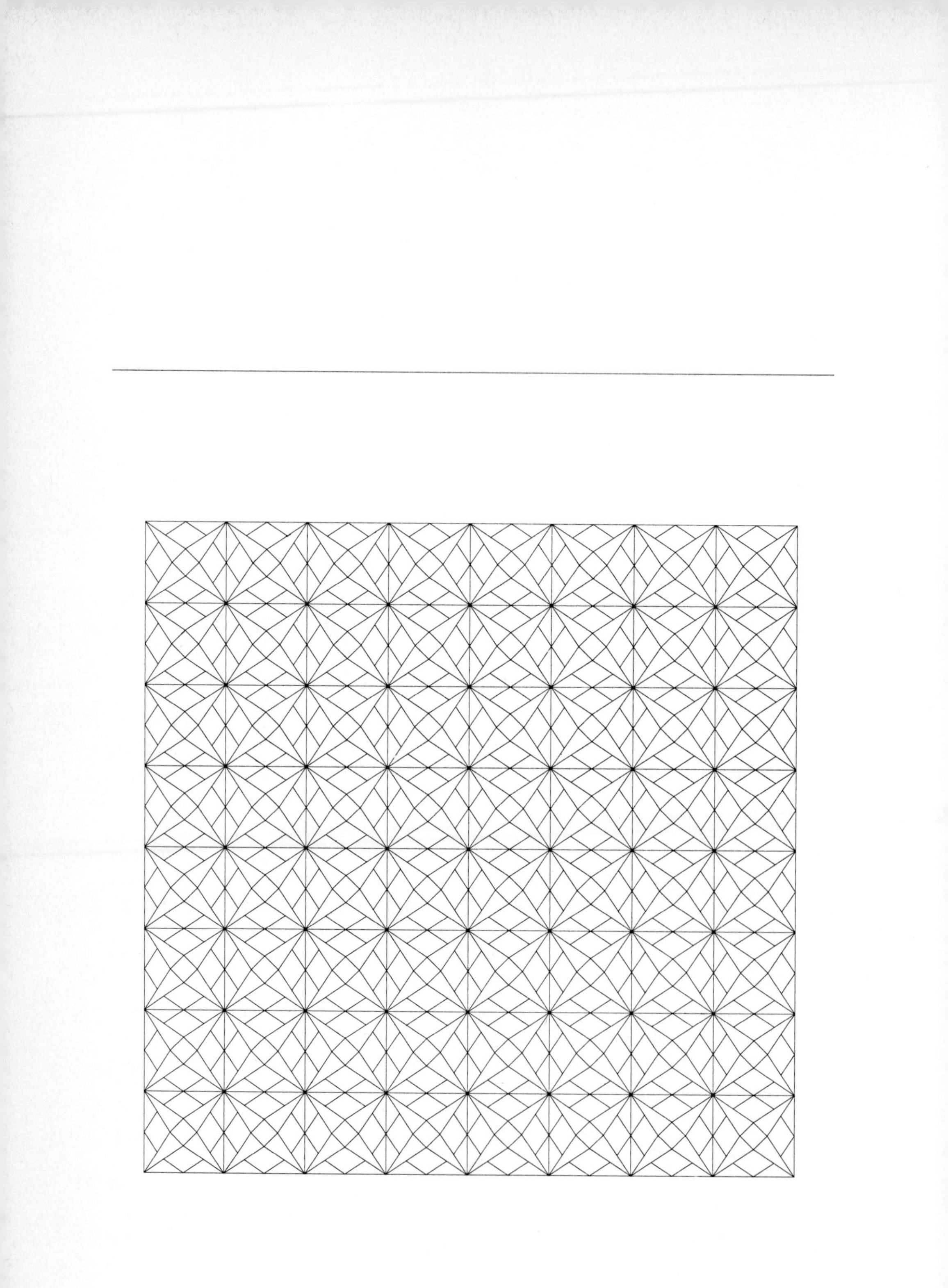

1

Introduction

The objective of this book is to study a broad variety of important and useful *algorithms*: methods for solving problems which are suited for computer implementation. We'll deal with many different areas of application, always trying to concentrate on "fundamental" algorithms that are important to know and interesting to study. Because of the large number of areas and algorithms to be covered, we won't be able to study many of the methods in great depth. However, we will try to spend enough time on each algorithm to understand its essential characteristics and to respect its subtleties. In short, our goal is to learn a large number of the most important algorithms used on computers today, well enough to be able to use and appreciate them.

To learn an algorithm well, one must implement and run it. Accordingly, the recommended strategy for understanding the programs presented in this book is to implement and test them, experiment with variants, and try them out on real problems. We will use the Pascal programming language to discuss and implement most of the algorithms; since, however, we use a relatively small subset of the language, our programs can easily be translated into many other modern programming languages.

Readers of this book are expected to have at least a year's experience in programming in high- and low-level languages. Also, some exposure to elementary algorithms on simple data structures such as arrays, stacks, queues, and trees might be helpful, though this material is reviewed in some detail in Chapters 3 and 4. An elementary acquaintance with machine organization, programming languages, and other basic computer science concepts is also assumed. (We'll review such material briefly when appropriate, but always within the context of solving particular problems.) A few of the applications areas we deal with require knowledge of elementary calculus. We'll also be using some very basic material involving linear algebra, geometry, and discrete mathematics, but previous knowledge of these topics is not necessary.

Algorithms

In writing a computer program, one is generally implementing a method of solving a problem that has been devised previously. This method is often independent of the particular computer to be used: it's likely to be equally appropriate for many computers. In any case, it is the method, not the computer program itself, which must be studied to learn how the problem is being attacked. The term *algorithm* is used in computer science to describe a problem-solving method suitable for implementation as computer programs. Algorithms are the "stuff" of computer science: they are central objects of study in many, if not most, areas of the field.

Most algorithms of interest involve complicated methods of organizing the data involved in the computation. Objects created in this way are called *data structures*, and they also are central objects of study in computer science. Thus algorithms and data structures go hand in hand; in this book we take the view that data structures exist as the byproducts or endproducts of algorithms, and thus need to be studied in order to understand the algorithms. Simple algorithms can give rise to complicated data structures and, conversely, complicated algorithms can use simple data structures. We will study detailed properties of many data structures in this book; indeed, it might well have been called *Algorithms and Data Structures*.

When a very large computer program is to be developed, a great deal of effort must go into understanding and defining the problem to be solved, managing its complexity, and decomposing it into smaller subtasks that can be easily implemented. It is often true that many of the algorithms required after the decomposition are trivial to implement. However, in most cases there are a few algorithms whose choice is critical because most of the system resources will be spent running those algorithms. In this book we will study a variety of fundamental algorithms basic to large programs in many applications areas.

The sharing of programs in computer systems is becoming more widespread, so that while serious computer users will *use* a large fraction of the algorithms in this book, they may need to *implement* only a somewhat smaller fraction of them. However, implementing simple versions of basic algorithms helps us to understand them better and thus use advanced versions more effectively. Also, mechanisms for sharing software on many computer systems often make it difficult to tailor standard programs to perform effectively on specific tasks, so that the opportunity to reimplement basic algorithms frequently arises.

Computer programs are often over-optimized. It may not be worthwhile to take pains to ensure that an implementation is the most efficient possible unless an algorithm is to be used for a very large task or is to be used many times. Otherwise, a careful, relatively simple implementation will suffice: one can have some confidence that it will work, and it is likely to run perhaps five or ten times slower than the best possible version, which means that it may run for an extra few seconds. By contrast, the proper choice of algorithm in the first place can make a difference of a factor of a hundred or a thousand or more, which might translate to minutes, hours, or even more in running time. In this book, we concentrate on the simplest reasonable implementations of the best algorithms.

Often several different algorithms (or implementations) are available to solve the same problem. The choice of the very best algorithm for a particular task can be a very complicated process, often involving sophisticated mathematical analysis. The branch of computer science which studies such questions is called *analysis of algorithms*. Many of the algorithms that we will study have been shown through analysis to have very good performance, while others are simply known to work well through experience. We will not dwell on comparative performance issues: our goal is to learn some reasonable algorithms for important tasks. But one should not use an algorithm without having some idea of what resources it might consume, so we will be aware of how our algorithms might be expected to perform.

Outline of Topics

Below are brief descriptions of the major parts of the book, giving some of the specific topics covered as well as some indication of our general orientation towards the material. This set of topics is intended to touch on as many fundamental algorithms as possible. Some of the areas covered are "core" computer science areas we'll study in some depth to learn basic algorithms of wide applicability. Other areas are advanced fields of study within computer science and related fields, such as numerical analysis, operations research, compiler construction, and the theory of algorithms—in these cases our treatment will serve as an introduction to these fields through examination of some basic methods.

FUNDAMENTALS in the context of this book are the tools and methods used throughout the later chapters. A short discussion of Pascal is included, followed by an introduction to basic data structures, including arrays, linked lists, stacks, queues, and trees. We discuss practical uses of recursion, and cover our basic approach towards analyzing and implementing algorithms.

SORTING methods for rearranging files into order are of fundamental importance and are covered in some depth. A variety of methods are developed, described, and compared. Algorithms for several related problems are treated, including priority queues, selection, and merging. Some of these algorithms are used as the basis for other algorithms later in the book.

SEARCHING methods for finding things in files are also of fundamental importance. We discuss basic and advanced methods for searching using trees and digital key transformations, including binary search trees, balanced trees, hashing, digital search trees and tries, and methods appropriate for very large files. Relationships among these methods are discussed, and similarities to sorting methods are pointed out.

STRING PROCESSING algorithms include a range of methods for dealing with leads to parsing. File compression techniques and cryptology are also considered. Again, an introduction to advanced topics is given through treatment of some elementary problems that are important in their own right.

GEOMETRIC ALGORITHMS are a collection of methods for solving problems involving points and lines (and other simple geometric objects) that have only recently come into use. We consider algorithms for finding the convex hull of a set of points, for finding intersections among geometric objects, for solving closest-point problems, and for multidimensional searching. Many of these methods nicely complement more elementary sorting and searching methods.

GRAPH ALGORITHMS are useful for a variety of difficult and important problems. A general strategy for searching in graphs is developed and applied to fundamental connectivity problems, including shortest path, minimum spanning tree, network flow, and matching. A unified treatment of these algorithms shows that they are all based on the same procedure, and this procedure depends on a basic data structure developed earlier.

MATHEMATICAL ALGORITHMS include fundamental methods from arithmetic and numerical analysis. We study methods for arithmetic with integers, polynomials, and matrices as well as algorithms for solving a variety of mathematical problems that arise in many contexts: random number generation, solution of simultaneous equations, data fitting, and integration. The emphasis is on algorithmic aspects of the methods, not the mathematical basis.

ADVANCED TOPICS are discussed for the purpose of relating the material in the book to several other advanced fields of study. Special-purpose hardware, dynamic programming, linear programming, exhaustive search, and NP-completeness are surveyed from an elementary viewpoint to give the reader some appreciation for the interesting advanced fields of study suggested by the elementary problems confronted in this book.

The study of algorithms is interesting because it is a new field (almost all of the algorithms we will study are less than twenty-five years old) with a rich tradition (a few algorithms have been known for thousands of years). New discoveries are constantly being made, and few algorithms are completely understood. In this book we will consider intricate, complicated, and difficult algorithms as well as elegant, simple, and easy algorithms. Our challenge is to understand the former and appreciate the latter in the context of many different potential applications. In doing so, we will explore a variety of useful tools and develop a way of "algorithmic thinking" that will serve us well in computational challenges to come.

□

2

Pascal

The programming language used throughout this book is Pascal. All languages have their good and bad points, and thus the choice of any particular language for a book like this has advantages and disadvantages. But many modern programming languages are similar, so by using relatively few language constructs and avoiding implementation decisions based on peculiarities of Pascal, we develop programs that are easily translatable to other languages. Our goal is to present the algorithms in as simple and direct form as possible; Pascal allows us to do this.

Algorithms are often described in textbooks and research reports in terms of imaginary languages—unfortunately, this often allows details to be omitted and leaves the reader rather far from a useful implementation. In this book we take the view that one of the best ways to understand an algorithm and to validate its utility is through experience with an actual implementation. Pascal has the virtue of being widely available, and the reader is encouraged to become conversant with a local Pascal programming environment. The Pascal implementations in this book are working programs that are intended to be run, experimented with, modified, and *used*.

The advantage of using Pascal is that it is widely used and has basic features similar to those in other modern programming languages; the disadvantage is that it lacks many features needed by sophisticated algorithms (and available in other programming environments). Some of the programs that we will encounter can be simplified by using more advanced language features (some not available in Pascal), but this is true less often than one might think. When appropriate, the discussion of such programs will cover relevant language issues.

A concise description of the Pascal language is given in the Wirth and Jensen *Pascal User Manual and Report* that serves as the definition for the language. Our purpose in this chapter is not to repeat information from that book but rather to examine the implementation of a simple (but classic) algorithm that illustrates some of the basic features of the language and style we'll be using.

Example: Euclid's Algorithm

To begin, we'll consider a Pascal program to solve a classic elementary problem: "Reduce a given fraction to its lowest terms." We want to write 2/3, not 4/6, 200/300, or 178468/267702. Solving this problem is equivalent to finding the *greatest common divisor* (gcd) of the numerator and the denominator: the largest integer which divides them both. A fraction is reduced to lowest terms by dividing both numerator and denominator by their greatest common divisor. An efficient method for finding the greatest common divisor was discovered by the ancient Greeks over two thousand years ago: it is called *Euclid's algorithm* because it is spelled out in detail in Euclid's famous treatise *Elements.*

Euclid's method is based on the fact that if u is greater than v then the greatest common divisor of u and v is the same as the greatest common divisor of v and $u - v$. This observation leads to the following implementation in Pascal:

```
program euclid(input, output);
var x, y: integer;
function gcd(u, v: integer): integer;
   var t: integer;
   begin
   repeat
      if u<v then
         begin t:=u; u:=v; v:=t end;
      u:=u-v
   until u=0;
   gcd:=v
   end;
begin
while not eof do
   begin
   readln(x, y);
   if (x>0) and (y>0) then writeln(x, y, gcd(x, y))
   end;
end.
```

First, we consider the properties of the language exhibited by this code. Pascal has a rigorous high-level syntax that allows easy identification of the main features of the program. The variables (**var**) and functions (**function**) used by the program are declared first, followed by the body of the program. (Other major program parts not used in the program above, which are declared before the program body, are **const**ants and **type**s.) Functions have the same format as the main program except that they return a value, which is set by assigning something to the function name within the body of the function. (The **function** is one type of "subroutine" in Pascal—the other type, which does not involve a return value, is the **procedure**.) The built-in function *readln* reads a line from the input and assigns the values found to the variables given as arguments; *writeln* is similar. A standard built-in predicate, *eof*, is set to **true** when there is no more input. (Input and output within a line are possible with *read*, *write*, and *eoln*.) The declaration of *input* and *output* in the program statement indicates that the program is using the "standard" input and output streams.

The body of the program above is trivial: it reads pairs of numbers from the input, then, if they are both positive, writes them and their greatest common divisor on the output. (What would happen if *gcd* were called with u or v negative or zero?)

The *gcd* function implements Euclid's algorithm itself: the program is a loop that first ensures that $u \geq v$ by exchanging them, if necessary, and then replaces u by $u - v$. The greatest common divisor of the variables u and v is always the same as the greatest common divisor of the original values presented to the procedure: eventually the process terminates with $u = v$, both equal to the greatest common divisor of the original (and all intermediate) values of u and v.

This example is written as a complete Pascal program that the reader may wish to use to become familiar with some Pascal programming system. The "driver" part of the program reads in pairs of numbers, then writes them out along with their greatest common divisor. This example is included to show how the algorithm might be exercised, and to underscore the point that the algorithms in this book are best understood when they are implemented and run on some sample input values. Depending on the quality of the debugging environment available, the reader might wish to instrument the programs further. For example, the intermediate values taken on by u and v in the **repeat** loop are of interest in the program above.

Though our topic in the present section is the language, not the algorithm, we must do justice to the classic Euclid's algorithm: the implementation above can be improved by noting that, once $u > v$, we continue to subtract off multiples of v from u until reaching a number less than v. But this number is exactly the same as the remainder left after dividing u by v, which is what the **mod** function computes: the greatest common divisor of u and v is the same as the greatest common divisor of v and $u \bmod v$. For example, the greatest common divisor of 461952 and 116298 is 18, as exhibited by the sequence

$$461952, 116298, 113058, 3240, 2898, 342, 162, 18.$$

(Each item in this sequence is the remainder left after dividing the previous two.) The reader may wish to modify the above implementation to use the **mod** operator and to note how much more efficient the modification is when, for example, finding the greatest common divisor of a very large number and a very small number. It turns out that this algorithm always uses a relatively small number of steps.

Types of Data

Most of the algorithms in this book operate on simple data types: integers, real numbers, characters, or strings of characters. One of the most important features of Pascal is its provision for building more complex data types from these elementary building blocks. This is one of the "advanced" features that we avoid using, to keep our examples simple and our focus on the dynamics of the algorithms rather than properties of their data. We strive to do this without loss of generality: indeed, the very availability of advanced capabilities such as Pascal provides makes it easy to transform an algorithm from a "toy" that operates on simple data types into a workhorse that operates on complex **records**. When the basic methods are best described in terms of user-defined types, we do so. For example, the geometric methods in Chapters 24–28 are based on types for points, lines, polygons, etc.

It is sometimes the case that the proper low-level representation of data is the key to performance. Ideally, the way that a program works shouldn't depend on how numbers are represented or how characters are packed (to pick two examples), but the price one must pay in performance through pursuit of this ideal is often too high. Programmers in the past responded to this situation by taking the drastic step of moving to assembly language or machine language, where there are few constraints on the representation. Fortunately, modern high-level languages provide mechanisms for creating sensible representations without going to such extremes. This allows us to do justice to some important classical algorithms. Of course, such mechanisms are necessarily machine-dependent, and we will not consider them in much detail, except to point out when they are appropriate. This issue is discussed in more detail in Chapters 10, 17 and 22, where algorithms based on binary representations of data are considered.

We also try to avoid dealing with machine-dependent representation issues when considering algorithms that operate on characters and character strings. Frequently, we simplify our examples by working only with the upper-case letters A through Z, using a simple code with the ith letter of the alphabet represented by the integer i. Representation of characters and character strings is such a fundamental part of the interface among the programmer, the programming language, and the machine that one should be sure to understand it fully before implementing algorithms for processing such data—the methods given in this book based on simplified representations are then easily adapted.

We use *integers* whenever possible. Programs that process *real* numbers fall in the domain of *numerical analysis*. Typically, their performance is intimately tied to mathematical properties of the representation. We return to this issue in Chapters 37, 38, 39, 41, and 43, where some fundamental numerical algorithms are discussed. In the meantime, we stick to integers even when real numbers might seem more appropriate, to avoid the inefficiency and inaccuracy normally associated with "floating point" representations.

Input/Output

Another area of significant machine dependency is the interaction between the program and its data, normally referred to as *input-output*. In operating systems, this term refers to the transfer of data between the computer and physical media such as magnetic tape or disk: we touch on such matters only in Chapters 13 and 18. Most often, we simply seek a systematic way to get data to and derive results from implementations of algorithms, such as *gcd* above.

When "reading" and "writing" is called for, we use standard Pascal features but invoke as few of the extra formatting facilities available as possible. Again, this is to keep the programs concise, portable, and easily translatable: one way in which the reader might wish to modify the programs is to embellish their interface with the programmer. Few modern Pascal or other programming environments actually take *read* or *write* to refer to an external medium: instead, they normally refer to "logical devices" or "streams" of data. Thus, the output of one program can be used as the input to another, without any physical reading or writing. Our tendency to streamline the input/output in our implementations makes them more useful in such environments.

Actually, in many modern programming environments it is appropriate and rather easy to use graphical representations such as those used in the figures throughout the book. (As described in the Epilog, these figures were actually produced by the programs themselves, with a very significantly embellished interface.)

Many of the methods we will discuss are intended for use within larger applications systems, so a more appropriate way for them to get their data is through parameters. This is the method used for the *gcd* procedure above. Also, several of the implementations in the later chapters of the book use programs from earlier chapters. Again, to avoid diverting our attention from the algorithms themselves, we resist the temptation to "package" the implementations for use as general utility programs. Certainly, many of the implementations that we study are quite appropriate as a starting point for such utilities, but a large number of system- and application-dependent questions that we ignore here must be satisfactorily addressed in developing such packages.

Often we write algorithms to operate on "global" data, to avoid excessive parameter passing. For example, the *gcd* function could operate directly on x and y, rather than bothering with the parameters u and v. This is not justified in this case because *gcd* is a well-defined function in terms of its two inputs. On the other hand, when several algorithms operate on the same data, or when a large amount of data is passed, we use global variables for economy in expressing the algorithms and to avoid moving data unnecessarily. Advanced features are available in Pascal and other languages and systems to allow this to be done more cleanly, but, again, our tendency is to avoid such language dependencies when possible.

Concluding Remarks

Many other examples similar to the *euclid* program above are given in the *Pascal User Manual and Report* and in the chapters that follow. The reader is encouraged to scan the manual, implement and test some simple programs and then read the manual carefully to become reasonably comfortable with most of the features of Pascal.

The Pascal programs given in this book are intended to serve as precise descriptions of algorithms, as examples of full implementations, and as starting points for practical programs. As mentioned above, readers conversant with other languages should have little difficulty reading the algorithms as presented in Pascal and then implementing them in another language. For example, the following is an implementation of Euclid's algorithm in C:

```
#include <stdio.h>
main()
{
  int x,y;
  while (scanf("%d %d",&x,&y)!=EOF)
    if ((x>0) && (y>0))
      printf("%d %d %d\n",x,y,gcd(x,y));
}
int gcd(u,v)
int u,v;
{
  int t;
  do {
    if (u<v)
      { t = u; u = v; v = t;};
    u = u-v;
  } while (u!=v);
  return(u);
}
```

For this algorithm, there is nearly a one-to-one correspondence between Pascal and C statements, as intended, though there are more concise implementations in both languages.

□

Exercises

1. Implement the classical version of Euclid's algorithm as described in the text.
2. Check what values your Pascal system computes for *u* **mod** *v* when *u* and *v* are not necessarily positive.
3. Implement a procedure to reduce a given fraction to lowest terms, using a **type** *fraction* = **record** *numerator, denominator: integer* **end**.
4. Write a **function** *convert: integer* that reads a decimal number one character (digit) at a time, terminated by a blank, and returns the value of that number.
5. Write a **procedure** *binary(x: integer)* that prints out the binary equivalent of a number.
6. Give all the values that *u* and *v* take on when *gcd* is invoked with the initial call *gcd(12345,56789)*.
7. Exactly how many Pascal statements are executed for the call in the previous exercise?
8. Write a program to compute the greatest common divisor of *three* integers *u*, *v*, and *w*.
9. Find the largest pair of numbers representable as *integer*s in your Pascal system whose greatest common divisor is 1.
10. Implement Euclid's algorithm in FORTRAN or BASIC.

3

Elementary Data Structures

☐ In this chapter, we discuss basic ways of organizing data for processing by computer programs. For many applications, the choice of the proper data structure is really the only major decision involved in the implementation: once the choice has been made, only very simple algorithms are needed. For the same data, some data structures require more or less space than others; for the same operations on the data, some data structures lead to more or less efficient algorithms than others. This theme will recur frequently throughout this book, as the choice of algorithm and data structure is closely intertwined, and we continually seek ways of saving time or space by making this choice properly.

A data structure is not a passive object: we also must consider the operations to be performed on it (and the algorithms used for these operations). This concept is formalized in the notion of an *abstract data type* which we discuss at the end of this chapter. But our primary interest is in concrete implementations, and we'll focus on specific representations and manipulations.

We're going to be dealing with arrays, linked lists, stacks, queues, and other simple variants. These are classical data structures with widespread applicability: along with trees (see Chapter 4), they form the basis for virtually all of the algorithms considered in this book. In this chapter, we consider basic representations and fundamental methods for manipulating these structures, work through some specific examples of their use, and discuss related issues such as storage management.

Arrays

Perhaps the most fundamental data structure is the *array*, which is defined as a primitive in Pascal and most other programming languages. An array is a fixed number of data items which are stored contiguously and which are accessible by an index. We refer to the ith element of an array a as $a[i]$. It is the responsibility of the programmer to store something meaningful in an array position *a*[*i*] before referring to it; neglecting this is one of the most common programming mistakes.

A simple example of the use of an array is given by the following program, which prints out all the prime numbers less than 1000. The method used, which dates back to the 3rd century B.C., is called the "sieve of Eratosthenes":

```
program primes(input.output);
const N=1000;
var a: array[1..N] of boolean;
    i,j: integer;
begin
a[1]:=false; for i:=2 to N do a[i]:=true;
for i:=2 to N div 2 do
    for j:=2 to N div i do a[i*j]:=false;
for i:=1 to N do
    if a[i] then write(i:4);
end.
```

This program uses an array consisting of the very simplest type of elements, boolean values. The goal of the program is to set *a*[*i*] to **true** if *i* is prime, **false** if it is not. It does this by, for each *i*, setting the array element corresponding to each multiple of *i* to **false**, since any number that is a multiple of any other number cannot be prime. Then it goes through the array once more, printing out the primes. (This program can be made somewhat more efficient by adding the test **if** *a*[*i*] **then** before the **for** loop involving *j*, since if *i* is not prime, the array elements corresponding to all of its multiples must already have been marked.) Note that the array is first "initialized" to indicate that no numbers are known to be nonprime: the algorithm sets to **false** array elements corresponding to indices that are known to be nonprime.

The sieve of Eratosthenes is typical of algorithms that exploit the fact that any item of an array can be efficiently accessed. The algorithm also accesses the items of the array sequentially, one after the other. In many applications, sequential ordering is important; in other applications sequential ordering is used because it is as good as any other. But the primary feature of arrays is that *if the index is known*, any item can be accessed in constant time.

The size of the array must be known beforehand; in Pascal, it must be known at compile time. To run the above program for a different value of *N*, it is necessary to change the constant *N*, then compile and execute. In some programming environments, it is possible to declare the size of an array at execution time (so that one could, for example, have a user type in the value of *N*, and then respond with the primes less than *N* without wasting memory by declaring an array as large as any value the user is allowed to type), but it is still a fundamental property of arrays that their size is fixed and must be known before they are used.

Arrays are fundamental data structures in that they have a direct correspondence with memory systems on virtually all computers. To retrieve the contents of a word from the memory in machine language, we provide an address. Thus, we could think of the entire computer memory as an array, with the memory addresses corresponding to array indices. Most computer language processors translate programs that involve arrays into rather efficient machine-language programs that access memory directly.

Another familiar way to structure information is to use a two-dimensional table of numbers organized into rows and columns. For example, a table of students' grades in a course might have one row for each student, one column for each assignment. On a computer, such a table would be represented as a *two-dimensional* array with two indices, one for the row and one for the column. Various algorithms on such structures are straightforward: for example, to compute the average grade on an assignment, we sum together the elements in a column and divide by the number of rows; to compute a particular student's average grade in the course, we sum together the elements in a row and divide by the number of columns. Two-dimensional arrays are widely used in applications of this type. Actually, on a computer, it is often convenient and rather straightforward to use more than two dimensions: an instructor might use a third index to keep student grade tables for a sequence of years.

Arrays also correspond directly to *vectors*, the mathematical term for indexed lists of objects. Similarly, two-dimensional arrays correspond to *matrices*. We study algorithms for processing these mathematical objects in Chapters 36 and 37.

Linked Lists

The second elementary data structure to consider is the *linked list*, which is defined as a primitive in some programming languages (notably in Lisp) but not in Pascal. However, Pascal does provide some basic primitive operations that make it easy to use linked lists.

The primary advantage of linked lists over arrays is that they can grow and shrink in size over their lifetime. In particular, their maximum size need not be known in advance. In practical applications, this often makes it possible to have several data structures share the same space, without paying particular attention to their relative sizes at any time.

A second advantage of linked lists is that they provide flexibility in allowing the items to be rearranged efficiently. This flexibility is gained at the expense of quick access to any arbitrary item in the list. This will become more apparent below, after we have examined some of the basic properties of linked lists and some of the fundamental operations we perform on them.

A linked list is a set of items organized sequentially, just like an array. In an array, the sequential organization is provided implicitly (by the position in the array); in a linked list, we use an explicit arrangement in which each item is part of

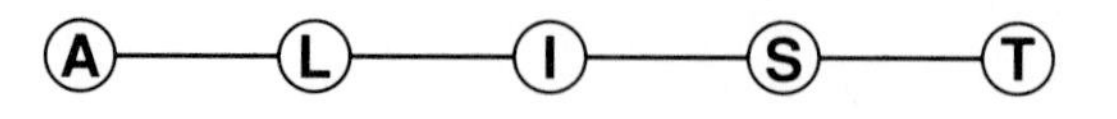

Figure 3.1 A linked list.

a "node" that also contains a "link" to the next node. Figure 3.1 shows a linked list, with items represented by letters, nodes by circles and links by lines connecting the nodes. We look in detail below at how lists are represented within the computer; for now we'll talk simply in terms of nodes and links.

Even the simple representation of Figure 3.1 exposes two details we must consider. First, every node has a link, so the link in the last node of the list must specify some "next" node. Our convention will be to have a "dummy" node, which we'll call z, for this purpose: the last node of the list will point to z, and z will point to itself. In addition, we normally will have a dummy node at the other end of the list, again by convention. This node, which we'll call *head*, will point to the first node in the list. The main purpose of the dummy nodes is to make certain manipulations with the links, especially those involving the first and last nodes on the list, more convenient. Other conventions are discussed below. Figure 3.2 shows the list structure with these dummy nodes included.

Now, this explicit representation of the ordering allows certain operations to be performed much more efficiently than would be possible for arrays. For example, suppose that we want to move the T from the end of the list to the beginning. In an array, we would have to move every item to make room for the new item at the beginning; in a linked list, we just change three links, as shown in Figure 3.3. The two versions shown in Figure 3.3 are equivalent; they're just drawn differently. We make the node containing T point to A, the node containing S point to z, and *head* point to T. Even if the list were very long, we could make this structural change by changing just three links.

More important, we can talk of "inserting" an item into a linked list (which makes it grow by one in length), an operation that is unnatural and inconvenient in an array. Figure 3.4 shows how to insert X into our example list by putting it in a node that points to S, then making the node containing I point to the new node. Only two links need to be changed for this operation, no matter how long the list is.

Similarly, we can speak of "deleting" an item from a linked list (which makes it shrink by one in length). For example, the third list in Figure 3.4 shows how to delete X from the second list simply by making the node containing I point to S,

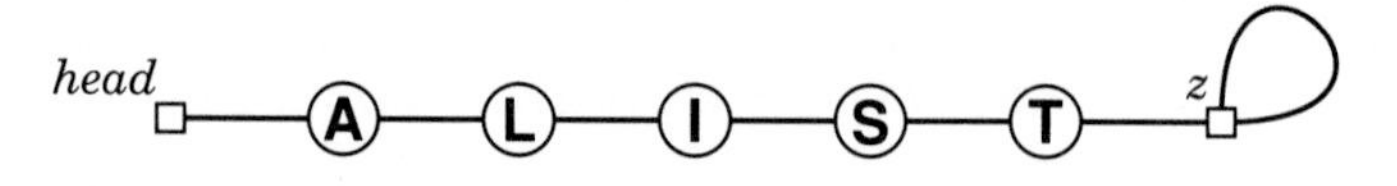

Figure 3.2 A linked list with its dummy nodes.

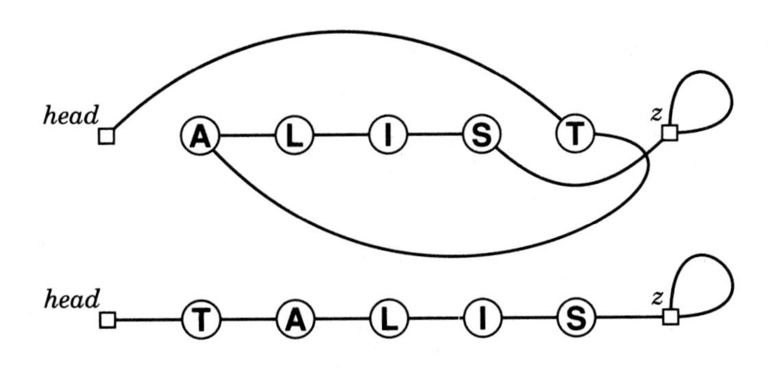

Figure 3.3 Rearranging a linked list.

skipping X. Now, the node containing X still exists (in fact it still points to S), and perhaps should be disposed of in some way — the point is that it is no longer part of this list, and cannot be accessed by following links from *head*. We will return to this issue below.

On the other hand, there are other operations for which linked lists are *not* well-suited. The most obvious of these is "find the kth item" (find an item given its index): in an array this is simply done with the access $a[k]$, but in a list we have to travel through k links.

Another operation that is unnatural on linked lists is "find the item *before* a given item". If all we have is the link to T in our sample list, then the only way

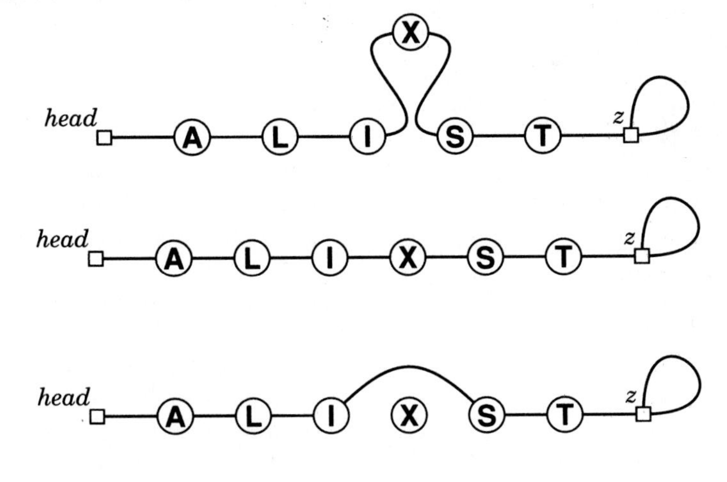

Figure 3.4 Insertion into and deletion from a linked list.

we can find the link to S is to start at *head* and travel through to find the node that points to T. As a matter of fact, this operation is necessary if we want to be able to delete a given node from a linked list: how else do we find the node whose link must be changed? In many applications, we can get around this problem by redesigning the fundamental deletion operation to be "delete the *next* node". A similar problem can be avoided for insertion by making the fundamental insertion operation "insert a given item *after* a given node" in the list.

Pascal provides primitive operations that allow linked lists to be implemented directly. The following code fragment is a sample implementation of the basic functions that we have discussed to this point.

```
type link=↑node;
      node= record key: integer; next: link end;
var head,z,t: link;
procedure listinitialize;
   begin
   new(head); new(z);
   head↑.next:=z; z↑.next:=z
   end;
procedure deletenext(t: link);
   begin
   t↑.next:=t↑.next↑.next;
   end;
procedure insertafter(v: integer; t: link);
   var x: link;
   begin
   new(x);
   x↑.key:=v; x↑.next:=t↑.next;
   t↑.next:=x
   end;
```

The precise format of the lists is described in the **type** declaration: the lists are made up of *nodes*, each node containing an integer and a *link* to the next node on the list. The *key* is an integer here only for simplicity, and could be arbitrarily complex—the *link* is the key to the list. The variable *head* is a *link* to the first node on a list: we can examine the other nodes in order by following *links* until reaching *z*, the *link* that points to the dummy node representing the end of the list. The Pascal syntax for following links is the "up-arrow": we write a reference to a link followed by this symbol to indicate a reference to the node pointed to by that link. For example, the reference *head↑.next↑.key* refers to the first item on a list, and *head↑.next↑.next↑.key* refers to the second.

The **type** declaration merely describes the formats of the nodes; nodes can be created only when the built-in procedure *new* is called. For example, the call *new*(*head*) creates a new node, putting a pointer to it in *head*. The purpose of this procedure is to relieve the programmer of the burden of "allocating" storage for the nodes as the list grows. (We discuss this mechanism in some detail below.) There is a corresponding built-in procedure *dispose* for deletion, which might be used by the calling routine, or perhaps the node, though deleted from one list, is to be added to another.

The reader is encouraged to check these Pascal implementations against the English-language descriptions given above. In particular, it is instructive at this stage to consider why the dummy nodes are useful. First, if the convention were to have *head* point to the beginning of the list rather than having a *head* node, then the *insert* procedure would need a special test for insertion at the beginning of the list. Second, the convention for z protects the *delete* procedure from (for example) a call to delete an item from an empty list.

Another common convention for terminating a list is to make the last node point to the first, rather than using either of the dummy nodes *head* or z. This is called a *circular list*: it allows a program to go around and around the list as long as there is something in it. Using one dummy node to mark the beginning (and the end) of the list and to help handle the case of an empty list is sometimes convenient.

It is possible to support the operation "find the item *before* a given item" by using a *doubly linked list* in which we maintain two links for each node, one to the item before, one to the item after. The cost of providing this extra capability is doubling the number of link manipulations per basic operation, so it is not normally used unless specifically called for. As mentioned above, however, if a node is to be deleted and only a link to it is available (perhaps it is also part of some other data structure), double linking may be called for.

We'll see many examples of applications of these and other basic operations on linked lists in later chapters. Since the operations involve only a few statements, we normally manipulate the lists directly rather than use the precise procedures above. As an example, we consider next a program for solving the so-called "Josephus problem" in the spirit of the sieve of Eratosthenes.

For the Josephus problem, we imagine that N people have decided to commit mass suicide by arranging themselves in a circle and killing the Mth person around the circle, closing ranks as each person drops out of the circle. The problem is to find out which person is the last to die (though perhaps that person would have a change of heart at the end!), or, more generally, to find the order in which the people are executed. For example, if $N = 9$ and $M = 5$, the people are killed in the order 5 1 7 4 3 6 9 2 8. The following program reads in N and M and prints out this ordering:

```
program josephus(input, output);
type link =↑node;
      node = record key: integer; next: link end;
var i, N, M: integer;
     t, x: link;
begin
read(N, M);
new(t); t↑.key:=1; x:=t;
for i:=2 to N do
   begin new(t↑.next); t:=t↑.next; t↑.key:=i end;
t↑.next:=x;
while t<>t↑.next do
   begin
   for i:=1 to M-1 do t:=t↑.next;
   write(t↑.next↑.key);
   x:=t↑.next;
   t↑.next:=t↑.next↑.next;
   dispose(x);
   end;
writeln(t↑.key);
end.
```

The program uses a circular linked list to simulate the sequence of executions directly. First, the list is built with keys from 1 to N: the variable x holds onto the beginning of the list as it is built, then the pointer in the last node in the list is set to x. Then, the program proceeds through the list, counting through $M - 1$ items and deleting the next, until only one is left (which then points to itself). Note the call to *dispose* for the delete, which corresponds to an execution: this is the opposite of *new* as mentioned above.

Storage Allocation

Pascal's pointers provide a convenient way to implement lists, as shown above, but there are alternatives. In this section we discuss how to use *arrays* to implement linked lists and how this is related to the actual representation of the lists in a Pascal program. As mentioned above, arrays are a rather direct representation of the memory of the computer, so analysis of how a data structure is implemented as an array will give some insight into how it might be represented at a low level in the computer. In particular, we're interested in seeing how several lists might be represented simultaneously.

In a direct-array representation of linked lists, we use indices instead of links. One way to proceed would be to define an array of records like those above but using *integers* (for array indices) rather than *links* for the *next* field. An alternative, which often turns out to be more convenient, is to use "parallel arrays": we keep the items in an array *key*[*1..N*] and the links in an array *next*[*1..N*]. Thus, *key*[*next*[*head*]] refers to the first item on the list, *key*[*next*[*next*[*head*]]] to the second, etc. The advantage of using parallel arrays is that the structure can be built "on top of" the data: the array *key* contains data and only data—all the structure is in the parallel array *next*. For example, another list can be built using the same data array and a different parallel "link" array, or more data can be added with more parallel arrays. The following code implements the basic list operations using parallel arrays:

```
var key, next: array[0..N] of integer;
    x, head, z: integer;
procedure listinitialize;
  begin
  head:=0; z:=1; x:=1;
  next[head]:=z; next[z]:=z
  end;
procedure deletenext(t: integer);
  begin
  next[t]:=next[next[t]]
  end;
procedure insertafter(v: integer; t: integer);
  begin
  x:=x+1;
  key[x]:=v; next[x]:=next[t];
  next[t]:=x
  end;
```

The "pointer" *x* replaces the storage allocation function *new*: it keeps track of the next unused position in the array.

Figure 3.5 shows how our sample list might be represented in parallel arrays, and how this representation relates to the graphical representation that we have been using. The *key* and *next* arrays are shown on the left, as they appear (for example) if S L A I T are inserted into an initially empty list, with S, L, and A inserted after *head*; I after L, and T after S. Position *0* is *head* and position *1* is *z* (these are set by *listinitialize*)—since *next[0]* is *4*, the first item on the list is *key[4]* (A); since *next[4]* is *3*, the second item on the list is *key[3]* (L), etc. In the second diagram from the left, the indices for the *next* array are replaced by lines—instead of putting a "4" at *next[0]*, we draw a line from node 0 down to node 4, etc. In the third diagram, we untangle the links to arrange list elements one after the other; then at the right, we simply draw the nodes in our usual graphical representation.

The crux of the matter is to consider how the built-in procedures *new* and *dispose* might be implemented. We presume that the only space for nodes and links are the arrays we've been using; this presumption puts us in the situation the system is in when it has to provide the capability to grow and shrink a data structure with a fixed data structure (the memory itself). For example, suppose that the node containing A is to be deleted from the example in Figure 3.5 and then disposed of. It is one thing to rearrange the links so that node is no longer hooked into the list, but what do we do with the space occupied by that node? And how do we find space for a node when *new* is called and more space is needed?

On reflection, the reader will see that the solution is clear: a linked list should be used to keep track of the free space! We refer to this list as the "free list". Then, when we *delete* a node from our list we dispose of it by *inserting* it onto the free list, and when we need a *new* node, we get it by *deleting* it from the free list. This mechanism allows several different lists to occupy the same array.

A simple example with two lists (but no free list) is shown in Figure 3.6. There are two list header nodes *hd1* = *0* and *hd2* = *6*, but both lists can share the same *z*. (To build multiple lists, the *listinitialize* procedure above would have to be modified to manage more than one *head*.) Now, *next[0]* is *4*, so the first item on the first list is *key[4]* (O); since *next[6]* is 7, the first item on the second list is *key[7]* (T), etc. The other diagrams in Figure 3.6 show the result of replacing *next* values by lines, untangling the nodes, and changing to our simple graphical representation, just as in Figure 3.5. This same technique could be used to maintain several lists in the same array, one of which would be a free list, as described above.

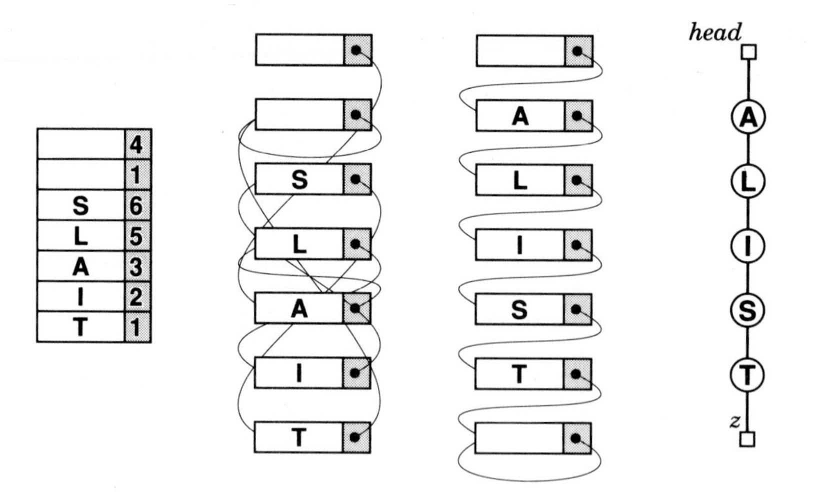

Figure 3.5 Array implementation of a linked list.

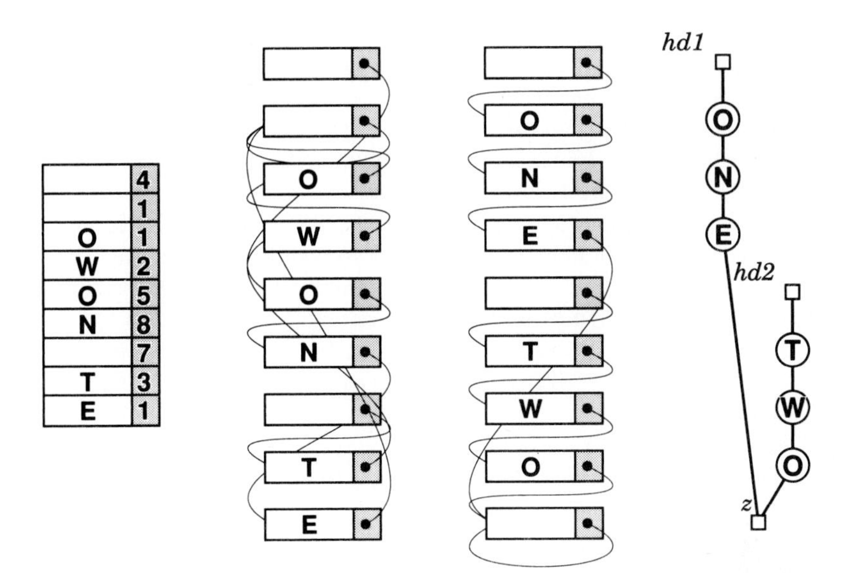

Figure 3.6 Two lists sharing the same space.

When storage management is provided by the system, as in Pascal, there is no reason to override it in this way. The description above is intended to indicate how the storage management is done by the system. (If the reader's system does not do storage management, the description above provides a starting point for an implementation.) The actual problem faced by the system is rather more complex, because not all nodes need be of the same size. Also, some systems relieve the user of the need to explicitly *dispose* nodes by using "garbage-collection" algorithms to remove any nodes not referenced by any link. A number of rather clever storage management algorithms have been developed to handle these two situations.

Pushdown Stacks

We have been concentrating on structuring data in order to insert, delete, or access items arbitrarily. Actually, it turns out that for many applications, it suffices to consider various (rather stringent) restrictions on how the data structure is accessed. Such restrictions are beneficial in two ways: first, they can alleviate the need for the program using the data structure to be concerned with its details (for example, keeping track of links to or indices of items); second, they allow simpler and more flexible implementations, since fewer operations need be supported.

The most important restricted-access data structure is the *pushdown stack*. Only two basic operations are involved: one can *push* an item onto the stack (insert it at the beginning) and *pop* an item (remove it from the beginning). A stack operates somewhat like a busy executive's "in" box: work piles up in a stack, and whenever the executive is ready to do some work, he takes it off the top. This might mean that something gets stuck in the bottom of the stack for some time, but a good executive would presumably manage to get the stack emptied periodically. It turns out that a computer program sometimes is naturally organized in this way, postponing some tasks while doing others, and thus pushdown stacks appear as the fundamental data structure for many algorithms.

We'll see a great many applications of stacks in the chapters that follow: for an introductory example, let's look at using stacks in evaluating arithmetic expressions. Suppose that one wants to find the value of a simple arithmetic expression involving multiplication and addition of integers, such as

$$5 * (((9 + 8) * (4 * 6)) + 7).$$

A stack is the ideal mechanism for saving intermediate results in such a calculation. The above example might be computed with the calls:

```
push(5);
push(9);
push(8);
push(pop+pop);
push(4);
push(6);
push(pop*pop);
push(pop*pop);
push(7);
push(pop+pop);
push(pop*pop);
writeln(pop);
```

The order in which the operations are performed is dictated by the parentheses in the expression, and by the convention that we proceed from left to right. Other conventions are possible; for example $4 * 6$ could be computed before $9 + 8$ in the example above.

Some calculators and some computing languages base their method of calculation on stack operations explicitly in this way: every operation pops its arguments from the stack and returns its results to the stack. As we'll see in Chapter 5, stacks often arise implicitly even when not used explicitly.

The basic stack operations are easy to implement using linked lists, as in the following implementation:

```
type link=↑node;
     node= record key: integer; next: link end;
var head,z: link;
procedure stackinit;
  begin
  new(head); new(z);
  head↑.next:=z; z↑.next:=z
  end;
procedure push(v: integer);
  var t: link;
  begin
  new(t);
  t↑.key:=v; t↑.next:=head↑.next;
  head↑.next:=t
  end;
function pop: integer;
  var t: link;
  begin
  t:=head↑.next;
  pop:=t↑.key;
  head↑.next:=t↑.next;
  dispose(t)
  end;
function stackempty: boolean;
  begin stackempty:=(head↑.next=z) end;
```

(This implementation also includes code to initialize a stack and to test it if is empty.) In an application in which only one stack is used, we can assume that the global variable *head* is the link to the stack; otherwise, the implementations can be modified to also pass around a link to the stack.

The order of calculation in the arithmetic example above requires that the operands appear *before* the operator so that they can be on the stack when the operator is encountered. Any arithmetic expression can be rewritten in this way—the example above corresponds to the expression

$$5\ 9\ 8\ +\ 4\ 6\ *\ *\ 7\ +\ *.$$

This is called *reverse Polish* notation (because it was introduced by a Polish logician), or *postfix*. The customary way of writing arithmetic expressions is called *infix*. One interesting property of postfix is that parentheses are not required; in infix they are needed to distinguish, for example, $5 * (((9 + 8) * (4 * 6)) + 7)$ from $((5*9)+8)*((4*6)+7)$. The following program converts a legal fully parenthesized infix expression into a postfix expression:

```
stackinit;
repeat
   repeat read(c) until c<>' ';
   if c=')' then write(chr(pop));
   if c='+' then push(ord(c));
   if c='*' then push(ord(c));
   while (c>='0') and (c<='9') do
      begin write(c); read(c) end;
   if c<>'(' then write (' ');
until eoln;
```

Arguments are simply passed through, since they appear in the postfix expression in the same order as in the infix expression. Then each right parenthesis indicates that both arguments for the last operator have been output, so the operator itself can be popped and written out. (This program does not check for errors in the input and requires spaces between operators, parentheses and operands.)

The chief reason to use postfix is that evaluation can be done in a very straightforward manner with a stack, as in the following program:

```
stackinit;
repeat
   x:=0;
   repeat read(c) until c<>' ';
   if c='*' then x:=pop*pop;
   if c='+' then x:=pop+pop;
   while (c>='0') and (c<='9') do
      begin x:=10*x+(ord(c)-ord('0')); read(c) end;
   push(x);
until eoln;
writeln(pop);
```

This program reads any postfix expression involving multiplication and addition of integers, then prints the value of the expression. Blanks are ignored, and the **while** loop converts integers from character format to numbers for calculation. Otherwise, the operation of the program is straightforward. Integers (operands) are pushed onto the stack and multiplication and addition replace the top two items on the stack by the result of the operation.

If the maximum size of a stack can be predicted in advance, it may be appropriate to use an array representation rather than a linked list, as in the following implementation:

```
const maxP=100;
var stack: array[0..maxP] of integer;
    p: integer;
procedure push(v: integer);
  begin stack[p]:=v; p:=p+1 end;
function pop: integer;
  begin p:=p-1; pop:=stack[p] end;
procedure stackinit;
  begin p:=0 end;
function stackempty: boolean;
  begin stackempty:=(p<=0) end;
```

The variable p is a global variable that keeps track of the location of the top of the stack. This is a very simple implementation that avoids the use of extra space for links, at the cost of perhaps wasting space by reserving room for the maximum size stack.

Figure 3.7 shows how a sample stack evolves through the series of *push* and *pop* operations represented by the sequence:

$$A * S A * M * P * L * E S * T * * * A * C K * * .$$

The appearance of a letter in this list means "push" (the letter); the asterisk means "pop".

Typically, a large number of operations will require only a small stack. If one is confident this is the case, then an array representation is called for. Otherwise, a linked list might allow the stack to grow and shrink gracefully, especially if it is one of many such data structures.

Queues

Another fundamental restricted-access data structure is called the *queue*. Again, only two basic operations are involved: one can *insert* an item into the queue at the beginning and *remove* an item from the end. Perhaps our busy executive's "in" box *should* operate like a queue, since then work that arrives first would get done

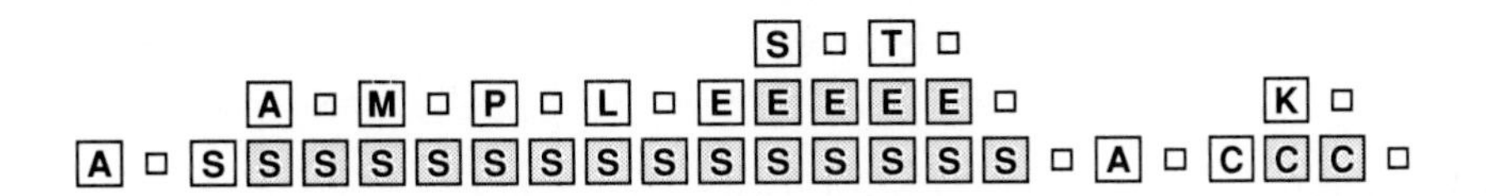

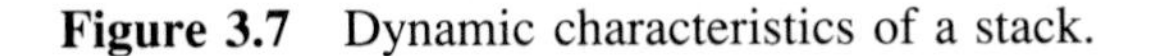

Figure 3.7 Dynamic characteristics of a stack.

first. In a stack, something can get buried at the bottom, but in a queue everything is processed in the order received.

Although stacks are encountered more often than queues because of their fundamental relationship with recursion (see Chapter 5), we will encounter algorithms for which the queue is the natural data structure. Stacks are sometimes referred to as obeying a "last in, first out" (LIFO) discipline; queues obey a "first in, first out" (FIFO) discipline.

The linked-list implementation of the queue operations is straightforward and left as an exercise for the reader. As with stacks, an array can also be used if one can estimate the maximum size, as in the following implementation:

```
const max=100;
var queue: array[0..max] of integer;
    head, tail: integer;
procedure put(v: integer);
  begin
  queue[tail]:=v; tail:=tail+1;
  if tail>max then tail:=0
  end;
function get: integer;
  begin
  get:=queue[head]; head:=head+1;
  if head>max then head:=0
  end;
procedure queueinitialize;
  begin head:=0; tail:=0 end;
function queueempty: boolean;
  begin queueempty:=(head=tail) end;
```

It is necessary to maintain two indices, one to the beginning of the queue (*head*) and one to the end (*tail*). The contents of the queue are all the elements in the array between *head* and *tail*, taking into account the "wraparound" back to 0 when the end of the array is encountered. If *head* = *tail* then the queue is defined to be empty; if *head* = *tail+1*, or *tail* = *max* and *head* = *0*, it is defined to be full.

Figure 3.8 shows how a sample queue evolves through the series of *get* and *put* operations represented by the sequence:

*A * S A * M * P * L E * Q * * * U * E U * * E * .*

The appearance of a letter in this list means "put" (the letter); the asterisk means "get".

In Chapter 20 we encounter a *deque* (or "double-ended queue"), which is a combination of a stack and a queue, and in Chapters 4 and 30 we discuss rather

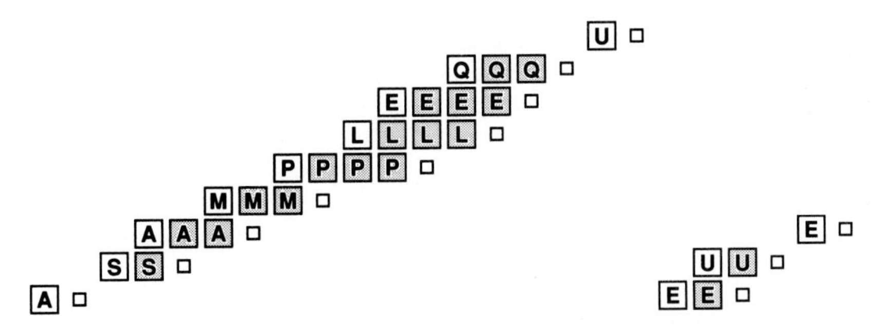

Figure 3.8 Dynamic characteristics of a queue.

fundamental examples involving the application of a queue as a mechanism to allow exploration of trees and graphs.

Abstract Data Types

We've seen above that it is often convenient to describe algorithms and data structures in terms of the operations performed, rather than in terms of details of implementation. When a data structure is defined in this way, it is called an *abstract data type*. The idea is to separate the "concept" of what the data structure should do from any particular implementation.

The defining characteristic of an abstract data type is that nothing outside of the definitions of the data structure and the algorithms operating on it should refer to anything inside, except through function and procedure calls for the fundamental operations. The main motivation for the development of abstract data types has been as a mechanism for organizing large programs. They provide a way to limit the size and complexity of the interface between (potentially complicated) algorithms and associated data structures and (a potentially large number of) programs that use the algorithms and data structures. This makes it easier to understand the large program, and makes it more convenient to change or improve the fundamental algorithms.

Stacks and queues are classic examples of abstract data types: most programs need be concerned only about a few well-defined basic operations, not details of links and indices.

Arrays and linked lists can in turn be thought of as refinements of a basic abstract data type called the *linear list.* Each of them can support operations such as *insert*, *delete*, and *access* on a basic underlying structure of sequentially ordered items. These operations suffice to describe the algorithms, and the linear list abstraction can be useful in the initial stages of algorithm development. But as we've seen, it is in the programmer's interest to define carefully which operations will be used, for the different implementations can have quite different performance characteristics. For example, using a linked list instead of an array for the sieve of Eratosthenes would be costly because the algorithm's efficiency depends on being able to get from any array position to any other quickly, and using an array instead of a linked list for the Josephus problem would be costly because the algorithm's efficiency depends on the disappearance of deleted elements.

Many more operations suggest themselves on linear lists that require much more sophisticated algorithms and data structures to support efficiently. The two most important are *sorting* the items in increasing order of their keys (the subject of Chapters 8-13), and *searching* for an item with a particular key (the subject of Chapters 14-18).

One abstract data type can be used to define another: we use linked lists and arrays to define stacks and queues. Indeed, we use the "pointer" and "record" abstractions provided by Pascal to build linked lists, and the "array" abstraction provided by Pascal to build arrays. In addition, we saw above that we can build linked lists with arrays, and we'll see in Chapter 36 that arrays should sometimes be built with linked lists! The real power of the abstract data type concept is that it allows us conveniently to construct large systems on different levels of abstraction, from the machine-language instructions provided by the computer, to the various capabilities provided by the programming language, to sorting, searching and other higher-level capabilities provided by algorithms as discussed in this book, to the even higher levels of abstraction the application may suggest.

In this book, we deal with relatively small programs that are rather tightly integrated with their associated data structures. While it is possible to talk of abstraction at the interface between our algorithms and their data structures, it is really more appropriate to focus on higher levels of abstraction (closer to the application): the concept of abstraction should not distract us from finding the most efficient solution to a particular problem. We take the view here that performance does matter! Programs developed with this in mind can then be used with some confidence in developing higher levels of abstraction for large systems.

Whether or not abstract data types are explicitly used, we are not freed from the obligation of stating precisely what our algorithms do. Indeed, it is often convenient to define the interfaces to the algorithms and data structures provided here as abstract data types; examples of this are found in Chapters 11 and 14. Moreover, the user of the algorithms and data structures is obliged to state clearly what he expects them to do—proper communication between the user of an algorithm and the person who implements it (even if they are the same person) is the key to success in building large systems. Programming environments which support the development of large systems have facilities which allow this to be done in a systematic way.

As mentioned above, real data structures rarely consist simply of integers and links. Nodes often contain a great deal of information and may belong to multiple independent data structures. For example, a file of personnel data may contain records with names, addresses, and various other pieces of information about employees, and each record may need to belong to one data structure for searching for particular employees, another for answering statistical queries, etc. It is possible to build up quite complex structures even using just the simple data structures described in this chapter: the records may be larger and more complex, but the algorithms are the same. Still, we need to be careful that we do not develop algorithms good for small records only: we return to this issue at the end of Chapter 8 and at the beginning of Chapter 14.

□

Exercises

1. Write a program to fill in a two-dimensional array of boolean values by setting $a[i,j]$ to **true** if the greatest common divisor of i and j is 1 and to **false** otherwise.

2. Implement a procedure *movenexttofront*(*t: link*) for a linked list that moves the node following the node pointed to by t to the beginning of the list. (Figure 3.3 is an example of this for the special case when t points to the next-to-last node in the list.)

3. Implement a procedure *exchange*(*t,u: link*) for a linked list that exchanges the positions of the nodes pointed to by t and u.

4. Write a program to solve the Josephus problem, using an array instead of a linked list.

5. Write procedures for insertion and deletion in a doubly linked list.

6. Write procedures for a linked list implementation of pushdown stacks, but using parallel arrays.

7. Give the contents of the stack after each operation in the sequence $E\ A\ S * Y * * Q\ U\ E * * * S\ T * * * I * O\ N * *$. Here a letter means "push" (the letter) and "*" means "pop."

8. Give the contents of the queue after each operation in the sequence $E\ A\ S * Y * * Q\ U\ E * * * S\ T * * * I * O\ N * *$. Here a letter means "put" (the letter) and "*" means "get."

9. Give a sequence of calls to *listinitialize*, *deletenext*, and *insertafter* that could have produced Figure 3.5.

10. Using a linked list, implement the basic operations for a queue.

4

Trees

The structures discussed in Chapter 3 are inherently one-dimensional: one item follows the other. In this chapter we consider two-dimensional linked structures called *trees*, which lie at the heart of many of our most important algorithms. A full discussion of trees could fill an entire book, for they arise in many applications outside of computer science and have been studied extensively as mathematical objects. Indeed, it might be said that *this* book is filled with a discussion of trees, for they are present, in a fundamental way, in every one of its sections. In this chapter, we consider the basic definitions and terminology associated with trees, examine some important properties, and look at ways of representing them within the computer. In later chapters, we shall see many algorithms that operate on these fundamental data structures.

Trees are encountered frequently in everyday life, and the reader is surely rather familiar with the basic concept. For example, many people keep track of ancestors and/or descendants with a family tree: as we'll see, much of our terminology is derived from this usage. Another example is found in the organization of sports tournaments; this usage, which we'll encounter in Chapter 11, was studied by Lewis Carroll. A third example is found in the organizational chart of a large corporation; this usage is suggestive of the "hierarchical decomposition" found in many computer science applications. A fourth example is a "parse tree" of an English sentence into its constituent parts; this is intimately related to the processing of computer languages, as discussed further in Chapter 21. Other examples will be touched on throughout the book.

Glossary

We begin our discussion of trees here by defining them as abstract objects and introducing most of the basic associated terminology. There are a number of equivalent ways to define trees, and a number of mathematical properties that imply this equivalence; these are discussed in more detail in the next section.

A *tree* is a nonempty collection of *vertices* and *edges* that satisfies certain requirements. A vertex is a simple object (also referred to as a *node*) that can have a name and can carry other associated information; an edge is a connection between two vertices. A *path* in a tree is a list of distinct vertices in which successive vertices are connected by edges in the tree. One node in the tree is designated as the *root*—the defining property of a tree is that there is exactly one path between the root and each of the other nodes in the tree. If there is more than one path between the root and some node, or if there is no path between the root and some node, then what we have is a graph (see Chapter 29), not a tree. Figure 4.1 shows an example of a tree.

Though the definition implies no "direction" on the edges, we normally think of the edges as all pointing away from the root (down in Figure 4.1) or towards the root (up in Figure 4.1) depending upon the application. We usually draw trees with the root at the top (even though this seems unnatural at first), and we speak of node y as being *below* node x (and x as *above* y) if x is on the path from y to the root (that is, if y is below x as drawn on the page and is connected to x by a path that does not pass through the root). Each node (except the root) has exactly one node above it which is called its *parent*; the nodes directly below a node are called its *children*. We sometimes carry the analogy to family trees further and refer to the "grandparent" or the "sibling" of a node: in Figure 4.1, P is the grandchild of R and has three siblings.

Nodes with no children are sometimes called *leaves*, or *terminal* nodes. To correspond to the latter usage, nodes with at least one child are sometimes called *nonterminal* nodes. Terminal nodes are often different from nonterminal nodes: for example, they may have no name or associated information. Especially in such situations, we refer to nonterminal nodes as *internal* nodes and terminal nodes as *external* nodes.

Any node is the root of a *subtree* consisting of it and the nodes below it. In the tree shown in Figure 4.1, there are seven one-node subtrees, one three-node subtree, one five-node subtree, and one six-node subtree. A set of trees is called

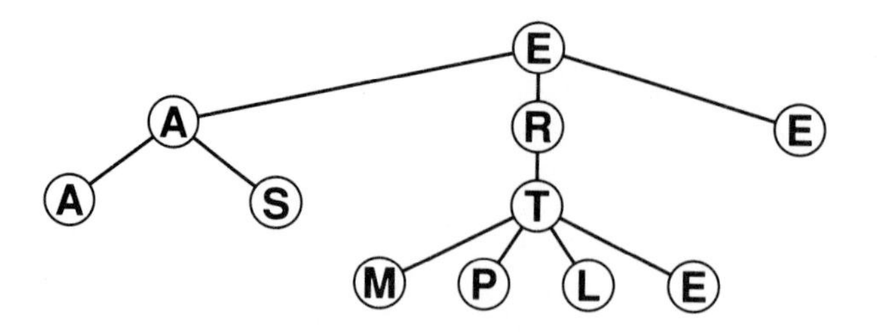

Figure 4.1 A sample tree.

a *forest*: for example, if we remove the root and the edges connecting it from the tree in Figure 4.1, we are left with a forest consisting of three trees rooted at A, R, and E.

Sometimes the way in which the children of each node are ordered is significant, sometimes it is not. An *ordered* tree is one in which the order of the children at every node is specified. Of course, the children are placed in some order when we draw a tree, and clearly there are many different ways to draw trees that are not ordered. As we will see below, this becomes significant when we consider representing trees in a computer, since there is much less flexibility in how to represent ordered trees. It is normally obvious from the application which type of tree is called for.

The nodes in a tree divide themselves into *levels*: the level of a node is the number of nodes on the path from it to the root (not including itself). Thus, for example, in Figure 4.1, R is on level 1 and S is on level 2. The *height* of a tree is the maximum level among all nodes in the tree (or the maximum distance to the root from any node). The *path length* of a tree is the sum of the levels of all the nodes in the tree (or the sum of the lengths of the paths from each node to the root). The tree in Figure 4.1 is of height 3 and path length 21. If internal nodes are distinguished from external nodes, we speak of *internal path length* and *external path length*.

If each node *must* have a specific number of children appearing in a specific order, then we have a *multiway* tree. In such a tree, it is appropriate to define special external nodes which have no children (and usually no name or other associated information). Then the external nodes act as "dummy" nodes for reference by nodes which do not have the specified number of children.

In particular, the simplest type of multiway tree is the *binary tree*. A binary tree is an ordered tree consisting of two types of nodes: external nodes (with no children) and internal nodes with exactly two children. An example of a binary tree is shown in Figure 4.2. Since the two children of each internal node are ordered, we refer to the *left child* and the *right child* of internal nodes: every internal node

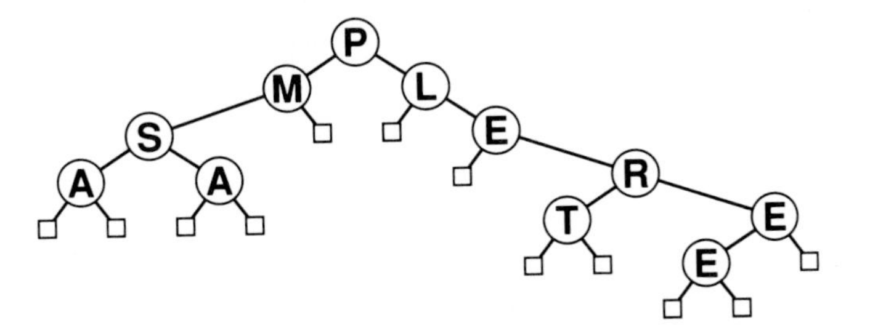

Figure 4.2 A sample binary tree.

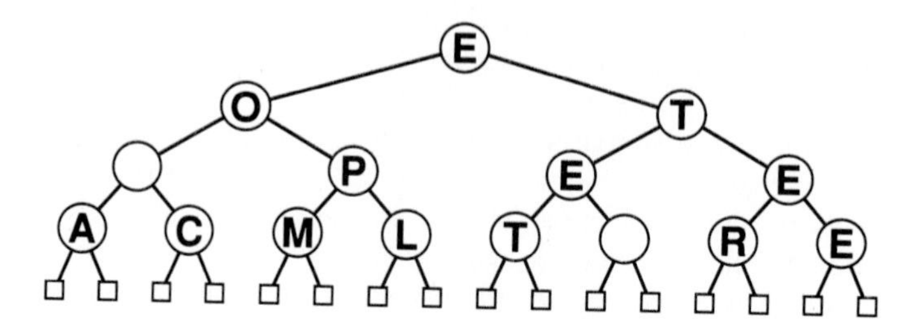

Figure 4.3 A complete binary tree.

must have both a left and a right child, though one or both of them might be an external node.

The purpose of the binary tree is to structure the internal nodes; the external nodes serve only as placeholders. We include them in the definition because the most commonly used representations for binary trees must account for each external node. A binary tree could be "empty", consisting of no internal nodes and one external node.

A *full* binary tree is one in which internal nodes completely fill every level, except possibly the last. A *complete* binary tree is a full binary tree where the internal nodes on the bottom level all appear to the left of the external nodes on that level. Figure 4.3 shows an example of a complete binary tree. As we shall see, binary trees appear extensively in computer applications, and performance is best when they are full (or nearly full). In Chapter 11 we will examine an important data structure based on complete binary trees.

The reader should note carefully that, though every binary tree is a tree, not every tree is a binary tree. Even considering only ordered trees in which every node has 0, 1, or 2 children, each such tree might correspond to many binary trees, because nodes with 1 child could be either left or right in a binary tree.

Trees are intimately connected with recursion, as we will see in the next chapter. In fact, perhaps the simplest way to define trees is recursively, as follows: "a tree is either a single node or a root node connected to a set of trees" and "a binary tree is either an external node or a root (internal) node connected to a left binary tree and a right binary tree."

Properties

Before considering representations, we continue in a mathematical vein by considering a number of important properties of trees. Again, there are a vast number of possible properties to consider—our purpose is to consider those which are particularly relevant to the algorithms to be considered later in this book.

Property 4.1 *There is exactly one path connecting any two nodes in a tree.*

Any two nodes have a *least common ancestor*: a node that is on the path from both nodes to the root, but with none of its children having the same property. For example, O is the least common ancestor of C and L in the tree of Figure 4.3. The least common ancestor must always exist because either the root is the least common ancestor, or both of the nodes are in the subtree rooted at one of the children of the root; in the latter case either that node is the least common ancestor, or both of the nodes are in the subtree rooted at one of its children, etc. There is a path from each of the nodes to the least common ancestor—patching these two paths together gives a path connecting the two nodes. ■

An important implication of Property 1 is that *any* node can be the root: each node in a tree has the property that there is exactly one path connecting that node with every other node in the tree. Technically, our definition, in which the root is identified, pertains to a *rooted tree* or *oriented tree*; a tree in which the root is not identified is called a *free tree*. The reader need not be concerned about making this distinction: either the root is identified, or it is not.

Property 4.2 *A tree with N nodes has $N - 1$ edges.*

This property follows directly from the observations that each node, except the root, has a unique parent, and every edge connects some node to its parent. We can also prove this fact by induction from the recursive definition. ■

The next two properties that we consider pertain to binary trees. As mentioned above, these structures occur quite frequently throughout this book, so it is worthwhile to devote some attention to their characteristics. This lays the groundwork for understanding the performance characteristics of various algorithms we will encounter.

Property 4.3 *A binary tree with N internal nodes has $N + 1$ external nodes.*

This property can be proven by induction. A binary tree with no internal nodes has one external node, so the property holds for $N = 0$. For $N > 0$, any binary tree with N internal nodes has k internal nodes in its left subtree and $N - 1 - k$ internal nodes in its right subtree for some k between 0 and $N - 1$, since the root is an internal node. By the inductive hypothesis, the left subtree has $k + 1$ external nodes and the right subtree has $N - k$ external nodes, for a total of $N + 1$. ■

Property 4.4 *The external path length of any binary tree with N internal nodes is $2N$ greater than the internal path length.*

This property can also be proven by induction, but an alternate proof is also instructive. Observe that any binary tree can be constructed by the following process: start with the binary tree consisting of one external node. Then repeat the following N times: pick some external node and replace it by a new internal node with two external nodes as children. If the external node chosen is at level k, the internal path length is increased by k, but the external path length is increased by $k + 2$

(one external node at level k is removed, but two at level $k+1$ are added). The process starts with a tree with internal and external path length both 0 and, for $N-1$ steps, increases the external path length by 2 more than the internal path length. ■

Finally, we consider simple properties of the "best" kind of binary trees, full trees. These trees are of interest because their height is guaranteed to be low, so we never have to do much work to get from the root to any node or vice versa.

Property 4.5 *The height of a full binary tree with N internal nodes is about* $\log_2 N$.

Referring to Figure 4.3, if the height is n, then we must have

$$2^{n-1} < N+1 \leq 2^n,$$

since there are $N+1$ external nodes. This implies the property stated. (Actually, the height is exactly equal to $\log_2 N$ rounded up to the nearest integer, but we will refrain from being quite so precise, as discussed in Chapter 6.) ■

Further mathematical properties of trees will be discussed as needed in the chapters which follow. At this point, we're ready to move on to the practical matter of representing trees in the computer and manipulating them in an efficient fashion.

Representing Binary Trees

The most prevalent representation of binary trees is a straightforward use of records with *two* links per node. Normally, we will use the link names l and r (abbreviations for "left" and "right") to indicate that the ordering chosen for the representation corresponds to the way the tree is drawn on the page. For some applications, it may be appropriate to have two different types of records, one for internal nodes, one for external nodes; for others, it may be appropriate to use just one type of node and to use the links in external nodes for some other purpose.

As an example in using and constructing binary trees, we'll continue with the simple example from the previous chapter, processing arithmetic expressions. There is a fundamental correspondence between arithmetic expressions and trees, as shown in Figure 4.4.

We use single-character identifiers rather than numbers for the arguments; the reason for this will become plain below. The *parse tree* for an expression is defined by the simple recursive rule: "put the operator at the root and then put the tree for the expression corresponding to the first operand on the left and the tree corresponding to the expression for the second operand on the right." Figure 4.4 is also the parse tree for $A\ B\ C + D\ E * * F + *$ (the same expression in postfix)—infix and postfix are two ways to represent arithmetic expressions, parse trees are a third.

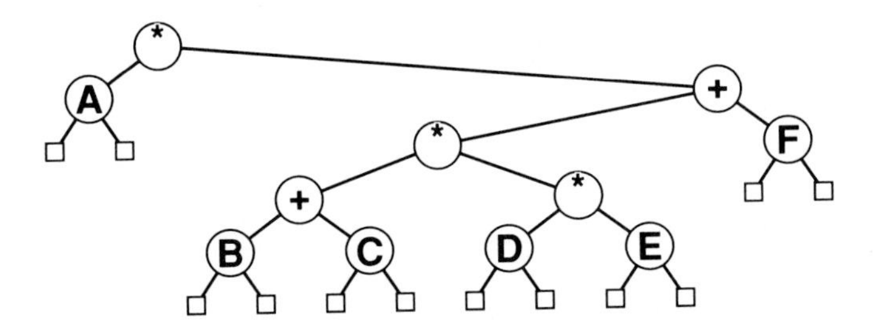

Figure 4.4 Parse tree for A*(((B+C)*(D*E))+F).

Since the operators take exactly two operands, a binary tree is appropriate for this kind of expression. More complicated expressions might require a different type of tree. We will revisit this issue in greater detail in Chapter 21; our purpose here is simply to construct a tree representation of an arithmetic expression.

The following code builds a parse tree for an arithmetic expression from a postfix input representation. It is a simple modification of the program given in the previous chapter for evaluating postfix expressions using a stack. Rather than saving the results of intermediate calculations on the stack, we save the expression trees, as in the following implementation:

```
type link=↑node;
       node=record info: char; l,r: link end;
var x,z: link;
      c: char;
begin
stackinit;
new(z); z↑.l:=z; z↑.r:=z;
repeat
  repeat read(c) until c<>' ';
  new(x); x↑.info:=c;
  if (c='*') or (c='+')
    then begin x↑.r:=pop; x↑.l:=pop end
    else  begin x↑.r:=z; x↑.l:=z end;
  push(x)
until eoln;
```

The procedures *stackinit*, *push*, and *pop* here refer to the pushdown stack code from Chapter 3, modified to put *links* on the stack rather than integers. The code for these is omitted here. Every node has a character and two links to other nodes. Each time a new nonblank character is encountered, a node is created for it using the Pascal primitive *new*. If it is an operator, subtrees for its operands are at the top of the stack, just as for postfix evaluation. If it is an operand, then its links are null. Rather than using null links, as with lists, we use a dummy node z whose links point to itself. In Chapter 14, we examine in detail how this makes certain operations on trees more convenient. Figure 4.5 shows the intermediate stages in the construction of the tree in Figure 4.4.

This rather simple program can be modified to handle more complicated expressions involving single-argument operators such as exponentiation. But the mechanism is very general: exactly the same mechanism is used, for example, to parse and compile Pascal programs. Once the parse tree has been created, then it can be used for many things, such as evaluating the expression or creating computer programs to evaluate the expression. Chapter 21 discusses general procedures for building parse trees. Below we shall see how the tree itself can be used to evaluate the expression. For the purposes of this chapter, however, we are most interested in the mechanics of the construction of the tree.

As with linked lists, there is always the alternative of using parallel arrays rather than pointers and records to implement the binary tree data structure. As before, this is especially useful when the number of nodes is known in advance. Also as before, the particular special case where the nodes need to occupy an array for some other purpose calls for this alternative.

The two-link representation for binary trees used above allows going *down* the tree but provides no way to move *up* the tree. The situation is analogous to singly-linked lists vs. doubly-linked lists: one can add another link to each node to allow more freedom of movement, but at the cost of a more complicated implementation. Various other options are available in advanced data structures to facilitate moving around in the tree, but for the algorithms in this book, the two-link representation generally suffices.

In the program above, we used a "dummy" node in lieu of external nodes. As

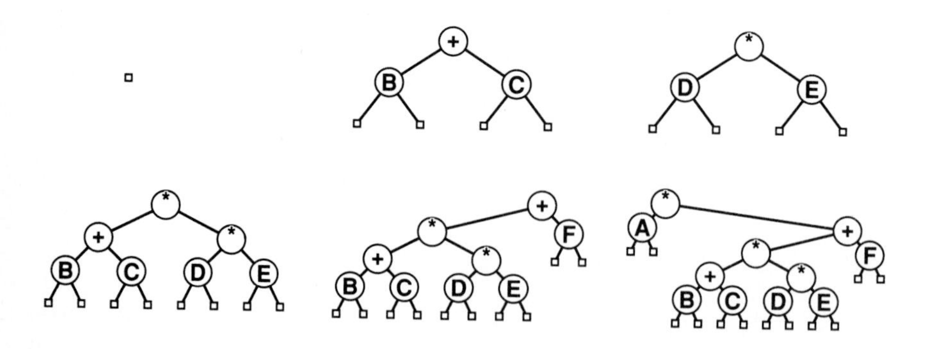

Figure 4.5 Building the parse tree for $A\ B\ C + D\ E * * F + *$.

with linked lists, this turns out to be convenient in most situations, but is not always appropriate, and there are two other commonly used solutions. One option is to use a different type of node for external nodes, one with no links. Another method is to mark the links in some way (to distinguish them from other links in the tree), then have them point elsewhere in the tree; one option for this is discussed below. We will revisit this issue in Chapters 14 and 17.

Representing Forests

Binary trees have two links below each internal node, so the representation used above for them is immediate. But what do we do for general trees, or forests, in which each node might require any number of links to the nodes below? It turns out that there are two relatively simple ways out of this dilemma.

First, in many applications, we don't need to go down the tree, only *up*! In such cases, we only need one link for each node, to its parent. Figure 4.6 shows this representation for the tree in Figure 4.1: the array a contains the information associated with each record and the array *dad* contains the parent links. Thus the information associated with the parent of $a[i]$ is in $a[dad[i]]$, etc. By convention, the root is set to point to itself. This is a rather compact representation that is definitely recommended if working up the tree is appropriate. We'll see examples of the use of this representation in Chapters 22 and 30.

To represent a forest for top-down processing, we need a way to handle the children of each node without preallocating a specific number for any node. But this is exactly the type of constraint that linked lists are designed to remove. Clearly, we should use a linked list for the children of each node. Then each node contains two links, one for the linked list connecting it to its siblings, the other for the linked list of its children. Figure 4.7 shows this representation for the tree of Figure 4.1. Rather than use a dummy node to terminate each list, we simply make it point back to the parent; this gives a way to move up the tree as well as down. (These links may be marked to distinguish them from "sibling" links; alternatively, we can scan through the children of a node by marking or saving the name of the parent so that the scan can be stopped when the parent is revisited.)

But in this representation, each node has exactly two links (one to its sibling on the right, the other to its leftmost child) One might then wonder whether there is a difference between this data structure and a binary tree. The answer is that

k	1	2	3	4	5	6	7	8	9	10	11
a[k]	Ⓐ	Ⓢ	Ⓐ	Ⓜ	Ⓟ	Ⓛ	Ⓔ	Ⓣ	Ⓡ	Ⓔ	Ⓔ
dad[k]	3	3	10	8	8	8	8	9	10	10	10

Figure 4.6 Parent link representation of a tree.

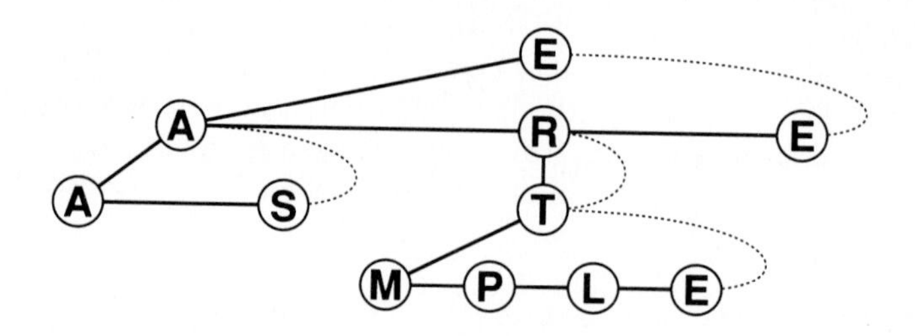

Figure 4.7 Leftmost child, right sibling representation of a tree.

there is not, as shown in Figure 4.8, the binary tree representation of the tree in Figure 4.1. That is, any forest can be represented as a binary tree by making the left link of each node point to its leftmost child, and the right link of each node point to its sibling on the right. (This fact is often surprising to the novice.)

Thus we may as well use forests whenever convenient in algorithm design. When working from the bottom up, the parent link representation makes forests easier to deal with than nearly any other kind of tree, and when working from the top down, they are essentially equivalent to binary trees.

Traversing Trees

Once a tree has been constructed, the first thing one needs to know is how to *traverse* it: how to systematically visit every node. This operation is trivial for linear lists by their definition, but for trees, there are a number of different ways to proceed. The methods differ primarily in the *order* in which they visit the nodes. As we'll see, different node orderings are appropriate for different applications.

For the moment, we'll concentrate on traversing binary trees. Because of the equivalence between forests and binary trees, the methods are useful for forests as well, but we also mention later how the methods apply directly to forests.

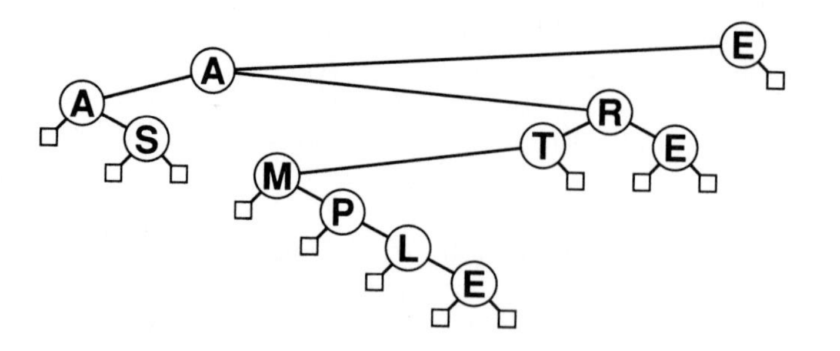

Figure 4.8 Binary tree representation of a tree.

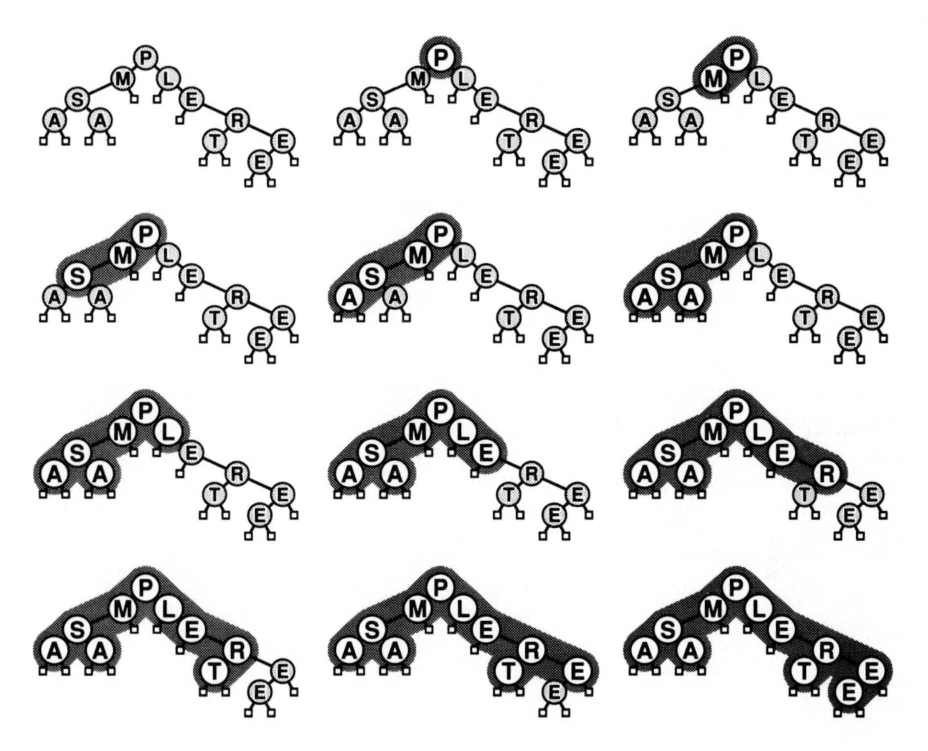

Figure 4.9 Preorder traversal.

The first method to consider is *preorder* traversal, which can be used, for example, to write out the expression represented by the tree in Figure 4.4 in prefix. The method is defined by the simple recursive rule: "visit the root, then visit the left subtree, then visit the right subtree". The simplest implementation of this method, a recursive one, is shown in the next chapter to be closely related to the following stack-based implementation:

```
procedure traverse(t: link);
  begin
  push(t);
  repeat
    t:=pop;
    visit(t);
    if t<>z then push(t↑.r);
    if t<>z then push(t↑.l);
  until stackempty;
  end;
```

(The stack is assumed to be initialized outside this procedure.) Following the rule, we "visit a subtree" by visiting the root first. Then, since we can't visit both subtrees at once, we save the right subtree on a stack and visit the left subtree. When the left subtree has been visited, the right subtree will be at the top of the stack; it can then be visited. Figure 4.9 shows this program in operation when applied to the binary tree in Figure 4.2: the order in which the nodes are visited is P M S A A L E R T E E.

To prove that this program actually visits the nodes of the tree in preorder, one can use induction with the inductive hypothesis that the subtrees are visited in preorder *and* that the contents of the stack just before visiting a subtree are the same as the contents of the stack just after.

Second, we consider *inorder* traversal, which can be used, for example, to write out arithmetic expressions corresponding to parse trees in infix (with some extra work to get the parentheses right). In a manner similar to preorder, inorder is defined with the recursive rule "visit the left subtree, then visit the root, then visit the right subtree." This is also sometimes called *symmetric order*, for obvious reasons. The implementation of a stack-based program for inorder is almost identical to

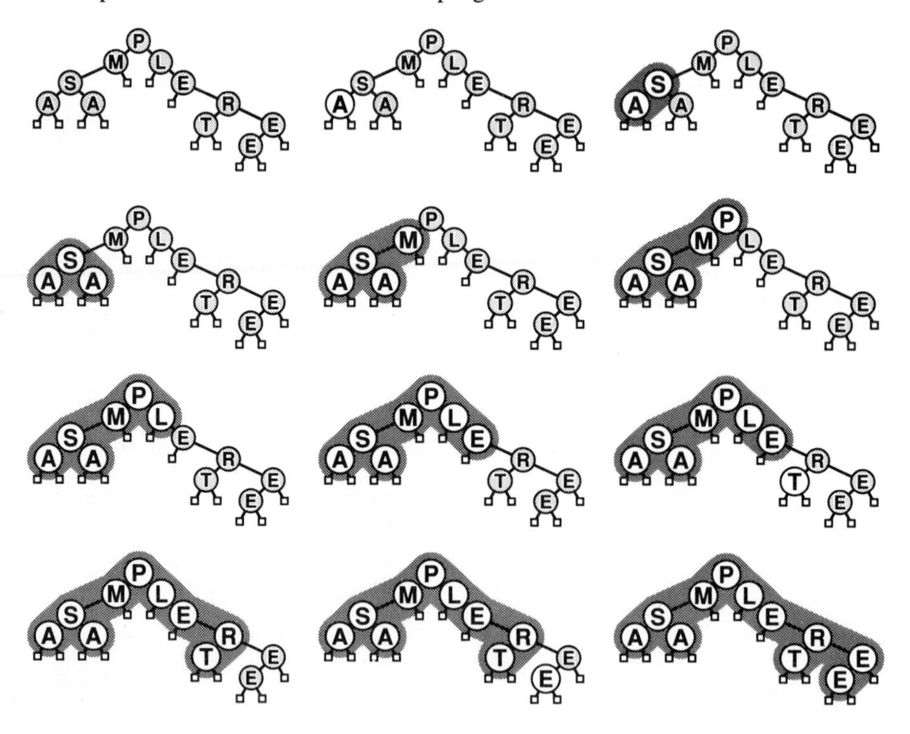

Figure 4.10 Inorder traversal.

the above program; we will omit it here because it is a main topic of the next chapter. Figure 4.10 shows how the nodes in the tree in Figure 4.2 are visited in inorder: the nodes are visited in the order A S A M P L E T R E E. This method of traversal is probably the most widely used: for example, it plays a central role in the applications of Chapters 14 and 15.

The third type of recursive traversal, called *postorder*, is defined, of course, by the recursive rule "visit the left subtree, then visit the right subtree, then visit the root." Figure 4.11 shows how the nodes of the tree in Figure 4.2 are visited in postorder: the nodes are visited in the order A A S M T E E R E L P. Visiting the expression tree of Figure 4.4 in postorder gives the expression $A\ B\ C + D\ E * * F + *$, as expected. Implementation of a stack-based program for postorder is more complicated than for the other two because one must arrange for the root and the right subtree to be saved while the left subtree is visited *and* for the root to be saved while the right subtree is visited. The details of this implementation are left as an exercise for the reader.

The fourth traversal strategy that we consider is not recursive at all—we simply visit the nodes as they appear on the page, reading down from top to bottom and

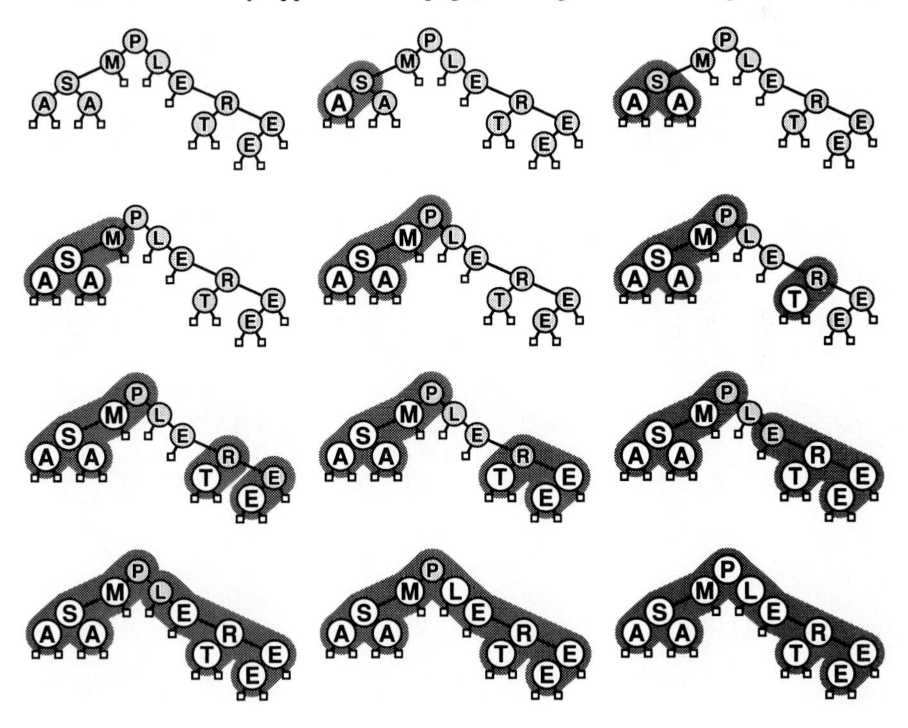

Figure 4.11 Postorder traversal.

from left to right. This is called *level-order* traversal because all the nodes on each level appear together, in order. Figure 4.12 shows how the nodes of the tree in Figure 4.2 are visited in level order.

Remarkably, level-order traversal can be achieved by using the program above for preorder, with a queue instead of a stack:

```
procedure traverse(t: link);
  begin
  put(t);
  repeat
    t:=get;
    visit(t);
    if t<>z then put(t↑.l);
    if t<>z then put(t↑.r);
  until queueempty;
  end;
```

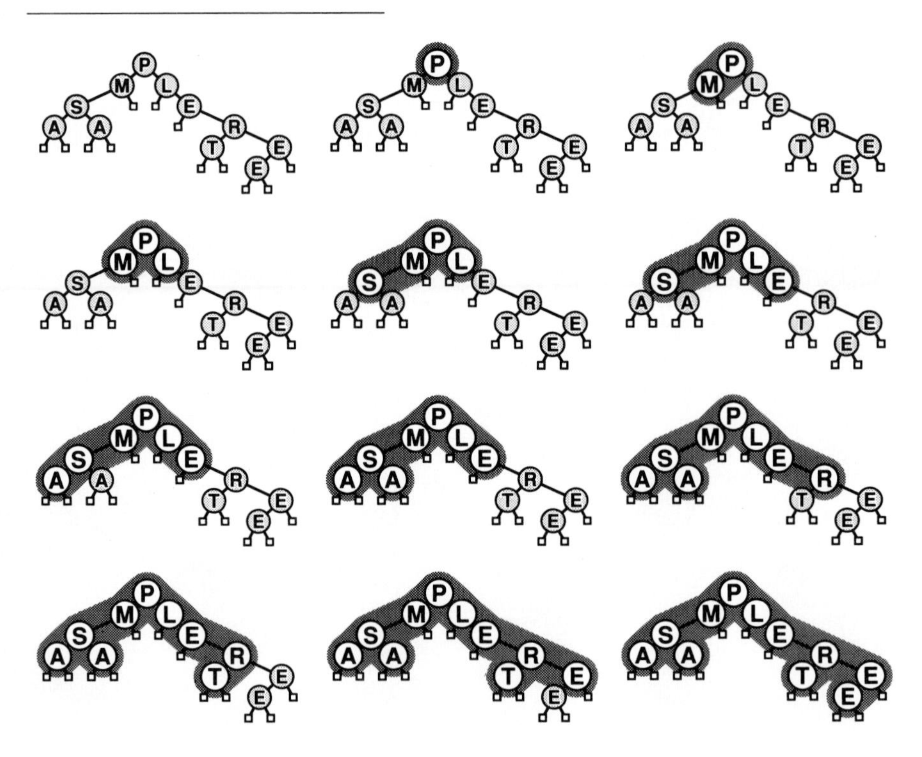

Figure 4.12 Level order traversal.

On the one hand, this program is virtually identical to the one above—the only difference is its use of a FIFO data structure where the other uses a LIFO data structure. On the other hand, these programs process trees in fundamentally different ways. These programs merit careful study, for they expose the essence of the difference between stacks and queues. We shall return to this issue in Chapter 30.

Preorder, postorder and level order are well defined for forests as well. To make the definitions consistent, think of a forest as a tree with an imaginary root. Then the preorder rule is "visit the root, then visit each of the subtrees," the postorder rule is "visit each of the subtrees, then visit the root." The level-order rule is the same as for binary trees. Note that preorder for a forest is the same as preorder for the corresponding binary tree as defined above, and that postorder for a forest is the same as inorder for the binary tree, but the level orders are *not* the same. Direct implementations using stacks and queues are straightforward generalizations of the programs given above for binary trees.

□

Exercises

1. Give the order in which the nodes are visited when the tree in Figure 4.3 is visited in preorder, inorder, postorder, and level order.
2. What is the height of a complete 4-way tree with N nodes?
3. Draw the parse tree for the expression (A+B)*C+(D+E).
4. Consider the tree of Figure 4.2 as a forest that is to be represented as a binary tree. Draw that representation.
5. Give the contents of the stack each time a node is visited during the preorder traversal depicted in Figure 4.9.
6. Give the contents of the queue each time a node is visited during the level order traversal depicted in Figure 4.12.
7. Give an example of a tree for which the stack in a preorder traversal uses more space than the queue in a level-order traversal.
8. Give an example of a tree for which the stack in a preorder traversal uses less space than the queue in a level-order traversal.
9. Give a stack-based implementation of postorder traversal of a binary tree.
10. Write a program to implement level-order traversal of a forest represented as a binary tree.

5

Recursion

□ Recursion is a fundamental concept in mathematics and computer science. The simple definition is that a recursive program is one that calls itself (and a recursive function is one that is defined in terms of itself). Yet a recursive program can't call itself always, or it would never stop (and a recursive function can't be defined in terms of itself always, or the definition would be circular); another essential ingredient is that there must be a *termination condition* when the program can cease to call itself (and when the function is not defined in terms of itself). All practical computations can be couched in a recursive framework.

Our primary purpose in this chapter is to examine recursion as a practical tool. First, we show some examples in which it is *not* practical, while showing the relationship between simple mathematical recurrences and simple recursive programs. Next we show a prototype example of a "divide-and-conquer" recursive program of the type that we use to solve fundamental problems in several later sections of this book. Finally, we discuss how recursion can be removed from any recursive program, and show a detailed example of removing recursion from a simple recursive tree traversal algorithm to get a simple nonrecursive stack-based algorithm.

As we shall see, many interesting algorithms are quite simply expressed with recursive programs, and many algorithm designers prefer to express methods recursively. But it is also very often the case that an equally interesting algorithm lies hidden in the details of a (necessarily) nonrecursive implementation—in this chapter we discuss techniques for finding such algorithms.

Recurrences

Recursive definitions of functions are quite common in mathematics—the simplest type, involving integer arguments, are called *recurrence relations*. Perhaps the most familiar such function is the *factorial* function, defined by the formula

$$N! = N \cdot (N-1)!, \qquad \text{for } N \geq 1 \text{ with } 0! = 1.$$

This corresponds directly to the following simple recursive program:

```
function factorial(N: integer): integer;
  begin
  if N=0
    then factorial:=1
    else factorial:=N*factorial(N-1);
  end;
```

On the one hand, this program illustrates the basic features of a recursive program: it calls itself (with a smaller value of its argument), and it has a termination condition in which it directly computes its result. On the other hand, there is no masking the fact that this program is nothing more than a glorified **for** loop, so it is hardly a convincing example of the power of recursion. Also, it is important to remember that it is a *program*, not an equation: for example, neither the equation nor the program above "works" for negative N, but the negative effects of this oversight are perhaps more noticeable with the program than with the equation. The call *factorial(−1)* results in an infinite recursive loop; this is in fact a common bug which can appear in more subtle forms in more complicated recursive programs.

A second well-known recurrence relation is the one that defines the *Fibonacci numbers*:

$$F_N = F_{N-1} + F_{N-2}, \qquad \text{for } N \geq 2 \text{ with } F_0 = F_1 = 1.$$

This defines the sequence

$$1, 1, 2, 3, 5, 8, 13, 21, 34, 55, 89, 144, 233, 377, 610, \ldots.$$

Again, the recurrence corresponds directly to the simple recursive program:

```
function fibonacci(N: integer): integer;
  begin
  if N<=1
    then fibonacci:=1
    else fibonacci:=fibonacci(N-1)+fibonacci(N-2);
  end;
```

This is an even less convincing example of the "power" of recursion; indeed, it *is* a convincing example that recursion should not be used blindly, or dramatic inefficiencies can result. The problem here is that the recursive calls indicate that F_{N-1} and F_{N-2} should be computed independently, when, in fact, one certainly would use F_{N-2} (and F_{N-3}) to compute F_{N-1}. It is actually easy to figure out the exact number of calls on the procedure *fibonacci* above that are required to compute

F_N: the number of calls needed to compute F_N is the number of calls needed to compute F_{N-1} plus the number of calls needed to compute F_{N-2} unless $N = 0$ or $N = 1$, when only one call is needed. But this fits the recurrence relation defining the Fibonacci numbers exactly: the number of calls on *fibonacci* to compute F_N is exactly F_N. It is well known that F_N is about ϕ^N, where $\phi = 1.61803\ldots$ is the "golden ratio": the awful truth is that the above program is an *exponential-time* algorithm for computing the Fibonacci numbers!

By contrast, it is very easy to compute F_N in linear time, as follows:

```
procedure fibonacci;
  const max=25;
  var i: integer;
      F: array[0..max] of integer;
  begin
  F[0]:=1; F[1]:=1;
  for i:=2 to max do F[i]:=F[i-1]+F[i-2]
  end;
```

This program computes the first *max* Fibonacci numbers, using an array of size *max*. (Since the numbers grow exponentially, *max* will be small.)

In fact, this technique of using an array to store previous results is typically the method of choice for evaluating recurrence relations, for it allows rather complex equations to be handled in a uniform and efficient manner. Recurrence relations often arise when we try to determine performance characteristics of recursive programs, and we'll see several examples in this book. For example, in Chapter 9 we encounter the equation

$$C_N = N - 1 + \frac{1}{N} \sum_{1 \le k \le N} (C_{k-1} + C_{N-k}). \qquad \text{for } N \ge 1 \text{ with } C_0 = 1.$$

The value of C_N can be rather easily computed using an array as in the program above. In Chapter 9, we discuss how this formula can be handled mathematically, and several recurrences which occur more frequently are discussed in Chapter 6.

Thus, the relationship between recursive programs and recursively defined functions is often more philosophical than practical. Strictly speaking, the problems pointed out above are associated not with the concept of recursion itself, but with the implementation: a (very smart) compiler might discover that the factorial function really could be implemented with a loop and that the Fibonacci function is better handled by storing all precomputed values in an array. Below, we'll look in more detail at the mechanics of implementing recursive programs.

Divide-and-Conquer

Most of the recursive programs we consider in this book use two recursive calls, each operating on about half the input. This is the so-called "divide and conquer" paradigm for algorithm design, which is often used to achieve significant economies. Divide-and-conquer programs normally do not reduce to trivial loops like the factorial program above, because they have two recursive calls; they normally do not lead to excessive recomputing as in the program for Fibonacci numbers above, because the input is divided without overlap.

As an example, let us consider the task of drawing the markings for each inch on a ruler: there is a mark at the 1/2" point, slightly shorter marks at 1/4" intervals, still shorter marks at 1/8" intervals, etc., as shown (in magnified form) in Figure 5.1. As we'll see there are many ways to carry out this task, and it is a prototype of simple divide-and-conquer computations.

If the desired resolution is $1/2^n$" we simplify things by multiplying everything by 2^n, so that our task is to put a mark at every point between 0 and 2^n, endpoints not included. We assume that we have at our disposal a procedure *mark*(x,h) to make a mark h units high at position x. The middle mark should be n units high, the marks in the middle of the left and right halves should be $n-1$ units high, etc. The following "divide-and-conquer" recursive program accomplishes our objective:

```
procedure rule(l,r,h: integer);
  var m: integer;
  begin
  if h>0 then
    begin
    m:=(l+r) div 2;
    mark(m,h);
    rule(l,m,h-1);
    rule(m,r,h-1)
    end
  end;
```

For example, the call *rule*(0,64,6) will yield Figure 5.1, with appropriate scaling. The idea behind the method is the following: to make the marks in an interval, first make the long mark in the middle. This divides the interval into two equal halves. Make the (shorter) marks in the first half, using the same procedure, then

Figure 5.1 A ruler.

make the (shorter) marks in the second half, using the same procedure.

It is normally prudent to pay special attention to the termination condition of a recursive program—otherwise it may not terminate! In the above program, we terminate when the length of the mark to be made is 0. Figure 5.2 shows the process in detail, giving the list of procedure calls and marks resulting from the call *rule(0,8,3)*. We mark the middle and call *rule* for the left half, then do the same for the left half, and so forth, until a mark of length 0 is called for. Eventually we return from *rule* and mark right halves in the same way.

For this problem, the order in which the marks are drawn is not particularly relevant. We could just as well have put the *mark* call *between* the two recursive calls, in which case the points for our example would be plotted in a completely different order, as shown in Figure 5.3. Here the marks are drawn in order from left to right.

The collection of marks drawn by these two procedures is the same; but the

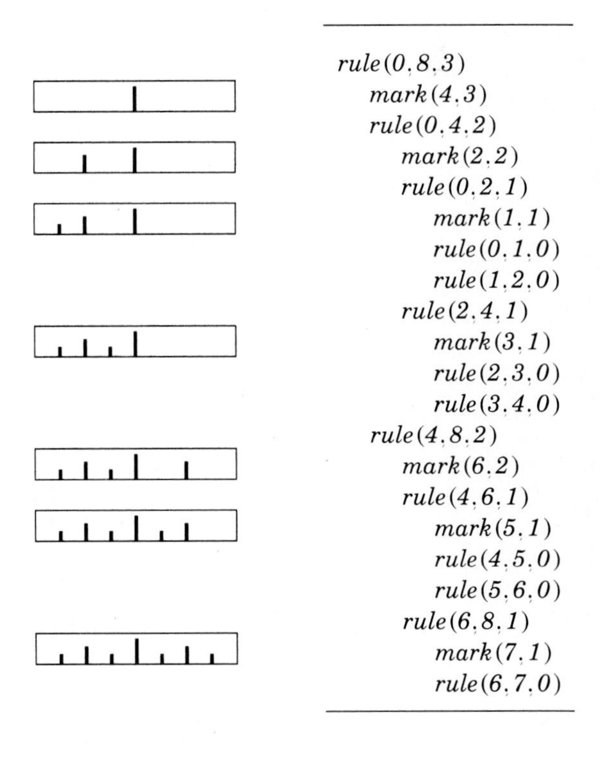

Figure 5.2 Drawing a ruler.

ordering is quite different. This may be explained by the tree diagram shown in Figure 5.4. This diagram has a node for each call to *rule*, labeled with the parameters used in the call. The children of each node correspond to the (recursive) calls to *rule*, along with their parameters, for that invocation. A tree like this can be drawn to illustrate the dynamic characteristics of any collection of procedures. Now, Figure 5.2 corresponds to traversing this tree in preorder (where "visiting" a node corresponds to making the indicated call to *mark*); Figure 5.3 corresponds to traversing it in inorder.

In general, divide-and-conquer algorithms involve doing some work to split the input into two pieces, or to merge the results of processing two independent "solved" portions of the input, or to help things along after half of the input has been processed. That is, there may be code before, after, or in between the two recursive calls. We'll see many examples of such algorithms later, especially in Chapters 9, 12, 27, 28, and 41. We also encounter algorithms in which it is not

```
rule(0,8,3)
   rule(0,4,2)
      rule(0,2,1)
         rule(0,1,0)
         mark(1,1)
         rule(1,2,0)
      mark(2,2)
      rule(2,4,1)
         rule(2,3,0)
         mark(3,1)
         rule(3,4,0)
   mark(4,3)
   rule(4,8,2)
      rule(4,6,1)
         rule(4,5,0)
         mark(5,1)
         rule(5,6,0)
      mark(6,2)
      rule(6,8,1)
         rule(6,7,0)
         mark(7,1)
```

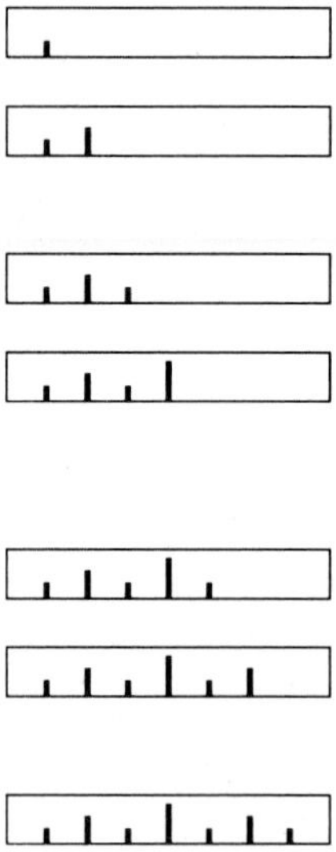

Figure 5.3 Drawing a ruler (inorder version).

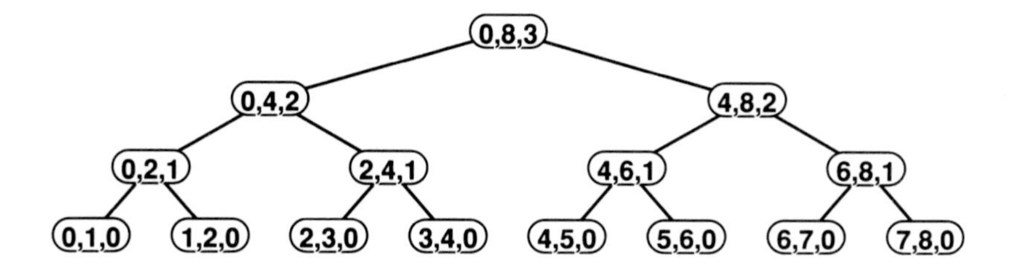

Figure 5.4 Recursive call tree for drawing a ruler.

possible to follow the divide-and-conquer regimen completely: perhaps the input is split into unequal pieces or into more than two pieces, or there is some overlap among the pieces. In the next chapter, we study the recurrence relations which help us determine precisely how much time and space various divide-and-conquer strategies can save.

It is also easy to develop nonrecursive algorithms for this task. The most straightforward method is to simply draw them in order, as in Figure 5.3, but with the direct loop **for** i:= *1* **to** $N-1$ **do** *mark*(*i*,*height*(*i*)). The function *height*(*i*) needed for this turns out to be not hard to compute: it is the number of trailing 0 bits in the binary representation of i. As mentioned in Chapter 2 and discussed in more detail in Chapter 10, working with the binary representation of numbers is possible, though not always convenient, in Pascal. It is actually possible to derive this method directly from the recursive version, through a laborious "recursion removal" process that we examine in detail below, for another problem.

Another nonrecursive algorithm, which does not correspond to any recursive implementation, is to draw the shortest marks first, then the next shortest, etc., as in the following rather compact program:

```
procedure rule(l,r,h: integer);
  var i,j: integer;
  begin
  j:=1;
  for i:=1 to h do
    begin
    for x:=0 to (l+r) div j do
      mark(l+j+x*(j+j),i);
    j:=j+j;
    end;
  end;
```

Figure 5.5 shows how this program draws the marks. This process corresponds to

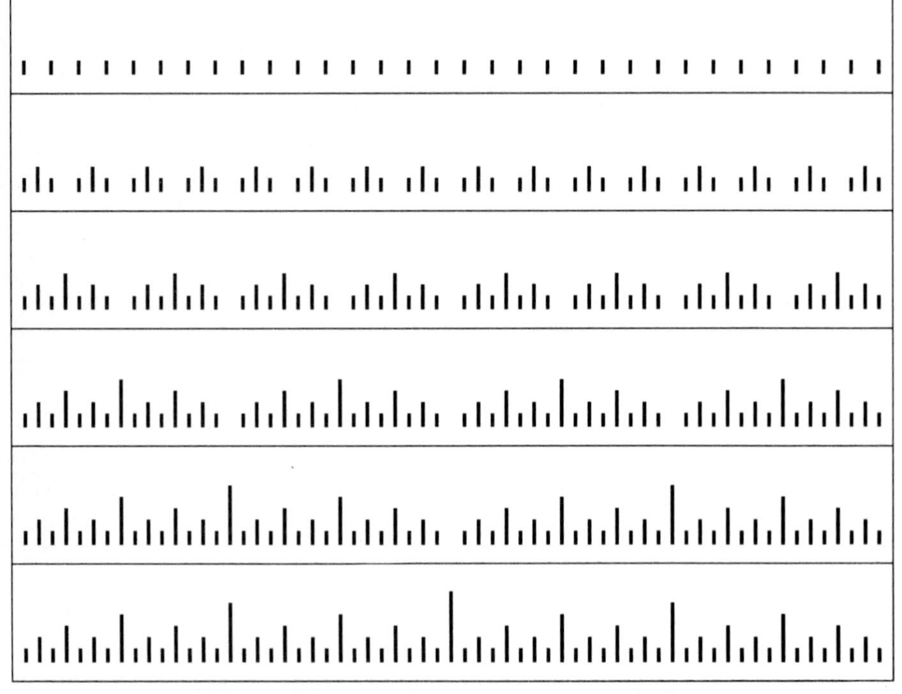

Figure 5.5 Drawing a ruler nonrecursively.

traversing the tree of Figure 5.4 in level order (from the bottom up), but it is not recursive.

This corresponds to the general method of algorithm design where we solve a problem by first solving trivial subproblems, then combining those solutions to solve slightly bigger subproblems, etc. until the whole problem is solved. This approach might be called "combine and conquer." While it is always possible to get an equivalent nonrecursive implementation of any recursive program, it is not always possible to rearrange the computations in this way—many recursive programs depend on the subproblems being solved in a particular order. This is a *bottom-up* approach as contrasted with the *top-down* orientation of divide-and-conquer. We'll encounter several examples of this: the most important is in Chapter 12. A generalization of the method is discussed in Chapter 42.

Figure 5.6 shows a two-dimensional pattern that illustrates how a simple recursive description can lead to a computation that appears to be rather complex. The program that produces the pattern on the left is actually just a slight generalization of *rule*:

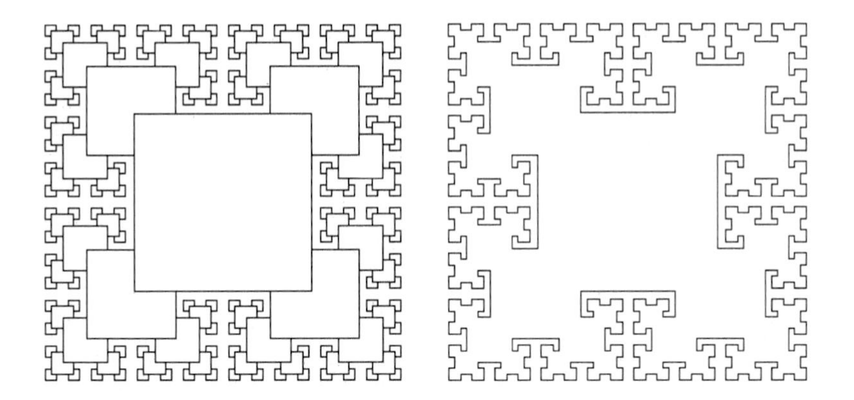

Figure 5.6 A fractal star, drawn with boxes (left) and outline only (right).

```
procedure star(x,y,r: integer);
  begin
  if r>0 then
    begin
    star(x−r,y+r,r div 2);
    star(x+r,y+r,r div 2);
    star(x−r,y−r,r div 2);
    star(x+r,y−r,r div 2);
    box(x,y,r);
    end
  end;
```

The drawing primitive used is simply a program which draws a square of size $2r$ centered at (x,y). Thus the pattern on the left in Figure 5.6 is simple to generate with a recursive program—the reader may be amused to try to find a recursive method for drawing the outline of the pattern shown on the right. The pattern on the left is also easy to generate with a bottom-up method like the one represented by Figure 5.5: draw the smallest squares, then the second smallest, etc. The reader may also be amused to try to find a nonrecursive method for drawing the outline.

Recursively defined geometric patterns like Figure 5.6 are sometimes called *fractals*. If more complicated drawing primitives are used, and more complicated recursive invocations (especially including recursively-defined functions on reals and in the complex plane), patterns of remarkable diversity and complexity can be developed.

Recursive Tree Traversal

As indicated in Chapter 4, perhaps the simplest way to traverse the nodes of a tree is with a recursive implementation. For example, the following program visits the nodes of a binary tree in inorder.

```
procedure traverse(t: link);
  begin
  if t<>z then
    begin
    traverse(t↑.l);
    visit(t);
    traverse(t↑.r)
    end
  end;
```

The implementation precisely mirrors the definition of inorder: "if the tree is nonempty, first traverse the left subtree, then visit the root, then traverse the right subtree." Obviously, preorder can be implemented by putting the call to *visit* before the two recursive calls, and postorder can be implemented by putting the call to *visit* after the two recursive calls.

This recursive implementation of tree traversal is more natural than a stack-based implementation both because trees are recursively defined structures and because preorder, inorder and postorder are recursively defined processes. By contrast, note that there is no convenient way to implement a recursive procedure for level-order traversal (for example). We will return to this issue in Chapters 29 and 30 when we consider traversal algorithms for graphs, much more complicated structures than trees.

Simple modifications to the recursive program above and appropriate implementation of *visit* can lead to programs that compute various properties of trees in a straightforward manner. For example, the following program shows how the coordinates for placing the binary tree nodes in the figures in this book might be computed. Suppose that the record for nodes includes two integer fields for the x and y coordinates of the node on the page. (To avoid details of scaling and translation, these are assumed to be relative coordinates: if the tree has N nodes and is of height h, the x coordinate runs left-to-right from 1 to N and the y coordinate runs top-to-bottom from 1 to h.) The following program fills in these fields with appropriate values for each node:

```
procedure visit(t: link);
  begin x:=x+1; t↑.x:=x; t↑.y:=y end;
procedure traverse(t: link);
  begin
  y:=y+1;
  if t<>z then
    begin
    traverse(t↑.l);
    visit(t);
    traverse(t↑.r)
    end
  y:=y-1;
  end;
```

The program uses two global variables, x and y, both assumed to be initialized to 0. The variable x keeps track of the number of nodes that have been visited in inorder; the variable y keeps the height of the tree. Each time *traverse* goes down in the tree it is incremented by one, and each time it goes up in the tree it is decremented by one.

In a similar manner, one could implement recursive programs to compute the path length of a tree, to implement another way to draw a tree, to evaluate an expression represented by an expression tree, etc.

Removing Recursion

But what is the relationship between the implementation above (recursive) and the implementation in Chapter 4 (nonrecursive) for tree traversal? Certainly they are strongly related, since, given any tree, the two programs produce precisely the same sequence of calls to *visit*. In this section, we study this question in detail by "mechanically" removing the recursion from the preorder traversal program given above to get a nonrecursive implementation.

This is the same task that a compiler is faced with when given the task of translating a recursive program into machine language. Our purpose is not primarily to study compilation techniques (though we do gain some insight into the problems faced by a compiler), but rather to study the relationship between recursive and nonrecursive implementations of algorithms. This theme arises again throughout the book.

To begin, we start with a recursive implementation of preorder traversal, exactly as described above:

```
procedure traverse(t: link);
  begin
  if t<>z then
    begin
    visit(t);
    traverse(t↑.l);
    traverse(t↑.r)
    end
  end;
```

First, the second recursive call can be easily removed because there is no code following it. Whenever the second call is to be executed, *traverse* is to be called (with the argument $t\uparrow.r$); then, when that call is complete, the *current* invocation of *traverse* is also complete. But this same sequence of events can be implemented with a **goto** rather than a recursive call, as follows:

```
procedure traverse(t: link);
  label 0,1;
  begin
0: if t=z then goto 1;
  visit(t);
  traverse(t↑.l);
  t:=t↑.r; goto 0;
1: end;
```

This is a well-known technique called *end-recursion removal* which is implemented on many compilers. Recursive programs are less viable on systems without this capability, because dramatic and unnecessary inefficiencies such as those arising with *factorial* and *fibonacci* above can arise. In Chapter 9, we shall study an important practical example.

Removing the other recursive call requires more work. In general, most compilers produce code which goes through the same sequence of actions for *any* procedure call: "push the values of local variables and the address of the next instruction on a stack, set the values of parameters to the procedure and **goto** the beginning of the procedure." Then, when a procedure completes, it must "pop the return address and values of local variables from the stack, reset the variables, and **goto** the return address." Of course, things are more complex for the general situation which must be faced by a real compiler; nevertheless, in this spirit, we can remove the second recursive call from our program as follows:

```
procedure traverse(t: link);
  label 0,1,2,3;
  begin
0: if t=z then goto 1;
  visit(t);
  push(t); t:=t↑.l; goto 0;
3: t:=t↑.r; goto 0;
1: if stackempty then goto 2;
  t:=pop; goto 3;
2: end;
```

There is only one local variable, *t*, so we push that on the stack and **goto** the beginning. There is only one return address, *3*, which is fixed, so we don't put it on the stack. At the end of the procedure, we set *t* from the stack and **goto** the return address *3*. When the stack is empty, we return from the first call to *traverse*.

Now, the recursion is removed, but we are left with a morass of **goto**s which comprise a rather opaque program. But these, too, can be "mechanically" removed to yield a more structured piece of code. First, the piece of code between the label *3* and the second **goto** *0* is surrounded by **goto**s and can simply be moved, eliminating the label *3* and associated **goto**. Next, the code between the label *0* and the first **goto** *0* is nothing more than a **while** loop. This leaves us with:

```
procedure traverse(t: link);
  label 0,2;
  begin
0: while t<>z do
    begin
    visit(t);
    push(t↑.r); t:=t↑.l
    end;
  if stackempty then goto 2;
  t:=pop; goto 0;
2: end;
```

Now we have another loop, which can be transformed into a **repeat** loop by adding an extra stack push (of the initial argument *t* on entry to traverse), leaving us with a **goto**-less program:

```
procedure traverse(t: link);
  begin
  push(t);
  repeat
    t:=pop;
    while t<>z do
      begin
      visit(t);
      push(t↑.r);
      t:=t↑.l
      end;
  until stackempty;
  end;
```

This version is the "standard" nonrecursive tree traversal method. It is a worthwhile exercise to forget for the moment how this program was derived and directly convince oneself that this program does preorder tree traversal as advertised.

Actually, the loop-within-a-loop structure of this program can be simplified (at the cost of some stack pushes):

```
procedure traverse(t: link);
  begin
  push(t);
  repeat
    t:=pop;
    if t<>z then
      begin
      visit(t);
      push(t↑.r);
      push(t↑.l);
      end;
  until stackempty;
  end;
```

This is strikingly similar to our original recursive preorder algorithm, but the two programs really are *quite* different. One primary difference is that this program can be run in virtually any programming environment, while the recursive implementation obviously requires one that supports recursion. Even in such an environment, this stack-based method is likely to be rather more efficient.

Finally, we note that this program puts null subtrees on the stack, a direct result of the decision in the original implementation to test whether the subtree is null as the first act of the recursive procedure. (Alternatively, the recursive implementation could test $t\uparrow.l$ and $t\uparrow.r$ and make the recursive call only for non-null subtrees.) Changing the above program to avoid putting null subtrees on the stack leads to the stack-based preorder traversal algorithm of Chapter 4.

```
procedure traverse(t: link);
  begin
  push(t);
  repeat
    t:=pop;
    visit(t);
    if t↑.r<>z then push(t↑.r);
    if t↑.l<>z then push(t↑.l);
  until stackempty;
  end;
```

Any recursive algorithm can be manipulated as above to remove the recursion; indeed, this is a primary task of the compiler. "Manual" recursion removal as described here, though a complicated task, often leads to both an efficient nonrecursive implementation and a better understanding of the nature of the recursive computation.

Perspective

It is certainly impossible to do justice to a topic as fundamental as recursion in so brief a discussion. Many of the best examples of recursive programs appear throughout the book—divide-and-conquer algorithms have been devised for a wide variety of problems. For many applications, there is no reason to go beyond a simple, direct recursive implementation; for others, we will consider the result of recursion removal as described in this chapter or derive alternate nonrecursive implementations directly.

Recursion lies at the heart of early theoretical studies into the very nature of computation. Recursive functions and programs play a central role in mathematical studies that attempt to separate problems that can be solved by a computer from problems which cannot.

In Chapter 44, we look at the use of recursive programs (and other techniques) for solving difficult problems in which a large number of possible solutions must be examined. As we shall see, recursive programming can be a quite effective means of organizing a complicated search through the set of possibilities.

□

Exercises

1. Write a recursive program to draw a binary tree so that the root appears at the center of the page, the root of the left subtree is at the center of the left half of the page, etc.
2. Write a recursive program to compute the external path length of a binary tree.
3. Write a recursive program to compute the external path length of a tree represented as a binary tree.
4. Give the coordinates produced when the recursive tree-drawing procedure given in the text is applied to the binary tree in Figure 4.2.
5. Mechanically remove the recursion from the *fibonacci* program given in the text to get a nonrecursive implementation.
6. Mechanically remove the recursion from the recursive *inorder* tree traversal algorithm to get a nonrecursive implementation.
7. Mechanically remove the recursion from the recursive *postorder* tree traversal algorithm to get a nonrecursive implementation.
8. Write a recursive "divide-and-conquer" program to draw an approximation to the line segment connecting two points (x_1, y_1) and (x_2, y_2) by drawing points using only integer coordinates. (Hint: first draw a point close to the middle.)
9. Write a recursive program for solving the Josephus problem (see Chapter 3).
10. Write a recursive implementation of Euclid's algorithm (see Chapter 1).

6

Analysis of Algorithms

☐ For most problems, many different algorithms are available. How is one to choose the best implementation? This is actually a well-developed area of study in computer science. We'll frequently have occasion to call on research results describing the performance of fundamental algorithms. However, comparing algorithms can be challenging indeed, and certain general guidelines will be useful.

Usually the problems we solve have a natural "size" (typically the amount of data to be processed) which we'll normally call N. We would like to describe the resources used (most often the amount of time taken) as a function of N. We're interested in the *average case*, the amount of time a program might be expected to take on "typical" input data, and in the *worst case*, the amount of time a program would take on the worst possible input configuration.

Some of the algorithms in this book are very well understood, to the point that accurate mathematical formulas are known for the average- and worst-case running time. Such formulas are developed by carefully studying the program, to find the running time in terms of fundamental mathematical quantities, and then doing a mathematical analysis of the quantities involved. On the other hand, the performance properties of other algorithms in this book are not understood at all — perhaps their analysis leads to unsolved mathematical questions, or perhaps known implementations are too complex for a detailed analysis to be reasonable, or (most likely) perhaps the types of input they encounter cannot be adequately characterized. Most algorithms fall somewhere in between these extremes: some facts are known about their performance, but they have really not been fully analyzed.

Several important factors go into this analysis which are usually outside a given programmer's domain of influence. First, Pascal programs are translated into machine code for a given computer, and it can be a challenging task to figure out exactly how long even one Pascal statement might take to execute (especially in an environment where resources are being shared, so that even the same program can have varying performance characteristics). Second, many programs are extremely sensitive to their input data, and performance might fluctuate wildly depending on the input. The average case might be a mathematical fiction that is not representative of the actual data on which the program is being used, and the worst case might be a bizarre construction that would never occur in practice. Third, many programs of interest are not well understood, and specific mathematical results may not be available. Finally, it is often the case that programs are not comparable at all: one runs much more efficiently on one particular kind of input, the other runs efficiently under other circumstances.

The above comments notwithstanding, it is often possible to predict precisely how long a particular program will take, or to know that one program will do better than another in particular situations. It is the task of the algorithm analyst to discover as much information as possible about the performance of algorithms; it is the task of the programmer to apply such information in selecting algorithms for particular applications. In this chapter we concentrate on the rather idealized world of the analyst; in the next we discuss practical considerations of implementation.

Framework

The first step in the analysis of an algorithm is to characterize the data which is to be used as input to the algorithm and to decide what type of analysis is appropriate. Ideally, we would like to be able to derive, for any given distribution of the probability of occurrence of the possible inputs, the corresponding distribution of possible running times of the algorithm. We're not able to achieve this ideal for any nontrivial algorithm, so we normally concentrate on bounding the performance statistic by trying to prove that the running time is always less than some "upper bound" *no matter what the input*, and on trying to derive the *average* running time for a "random" input.

The second step in the analysis of an algorithm is to identify abstract operations upon which the algorithm is based, in order to separate the analysis from the implementation. Thus, for example, we separate the study of how many comparisons a sorting algorithm makes from the determination of how many microseconds a particular computer takes to execute whatever machine code a particular compiler produces for the code fragment **if** $a[i]<v$ **then**... . Both these elements are required to determine the actual running time of a program on a particular computer. The former is determined by properties of the algorithm; the latter by properties of the computer. This separation often allows us to make comparisons of algorithms which are at least somewhat independent of particular implementations or particular computers.

While the number of abstract operations involved can in principle be large, it is usually the case that the performance of the algorithms we consider depends on only a few quantities. In general it is rather straightforward to identify the relevant quantities for a particular program—one way to do so is to use a "profiling" option (available in many Pascal implementations) to give instruction frequency counts for some sample runs. In this book, we concentrate on the most important such quantities for each program.

Third, we proceed to the mathematical analysis itself, with the goal of finding average- and worst-case values for each of the fundamental quantities. It is not difficult to find an upper bound on the running time of a program—the challenge is to find the *best* upper bound, one which could actually be achieved if the worst input were encountered. This gives the worst case: the average case normally requires a rather sophisticated mathematical analysis. Once such analyses have been successfully performed on the fundamental quantities, the time associated with each quantity can be determined and expressions for the total running time obtained.

In principle, the performance of an algorithm often can be analyzed to an extremely precise level of detail, limited only by uncertainty about the performance of the computer or by the difficulty of determining the mathematical properties of some of the abstract quantities. However, it is rarely worthwhile to do a complete detailed analysis, so we are always interested in *estimating* in order to suppress detail. (Actually, estimates that seem very rough often turn out to be rather accurate.) Such rough estimates are quite often easy to obtain via the old programming saw "90% of the time is spent in 10% of the code." (This has been quoted in the past for many different values of "90%.")

Analysis of an algorithm is a cyclic process of analyzing, estimating and refining the analysis until an answer to the desired level of detail is reached. Actually, as discussed in the next chapter, the process should also include improvements in the implementation, and indeed such improvements are often suggested by the analysis.

With these caveats in mind, our *modus operandi* will be to look for rough estimates for the running time of our programs for purposes of classification, secure in the knowledge that a fuller analysis can be done for important programs when necessary.

Classification of Algorithms

As mentioned above, most algorithms have a primary parameter N, usually the number of data items to be processed, which affects the running time most significantly. The parameter N might be the degree of a polynomial, the size of a file to be sorted or searched, the number of nodes in a graph, etc. Virtually all of the algorithms in this book have running time proportional to one of the following functions:

1 Most instructions of most programs are executed once or at most only a few times. If all the instructions of a program have this property, we say that its running time is *constant*. This is obviously the situation to strive for in algorithm design.

$\log N$ When the running time of a program is *logarithmic,* the program gets slightly slower as N grows. This running time commonly occurs in programs that solve a big problem by transforming it into a smaller problem, cutting the size by some constant fraction. For our range of interest, the running time can be considered to be less than a "large" constant. The base of the logarithm changes the constant, but not by much: when N is a thousand, $\log N$ is 3 if the base is 10, 10 if the base is 2; when N is a million, $\log N$ is doubled. Whenever N doubles, $\log N$ increases by a constant, but $\log N$ doesn't double until N increases to N^2.

N When the running time of a program is *linear,* it generally is the case that a small amount of processing is done on each input element. When N is a million, then so is the running time. Whenever N doubles, then so does the running time. This is the optimal situation for an algorithm that must process N inputs (or produce N outputs).

$N \log N$ This running time arises for algorithms that solve a problem by breaking it up into smaller subproblems, solving them independently, and then combining the solutions. For lack of a better adjective (*linearithmic*?), we'll say that the running time of such an algorithm is "$N \log N$." When N is a million, $N \log N$ is perhaps twenty million. When N doubles, the running time more than doubles (but not much more).

N^2 When the running time of an algorithm is *quadratic,* it is practical for use only on relatively small problems. Quadratic running times typically arise in algorithms that process all pairs of data items (perhaps in a double nested loop). When N is a thousand, the running time is a million. Whenever N doubles, the running time increases fourfold.

N^3 Similarly, an algorithm that processes triples of data items (perhaps in a triple-nested loop) has a *cubic* running time and is practical for use only on small problems. When N is a hundred, the running time is a million. Whenever N doubles, the running time increases eightfold.

2^N Few algorithms with *exponential* running time are likely to be appropriate for practical use, though such algorithms arise naturally as "brute-force" solutions to problems. When N is twenty, the running time is a million. Whenever N doubles, the running time squares!

The running time of a particular program is likely to be some constant times one of these terms (the "leading term") plus some smaller terms. The values of the constant coefficient and the terms included depend on the results of the analysis and on implementation details. Roughly, the coefficient of the leading term has to do with the number of instructions in the inner loop: at any level of algorithm design it's prudent to limit the number of such instructions. For large N the effect of the leading term dominates; for small N or for carefully engineered algorithms, more terms may contribute and comparisons of algorithms

$\lg N$	$\lg^2 N$	$\sqrt{N}$	N	$N \lg N$	$N \lg^2 N$	$N^{3/2}$	N^2
3	9	3	10	30	90	30	100
6	36	10	100	600	3,600	1,000	10,000
9	81	31	1,000	9,000	81,000	31,000	1,000,000
13	169	100	10,000	130,000	1,690,000	1,000,000	100,000,000
16	256	316	100,000	1,600,000	25,600,000	31,600,000	*ten billion*
19	361	1,000	1,000,000	19,000,000	361,000,000	*one billion*	*one trillion*

Figure 6.1 Approximate relative values of functions.

are more difficult. In most cases, we'll simply refer to the running time of programs as "linear," "$N \log N$," "cubic," etc., with the implicit understanding that more detailed analysis or empirical studies must be done in cases where efficiency is very important.

A few other functions do arise. For example, an algorithm with N^2 inputs that has a running time cubic in N is more properly classed as an $N^{3/2}$ algorithm. Also, some algorithms have two stages of subproblem decomposition, which leads to a running time proportional to $N \log^2 N$. Both these functions should be considered to be much closer to $N \log N$ than to N^2 for large N.

One further note on the "log" function. As mentioned above, the base of the logarithm changes things only by a constant factor. Since we often deal with analytic results only to within a constant factor, it doesn't matter much what the base is, so we refer to "$\log N$," etc. On the other hand, it is sometimes the case that concepts can be explained more clearly when some specific base is used. In mathematics, the *natural logarithm* (base $e = 2.718281828\ldots$) arises so frequently that a special abbreviation is commonly used: $\log_e N \equiv \ln N$. In computer science, the *binary logarithm* (base 2) arises so frequently that the abbreviation $\log_2 N \equiv \lg N$ is commonly used. For example, $\lg N$ rounded up to the nearest integer is the number of bits required to represent N in binary.

Figure 6.1 indicates the relative size of some of these functions: approximate values of $\lg N$, $\lg^2 N$, $\sqrt{N}$, N, $N \lg N$, $N \lg^2 N$, $N^{3/2}$, N^2 are given for various N. The quadratic function clearly dominates, especially for large N, and differences among smaller functions may not be as expected for small N. For example, $N^{3/2}$ should be greater than $N \lg^2 N$ for very large N, but not for the smaller values which might occur in practice. This table is not intended to give a literal comparison of the functions for all N—numbers, tables and graphs relating to specific algorithms can do that. But it does give a realistic first impression.

Computational Complexity

One approach to studying the performance of algorithms is to study the *worst-case* performance, ignoring constant factors, in order to determine the functional dependence of the running time (or some other measure) on the number of inputs (or some other variable). This approach is attractive because it allows one to *prove* precise mathematical statements about the running time of programs: for example, one can say that the running time of Mergesort (see Chapter 11) is *guaranteed* to be proportional to $N \log N$.

The first step in the process is to make the notion of "proportional to" mathematically precise, while at the same time separating the analysis of an algorithm from any particular implementation. The idea is to ignore constant factors in the analysis: in most cases, if we want to know whether the running time of an algorithm is proportional to N or proportional to $\log N$, it matters not whether the algorithm is to be run on a microcomputer or on a supercomputer, and it matters not whether the inner loop has been carefully implemented with only a few instructions or badly implemented with many instructions. From a mathematical point of view, these are equivalent.

The mathematical artifact for making this notion precise is called the *O-notation*, or "big-Oh notation," defined as follows:

Notation. *A function $g(N)$ is said to be $O(f(N))$ if there exist constants c_0 and N_0 such that $g(N)$ is less than $c_0 f(N)$ for all $N > N_0$.*

Informally, this encapsulates the notion of "is proportional to" and frees the analyst from considering the details of particular machine characteristics. Furthermore, the statement that the running time of an algorithm is $O(f(N))$ is independent of the algorithm's input. Since we're interested in studying the *algorithm*, not the input or the implementation, the O-notation is a useful way to state upper bounds on running time which are independent of both inputs and implementation details.

The O-notation has been extremely useful in helping analysts to classify algorithms by performance and in guiding algorithm designers in the search for the "best" algorithms for important problems. The goal of the study of the *computational complexity* of an algorithm is to show that its running time is $O(f(N))$ for some function f, *and* that there can be no algorithm with a running time of $O(g(N))$ for any "smaller" function $g(N)$ (a function with $\lim_{N\to\infty} g(N)/f(N) = 0$). We try to provide both an "upper bound" and a "lower bound" on the worst-case running time. Proving upper bounds is often a matter of counting and analyzing statement frequencies (we will see many examples in the chapters that follow); proving lower bounds is a difficult matter of carefully constructing a machine model and determining which fundamental operations must be performed by any algorithm to solve a problem (we rarely touch upon this). When computational studies show that the upper bound of an algorithm matches its lower bound, then we have some confidence that it's fruitless to try to design an algorithm which is fundamentally faster and we can start to concentrate on the implementation. This point of view has proven very helpful to algorithm designers in recent years.

However, one must be extremely careful of interpreting results expressed using the O-notation, for at least four reasons: first, it is an "upper bound" and the quantity in question might be much lower; second, the input that causes the worst case may be unlikely to occur in practice; third, the constant c_0 is unknown and need not be small; and fourth, the constant N_0 is unknown and need not be small. We consider each of these in turn.

The statement that the running time of an algorithm is $O(f(N))$ does *not* imply that the algorithm ever takes that long: it says only that the analyst has been able to prove that it never takes longer. The actual running time might always be much lower. Better notation has been developed to cover the situation where it is also known that there exists some input for which the running time is $O(f(N))$, but there are many algorithms for which it is rather difficult to construct a worst-case input.

Even if the worst-case input is known, it can be the case that the inputs actually encountered in practice lead to much lower running times. Many extremely useful algorithms have a bad worst case. For example, perhaps the most widely used sorting algorithm, Quicksort, has a running time of $O(N^2)$, but it is possible to arrange things so that the running time for inputs encountered in practice is proportional to $N \log N$.

The constants c_0 and N_0 implicit in the O-notation often hide implementation details which are important in practice. Obviously, to say that an algorithm has running time $O(f(N))$ says nothing about the running time if N happens to be less than N_0, and c_0 might be hiding a large amount of "overhead" designed to avoid a bad worst case. We would prefer an algorithm using N^2 nanoseconds over one using $\log N$ centuries, but we couldn't make this choice on the basis of the O-notation. Figure 6.2 shows the situation for two typical functions, with more realistic values of the constants, in the range $0 \le N \le 1{,}000{,}000$. The $N^{3/2}$ function, which might have been mistakenly assumed to be the largest of the four since it is asymptotically the largest, is actually among the smallest for small N, and is less than $N \lg^2 N$ until N runs well into the tens of thousands. Programs whose

N	$\frac{1}{4}N \lg^2 N$	$\frac{1}{2}N \lg^2 N$	$N \lg^2 N$	$N^{3/2}$
10	22	45	90	30
100	900	1,800	3,600	1,000
1,000	20,250	40,500	81,000	31,000
10,000	422,500	845,000	1,690,000	1,000,000
100,000	6,400,000	12,800,000	25,600,000	31,600,000
1,000,000	90,250,000	180,500,000	361,000,000	1,000,000,000

Figure 6.2 Significance of constant factors in comparing functions.

running times depend on functions such as these can't be intelligently compared without careful attention to constant factors and implementation details.

One should definitely think twice before, for example, using an algorithm with running time $O(N^2)$ in favor of one with running time $O(N)$, but neither should one blindly follow the complexity result expressed in O-notation. For practical implementations of algorithms of the type considered in this book, complexity proofs often are too general and the O-notation is too imprecise to be helpful. Computational complexity must be considered the very first step in a progressive process of refining the analysis of an algorithm to reveal more details about its properties. In this book we concentrate on later steps, closer to actual implementations.

Average-Case Analysis

Another approach to studying the performance of algorithms is to examine the *average case.* In the simplest situation, we can precisely characterize the inputs to the algorithm: for example, a sorting algorithm might operate on an array of N random integers, or a geometric algorithm might process a set of N random points in the plane with coordinates between 0 and 1. Then, we calculate the average number of times each instruction is executed, and calculate the average running time of the program by multiplying each instruction frequency by the time required for the instruction and adding them all together. There are, however, at least three difficulties with this approach, which we consider in turn.

First, on some computers, it may be rather difficult to determine precisely the amount of time required for each instruction. Worse, this is subject to change, and a great deal of detailed analysis for one computer may not be relevant at all to the running time of the same algorithm on another computer. This is exactly the type of problem that computational complexity studies are designed to avoid.

Second, the average-case analysis itself often is a difficult mathematical challenge requiring intricate and detailed arguments. By its nature, the mathematics involved in proving upper bounds is normally less complex, because it need not be as precise. The average-case performance of many algorithms is unknown.

Third, and most serious, in average-case analysis the input model may not accurately characterize the inputs encountered in practice, or there may be no natural input model at all. How should one characterize the input to a program which processes English-language text? On the other hand, few would argue against the use of input models such as "randomly ordered file" for a sorting algorithm, or "random point set" for a geometric algorithm, and for such models it is possible to derive mathematical results which can accurately predict the performance of programs running on actual applications. Though the derivation of such results is normally beyond the scope of this book, we will give a few examples (see Chapter 9), and cite relevant results when appropriate.

Approximate and Asymptotic Results

Often, the results of a mathematical analysis are not exact but are approximate in a precise technical sense: the result might be an expression consisting of a sequence of decreasing terms. Just as we are most concerned with the inner loop of a program, we are most concerned with the *leading term* (the largest term) of a mathematical expression. It was for this type of application that the O-notation was originally developed, and, properly used, it allows one to make concise statements which give good approximations to mathematical results.

For example, suppose (after some mathematical analysis) we determine that a particular algorithm has an inner loop which is iterated $N \lg N$ times on the average (say), an outer section which is iterated N times, and some initialization code which is executed once. Suppose further that we determine (after careful scrutiny of the implementation) that each iteration of the inner loop requires a_0 microseconds, the outer section requires a_1 microseconds, and the initialization part a_2 microseconds. Then we know that the average running time of the program (in microseconds) is

$$a_0 N \lg N + a_1 N + a_2.$$

But it is also true that the running time is

$$a_0 N \lg N + O(N).$$

(The reader may wish to check this from the definition of $O(N)$.) This is significant because, if we're interested in an approximate answer, it says that, for large N, we may not need to find the values of a_1 or a_2. More important, there could well be other terms in the exact running time which may be difficult to analyze: the O-notation provides us with a way to get an approximate answer for large N without bothering with such terms.

Technically, we have no real assurance that small terms can be ignored in this way, because the definition of the O-notation says nothing whatever about the size of the constant c_0: it could be very large. But (though we don't usually bother) there are usually ways in such cases to put bounds on the constants that are small when compared to N, so we normally are justified in ignoring quantities represented by the O-notation when there is a well-specified leading (larger) term. When we do this, we are secure in the knowledge that we could carry out such a proof, if absolutely necessary, though we rarely do so.

In fact, when a function $f(N)$ is asymptotically large compared to another function $g(N)$, we use in this book the (decidedly nontechnical) terminology "about $f(N)$" to mean $f(N)+O(g(N))$. What we lose in mathematical precision we gain in clarity, for we're more interested in the performance of algorithms than in mathematical details. In such cases, the reader can rest assured that, for large N (if not for all N), the quantity in question will be rather close to $f(N)$. For example, even if we know that a quantity is $N(N-1)/2$, we may refer to it as being "about" $N^2/2$. This is more quickly understood and, for example, deviates from the truth only by a tenth of a percent for $N = 1000$. The precision lost in such cases pales by comparison with the precision lost in the more common usage $O(f(N))$. Our goal is to be both precise and concise when describing the performance of algorithms.

Basic Recurrences

As we'll see in the chapters that follow, a great many algorithms are based on the principle of recursively decomposing a large problem into smaller ones, using solutions to the subproblems to solve the original problem. The running time of such algorithms is determined by the size and number of the subproblems and the cost of the decomposition. In this section we look at basic methods for analyzing such algorithms and derive solutions to a few standard formulas which arise in the analysis of many of the algorithms we'll be studying. Understanding the mathematical properties of the formulas in this section will give insight into the performance properties of algorithms throughout the book.

The very nature of a recursive program dictates that its running time for input of size N will depend on its running time for smaller inputs: this translates to a mathematical formula called a *recurrence relation*. Such formulas precisely describe the performance of the corresponding algorithms: to derive the running time, we solve the recurrences. More rigorous arguments related to specific algorithms will come up when we get to the algorithms: here we're interested in the formulas, not the algorithms.

Formula 1. This recurrence arises for a recursive program that loops through the input to eliminate one item:

$$C_N = C_{N-1} + N, \qquad \text{for } N \geq 2 \text{ with } C_1 = 1.$$

Solution: C_N is about $N^2/2$. To solve such a recurrence, we "telescope" it by applying it to itself, as follows:

$$\begin{aligned} C_N &= C_{N-1} + N \\ &= C_{N-2} + (N-1) + N \\ &= C_{N-3} + (N-2) + (N-1) + N \\ &\vdots \\ &= C_1 + 2 + \cdots + (N-2) + (N-1) + N \\ &= 1 + 2 + \cdots + (N-2) + (N-1) + N \\ &= \frac{N(N+1)}{2}. \end{aligned}$$

Evaluating the sum $1 + 2 + \cdots + (N-2) + (N-1) + N$ is elementary: the result given above can be established by adding the same sum, but in reverse order, term by term. This result, twice the value sought, consists of N terms, each of which sums to $N + 1$.

Formula 2. This recurrence arises for a recursive program that halves the input in one step:

$$C_N = C_{N/2} + 1, \qquad \text{for } N \geq 2 \text{ with } C_1 = 0.$$

Solution: C_N is about $\lg N$. As written, this equation is meaningless unless N is even or we assume that $N/2$ is an integer division: for now, assume that $N = 2^n$, so that the recurrence is always well-defined. (Note that $n = \lg N$.) But then the recurrence telescopes even more easily than our first recurrence:

$$\begin{aligned} C_{2^n} &= C_{2^{n-1}} + 1 \\ &= C_{2^{n-2}} + 1 + 1 \\ &= C_{2^{n-3}} + 3 \\ &\vdots \\ &= C_{2^0} + n \\ &= n. \end{aligned}$$

It turns out that the precise solution for general N depends on properties of the binary representation of N, but C_N is about $\lg N$ for all N.

Formula 3. This recurrence arises for a recursive program that halves the input, but perhaps must examine every item in the input.

$$C_N = C_{N/2} + N, \qquad \text{for } N \geq 2 \text{ with } C_1 = 0.$$

Solution: C_N is about $2N$. This telescopes to the sum $N + N/2 + N/4 + N/8 + \ldots$ (as above, this is only precisely defined when N is a power of two). If the sequence were infinite, this is a simple geometric series that evaluates to exactly $2N$. For general N, the precise solution again involves the binary representation of N.

Formula 4. This recurrence arises for a recursive program that has to make a linear pass through the input, before, during, or after it is split into two halves:

$$C_N = 2C_{N/2} + N, \qquad \text{for } N \geq 2 \text{ with } C_1 = 0.$$

Solution: C_N is about $N \lg N$. This is our most widely cited solution, because it is prototypical of many standard divide-and-conquer algorithms.

$$\begin{aligned} C_{2^n} &= 2C_{2^{n-1}} + 2^n \\ \frac{C_{2^n}}{2^n} &= \frac{C_{2^{n-1}}}{2^{n-1}} + 1 \\ &= \frac{C_{2^{n-2}}}{2^{n-2}} + 1 + 1 \\ &\vdots \\ &= n. \end{aligned}$$

The solution is developed very much as in Formula 2, but with the additional trick of dividing both sides of the recurrence by 2^n at the second step to make the recurrence telescope.

Formula 5. This recurrence arises for a recursive program that splits the input into two halves with one step, such as our ruler-drawing program in Chapter 5.

$$C_N = 2C_{N/2} + 1, \qquad \text{for } N \geq 2 \text{ with } C_1 = 0.$$

Solution: C_N is about $2N$. This is derived in the same manner as Formula 4.

Minor variants of these formulas, involving different initial conditions or slight differences in the additive term, can be handled using the same solution techniques, though the reader is warned that some recurrences that seem similar actually may be rather difficult to solve. (There are a variety of advanced general techniques for dealing with such equations with mathematical rigor.) We will encounter a few more complicated recurrences in later chapters, but we defer discussion of their solution until they arise.

Perspective

Many of the algorithms in this book have been subjected to detailed mathematical analyses and performance studies far too complex to be discussed here. Indeed, it is on the basis of such studies that we are able to recommend many of the algorithms we discuss.

Not all algorithms are worthy of such intense scrutiny; indeed, during the design process, it is preferable to work with approximate performance indicators to guide the design process without extraneous detail. As the design becomes more refined, so must the analysis, and more sophisticated mathematical tools need to be applied. Often, the design process leads to detailed complexity studies that lead to "theoretical" algorithms rather far from any particular application. It is a common mistake to assume that rough analyses from complexity studies will translate immediately into efficient practical algorithms: this often leads to unpleasant surprises. On the other hand, computational complexity is a powerful tool for suggesting departures in design upon which important new methods can be based.

One should not use an algorithm without some indication of how it will perform: the approaches described in this chapter will help provide some indication of performance for a wide variety of algorithms, as we will see in the chapters that follow. In the next chapter we discuss other important factors that come into play when choosing an algorithm.

□

Exercises

1. Suppose it is known that the running time of one algorithm is $O(N \log N)$ and that the running time of another algorithm is $O(N^3)$. What does this say about the relative performance of the algorithms?

2. Suppose it is known that the running time of one algorithm is always about $N \log N$ and that the running time of another algorithm is $O(N^3)$. What does this say about the relative performance of the algorithms?

3. Suppose it is known that the running time of one algorithm is always about $N \log N$ and that the running time of another algorithm is always about N^3. What does this say about the relative performance of the algorithms?

4. Explain the difference between $O(1)$ and $O(2)$.

5. Solve the recurrence

$$C_N = C_{N/2} + N^2, \qquad \text{for } N \geq 2 \text{ with } C_1 = 0$$

   when N is a power of two.

6. For what values of N is $10N \lg N > 2N^2$?

7. Write a program to compute the exact value of C_N in Formula 2, as discussed in Chapter 5. Compare the results to $\lg N$.

8. Prove that the precise solution to Formula 2 is $\lg N + O(1)$.

9. Write a recursive program to compute the largest integer less than $\log_2 N$. (Hint: for $N > 1$, the value of this function for N **div** *2* is one greater than for N.)

10. Write an iterative program for the problem in the previous exercise. Then write a program that does the computation using Pascal library subroutines. If possible on your computer system, compare the performance of these three programs.

7

Implementation of Algorithms

☐ As mentioned in Chapter 1, our focus in this book is on the algorithms themselves—when discussing each algorithm, we treat it as if its performance is the crucial factor in the successful completion of some larger task. This point of view is justified both because such situations do arise for each algorithm and because the careful attention we give to finding an efficient way to solve a problem also often leads to a more elegant (and more efficient) algorithm. Of course, this narrow focus is rather unrealistic, since there are many other very real factors that must be taken into consideration when solving a complicated problem with a computer. In this chapter, we discuss issues related to making the rather idealized algorithms that we describe useful in practical applications.

The properties of the algorithm, after all, are only one side of the coin—a computer can be used to solve a problem effectively only if the problem itself is well understood. Careful consideration of properties of applications is beyond the scope of this book; our intention is to provide enough information about basic algorithms that one may make intelligent decisions about their use. Most of the algorithms we consider have proven useful for a variety of applications. The range of algorithms available to solve various problems owes to the range of needs of various applications. There is no "best" searching algorithm (to pick one example), but one method might be quite suitable for application in an airlines reservation system and another might be quite useful for use in the inner loop of a code-breaking program.

Algorithms rarely exist in a vacuum, except possibly in the minds of theoretical algorithm designers who invent methods without regard to any eventual implementation, or applications systems programmers who "hack in" *ad hoc* methods to solve problems that are otherwise well understood. Proper algorithm design involves putting some thought into the potential impact of design decisions on implementations, and proper applications programming involves putting some thought into performance properties of the basic methods used.

Selecting an Algorithm

As we'll see in the chapters that follow, there usually are a number of algorithms available to solve each problem, all with differing performance characteristics, ranging from a simple, "brute-force" (but probably inefficient) solution to a complex "well-tuned" (and maybe even optimal) solution. (It is not in general true that the more efficient an algorithm is, the more complicated the implementation must be, since some of our best algorithms are rather elegant and concise, but for the purposes of this discussion, let's assume that this rule holds.) As argued above, one cannot decide what algorithm to use for a problem without analyzing the needs of the problem. How often is the program to be run? What are the general characteristics of the computer system to be used? Is the algorithm a small part of a large application, or vice versa?

The first rule of implementation is that one should *first implement the simplest algorithm to solve a given problem.* If the particular problem instance that is encountered turns out to be easy, then the simple algorithm may solve the problem and nothing more need be done; if a more sophisticated algorithm is called for, then the simple implementation provides a correctness check for small cases and a baseline for evaluating performance characteristics.

If an algorithm is to be run only a few times on cases that are not too large, then it is certainly preferable to have the computer take a little extra time running a slightly less efficient algorithm than to have the programmer take a significant amount of extra time developing a sophisticated implementation. Of course, there is the danger that one could end up using the program more than originally envisioned, so one should always be prepared to start over and implement a better algorithm.

If the algorithm is to be implemented as part of a large system, the "brute-force" implementation provides the required functionality in a reliable manner, and performance can be upgraded in a controlled way by substituting a better algorithm later. Of course, one should take care not to foreclose options by implementing the algorithm in such a way that it is difficult to upgrade later, and one should take a very careful look at which algorithms are creating performance bottlenecks when studying the performance of the system as a whole. Also, in large systems it is often the case that design requirements of the system dictate from the start which algorithm is best. For example, perhaps a system-wide data structure is a particular form of linked list or tree, so that algorithms based on that particular structure are preferable. On the other hand, one should pay some attention to the algorithms to be used when making such system-wide decisions, because, in the end, it very often does turn out that performance of the whole system depends on the performance of some basic algorithm such as those discussed in this book.

If the algorithm is to be run only a few times, but on very large problems, then one would like to have some confidence that it produces meaningful output and some estimate of how long it will take. Again, a simple implementation can often be quite useful in setting up for a long run, including the development of instrumentation for checking the output.

The most common mistake made in selecting an algorithm is to ignore performance characteristics. Faster algorithms are often more complicated, and implementors are often willing to accept a slower algorithm to avoid having to deal with added complexity. But a faster algorithm is often not much more complicated, and dealing with slight added complexity is a small price to pay to avoid dealing with a slow algorithm. Users of a surprising number of computer systems lose substantial time waiting for simple quadratic algorithms to finish when only slightly more complicated $N \log N$ algorithms are available that could run in a fraction the time.

The second most common mistake made in selecting an algorithm is to pay too much attention to performance characteristics. An $N \log N$ algorithm might be only slightly more complicated than a quadratic algorithm for the same problem, but a better $N \log N$ algorithm might give rise to a substantial increase in complexity (and might actually be faster only for very large values of N). Also, many programs are really run only a few times: the time required to implement and debug an optimized algorithm might be substantially more than the time required simply to run a slightly slower one.

Empirical Analysis

As mentioned in Chapter 6, it is unfortunately all too often the case that mathematical analysis can shed very little light on how well a given algorithm can be expected to perform in a given situation. In such cases, we need to rely on *empirical analysis*, where we carefully implement an algorithm and monitor its performance on "typical" input. In fact, this should be done even when full mathematical results *are* available, in order to check their validity.

Given two algorithms to solve the same problem, there's no mystery in the method: run them both to see which one takes longer! This might seem too obvious to mention, but it is probably the most common omission in the comparative study of algorithms. The fact that one algorithm is ten times faster than another is very unlikely to escape the notice of someone who waits three seconds for one to finish and thirty seconds for the other to finish, but it is very easy to overlook as a small constant overhead factor in a mathematical analysis.

However, it is also easy to make mistakes when comparing implementations, especially if different machines, compilers, or systems are involved, or if very large programs with ill-specified inputs are being compared. Indeed, a factor that led to the development of the mathematical analysis of algorithms has been the tendency to rely on "benchmarks" whose performance is perhaps better understood through careful analysis.

The principal danger in comparing programs empirically is that one implementation may be more "optimized" than the other. The inventor of a proposed new algorithm is likely to pay very careful attention to every aspect of its implementation, and not to the details of implementing a classical competing algorithm. To be confident of the accuracy of an empirical study comparing algorithms, one must be sure that the same amount of attention is given to the implementations. Fortunately, this is often the case: many excellent algorithms are derived from relatively minor modifications to other algorithms for the same problem, and comparative studies really are valid.

An important special case arises when an algorithm is to be compared to another version of *itself*, or different implementation approaches are to be compared. An excellent way to check the efficacy of a particular modification or implementation idea is to run both versions on some "typical" input, then pay more attention to the faster one. Again, this seems almost too obvious to mention, but a surprising number of researchers involved in algorithm design never implement their designs, so let the user beware!

As outlined above and at the beginning of Chapter 6, the view taken here is that design, implementation, mathematical analysis, and empirical analysis all contribute in important ways to the development of good implementations of good algorithms. We want to use whatever tools are available to gain information about the properties of our programs, then modify or develop new programs on the basis of that information. On the other hand, one is not always justified in making large numbers of small changes in hopes of slight performance improvements. Next, we discuss this issue in more detail.

Program Optimization

The general process of making incremental changes to a program to produce another version that runs faster, is called *program optimization*. This is a misnomer because we're unlikely to see a "best" implementation—we can't optimize a program, but we can hope to improve it. Normally, program optimization refers to automatic techniques applied as part of the compilation process to improve the performance of compiled code. Here we use the term to refer to *algorithm-specific* improvements. Of course, the process is also rather dependent on the programming environment and machine used, so we consider only general issues here, not specific techniques.

This type of activity is justified only if one is sure that the program will be used many times or for a large input *and* if experimentation proves that effort put into improving the implementation will be rewarded with better performance. The best way to improve the performance of an algorithm is through a gradual process of transforming the program into better and better implementations. The recursion-removal example in Chapter 5 is an example of such a process, though preformance improvement was not our goal in that case.

The first step in implementing an algorithm is to develop a working version of the algorithm in its simplest form. This provides a baseline for refinements and improvements and, as mentioned above, is very often all that is needed. Any mathematical results available should be checked against the implementation: for example, if the analysis seems to say that the running time is $O(\log N)$ but the actual running time starts to run into seconds, then something is wrong with either the implementation or the analysis, and both should be studied more carefully.

The next step is to identify the "inner loop" and to try to minimize the number of instructions involved. Perhaps the easiest way to find the inner loop is to run the program and then check which instructions are executed most often. Normally, this is an extremely good indication of where the program should be improved. Every instruction in the inner loop should be scrutinized. Is it really necessary? Is there a more efficient way to accomplish the same task? For example, it usually pays to remove procedure calls from the inner loop. There are a number of other "automatic" techniques for doing this, many of which are implemented in standard compilers. Ultimately, the best performance is achieved by moving the inner loop into machine or assembly language, but this is usually the last resort.

Not all "improvements" actually result in performance gains, so it is extremely important to check the extent of the savings realized at each step. Moreover, as the implementation becomes more and more refined, it is wise to re-examine whether such careful attention to the details of the code is justified. In the past, computer time was so expensive that spending programmer time to save computing cycles was almost always justified, but the tables have turned in recent years.

For example, consider the preorder tree traversal algorithm discussed in Chapter 5. Actually, recursion removal is the first step in "optimizing" this algorithm, because it focuses on the inner loop. The nonrecursive version given is actually likely to be slower than the recursive version on many systems (the reader might wish to test this) because the inner loop is longer and includes four (albeit nonrecursive) procedure calls (to *pop*, *push*, *push* and *stackempty*) instead of two. If the calls to the stack procedures are replaced with the code for directly accessing the stack (using, say, an array implementation), this program is likely to be significantly faster than the recursive version. (One of the *push* operations is overhead from the algorithm, so the standard loop-within-a-loop program should probably be the basis for an optimized version.) Then it is plain that the inner loop involves incrementing the stack pointer, storing a pointer ($t\uparrow.r$) in the stack array, resetting the t pointer (to $t\uparrow.l$), and comparing it to z. On many machines, this could be implemented in four machine-language instructions, though a typical compiler is likely to produce twice as many or more. This program can be made to run perhaps four or five times faster than the straightforward recursive implementation without too much work.

Obviously, the issues under discussion here are extremely system- and machine-dependent. One cannot embark on a serious attempt to speed up a program without rather detailed knowledge of the operating system and the programming environment. The optimized version of a program can become rather fragile and difficult to change, and a new compiler or a new operating system (not to mention a new computer) might completely ruin a carefully optimized implementation. On the other hand, we do focus on efficiency in our implementations by paying attention to the inner loop at a high level and by ensuring that overhead from the algorithm is minimized. The programs in this book are tightly coded and amenable to further improvement in a straightforward manner for any particular programming environment.

Implementation of an algorithm is a cyclic process of developing a program, debugging it and learning its properties, then refining the implementation until a desired level of performance is reached. As discussed in Chapter 6, mathematical analysis can usually help in the process: first, to suggest which algorithms are promising candidates to perform well in a careful implementation; second, to help verify that the implementation is performing as expected. In some cases, this process can lead to the discovery of facts about the problem that suggest a new algorithm or substantial improvements in an old one.

Algorithms and Systems

Implementations of the algorithms in this book may be found in a wide variety of large programs, operating systems, and applications systems. Our intention is to describe the algorithms and to encourage to the reader to focus on their dynamic properties through experimentation with the implementations given. For some applications, the implementations may be quite useful exactly as given, but for other applications more work may be required.

First, as mentioned in Chapter 2, the programs in this book use only basic features of Pascal, rather than taking advantage of more advanced capabilities that are available in Pascal and other programming environments. Our purpose is to study algorithms, not systems programming or advanced features of programming languages. It is hoped that the essential features of the algorithms are best exposed through simple, direct implementations in a near-universal language.

The programming style we use is somewhat terse, with short variable names and few comments, so that the control structures stand out. The "documentation" of the algorithms is the accompanying text. It is expected that readers who use these programs in actual applications will flesh them out somewhat in adapting them for a particular use. A more "defensive" programming style is justified in building real systems: the programs must be implemented so that they can be changed easily, quickly read and understood by other programmers, and interface well with other parts of the system.

In particular, the data structures required for applications normally contain rather more information than those used in this book, though the algorithms that we consider are appropriate for more complex data structures. For example, we speak of searching through files containing integers or short character strings, while an application typically would require considering long character strings that are part of large records. But the basic methods available in both cases are the same. In such cases, we will discuss salient features of each algorithm and how they might relate to various application requirements.

Many of the comments above concerning improving the performance of a particular algorithm apply to improving performance in a large system as well. However, on this larger scale, a technique for improving the performance of the system might be to replace a module implementing one algorithm with a module implementing another. A basic principle of building large systems is that such changes should be possible. Typically, as a system evolves into being, more precise knowledge is gained about the specific requirements for particular modules. This more specific knowledge makes it possible to more carefully select the best algorithm for use to satisfy those needs; then one can concentrate on improving the performance of that algorithm, as described above. It is certainly the case that the vast majority of system code is only executed a few times (or not at all)—the primary concern of the system builder is to create a coherent whole. On the other hand, it also is very likely that when a system comes into use, many of its resources will be devoted to solving fundamental problems of the type discussed in this book, so that it is appropriate for the system builder to be cognizant of the basic algorithms that we discuss.

□

Exercises

1. How long does it take to count to 100,000? Estimate how long the program *j:=0;* **for** *i:= 1* **to** *100000* **do** *j:=j+1* should take on your programming environment, then run the program to test your estimate.
2. Answer the previous question using **repeat** and **while**.
3. By running on small values, estimate how long it would take the sieve of Eratosthenes implementation in Chapter 3 to run with N = 1,000,000 (if enough memory were available).
4. "Optimize" the sieve of Eratosthenes implementation in Chapter 3 to find the largest prime you can in ten seconds of computing.
5. Test the assertion in the text that removing recursion from the preorder tree traversal algorithm from Chapter 5 (with procedure calls for stack operations) makes the program slower.
6. Test the assertion in the text that removing recursion from the preorder tree traversal algorithm from Chapter 5 (and implementing stack operations inline) makes the program faster.
7. Examine the assembly-language program produced by the Pascal compiler in your local programming environment for the recursive preorder tree traversal algorithm for Chapter 5.
8. Design an experiment to test which of the linked list or array implementation of a pushdown stack is more efficient in your programming environment.
9. Which is more efficient, the nonrecursive or the recursive method for drawing a ruler given in Chapter 5?
10. Exactly how many extraneous stack pushes are used by the nonrecursive implementation given in Chapter 5 when traversing a complete tree of $2^n - 1$ nodes in preorder?

SOURCES for Fundamentals

There are a large number of introductory textbooks on programming and elementary data structures. Still, the best source for specific facts about Pascal and examples of Pascal programs, in the same spirit as those found in this book, is Wirth and Jensen's report on the language. The most comprehensive collection of information about properties of elementary data structures and trees is Knuth's Volume 1: Chapters 3 and 4 cover only a small fraction of the information there.

The classic reference on the analysis of algorithms based on asymptotic worst-case performance measures is Aho, Hopcroft, and Ullman's book. Knuth's books cover average-case analysis more fully and are the authoritative source on specific properties of various algorithms (for example, nearly fifty pages in Volume 2 are devoted to Euclid's algorithm.) Gonnet's book does both worst- and average-case analysis, and covers many recently-developed algorithms.

The book by Graham, Knuth and Patashnik covers the type of mathematics that commonly arises in the analysis of algorithms. For example, this book describes many techniques for solving recurrence relations like those given in Chapter 6 and the many more difficult ones that we encounter later on. Such material is also sprinkled liberally throughout Knuth's books.

The book by Roberts covers material related to Chapter 6, and Bentley's books take much the same point of view as Chapter 7 and later sections of this book. Bentley describes in detail a number of complete case studies on evaluating various approaches to developing algorithms and implementations for solving some interesting problems.

A. V. Aho, J. E. Hopcroft, and J. D. Ullman, *The Design and Analysis of Algorithms*, Addison-Wesley, Reading, MA, 1975.

J. L. Bentley, *Programming Pearls*, Addison-Wesley, Reading, MA, 1985; *More Programming Pearls*, Addison-Wesley, Reading, MA, 1988.

G. H. Gonnet, *Handbook of Algorithms and Data Structures*, Addison-Wesley, Reading, MA, 1984.

R. L. Graham, D. E. Knuth, and O. Patashnik, *Concrete Mathematics*, Addison-Wesley, Reading, MA, 1988.

K. Jensen and N. Wirth, *Pascal User Manual and Report*, Springer-Verlag, Berlin 1974.

D. E. Knuth, *The Art of Computer Programming. Volume 1: Fundamental Algorithms*, second edition, Addison-Wesley, Reading, MA, 1973; *Volume 2: Seminumerical Algorithms*, second edition, Addison-Wesley, Reading, MA, 1981; *Volume 3: Sorting and Searching*, second printing, Addison-Wesley, Reading, MA, 1975.

E. Roberts, *Thinking Recursively*, John Wiley & Sons, New York, 1986.

Sorting Algorithms

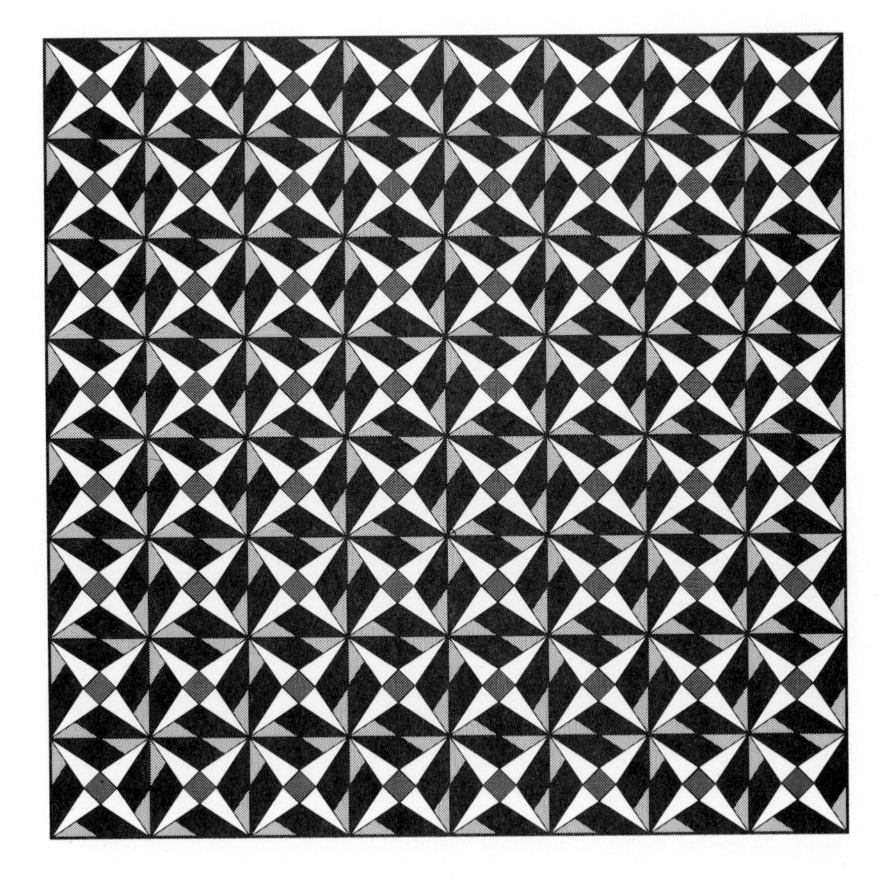

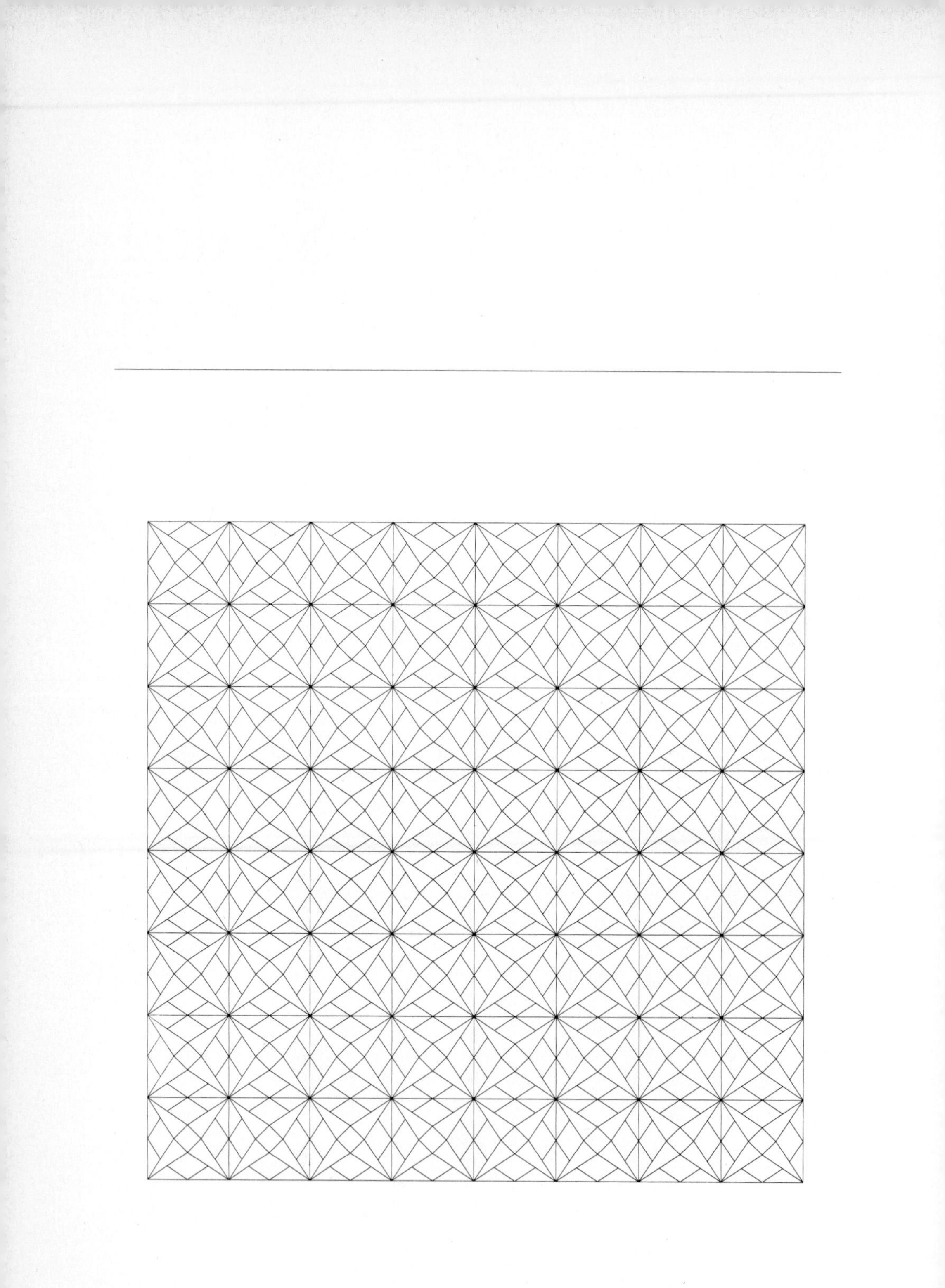

8

Elementary Sorting Methods

☐ As our first excursion into the area of sorting algorithms, we'll study some "elementary" methods which are appropriate for small files or files with some special structure. There are several reasons for studying these simple sorting algorithms in some detail. First, they provide a relatively painless way to learn terminology and basic mechanisms for sorting algorithms so that we get an adequate background for studying the more sophisticated algorithms. Second, in a great many applications of sorting it's better to use these simple methods than the more powerful general-purpose methods. Finally, some of the simple methods extend to better general-purpose methods or can be used to improve the efficiency of more powerful methods.

As mentioned above, there are several sorting applications in which a relatively simple algorithm may be the method of choice. Sorting programs are often used only once (or only a few times). If the number of items to be sorted is not too large (say, less than five hundred elements), it may well be more efficient just to run a simple method than to implement and debug a complicated method. Elementary methods are always suitable for small files (say, less than fifty elements); it is unlikely that a sophisticated algorithm would be justified for a small file, unless a very large number of such files are to be sorted. Other types of files that are relatively easy to sort are ones that are already almost sorted (or already sorted!) or ones that contain large numbers of equal keys. Simple methods can do much better on such well-structured files than general-purpose methods.

As a rule, the elementary methods that we'll be discussing take about N^2 steps to sort N randomly arranged items. If N is small enough, this may not be a problem, and if the items are not randomly arranged, some of the methods may run much faster than more sophisticated ones. However, it must be emphasized that these methods should *not* be used for large, randomly arranged files, with the notable exception of Shellsort, which is actually the sorting method of choice for a great many applications.

Rules of the Game

Before considering some specific algorithms, it will be useful to discuss some general terminology and basic assumptions for sorting algorithms. We'll be considering methods of sorting *files* of *records* containing *keys*. The keys, which are only part of the records (often a small part), are used to control the sort. The objective of the sorting method is to rearrange the records so that their keys are ordered according to some well-defined ordering rule (usually numerical or alphabetical order).

If the file to be sorted will fit into memory (or, in our context, if it will fit into a Pascal **array**), then the sorting method is called *internal*. Sorting files from tape or disk is called *external* sorting. The main difference between the two is that any record can easily be accessed in an internal sort, while an external sort must access records sequentially, or at least in large blocks. We'll look at a few external sorts in Chapter 13, but most of the algorithms that we'll consider are internal sorts.

As usual, the main performance parameter that we'll be interested in is the running time of our sorting algorithms. The first four methods that we'll examine in this chapter require time proportional to N^2 to sort N items, while more advanced methods can sort N items in time proportional to $N \log N$. (It can be shown that no sorting algorithm can use less than $N \log N$ comparisons between keys.) After examining the simple methods, we'll look at a more advanced method that can run in time proportional to $N^{3/2}$ or less, and we'll see that there are methods that use digital properties of keys to get a total running time proportional to N.

The amount of extra memory used by a sorting algorithm is the second important factor we'll be considering. Basically, the methods divide into three types: those that sort in place and use no extra memory except perhaps for a small stack or table; those that use a linked-list representation and so use N extra words of memory for list pointers; and those that need enough extra memory to hold another copy of the array to be sorted.

A characteristic of sorting methods which is sometimes important in practice is *stability*. A sorting method is called *stable* if it preserves the relative order of equal keys in the file. For example, if an alphabetized class list is sorted by grade, then a stable method produces a list in which students with the same grade are still in alphabetical order, but a non-stable method is likely to produce a list with no vestige of the original alphabetic order. Most of the simple methods are stable, but most of the well-known sophisticated algorithms are not. If stability is vital, it can be forced by appending a small index to each key before sorting or by lengthening the sort key in some other way. It is easy to take stability for granted: people often react to the unpleasant effects of instability with disbelief. Actually, few methods achieve stability without using significant extra time or space.

The following program, for sorting three records, is intended to illustrate the general conventions that we'll be using. In particular, the main program is a peculiar way to exercise a program that works only for $N = 3$; the point is that any sorting program could be substituted for *sort3* in this "driver" program.

```
program threesort(input,output);
const maxN=100;
var a: array[1..maxN] of integer;
      N,i: integer;
procedure sort3;
   var t: integer;
   begin
   if a[1]>a[2] then
      begin t:=a[1]; a[1]:=a[2]; a[2]:=t end;
   if a[1]>a[3] then
      begin t:=a[1]; a[1]:=a[3]; a[3]:=t end;
   if a[2]>a[3] then
      begin t:=a[2]; a[2]:=a[3]; a[3]:=t end;
   end;
begin
readln(N);
for i:=1 to N do read(a[i]);
if N=3 then sort3;
for i:=1 to N do write(a[i]);
writeln
end.
```

The three assignment statements following each **if** actually implement an "exchange" operation. We'll write out the code for such exchanges rather than use a procedure call because they're fundamental to many sorting programs and often fall in the inner loop.

In order to concentrate on algorithmic issues, we'll work with algorithms that simply sort arrays of integers into numerical order. It is generally straightforward to adapt such algorithms for use in a practical application involving large keys or records. Basically, sorting programs access records in one of two ways: either keys are accessed for comparison, or entire records are accessed to be moved. Most of the algorithms we will study can be recast in terms of these two operations on arbitrary records. If the records to be sorted are large, it is normally wise to avoid shuffling them around by doing an "indirect sort": here the records themselves are not necessarily rearranged, but rather an array of pointers (or indices) is rearranged so that the first pointer points to the smallest record, etc. The keys can be kept either with the records (if they are large) or with the pointers (if they are small). If necessary, the records can then be rearranged after the sort, as described later in this chapter.

The *sort3* program above uses an even more constrained access to the file: it is three instructions of the form "compare two records and exchange them if necessary to put the one with the smaller key first." Programs restricted to such instructions are interesting because they are well-suited for hardware implementation. We'll study this issue in more detail in Chapter 40.

By using programs which simply operate on a global array, we're ignoring "packaging problems" that can be troublesome in some programming environments. Should the array be passed to the sorting routine as a parameter? Can the same sorting routine be used to sort arrays of integers and arrays of reals (and arrays of arbitrarily complex records)? Even with our simple assumptions, we must circumvent the lack of dynamic array sizes in Pascal by predeclaring a maximum. Such concerns will be easier to deal with in programming environments of the future than in those of the past and present. For example, some modern languages have quite well-developed facilities for packaging programs together into large systems. On the other hand, such mechanisms are not really required for many applications: small programs which work directly on global arrays have many uses, and some operating systems make it quite easy to put together simple programs, like the one above, which serve as "filters" between their input and their output. Obviously, these comments apply to many of the other algorithms that we will examine, though the effects mentioned are perhaps felt most acutely for sorting algorithms.

Some of the programs use a few other global variables. Declarations which are not obvious will be included with the program code. Also, we'll sometimes assume that the array bounds go to 0 or $N+1$, to hold special keys used by some of the algorithms. We'll frequently use letters from the alphabet rather than numbers for examples: these are handled in the obvious way using Pascal's *ord* and *chr* "transfer functions" between integers and characters.

Selection Sort

One of the simplest sorting algorithms works as follows: first find the smallest element in the array and exchange it with the element in the first position, then find the second smallest element and exchange it with the element in the second position, and continue in this way until the entire array is sorted. This method is called *selection sort* because it works by repeatedly "selecting" the smallest remaining element, as shown in Figure 8.1. The first pass has no effect because there is no element in the array smaller than the A at the left. On the second pass, the second A is the smallest remaining element, so it is exchanged with the S in the second position. Then the first E is exchanged with the O in the third position on the third pass, then the second E is exchanged with the R in the fourth position on the fourth pass, etc.

The following program is a full implementation of this process. For each i from 1 to $N-1$, it exchanges the minimum element in $a[i..N]$ with $a[i]$:

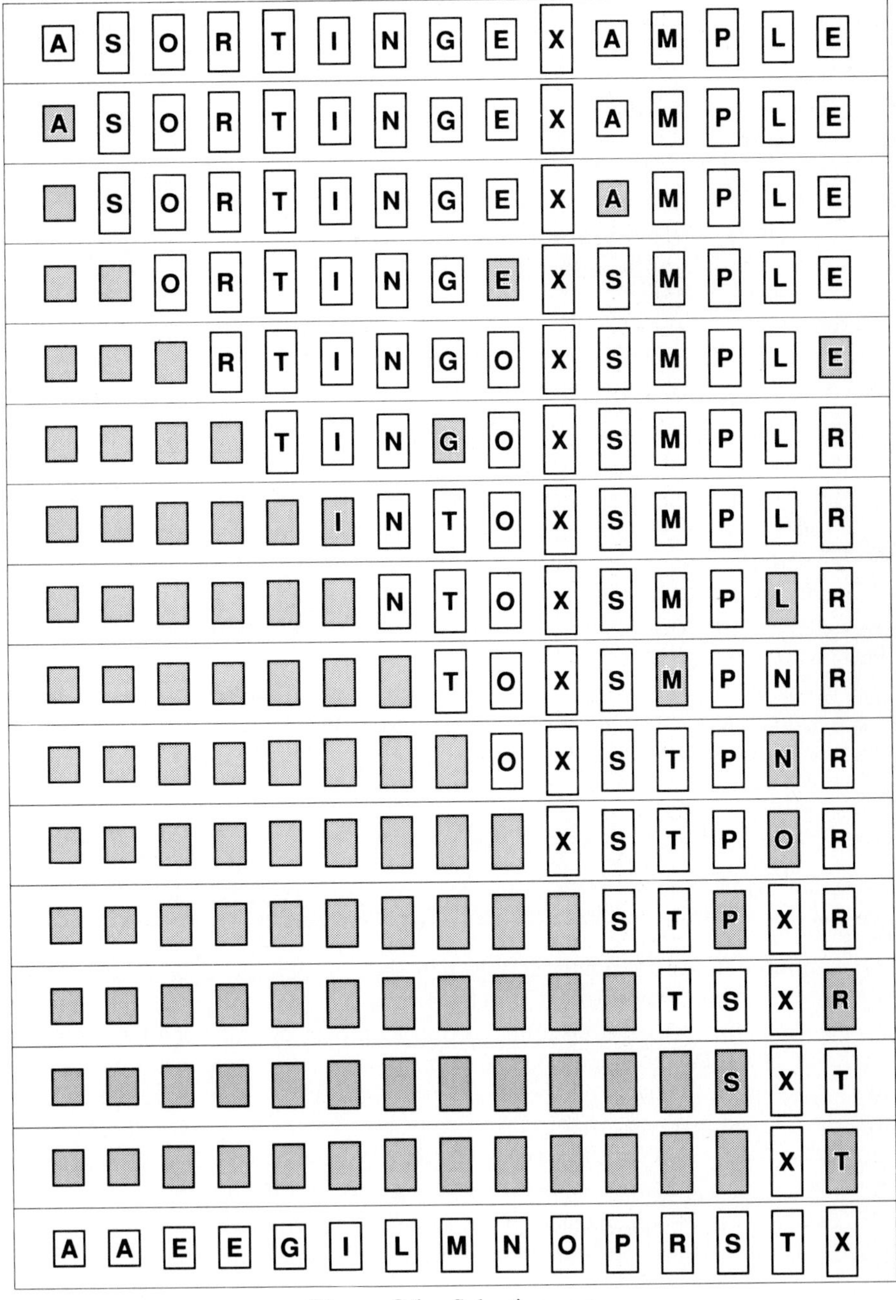

Figure 8.1 Selection sort.

```
procedure selection;
  var i,j,min,t: integer;
  begin
  for i:=1 to N-1 do
    begin
    min:=i;
    for j:=i+1 to N do
      if a[j]<a[min] then min:=j;
    t:=a[min]; a[min]:=a[i]; a[i]:=t
    end;
  end;
```

As the pointer i travels from left to right through the file, the elements to the left of the pointer are in their final position in the array (and will not be touched again), so that the array is fully sorted when the pointer reaches the right end.

This is among the simplest of sorting methods, and it will work very well for small files. The "inner loop" is the comparison $a[j]<a[min]$ (plus the code necessary to increment j and check that it does not exceed N), which could hardly be simpler. Below we discuss the number of times these instructions are likely to be executed.

Furthermore, despite its evident "brute-force" approach, selection sort actually has a quite important application: because each item is actually moved at most once, selection sort is the method of choice for sorting files with very large records and small keys. This is discussed in detail below.

Insertion Sort

An algorithm almost as simple as selection sort but perhaps more flexible is *insertion sort*. This is the method people often use to sort bridge hands: consider the elements one at a time, inserting each in its proper place among those already considered (keeping them sorted). The element being considered is inserted merely by moving larger elements one position to the right and then inserting the element into the vacated position, as shown in Figure 8.2. The S in the second position is larger than the A, so it doesn't have to be moved. When the O in the third position is encountered, it is exchanged with the S to put A O S in sorted order, etc.

This process is implemented in the following program. For each i from 2 to N, $a[i]$ is inserted into position among the elements in $a[1..i-1]$:

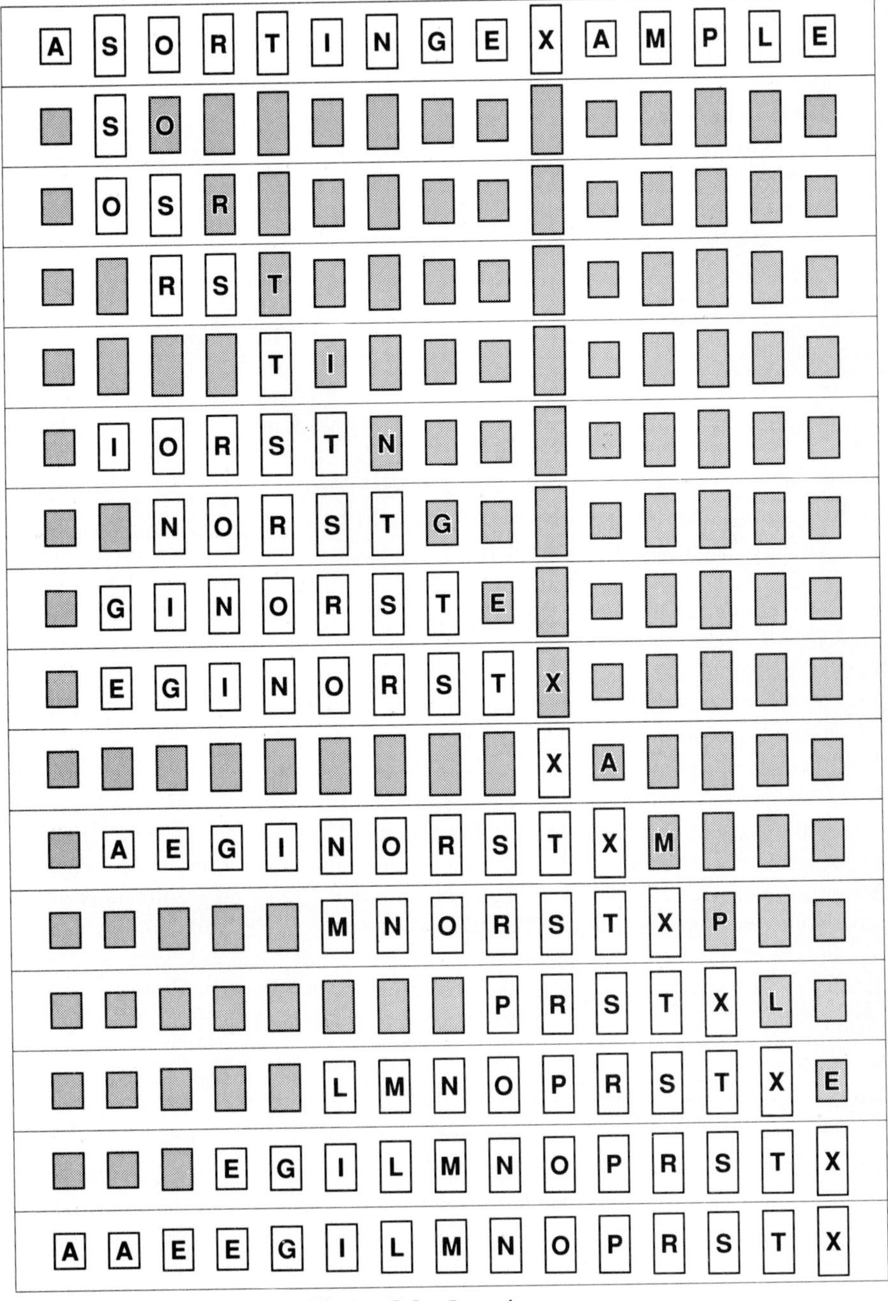

Figure 8.2 Insertion sort.

```
procedure insertion;
  var i,j,v: integer;
  begin
  for i:=2 to N do
    begin
    v:=a[i]; j:=i;
    while a[j-1]>v do
      begin a[j]:=a[j-1]; j:=j-1 end;
    a[j]:=v
    end
  end;
```

As in selection sort, the elements to the left of the pointer i are in sorted order during the sort, but they are not in their final position, as they may have to be moved to make room for smaller elements encountered later. However, the array is fully sorted when the pointer reaches the right end.

There is one more important detail to consider: As is, the procedure *insertion* doesn't work, because the **while** will run past the left end of the array when v is the smallest element in the array. To fix this, we put a "sentinel" key in $a[0]$, making it at least as small as the smallest element in the array. Sentinels are commonly used in situations like this to avoid including a test (in this case $j>1$) which almost always succeeds within the inner loop.

If for some reason it is inconvenient to use a sentinel and the array really must have the bounds $[1..N]$, standard Pascal does not offer a clean alternative, since it does not have a "conditional" **and** instruction: the situation seems to call for the test **while** $(j> 1)$ **and** $(a[j-1]> v)$, but this won't work because even when $j=1$, the second part of the **and** will be evaluated and will cause an out-of-bounds array access. A **goto** out of the loop seems to be required. (Some programmers prefer to goto some lengths to avoid **goto** instructions, for example by performing an action within the loop to ensure that the loop terminates. In this case, such a solution hardly seems justified, since it makes the program no clearer and adds overhead every time through the loop to guard against a rare event.)

Digression: Bubble Sort

An elementary sorting method that is often taught in introductory classes is *bubble sort*: keep passing through the file, exchanging adjacent elements, if necessary; when no exchanges are required on some pass, the file is sorted. An implementation of this method is given below.

```
procedure bubble;
  var i,j,t: integer;
  begin
  for i:=N downto 1 do
    for j:=2 to i do
      if a[j-1]>a[j] then
        begin t:=a[j-1]; a[j-1]:=a[j]; a[j]:=t end
  end;
```

It takes a moment's reflection to convince oneself that this works at all. To do so, note that whenever the maximum element is encountered during the first pass, it is exchanged with each of the elements to its right, until it gets into position at the right end of the array. Then on the second pass, the second largest element will be put into position, etc. Thus bubble sort operates as a type of selection sort, though it does much more work to get each element into position.

Performance Characteristics of Elementary Sorts

Direct illustrations of the operating characteristics of selection sort, insertion sort, and bubble sort are given in Figures 8.3, 8.4, and 8.5. These diagrams show the contents of the array a for each of the algorithms after the outer loop has been iterated $N/4$, $N/2$, and $3N/4$ times (starting with a random permutation of the integers 1 to N as input). In the diagrams, a square is placed at position (i,j) for $a[i]=j$. An unordered array is thus a random display of squares; in a sorted array each square appears above the one to its left. For clarity in the diagrams, we show *permutations* (rearrangements of the integers 1 to N), which, when sorted, have the squares all aligned along the main diagonal. The diagrams show how the different methods progress towards this goal.

Figure 8.3 shows how selection sort moves from left to right, putting elements in their final position without looking back. What is not apparent from this diagram

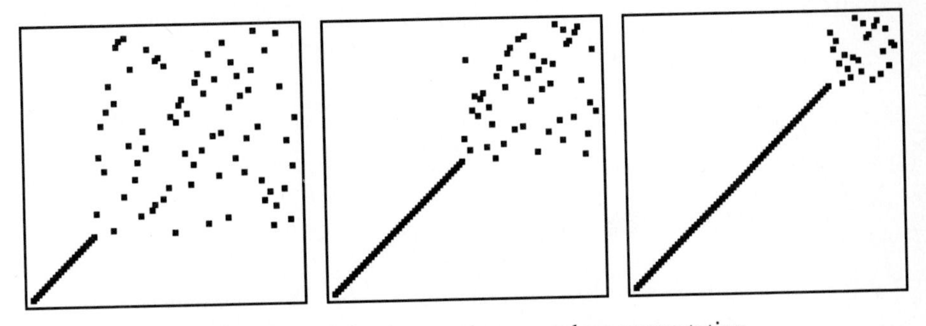

Figure 8.3 Selection sorting a random permutation.

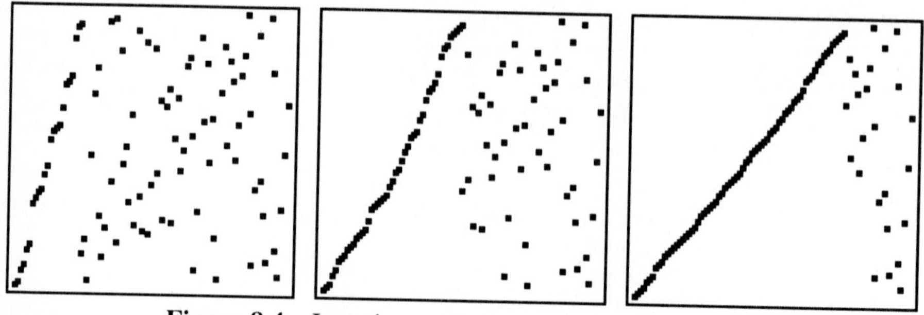

Figure 8.4 Insertion sorting a random permutation.

is the fact that selection sort spends most of its time trying to find the minimum element in the "unsorted" part of the array.

Figure 8.4 shows how insertion sort also moves from left to right, inserting newly encountered elements into position without looking any further forward. The left part of the array is continually changing.

Figure 8.5 shows the similarity between selection sort and bubble sort. Bubble sort "selects" the maximum remaining element at each stage, but wastes some effort imparting some order to the "unsorted" part of the array.

All of the methods are quadratic in both the worst and the average case, and none require extra memory. Thus, comparisons among them depend upon the length of the inner loops or on special characteristics of the input.

Property 8.1 *Selection sort uses about $N^2/2$ comparisons and N exchanges.*

This property is easy to see by examining Figure 8.1, which is an N-by-N table in which a letter corresponds to each comparison. But this is just about half the elements, those above the diagonal. The $N-1$ elements on the diagonal (not the last) each correspond to an exchange. More precisely: for each i from 1 to $N-1$, there is one exchange and $N-i$ comparisons, so there is a total of $N-1$

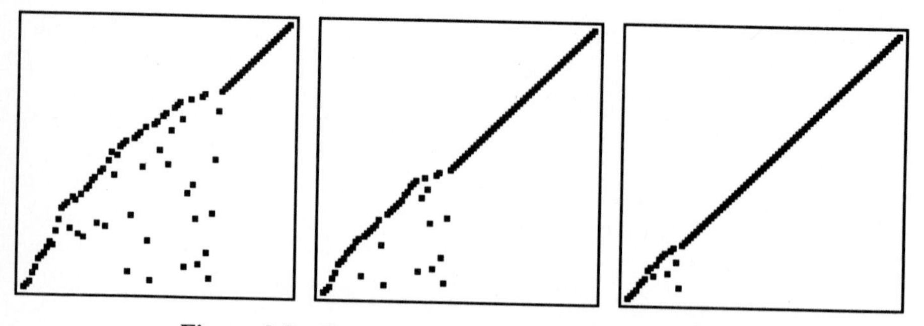

Figure 8.5 Bubble sorting a random permutation.

exchanges and $(N-1)+(N-2)+\cdots+2+1=N(N-1)/2$ comparisons. These observations hold no matter what the input data is: the only part of selection sort that does depend on the input is the number of times *min* is updated. In the worst case, this could also be quadratic, but in the average case, this quantity turns out to be only $O(N \log N)$, so we can expect the running time of selection sort to be quite insensitive to the input. ■

Property 8.2 *Insertion sort uses about $N^2/4$ comparisons and $N^2/8$ exchanges on the average, twice as many in the worst case.*

As implemented above, the number of comparisons and of "half-exchanges" (moves) is the same. As just argued, this quantity is easy to visualize in Figure 8.2, the N-by-N diagram which gives the details of the operation of the algorithm. Here, the elements below the diagonal are counted, all of them in the worst case. For random input, we expect each element to go about halfway back, on the average, so half of the elements below the diagonal should be counted. (It is not difficult to make these arguments more precise.) ■

Property 8.3 *Bubble sort uses about $N^2/2$ comparisons and $N^2/2$ exchanges on the average and in the worst case.*

In the worst case (file in reverse order), it is clear that the ith bubble sort pass requires $N-i$ comparisons *and* exchanges, so the proof goes as for selection sort. But the running time of bubble sort does depend on the input. For example, note that only one pass is required if the file is already in order (insertion sort is also fast in this case). It turns out that the average-case performance is not significantly better that the worst case, as stated, though this analysis is rather more difficult. ■

Property 8.4 *Insertion sort is linear for "almost sorted" files.*

Though the concept of an "almost sorted" file is necessarily rather imprecise, insertion sort works well for some types of non-random files that often arise in practice. General-purpose sorts are commonly misused for such applications; actually, insertion sort can take advantage of the order present in the file.

For example, consider the operation of insertion sort on a file which is already sorted. Each element is immediately determined to be in its proper place in the file, and the total running time is linear. The same is true for bubble sort, but selection sort is still quadratic. Even if a file is not completely sorted, insertion sort can be quite useful because its running time depends quite heavily on the order present in the file. The running time depends on the number of *inversions*: for each element count up the number of elements to its left which are greater. This is the distance the elements have to move when inserted into the file during insertion sort. A file which has some order in it will have fewer inversions in it than one which is arbitrarily scrambled.

Suppose one wants to add a few elements to a sorted file to produce a larger sorted file. One way to do so is to append the new elements to the end of the file, then call a sorting algorithm. Clearly, the the number of inversions is low in such a file: a file with only a constant number of elements out of place will have only a linear number of inversions. Another example is a file in which each element is only some constant distance from its final position. Files like this can be created in the initial stages of some advanced sorting methods: at a certain point it is worthwhile to switch over to insertion sort.

For such files, insertion sort will outperform even the sophisticated methods described in the next few chapters. ■

To compare the methods further, one needs to analyze the cost of comparisons and exchanges, a factor which in turn depends on the size of the records and keys. For example, if the records are one-word keys, as in the implementations above, then an exchange (two array accesses) should be about twice as expensive as a comparison. In such a situation, the running times of selection and insertion sort are roughly comparable, but bubble sort is twice as slow. (In fact, bubble sort is likely to be twice as slow as insertion sort under almost any circumstances!) But if the records are large in comparison to the keys, then selection sort will be best.

Property 8.5 *Selection sort is linear for files with large records and small keys.*

Suppose that the cost of a comparison is 1 time unit and the cost of an exchange is M time units. (For example, this might be the case with M-word records and 1-word keys.) Then selection sort takes about N^2 time for comparisons and about NM time for exchanges to sort a file of size NM. If $N = O(M)$, this is linear in the amount of data. ■

Sorting Files with Large Records

It is actually possible (and desirable) to arrange things so that *any* sorting method uses only N "exchanges" of full records, by having the algorithm operate indirectly on the file (using an array of indices) and then do the rearrangement afterwards.

Specifically, if the array $a[1..N]$ consists of large records, then we prefer to manipulate a "pointer array" $p[1..N]$, accessing the original array only for comparisons. If we define $p[i]=i$ initially, then the algorithms above (and all the algorithms in chapters that follow) need only be modified to refer to $a[p[i]]$ rather than $a[i]$ when using $a[i]$ in a comparison, and to refer to p rather than a when doing data movement. This produces an algorithm that will "sort" the index array so that $p[1]$ is the index of the smallest element in a, $p[2]$ is the index of the second smallest element in a, etc. and the cost of moving large records around excessively is avoided. The following code shows how insertion sort might be modified to work in this way.

```
procedure insertion;
  var i,j,v: integer;
  begin
  for i:=1 to N do p[i]:=i;
  for i:=2 to N do
    begin
    v:=p[i]; j:=i;
    while a[p[j-1]]>a[v] do
      begin p[j]:=p[j-1]; j:=j-1 end;
    p[j]:=v
    end
  end;
```

In this program, the array a is accessed only to compare keys of two records. Thus, it could be easily modified to handle files with very large records by modifying the comparison to access only a small field of a large record, or by making the comparison a more complicated procedure. Figure 8.6 shows how this process produces a permutation that specifies the order in which the array elements could be accessed to define a sorted list. For many applications, this will suffice (the data may not need to be moved at all). For example, one could print out the data in sorted order simply by referring to it indirectly through the index array, as in the sort itself.

But what if the data must actually be rearranged, as at the bottom of Figure 8.6? If there is enough extra memory for another copy of the array, this is trivial,

Before Sort

k	1	2	3	4	5	6	7	8	9	10	11	12	13	14	15
a[k]	A	S	O	R	T	I	N	G	E	X	A	M	P	L	E
p[k]	1	2	3	4	5	6	7	8	9	10	11	12	13	14	15

After Sort

k	1	2	3	4	5	6	7	8	9	10	11	12	13	14	15
a[k]	A	S	O	R	T	I	N	G	E	X	A	M	P	L	E
p[k]	1	11	9	15	8	6	14	12	7	3	13	4	2	5	10

After Permute

k	1	2	3	4	5	6	7	8	9	10	11	12	13	14	15
a[k]	A	A	E	E	G	I	L	M	N	O	P	R	S	T	X
p[k]	1	2	3	4	5	6	7	8	9	10	11	12	13	14	15

Figure 8.6 Rearranging a "sorted" array.

but what about the more normal situation when there isn't enough room for another copy of the file?

In our example, the first A is in its proper position, with *p[1]=1*, so nothing need be done with it. The first thing that we would like to do is to put the record with the next smallest key (the one with index *p[2]*) into the second position in the file. But before doing that, we need to save the record that is in that position, say in *t*. Now, after the move, we can consider there to be a "hole" in the file at position *p[2]*. But we know that the record at position *p[p[2]]* should eventually fill that hole. Continuing in this way, we eventually come to the point where we need the item originally in the second position, which we have been holding in *t*. In our example, this process leads to the series of assignments *t:=a[2]; a[2]:=a[11]; a[11]:=a[13]; a[13]:=a[2]; a[2]:=t;* . These assignments put the records with keys A, P, and S into their proper place in the file, which can be marked by setting *p[2]=2*, *p[11]=11*, and *p[13]=13*. (Any element with *p[i]=i* is in place and need not be touched again.) Now the process can be followed again for the next element which is not in place, etc., and this ultimately rearranges the entire file, moving each record only once, as in the following code:

```
procedure insitu;
  var i,j,k,t: integer;
  begin
  for i:=1 to N do
    if p[i]<>i then
      begin
      t:=a[i]; k:=i;
      repeat
        j:=k; a[j]:=a[p[j]];
        k:=p[j]; p[j]:=j;
      until k=i;
      a[j]:=t
      end;
  end;
```

The viability of this technique for particular applications of course depends on the relative size of records and keys in the file to be sorted. Certainly one would not go to such trouble for a file consisting of small records, because of the extra space required for the index array and the extra time required for the indirect comparisons. But for files consisting of large records, it is almost always desirable to use an indirect sort, and in many applications it may not be necessary to move the data at all. Of course, for files with very large records, plain selection sort is the method to use, as discussed above.

Because of the availability of this indirect approach, the conclusions we draw in this chapter and those which follow when comparing methods to sort files of integers are likely to apply to more general situations.

Shellsort

Insertion sort is slow because it exchanges only adjacent elements. For example, if the smallest element happens to be at the end of the array, N steps are needed to get it where it belongs. *Shellsort* is a simple extension of insertion sort which gains speed by allowing exchanges of elements that are far apart.

The idea is to rearrange the file to give it the property that taking every hth element (starting anywhere) yields a sorted file. Such a file is said to be *h-sorted*. Put another way, an h-sorted file is h independent sorted files, interleaved together. By h-sorting for some large values of h, we can move elements in the array long distances and thus make it easier to h-sort for smaller values of h. Using such a

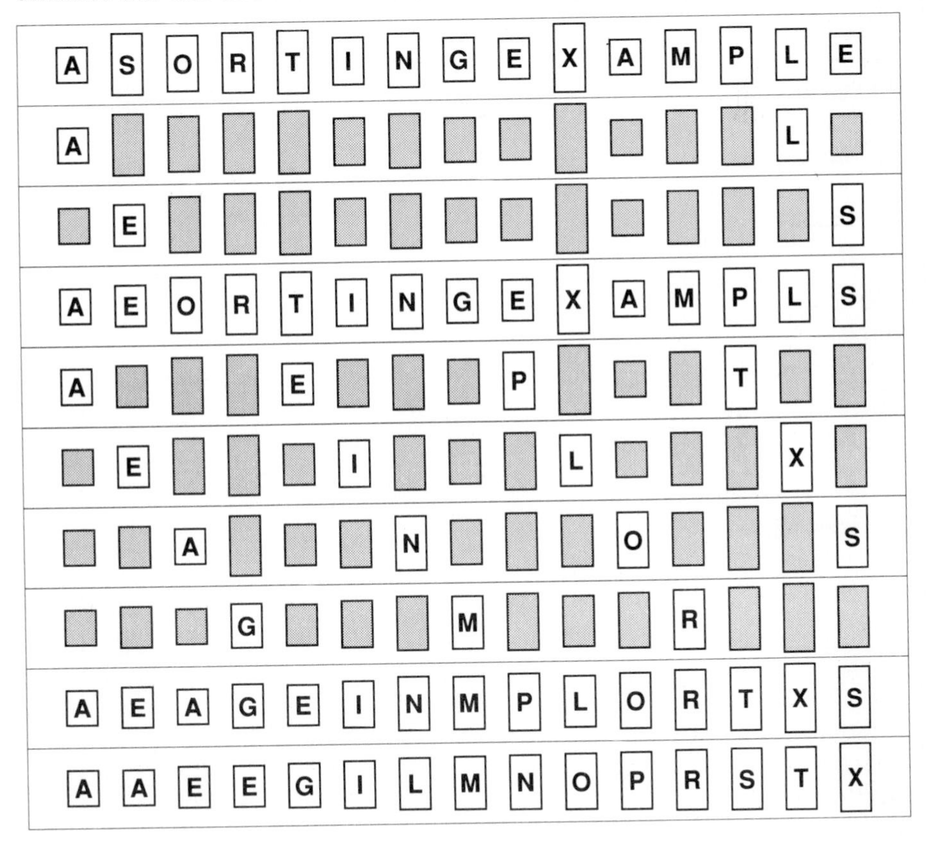

Figure 8.7 Shellsort.

procedure for any sequence of values of h which ends in 1 will produce a sorted file: this is Shellsort.

Figure 8.7 shows the operation of Shellsort on our sample file with the increments ..., 13, 4, 1. In the first pass, the A in position 1 is compared to the L in position 14, then the S in position 2 is compared (and exchanged) with the E in position 15. In the second pass, the A T E P in positions 1, 5, 9, and 13 are rearranged to put A E P T in those positions, and similarly for positions 2, 6, 10, and 14, etc. The last pass is just insertion sort, but no element has to move very far.

One way to implement Shellsort would be, for each h, to use insertion sort independently on each of the h subfiles. (Sentinels would not be used because there would have to be h of them, for the largest value of h used.) But it turns out to be much easier than that: If we replace every occurrence of "1" by "h" (and "2" by "$h+1$") in insertion sort, the resulting program h-sorts the file, as follows.

```
procedure shellsort;
  label 0;
  var i,j,h,v: integer;
  begin
  h:=1; repeat h:=3*h+1 until h>N;
  repeat
    h:=h div 3;
    for i:=h+1 to N do
    begin
      v:=a[i]; j:=i;
      while a[j-h]>v do
        begin
        a[j]:=a[j-h]; j:=j-h;
        if j<=h then goto 0
        end;
  0: a[j]:=v
    end
  until h=1;
  end;
```

This program uses the increment sequence ..., 1093, 364, 121, 40, 13, 4, 1. Other increment sequences might do about as well as this in practice, but some care must be exercised, as discussed below. Figure 8.8 shows this program in operation on a random permutation, by displaying the contents of the array a after each h-sort.

The increment sequence in this program is easy to use and leads to an efficient sort. Many other increment sequences lead to a more efficient sort (the reader might

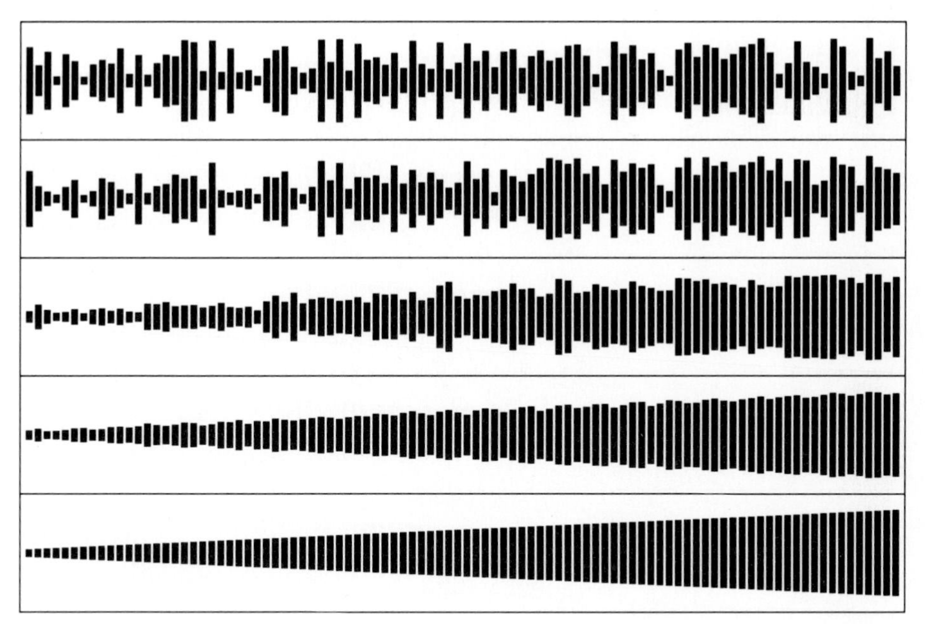

Figure 8.8 Shellsorting a Random Permutation..

be amused to try to discover one), but it is difficult to beat the above program by more than 20% even for relatively large N. (The possibility that much better increment sequences exist is still, however, quite real.) On the other hand, there are some bad increment sequences. Shellsort is sometimes implemented by starting at h=N (instead of initializing so as to ensure the same sequence is always used as above). This virtually ensures that a bad sequence will turn up for some N.

The above description of the efficiency of Shellsort is necessarily imprecise because no one has been able to analyze the algorithm. This makes it difficult not only to evaluate different increment sequences, but also to compare Shellsort with other methods analytically. Not even the functional form of the running time for Shellsort is known (furthermore, the form depends on the increment sequence). For the above program, two conjectures are $N(\log N)^2$ and $N^{1.25}$. The running time is not particularly sensitive to the initial ordering of the file, especially in contrast to, say, insertion sort, which is linear for a file already in order but quadratic for a file in reverse order. Figure 8.9 shows the operation of Shellsort on a file in reverse order.

Property 8.6 *Shellsort never does more than $N^{3/2}$ comparisons (for the increments 1, 4, 13, 40, 121, ...).*

The proof of this property is beyond the scope of this book, but the reader may not only appreciate its difficulty but also be convinced that Shellsort will run well

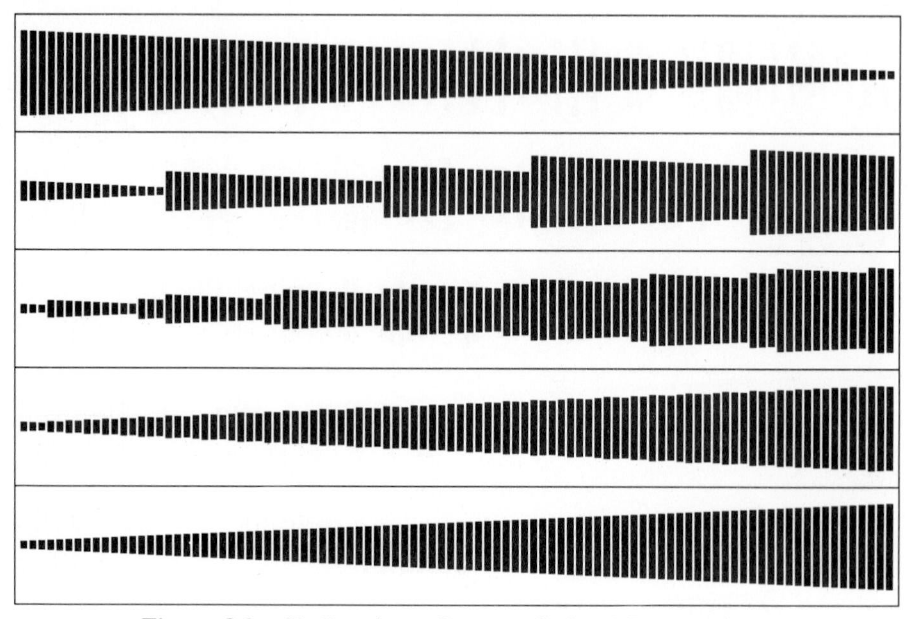

Figure 8.9 Shellsorting a Reverse-Ordered Permutation..

in practice by attempting to construct a file for which Shellsort runs slowly. As mentioned above, there are some bad increment sequences for which Shellsort may require a quadratic number of comparisons, but the $N^{3/2}$ bound has been shown to hold for a wide variety of sequences, including the one used above. Even better worst-case bounds are known for some special sequences. ■

Figure 8.10, showing a different view of Shellsort in operation, may be compared with Figures 8.3, 8.4, and 8.5. This figure shows the contents of the array after each h-sort (except the last, which completes the sort). In these diagrams, we might imagine a rubber band, fastened at the lower left and upper right corners, being stretched tighter to bring all the points toward the diagonal. The three diagrams in Figures 8.3, 8.4, and 8.5 each represent a significant amount of work by the algorithm illustrated; by contrast, each of the diagrams in Figure 8.10 represents only one h-sorting pass.

Shellsort is the method of choice for many sorting applications because it has acceptable running time even for moderately large files (say, less than 5000 elements) and requires only a very small amount of code that is easy to get working. We'll see methods that are more efficient in the next few chapters, but they're perhaps only twice as fast (if that much) except for large N, and they're significantly more complicated. In short, if you have a sorting problem, *use the above program*, then determine whether the extra effort required to replace it with a sophisticated method will be worthwhile.

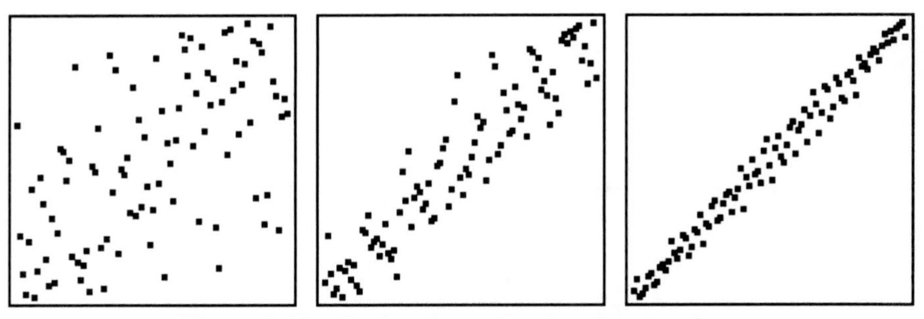

Figure 8.10 Shellsorting a Random Permutation..

Distribution Counting

A very special situation for which there is a simple sorting algorithm is the following: "sort a file of N records whose keys are distinct integers between 1 and N." This problem can be solved using a temporary array t with the statement **for** i:=*1* **to** N **do** $t[a[i]]$:=$a[i]$. (Or, as we saw above, it is possible, though more complicated, to solve this problem without an auxiliary array.)

A more realistic problem in the same spirit is: "sort a file of N records whose keys are integers between 0 and $M-1$." If M is not too large, an algorithm called *distribution counting* can be used to solve this problem. The idea is to count the number of keys with each value and then use the counts to move the records into position on a second pass through the file, as in the following code:

```
for j:=0 to M-1 do count[j]:=0;
for i:=1 to N do
  count[a[i]]:=count[a[i]]+1;
for j:=1 to M-1 do
  count[j]:=count[j-1]+count[j];
for i:=N downto 1 do
  begin
  b[count[a[i]]]:=a[i];
  count[a[i]]:=count[a[i]]-1
  end;
for i:=1 to N do a[i]:=b[i];
```

To see how this code works, consider the sample file of integers in the top row of Figure 8.11. The first **for** loop initializes the counts to 0; the second sets *count*[*1*]=*6*, *count*[*2*]=*4*, *count*[*3*]=*1*, and *count*[*4*]=*4* because there are six A's, four B's, etc. Then the third **for** loop adds these numbers to produce *count*[*1*]=*6*,

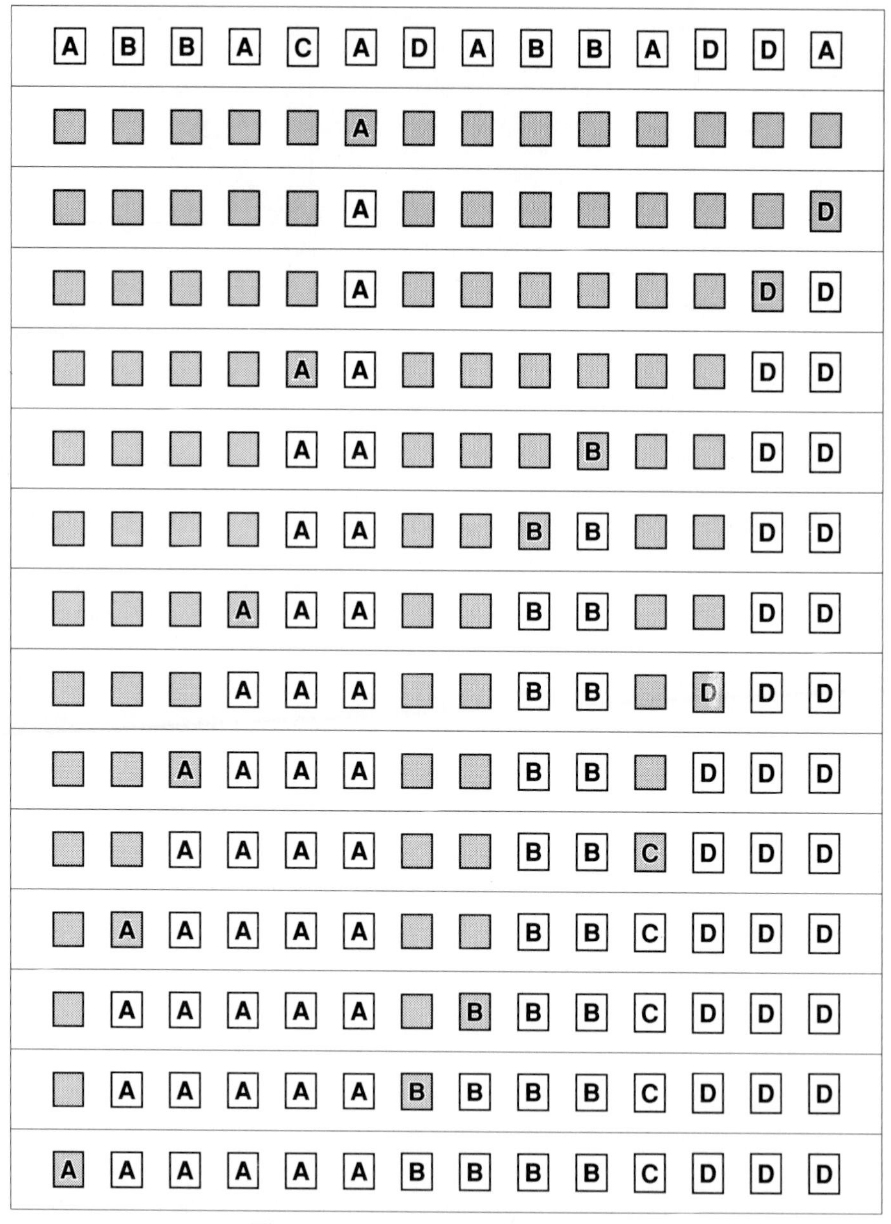

Figure 8.11 Distribution Counting.

count[*2*]=*10*, *count*[*3*]=*11*, and *count*[*4*]=*15*. That is, there are six keys less than or equal to A, ten keys less than or equal to B, etc.

Now, these can be used as addresses to sort the array, as shown in the figure. The original input array a is shown on the top line; the rest of the figure shows the temporary array t being filled. For example, when the A at the end of the file is encountered, it's put into location 6, since *count*[*1*] says that there are six keys less than or equal to A. Then *count*[*1*] is decremented, since there's now one fewer key less than or equal to A. Then the D from the next to last position in the file is put into location 14 and *count*[*4*] is decremented, etc. The inner loop goes from N down to 1 so that the sort will be stable. (The reader may wish to check this.)

This method will work very well for the type of files postulated. Furthermore, it can be extended to produce a much more powerful method that we'll examine in Chapter 10.

□

Exercises

1. Give a sequence of "compare-exchange" operations for sorting four records.
2. Which of the three elementary methods (selection sort, insertion sort, or bubble sort) runs fastest for a file which is already sorted?
3. Which of the three elementary methods runs fastest for a file in reverse order?
4. Test the hypothesis that selection sort is the fastest of the three elementary methods (for sorting integers), then insertion sort, then bubble sort.
5. Give a good reason why it might be inconvenient to use a sentinel key for insertion sort (aside from the one that comes up in the implementation of Shellsort).
6. How many comparisons are used by Shellsort to 7-sort, then 3-sort the keys E A S Y Q U E S T I O N?
7. Give an example to show why 8, 4, 2, 1 would not be a good way to finish off a Shellsort increment sequence.
8. Is selection sort stable? How about insertion sort and bubble sort?
9. Give a specialized version of distribution counting for sorting files where elements have only one of two values (either x or y).
10. Experiment with different increment sequences for Shellsort: find one that runs faster than the one given for a random file of 1000 elements.

9

Quicksort

☐ In this chapter, we'll study the sorting algorithm which is probably more widely used than any other, Quicksort. The basic algorithm was invented in 1960 by C. A. R. Hoare, and it has been studied by many people since that time. Quicksort is popular because it's not difficult to implement, it's a good "general-purpose" sort (it works well in a variety of situations), and it consumes fewer resources than any other sorting method in many situations.

The desirable features of the Quicksort algorithm are that it is in-place (uses only a small auxiliary stack), requires only about $N \log N$ operations on the average to sort N items, and has an extremely short inner loop. The drawbacks of the algorithm are that it is recursive (implementation is complicated if recursion is not available), it takes about N^2 operations in the worst case, and it is fragile: a simple mistake in the implementation can go unnoticed and can cause it to perform badly for some files.

The performance of Quicksort is very well understood. It has been subjected to a thorough mathematical analysis and very precise statements can be made about performance issues. The analysis has been verified by extensive empirical experience, and the algorithm has been refined to the point where it is the method of choice in a broad variety of practical sorting applications. This makes it worthwhile for us to look somewhat more carefully than for other algorithms at ways of efficiently implementing Quicksort. Similar implementation techniques are appropriate for other algorithms; with Quicksort we can use them with confidence because its performance is so well understood.

It is tempting to try to develop ways to improve Quicksort: a faster sorting algorithm is computer science's "better mousetrap." Almost from the moment Hoare first published the algorithm, "improved" versions have been appearing in the literature. Many ideas have been tried and analyzed, but it is easy to be deceived, because the algorithm is so well balanced that the effects of improvements in one part of the program can be more than offset by the effects of bad performance in another part of the program. We'll examine in some detail three modifications which do improve Quicksort substantially.

A carefully tuned version of Quicksort is likely to run significantly faster on most computers than any other sorting method. However, it must be cautioned that tuning any algorithm can make it more fragile, leading to undesirable and unexpected effects for some inputs. Once a version has been developed which seems free of such effects, this is likely to be the program to use for a library sort utility or for a serious sorting application. But if one is not willing to invest the effort to be sure that a Quicksort implementation is not flawed, Shellsort might well be a safer choice that will perform quite well for less implementation effort.

The Basic Algorithm

Quicksort is a "divide-and-conquer" method for sorting. It works by *partitioning* a file into two parts, then sorting the parts independently. As we will see, the exact position of the partition depends on the file, so the algorithm has the following recursive structure:

```
procedure quicksort(l,r: integer);
  var i: integer;
  begin
  if r>l then
    begin
    i:=partition(l,r);
    quicksort(l,i-1);
    quicksort(i+1,r)
    end
  end;
```

The parameters l and r delimit the subfile within the original file that is to be sorted; the call *quicksort*$(1,N)$ sorts the whole file.

The crux of the method is the *partition* procedure, which must rearrange the array to make the following three conditions hold:

(i) the element $a[i]$ is in its final place in the array for some i,

(ii) all the elements in $a[l], \ldots, a[i-1]$ are less than or equal to $a[i]$,

(iii) all the elements in $a[i+1], \ldots, a[r]$ are greater than or equal to $a[i]$.

This can be simply and easily implemented through the following general strategy. First, arbitrarily choose $a[r]$ to be the element that will go into its final position. Next, scan from the left end of the array until an element greater than $a[r]$ is found and scan from the right end of the array until an element less than $a[r]$ is found. The two elements which stopped the scans are obviously out of place in the final partitioned array, so exchange them. (Actually, it turns out, for reasons described

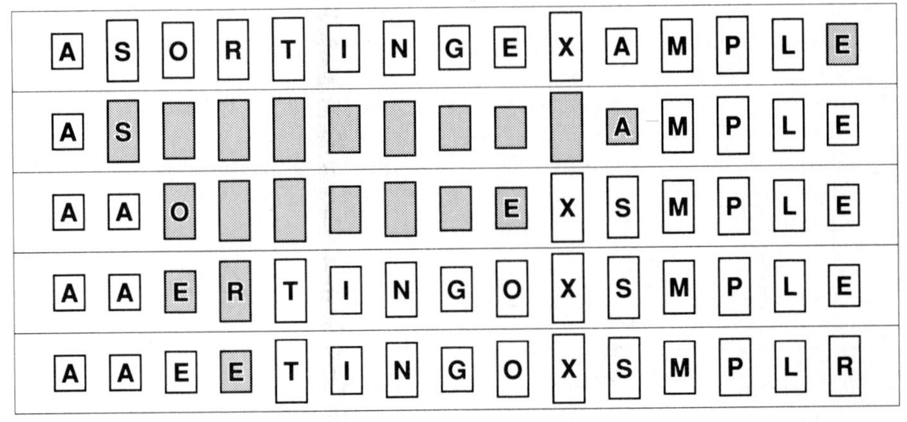

Figure 9.1 Partitioning.

below, to be best to also stop the scans for elements equal to $a[r]$, even though this might seem to involve some unnecessary exhanges.) Continuing in this way ensures that all array elements to the left of the left pointer are less than $a[r]$ and all array elements to the right of the right pointer are greater than $a[r]$. When the scan pointers cross, the partitioning process is nearly complete: all that remains is to exchange $a[r]$ with the leftmost element of the right subfile (the element pointed to by the left pointer).

Figure 9.1 shows how our sample file of keys is partitioned with this method. The rightmost element, E, is chosen as the partitioning element. First the scan from the left stops at the S, then the scan from the right stops at the A (as shown on the second line of the table), and then these two are exchanged. Next the scan from the left stops at the O, then the scan from the right stops at the E (as shown on the third line of the table), then these two are exchanged. Next the pointers cross. The scan from the left stops at the R, and the scan from the right stops at the E. The proper move at this point is to exchange the E at the right with the R, leaving the partitioned file shown on the last line of Figure 9.1.

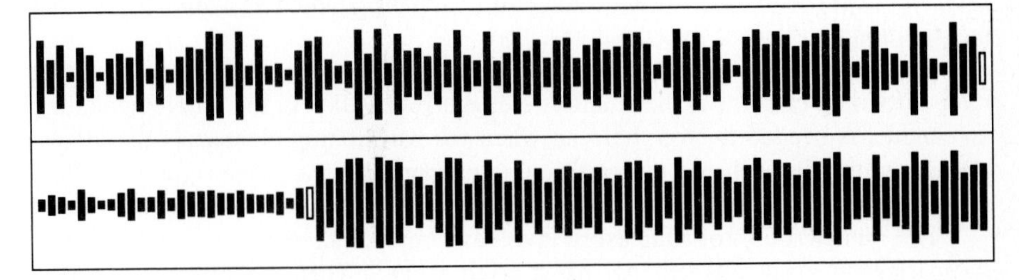

Figure 9.2 Partitioning a larger file.

Of course, the partitioning process is not stable, since any key might be moved past a large number of keys equal to it (which haven't even been examined yet) during any exchange.

Figure 9.2 shows the result of partitioning a larger file: with small elements on the left and large elements on the right, the partitioned file has considerably more "order" in it than the random file. The sort is finished by sorting the two subfiles on either side of the partitioning element (recursively). The following program gives a full implementation of the method.

```
procedure quicksort(l,r: integer);
  var v,t,i,j: integer;
  begin
  if r>l then
    begin
    v:=a[r]; i:=l-1; j:=r;
    repeat
      repeat i:=i+1 until a[i]>=v;
      repeat j:=j-1 until a[j]<=v;
      t:=a[i]; a[i]:=a[j]; a[j]:=t;
    until j<=i;
    a[j]:=a[i]; a[i]:=a[r]; a[r]:=t;
    quicksort(l,i-1);
    quicksort(i+1,r)
    end
  end;
```

In this implementation, the variable v holds the current value of the "partitioning element" $a[r]$ and i and j are the left and right scan pointers, respectively. An extra exchange of $a[i]$ with $a[j]$ is done with $j < i$ just after the pointers cross but before the crossing is detected and the outer **repeat** loop exited. (This could be avoided with a **goto**.) The three assignment statements following that loop implement the exchanges $a[i]$ with $a[j]$ (to undo the extra exchange) and $a[i]$ with $a[r]$ (to put the partitioning element into position).

As in insertion sort, a sentinel key is needed to stop the scan in the case that the partitioning element is the smallest element in the file. In this implementation, no sentinel is needed to stop the scan when the partitioning element is the largest element in the file, because the partitioning element itself is at the right end of the file to stop the scan. We'll shortly see an easy way to eliminate both sentinel keys.

The "inner loop" of Quicksort involves simply incrementing a pointer and comparing an array element against a fixed value. This is really what makes Quicksort quick: it's hard to imagine a simpler inner loop. The beneficial effect of sentinels is also underscored here, since adding just one superfluous test to the

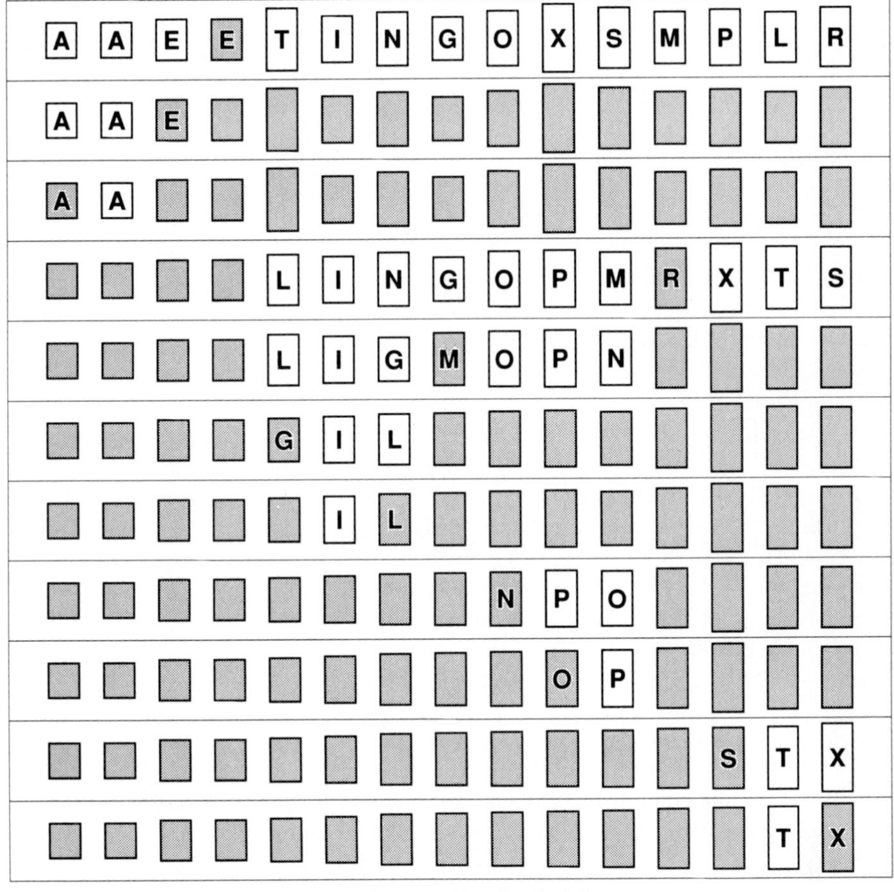

Figure 9.3 Subfiles in Quicksort.

inner loop will have a pronounced effect on performance.

Now the two subfiles are sorted recursively, finishing the sort. Figure 9.3 traces through these recursive calls. Each line depicts the result of partitioning the displayed subfile using the partitioning element (shaded in the diagram). If the initial test in the program were $r >= l$ rather than $r > l$, then every element would (eventually) be put into place by being used as a partitioning element; in the implementation as given, files of size 1 are not partitioned, as indicated in Figure 9.3. A generalization of this improvement is discussed in more detail below.

The most disturbing feature of the program above is that it is very inefficient on simple files. For example, if it is called with a file that is already sorted, the partitions will be degenerate, and the program will call itself N times, only knocking off one element for each call. This means not only that the time required will be about $N^2/2$, but also that the space required to handle the recursion will be about N (see below), which is unacceptable. Fortunately, there are relatively easy ways to ensure that this worst case doesn't occur in actual applications of the program.

When equal keys are present in the file, two subtleties become apparent. First, there is the question whether to have both pointers stop on keys equal to the partitioning element, or to have one pointer stop and the other scan over them, or to have both pointers scan over them. This question has actually been studied in some detail mathematically, and results show that it's best to have both pointers stop. This tends to balance the partitions in the presence of many equal keys. Second, there is the question of properly handling the pointer crossing in the presence of equal keys. Actually, the program above can be slightly improved by terminating the scans when $j< i$ and then using *quicksort*(l,j) for the first recursive call. This is an improvement because when $j=i$ we can put two elements into position with the partitioning by letting the loop iterate one more time. (This case would occur, for example, if R were E in the example above.) It is probably worth making this change because the program as given leaves a record with a key equal to the partitioning key in $a[r]$, and this makes the first partition in the call *quicksort*$(i+1,r)$ degenerate because its rightmost key is its smallest. The implementation of partitioning given above is a bit easier to understand, however, so we'll leave it as is in the discussions below, with the understanding that this change should be made when large numbers of equal keys are present.

Performance Characteristics of Quicksort

The best thing that could happen in Quicksort would be that each partitioning stage divides the file exactly in half. This would make the number of comparisons used by Quicksort satisfy the divide-and-conquer recurrence

$$C_N = 2C_{N/2} + N.$$

(The $2C_{N/2}$ covers the cost of sorting the two subfiles; the N is the cost of examining each element, using one partitioning pointer or the other.) From Chapter 6, we know that this recurrence has the solution

$$C_N \approx N \lg N.$$

Though things don't always go this well, it is true that the partition falls in the middle *on the average.* Taking into account the precise probability of each partition position makes the recurrence more complicated and more difficult to solve, but the final result is similar.

Property 9.1 *Quicksort uses about $2N \ln N$ comparisons on the average.*

The precise recurrence formula for the number of comparisons used by Quicksort for a random permutation of N elements is

$$C_N = N + 1 + \frac{1}{N} \sum_{1 \le k \le N} (C_{k-1} + C_{N-k}). \qquad \text{for } N \ge 2 \text{ with } C_1 = C_0 = 0.$$

The $N + 1$ term covers the cost of comparing the partitioning element with each of the others (two extra for where the pointers cross); the rest comes from the observation that each element k is likely to be the partitioning element with probability $1/k$, after which we are left with random files of size $k - 1$ and $N - k$.)

Though it looks rather complicated, this recurrence is actually easy to solve, in three steps. First, $C_0 + C_1 + \cdots + C_{N-1}$ is the same as $C_{N-1} + C_{N-2} + \cdots + C_0$, so we have

$$C_N = N + 1 + \frac{2}{N} \sum_{1 \le k \le N} C_{k-1}.$$

Second, we can eliminate the sum by multiplying both sides by N and subtracting the same formula for $N - 1$:

$$NC_N - (N - 1)C_{N-1} = N(N + 1) - (N - 1)N + 2C_{N-1}.$$

This simplifies to the recurrence

$$NC_N = (N + 1)C_{N-1} + 2N.$$

Third, dividing both sides by $N(N + 1)$ gives a recurrence that telescopes:

$$\frac{C_N}{N+1} = \frac{C_{N-1}}{N} + \frac{2}{N+1} = \frac{C_{N-2}}{N-1} + \frac{2}{N} + \frac{2}{N+1} = \cdots = \frac{C_2}{3} + \sum_{3 \le k \le N} \frac{2}{k+1}.$$

This exact answer is nearly equal to a sum that is easily approximated by an integral:

$$\frac{C_N}{N+1} \approx 2 \sum_{1 \le k \le N} \frac{1}{k} \approx 2 \int_1^N \frac{1}{x} dx = 2 \ln N,$$

which implies the stated result. Note that $2N \ln N \approx 1.38N \lg N$, so that the average number of comparisons is only about 38% higher than the best case. ■

Thus, the implementation above performs very well for random files, which makes it a very reasonable general-purpose sort for many applications. However, if the sort is to be used a great many times or if it is to be used to sort a very large file, then it might be worthwhile to implement several of the improvements discussed below which can make it much less likely that a bad case will occur, reduce the average running time by 20%, and easily eliminate the need for a sentinel key.

Removing Recursion

As in Chapter 5, we can remove recursion in the Quicksort program by using an explicit pushdown stack, which we think of as containing "work to be done" in the form of subfiles to be sorted. Any time we need a subfile to process, we pop the stack. When we partition, we create two subfiles to be processed which can be pushed on the stack. This leads to the following nonrecursive implementation of Quicksort:

```
procedure quicksort;
  var t,i,l,r: integer;
  begin
  l:=1; r:=N; stackinit;
  push(l); push(r);
  repeat
    if r>l then
      begin
      i:=partition(l,r);
      if (i-l)>(r-i)
        then begin push(l); push(i-1); l:=i+1 end
        else  begin push(i+1); push(r); r:=i-1 end;
      end
    else
      begin r:=pop; l:=pop end;
  until stackempty;
  end;
```

This program differs from the description above in two important ways. First, the two subfiles are not put on the stack in some arbitrary order, but their sizes are checked and the larger of the two is put on the stack first. Second, the smaller of the two subfiles is not put on the stack at all; the values of the parameters are simply reset. This is the "end-recursion-removal" technique discussed in Chapter 5. For Quicksort, the combination of end-recursion removal and the policy of processing the smaller of the two subfiles first turns out to ensure that the stack need contain room for only about $\lg N$ entries, since each entry on the stack after the top one must represent a subfile less than half the size of the previous entry.

This is in sharp contrast to the size of the stack in the worst case in the recursive implementation, which could be as large as N (for example, when the file is already sorted). This is a subtle but real difficulty with a recursive implementation of Quicksort: there's always an underlying stack, and a degenerate case on a large file could cause the program to terminate abnormally because of lack of memory, behavior obviously undesirable for a library sorting routine. Below we'll

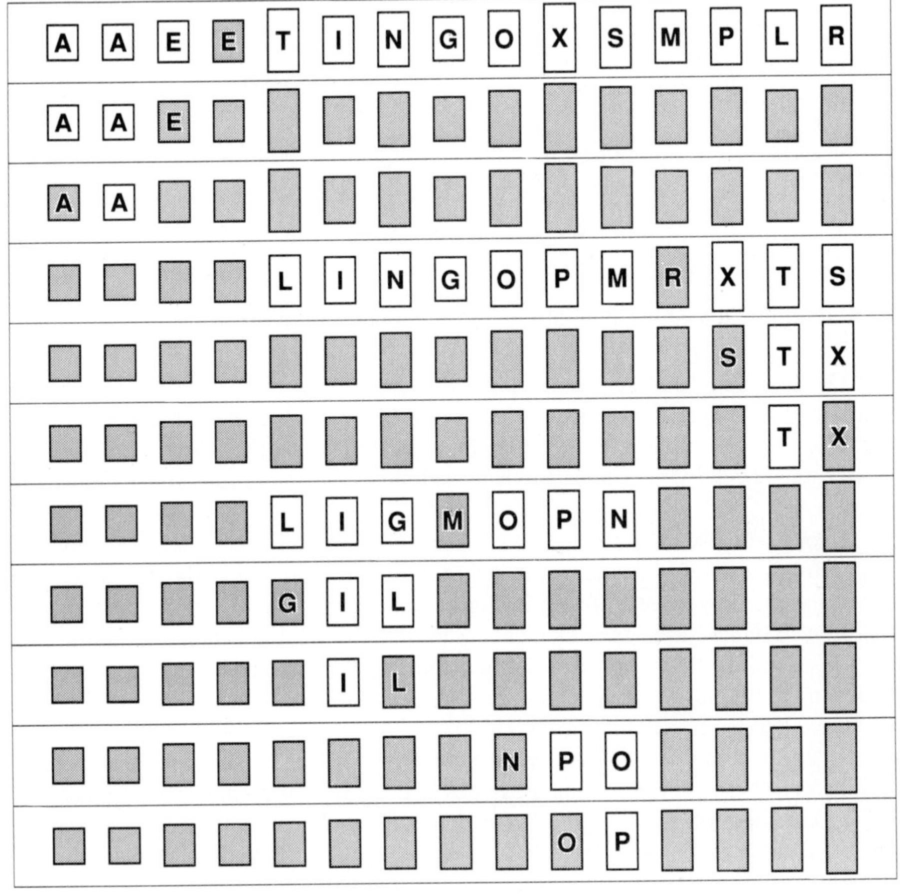

Figure 9.4 Subfiles in Quicksort (nonrecursive).

see ways to make degenerate cases extremely unlikely, but avoiding this problem in a recursive implementation is difficult without end-recursion removal. (Even switching the order in which subfiles are processed doesn't help.)

The simple use of an explicit stack in the program above leads to a far more efficient program than the direct recursive implementation, but there is still overhead that could be removed. The problem is that, if *both* subfiles have only one element, an entry with $r=l$ is put on the stack only to be immediately taken off and discarded. It is straightforward to change the program so that it puts no such files on the stack. This change is even more effective when the next improvement described below is included—this involves ignoring small subfiles in the same way, so the chances that both subfiles need to be ignored are much higher.

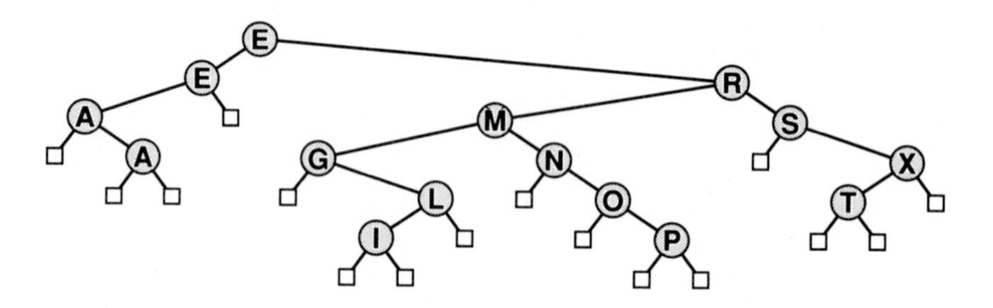

Figure 9.5 Tree diagram of the partitioning process in Quicksort.

Of course the nonrecursive method processes the same subfiles as the recursive method for any file; it just does them in a different order. Figure 9.4 shows the partitions for our example: the first three partitions are the same, but then the nonrecursive method partitions the right subfile of R first since it is smaller than the left, etc.

If we "collapse" Figures 9.3 and 9.4 and connect each partitioning element to the partitioning element used in its two subfiles, we get the static representation of the partitioning process shown in Figure 9.5. In this binary tree, each subfile is represented by its partitioning element (or itself if it is of size 1), and the subtrees of each node are the trees representing the subfiles after partitioning. Square external nodes in the tree represent null subfiles. (For clarity, the second A, the I, the P and the T are shown as having two null subfiles: as discussed above, the variants of the algorithm deal with null subfiles differently.) The recursive implementation of Quicksort corresponds to visiting the nodes of this tree in preorder; the nonrecursive implementation corresponds to a "visit the smaller subtree first" rule. In Chapter 14, we'll see how this tree leads to a direct relationship between Quicksort and a fundamental searching method.

Small Subfiles

The second improvement to Quicksort arises from the observation that a recursive program is guaranteed to call itself for many small subfiles, so it should use as good a method as possible when small subfiles are encountered. One obvious way to do this is to change the test at the beginning of the recursive routine from "**if** $r > l$ **then**" to a call on insertion sort (modified to accept parameters defining the subfile to be sorted): that is, "**if** $r-l <= M$ **then** $insertion(l, r)$." Here M is some parameter whose exact value depends upon the implementation. The value chosen for M need not be the best possible: the algorithm works about the same for M in the range from about 5 to about 25. The reduction in the running time is on the order of 20% for most applications.

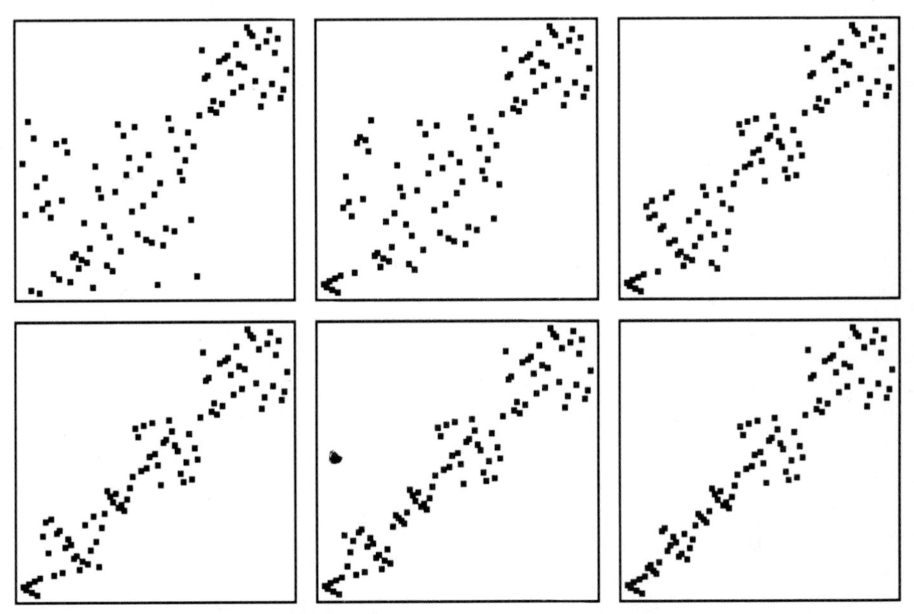

Figure 9.6 Quicksort (recursive implementation, $M = 12$).

A slightly easier way to handle small subfiles, which is also slightly more efficient, is to just change the test at the beginning to "**if** $r-l > M$ **then**": that is, simply ignore small subfiles during partitioning. In the nonrecursive implementation, this would be done by not putting any files of less than M on the stack. After partitioning, what is left is a file that is almost sorted. As mentioned in the previous chapter, however, insertion sort is the method of choice for such files. That is, insertion sort will work about as well for such a file as for the collection of little files that it would get if it were being used directly. This method should be used with caution, because the insertion sort is likely always to sort even if Quicksort has a bug which causes it not to work at all. The excessive cost may be the only sign that something went wrong.

Figure 9.6 gives a view of this process in a large, randomly ordered array. These diagrams graphically depict how each partition divides a subarray into two independent subproblems to be tackled independently. A subarray in these figures is shown as a square of randomly arranged dots; the partitioning process divides such a square into two smaller squares, with one element (the partitioning element) ending up on the diagonal. Elements not involved in partitioning end up quite close to the diagonal, leaving an array which is easily handled by insertion sort. As indicated above, the corresponding diagram for the nonrecursive implementation of Quicksort is similar, but partitions are done in a different order.

Median-of-Three Partitioning

The third improvement to Quicksort is to use a better partitioning element. There are several possibilities here. The safest choice to avoid the worst case would be a random element from the array for a partitioning element. Then the worst case will happen with negligibly small probability. This is a simple example of a "probabilistic algorithm," one which uses randomness to achieve good performance almost always, regardless of the arrangement of the input. Randomness can be a useful tool in algorithm design, especially if some bias in the input is suspected. However, for Quicksort it is probably overkill to put in a full random-number generator just for this purpose: an arbitrary number will do just as well (see Chapter 35).

A more useful improvement is to take three elements from the file, then use the median of the three for the partitioning element. If the three elements chosen are from the left, middle, and right of the array, then the use of sentinels can be avoided as follows: sort the three elements (using the three-exchange method in the previous chapter), then exchange the one in the middle with $a[r-1]$, and then run the partitioning algorithm on $a[l+1..r-2]$. This improvement is called the *median-of-three* partitioning method.

The median-of-three method helps Quicksort in three ways. First, it makes the worst case much more unlikely to occur in any actual sort. In order for the sort to take N^2 time, two out of the three elements examined must be among the largest or among the smallest elements in the file, and this must happen consistently through most of the partitions. Second, it eliminates the need for a sentinel key for partitioning, since this function is served by the three elements examined before partitioning. Third, it actually reduces the total average running time of the algorithm by about 5%.

The combination of a nonrecursive implementation of the median-of-three method with a cutoff for small subfiles can improve the running time of Quicksort over the naive recursive implementation by 25% to 30%. Further algorithmic improvements are possible (for example, the median of five or more elements could be used), but the amount of time gained will be marginal. More significant time savings can be realized (with less effort) by coding the inner loops (or the whole program) in assembly or machine language. Neither path is recommended except for experts with serious sorting applications.

Selection

One application related to sorting where a full sort may not always be necessary is the operation of finding the median of a set of numbers. This is a common computation in statistics and in various other data processing applications. One way to proceed would be to sort the numbers and look at the middle one, but it turns out that we can do better, using the Quicksort partitioning process.

The operation of finding the median is a special case of the operation of *selection*: find the kth smallest of a set of numbers. Since an algorithm cannot guarantee that a particular item is the kth smallest without having examined and identified the $k-1$ elements which are smaller and the $N-k$ elements which are larger, most selection algorithms can return all of the k smallest elements of a file without a great deal of extra calculation.

Selection has many applications in the processing of experimental and other data. The use of the median and other *order statistics* to divide a file up into smaller groups is very common. Often only a small part of a large file is to be saved for further processing; in such cases, a program which can select, say, the top ten percent of the elements of the file might be more appropriate than a full sort.

We've already seen an algorithm which can be directly adapted to selection. If k is very small, then *selection sort* will work very well, requiring time proportional to Nk: first find the smallest element, then find the second smallest by finding the smallest of the remaining items, etc. For slightly larger k, we'll see methods in Chapter 11 which can be immediately adapted to run in time proportional to $N \log k$.

An interesting method which adapts well to all values of k and runs in linear time on the average can be formulated from the partitioning procedure used in Quicksort. Recall that Quicksort's partitioning method rearranges an array $a[1..N]$ and returns an integer i such that $a[1], \ldots, a[i-1]$ are less than or equal to $a[i]$ and $a[i+1], \ldots, a[N]$ are greater than or equal to $a[i]$. If the kth smallest element in the file is sought and we have $k=i$, then we're done. Otherwise, if $k < i$ then we need to look for the kth smallest element in the left subfile, and if $k > i$ then we need to look for the $(k-i)$th smallest element in the right subfile. Adjusting this argument to apply to finding the kth smallest element in an array $a[l..r]$ leads immediately to the following recursive formulation.

```
procedure select(l,r,k: integer);
  var i;
  begin
  if r>l then
    begin
    i:=partition(l,r);
    if i>l+k-1 then select(l,i-1,k);
    if i<l+k-1 then select(i+1,r,k-i);
    end
  end;
```

This procedure rearranges the array so that $a[l], \ldots, a[k-1]$ are less than or equal to $a[k]$ and $a[k+1], \ldots, a[r]$ are greater than or equal to $a[k]$.

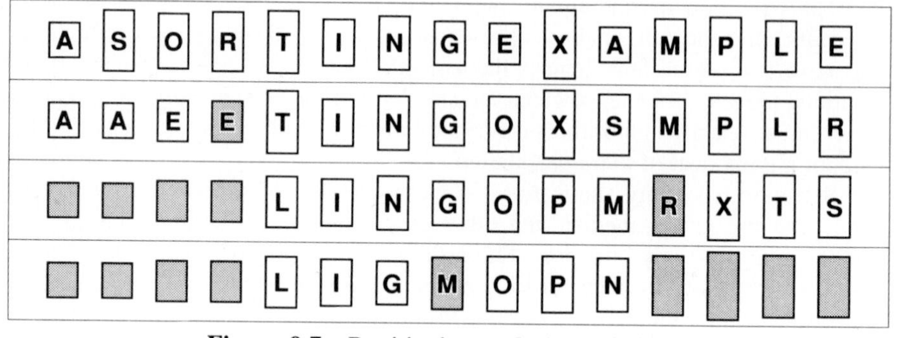

Figure 9.7 Partitioning to find the median.

For example, the call *select(1,N, (N+1)* **div** *2)* partitions the array on its median value. For the keys in our sorting example, this program uses only three recursive calls to find the median, as shown in Figure 9.7. The file is rearranged so that the median is in place with smaller elements to the left and larger elements to the right (equal elements could be on either side), but it is not fully sorted.

Since the *select* procedure always ends with only one call on itself, it is not really recursive in that no stack is needed to remove the recursion: when the time comes for the recursive call, we can simply reset the parameters and go back to the beginning, since there is nothing more to do. Also, we can eliminate the simple calculations involving k, as in the following implementation.

```
procedure select(k: integer);
  var v,t,i,j,l,r: integer;
  begin
  l:=1; r:=N;
  while r>l do
    begin
    v:=a[r]; i:=l-1; j:=r;
    repeat
      repeat i:=i+1 until a[i]>=v;
      repeat j:=j-1 until a[j]<=v;
      t:=a[i]; a[i]:=a[j]; a[j]:=t;
    until j<=i;
    a[j]:=a[i]; a[i]:=a[r]; a[r]:=t;
    if i>=k then r:=i-1;
    if i<=k then l:=i+1;
    end;
  end;
```

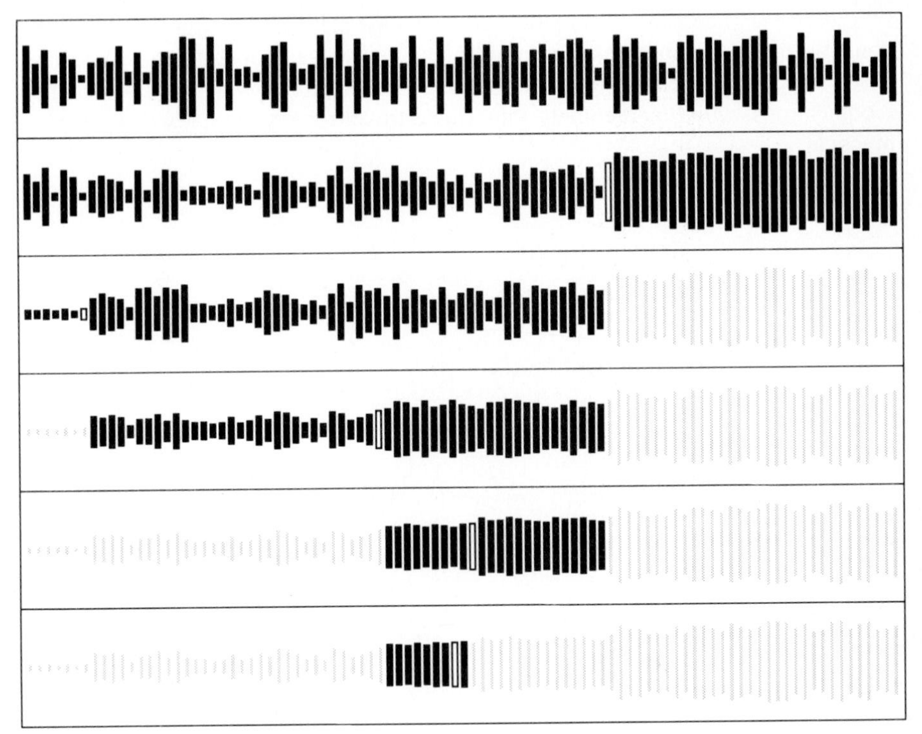

Figure 9.8 Finding the Median.

We use the identical partitioning procedure to Quicksort and, as with Quicksort, could change it slightly if many equal keys are expected.

Figure 9.8 shows the selection process on a larger (random) file. As with Quicksort, we can (very roughly) argue that, on a very large file, each partition should roughly split the array in half, so the whole process should require about $N + N/2 + N/4 + N/8 + \ldots = 2N$ comparisons. As with Quicksort, this rough argument is not too far from the truth.

Property 9.2 *Quicksort-based selection is linear-time on the average.*

An analysis similar to, but significantly more complex than, that given above for Quicksort leads to the result that the average number of comparisons is about $2N + 2k \ln(N/k) + 2(N - k) \ln(N/(N - k))$, which is linear for any allowed value of k. For $k = N/2$ (finding the median), this evaluates to about $(2 + 2\ln 2)N$ comparisons. ■

The worst case is about the same as for Quicksort: using this method to find the smallest element in an already sorted file would result in a quadratic running time. One could use an arbitrary or a random partitioning element, but some care should be exercised: for example, if the smallest element is sought, we probably don't want the file split near the middle. It is possible to modify this Quicksort-based selection procedure so that its running time is *guaranteed* to be linear. These modifications, while theoretically important, are extremely complex and not at all practical.

□

Exercises

1. Implement a recursive Quicksort with a cutoff to insertion sort for subfiles with less than M elements and empirically determine the value of M for which it runs fastest on a random file of 1000 elements.

2. Solve the previous problem for a nonrecursive implementation.

3. Solve the previous problem also incorporating the median-of-three improvement.

4. About how long will Quicksort take to sort a file of N equal elements?

5. What is the maximum number of times during the execution of Quicksort that the largest element can be moved?

6. Show how the file A B A B A B A is partitioned, using the two methods suggested in the text.

7. How many comparisons does Quicksort use to sort the keys E A S Y Q U E S T I O N?

8. How many "sentinel" keys are needed if insertion sort is called directly from within Quicksort?

9. Would it be reasonable to use a queue instead of a stack for a nonrecursive implementation of Quicksort? Why or why not?

10. Write a program to rearrange a file so that *all* the elements with keys equal to the median are in place, with smaller elements to the left and larger elements to the right.

10

Radix Sorting

☐ The "keys" used to define the order of the records for files for many sorting applications can be very complicated. (For example, consider the ordering function used in the telephone book or a library catalogue.) Because of this, it is reasonable to define sorting methods in terms of the basic operations of "comparing" two keys and "exchanging" two records. Most of the methods we have studied can be described in terms of these two fundamental operations. For many applications, however, it is possible to take advantage of the fact that the keys can be thought of as numbers from some restricted range. Sorting methods which take advantage of the digital properties of these numbers are called *radix sorts*. These methods do not just compare keys: they process and compare pieces of keys.

Radix-sorting algorithms treat the keys as numbers represented in a base-M number system, for different values of M (the *radix*), and work with individual digits of the numbers. For example, consider a clerk who must sort a pile of cards with three-digit numbers printed on them. One reasonable way for him to proceed is to make ten piles: one for the numbers less than 100, one for the numbers between 100 and 199, etc., place the cards in the piles, and then deal with the piles individually, by using the same method on the next digit or by using some easier method if there are only a few cards. This is a simple example of a radix sort with $M = 10$. We'll examine this and some other methods in detail in this chapter. Of course, with most computers it's more convenient to work with $M = 2$ (or some power of 2) rather than $M = 10$.

Anything that's represented inside a digital computer can be treated as a binary number, so many sorting applications can be recast to make feasible the use of radix sorts operating on keys which are binary numbers. Unfortunately, Pascal and many other languages intentionally make it difficult to write a program that depends on the binary representation of numbers. (The reason is that Pascal is intended to be a language for expressing programs in a machine-independent manner, and different computers may use different representations for the same numbers.)

This philosophy eliminates many types of "bit-flicking" techniques in situations admittedly better handled by fundamental Pascal constructs such as records and sets; however, radix sorting seems to have been a casualty of this progressive philosophy. Fortunately, it's not too difficult to use arithmetic operations to simulate the operations needed, and so here we'll be able to write (inefficient) Pascal programs to describe the algorithms that can be easily translated to efficient programs in programming languages that support bit operations on binary numbers.

Bits

Given a (key represented as a) binary number, the fundamental operation needed for radix sorts is extracting a contiguous set of bits from the number. Suppose we are to process keys which we know to be integers between 0 and 1000. We may assume that these are represented by ten-bit binary numbers. In machine language, bits are extracted from binary numbers by using bitwise "and" operations and shifts. For example, the leading two bits of a ten-bit number are extracted by shifting right eight bit positions, then doing a bitwise "and" with the mask 0000000011. In Pascal, these operations can be simulated with **div** and **mod**. For example, the leading two bits of a ten-bit number x are given by (x **div** *256*) **mod** *4*. In general, "shift x right k bit positions" can be simulated by computing x **div** 2^k, and "zero all but the j rightmost bits of x" can be simulated by computing x **mod** 2^j. In our description of the radix-sort algorithms, we'll assume the existence of a **function** *bits*(x,k,j: *integer*): *integer* which combines these operations to return the j bits which appear k bits from the right in x by computing (x **div** 2^k) **mod** 2^j. For example, the rightmost bit of x is returned by the call *bits*(x, *0*, *1*). This function can be made efficient by precomputing (or defining as constants) the powers of 2. Note that a program which uses only this function will do radix sorting whatever the representation of the numbers, though we can hope for much-improved efficiency if the representation is binary and the compiler is clever enough to notice that the computation can actually be done with machine-language "shift" and "and" instructions. Many Pascal implementations have extensions to the language which allow these operations to be specified somewhat more directly.

Armed with this basic tool, we'll consider two types of radix sorts which differ in the order in which they examine the bits of the keys. We assume that the keys are not short, so that it is worthwhile to go to the effort of extracting their bits. If the keys are short, then the distribution-counting method of Chapter 8 can be used. Recall that this method can sort N keys known to be integers between 0 and $M-1$ in linear time, using one auxiliary table of size M for counts and another of size N for rearranging records. Thus, if we can afford a table of size 2^b, then b-bit keys can easily be sorted in linear time. Radix sorting comes into play if the keys are sufficiently long (say $b = 32$) that this is not possible.

The first basic method for radix sorting that we'll consider examines the bits in the keys from left to right. It is based on the fact that the outcome of "comparisons" between two keys depends only on the value of the bits in the first position at which they differ (reading from left to right). Thus, all keys with leading bit 0 appear before all keys with leading bit 1 in the sorted file; among the keys with leading bit 1, all keys with second bit 0 appear before all keys with second bit 1, and so forth. The left-to-right radix sort, which is called *radix exchange sort*, sorts by systematically dividing up the keys in this way.

The second basic method that we'll consider, called *straight radix sort*, examines the bits in the keys from right to left. It is based on an interesting principle that reduces a sort on b-bit keys to b sorts on 1-bit keys. We'll see how this can be combined with distribution counting to produce a sort that runs in linear time under quite generous assumptions.

Radix Exchange Sort

Suppose we can rearrange the records of a file so that all those whose keys begin with a 0 bit come before all those whose keys begin with a 1 bit. This immediately defines a recursive sorting method: if the two subfiles are sorted independently, then the whole file is sorted. The rearrangement (of the file) is done very much as in the partitioning in Quicksort: scan from the left to find a key which starts with a 1 bit, scan from the right to find a key which starts with a 0 bit, exchange, and continue the process until the scanning pointers cross. This leads to a recursive sorting procedure that is very similar to Quicksort:

```
procedure radixexchange(l,r,b: integer);
  var t,i,j: integer;
  begin
  if (r>l) and (b>=0) then
    begin
    i:=l; j:=r;
    repeat
      while (bits(a[i],b,1)=0) and (i<j) do i:=i+1;
      while (bits(a[j],b,1)=1) and (i<j) do j:=j-1;
      t:=a[i]; a[i]:=a[j]; a[j]:=t;
    until j=i;
    if bits(a[r],b,1)=0 then j:=j+1;
    radixexchange(l,j-1,b-1);
    radixexchange(j,r,b-1);
    end
  end;
```

For simplicity, assume that $a[1..N]$ contains positive integers less than 2^{32} (so that they can be represented as 31-bit binary numbers). Then *radixexchange*(*1*,*N*,*30*)

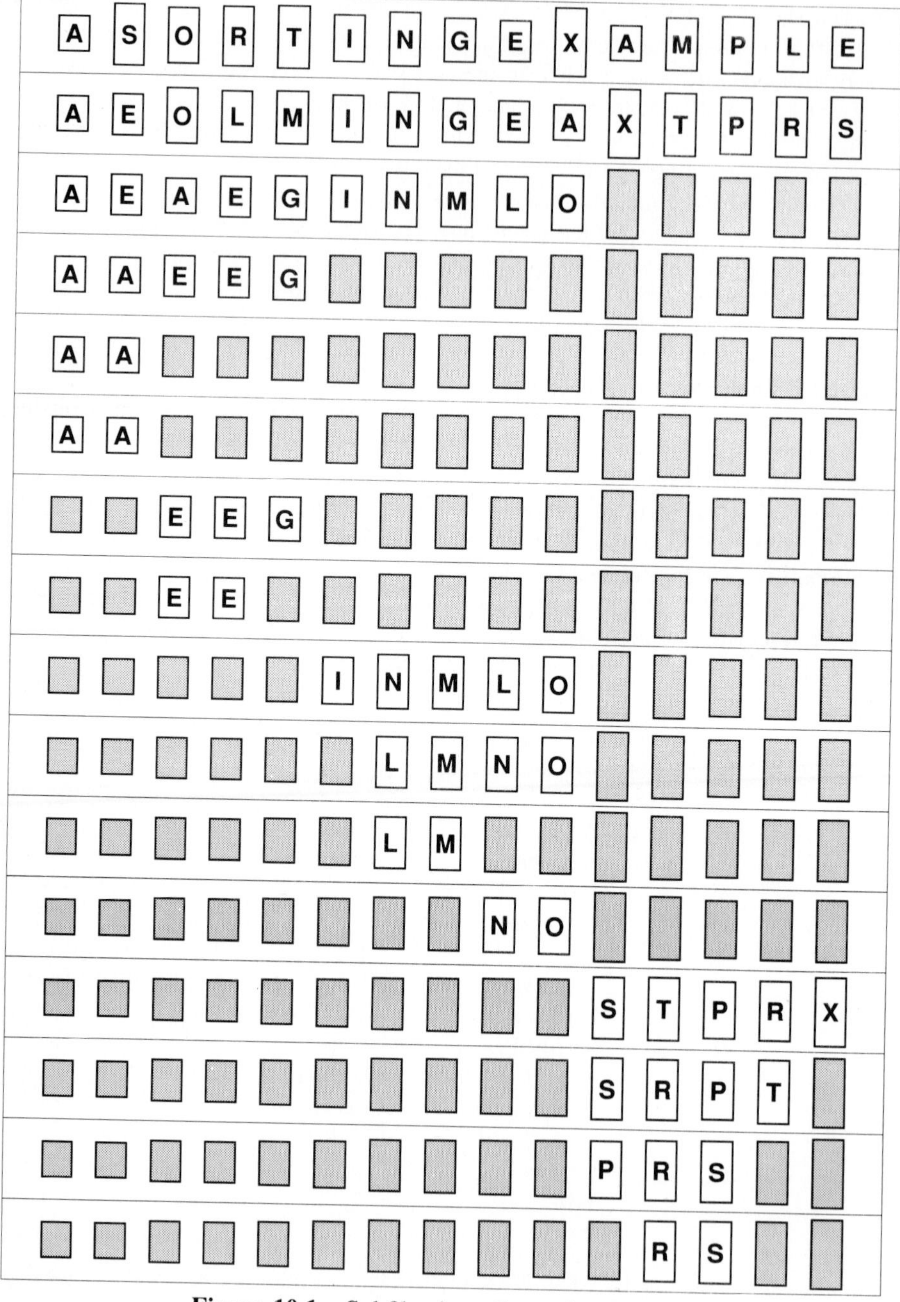

Figure 10.1 Subfiles in radix exchange sort.

will sort the array. The variable b keeps track of the bit being examined, ranging from 30 (leftmost) down to 0 (rightmost). (It is normally possible to adapt the implementation of *bits* to the machine representation of negative numbers so that negative numbers are handled in a uniform way in the sort.)

This implementation is obviously quite similar to the recursive implementation of Quicksort in Chapter 9. Essentially, the partitioning in radix-exchange sort is like partitioning in Quicksort except that the number 2^b is used as the partitioning element instead of some number from the file. Since 2^b may not be in the file, there can be no guarantee that an element is put into its final place during partitioning. Also, since only one bit is being examined, we can't rely on sentinels to stop the pointer scans; therefore the tests $(i< j)$ are included in the scanning loops. As with Quicksort, an extra exchange is done for the case $j=i$, but it is not necessary to undo this exchange outside the loop because the "exchange" is of $a[i]$ with itself. The partitioning stops with $j=i$ and all elements to the right of $a[i]$ having 1 bits in the bth position and all elements to the left of $a[i]$ having 0 bits in the bth position. The element $a[i]$ itself will have a 1 bit *unless* all keys in the file have a 0 in position b. The implementation above has an extra test just after the partitioning loop to cover this case.

Figure 10.1 shows how our sample file of keys is partitioned and sorted by this method. This can be compared with Figure 9.2 for Quicksort, though the operation of the partitioning method is completely opaque without the binary representation of the keys.

Figure 10.2 shows the partition in terms of the binary representation of the keys. A simple five-bit code is used, with the ith letter in the alphabet represented by the binary representation of the number i. This is a simplified version of real

A 0 0 0 0 1	A 0 0 0 0 1	A 0 0 0 0 1	A 0 0 0 0 1	A 0 0 0 0 1	A 0 0 0 0 1
S 1 0 0 1 1	E 0 0 1 0 1	E 0 0 1 0 1	A 0 0 0 0 1	A 0 0 0 0 1	A 0 0 0 0 1
O 0 1 1 1 1	O 0 1 1 1 1	A 0 0 0 0 1	E 0 0 1 0 1	E 0 0 1 0 1	E 0 0 1 0 1
R 1 0 0 1 0	L 0 1 1 0 0	E 0 0 1 0 1	E 0 0 1 0 1	E 0 0 1 0 1	E 0 0 1 0 1
T 1 0 1 0 0	M 0 1 1 0 1	G 0 0 1 1 1	G 0 0 1 1 1	G 0 0 1 1 1	
I 0 1 0 0 1	I 0 1 0 0 1	I 0 1 0 0 1	I 0 1 0 0 1		
N 0 1 1 1 0	N 0 1 1 1 0	N 0 1 1 1 0	N 0 1 1 1 0	L 0 1 1 0 0	L 0 1 1 0 0
G 0 0 1 1 1	G 0 0 1 1 1	M 0 1 1 0 1	M 0 1 1 0 1	M 0 1 1 0 1	M 0 1 1 0 1
E 0 0 1 0 1	E 0 0 1 0 1	L 0 1 1 0 0	L 0 1 1 0 0	N 0 1 1 1 0	N 0 1 1 1 0
X 1 1 0 0 0	A 0 0 0 0 1	O 0 1 1 1 1	O 0 1 1 1 1	O 0 1 1 1 1	O 0 1 1 1 1
A 0 0 0 0 1	X 1 1 0 0 0	S 1 0 0 1 1	S 1 0 0 1 1	P 1 0 0 0 0	
M 0 1 1 0 1	T 1 0 1 0 0	T 1 0 1 0 0	R 1 0 0 1 0	R 1 0 0 1 0	R 1 0 0 1 0
P 1 0 0 0 0	P 1 0 0 0 0	P 1 0 0 0 0	P 1 0 0 0 0	S 1 0 0 1 1	S 1 0 0 1 1
L 0 1 1 0 0	R 1 0 0 1 0	R 1 0 0 1 0	T 1 0 1 0 0		
E 0 0 1 0 1	S 1 0 0 1 1	X 1 1 0 0 0			

Figure 10.2 Radix exchange sort ("left-to-right" radix sort).

character codes, which use more bits (seven or eight) and represent more characters (upper/lower case letters, numbers, special symbols). By translating the keys in Figure 10.1 to this five-bit character code, compressing the table so that the subfile partitioning is shown "in parallel" rather than one per line, and then transposing rows and columns, we can show in Figure 10.2 how the leading bits of the keys control partitioning. In this figure, each partition is indicated by a white "0" subfile followed by a gray "1" subfile in the next diagram to the right, except that subfiles of size 1 drop out of the partitioning process as encountered.

One serious potential problem for radix sort not brought out in this example is that degenerate partitions (partitions with all keys having the same value for the bit being used) can happen frequently. This situation arises commonly in real files when small numbers (with many leading zeros) are being sorted. It also occurs for characters: for example, suppose that 32-bit keys are made up from four characters by encoding each in a standard eight-bit code and then putting them together. Then degenerate partitions are likely to occur at the beginning of each character position, since for example, lower-case letters all begin with the same bits in most character codes. Many other similar effects are obviously of concern when sorting encoded data.

It can be seen from Figure 10.2 that once a key is distinguished from all the other keys by its left bits, no further bits are examined. This is a distinct advantage in some situations, a disadvantage in others. When the keys are truly random bits, each key should differ from the others after about $\lg N$ bits, which could be many fewer than the number of bits in the keys. This is because, in a random situation, we expect each partition to divide the subfile in half. For example, sorting a file with 1000 records might involve only examining about ten or eleven bits from each key (even if the keys are, say, 32-bit keys). On the other hand, notice that all the bits of equal keys are examined. Radix sorting simply does not work well on files which contain many equal keys. Radix-exchange sort is actually slightly faster than Quicksort if the keys to be sorted are comprised of truly random bits,

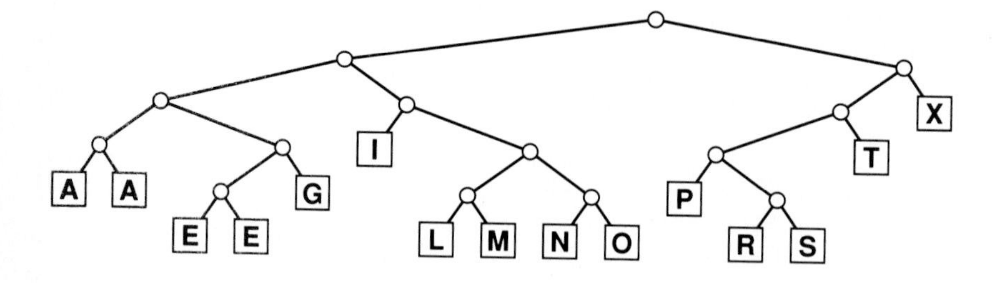

Figure 10.3 Tree diagram of the partitioning process in radix-exchange sort.

but Quicksort can adapt better to less random situations.

Figure 10.3, the tree that represents the partitioning process for radix exchange sort, may be compared with Figure 9.5. In this binary tree, the internal nodes represent the partitioning points, and the external nodes are the keys in the file, all of which end up in subfiles of size 1. In Chapter 17 we will see how this tree suggests a direct relationship between radix-exchange sort and a fundamental searching method.

The basic recursive implementation given above can be improved by removing recursion and treating small subfiles differently, just as we did for Quicksort.

Straight Radix Sort

An alternative radix-sorting method is to examine the bits from right to left. This is the method used by old computer-card-sorting machines: a deck of cards was run through the machine 80 times, once for each column, proceeding from right to left. Figure 10.4 shows how a right-to-left bit-by-bit radix sort works on our file of sample keys. The ith column in Figure 10.4 is sorted on the trailing i bits of the keys, and is derived from the $(i-1)$st column by extracting all the keys with a 0 in the ith bit, then all the keys with a 1 in the ith bit.

It's not easy to be convinced that the method works; in fact it doesn't work at all unless the one-bit partitioning process is stable. Once stability has been identified as being important, a trivial proof that the method works can be found: after putting keys with ith bit 0 before those with ith bit 1 (in a stable manner), we know that any two keys appear in proper order (on the basis of the bits so far examined) in the file either because their ith bits are different, in which case

A	00001	R	10010	T	10100	X	11000	P	10000	A	00001
S	10011	T	10100	X	11000	P	10000	A	00001	A	00001
O	01111	N	01110	P	10000	A	00001	A	00001	E	00101
R	10010	X	11000	L	01100	I	01001	R	10010	E	00101
T	10100	P	10000	A	00001	A	00001	S	10011	G	00111
I	01001	L	01100	I	01001	R	10010	T	10100	I	01001
N	01110	A	00001	E	00101	S	10011	E	00101	L	01100
G	00111	S	10011	A	00001	T	10100	E	00101	M	01101
E	00101	O	01111	M	01101	L	01100	G	00111	N	01110
X	11000	I	01001	E	00101	E	00101	X	11000	O	01111
A	00001	G	00111	R	10010	M	01101	I	01001	P	10000
M	01101	E	00101	N	01110	E	00101	L	01100	R	10010
P	10000	A	00001	S	10011	N	01110	M	01101	S	10011
L	01100	M	01101	O	01111	O	01111	N	01110	T	10100
E	00101	E	00101	G	00111	G	00111	O	01111	X	11000

Figure 10.4 Straight radix sort ("right-to-left" radix sort).

partitioning puts them in the proper order, or because their ith bits are the same, in which case they're in proper order because of stability. The requirement of stability means, for example, that the partitioning method used in the radix-exchange sort can't be used for this right-to-left sort.

The partitioning is like sorting a file with only two values, and the distribution counting sort in Chapter 8 is entirely appropriate for this. If we assume that $M = 2$ in the distribution counting program and replace *a[i]* by *bits(a[i],k,1)*, then that program becomes a method for sorting the elements of the array *a* on the bit *k* positions from the right and putting the result in a temporary array *t*. But there's no reason to use $M = 2$; in fact, we should make M as large as possible, realizing that we need a table of M counts. This corresponds to using m bits at a time during the sort, with $M = 2^m$. Thus, straight radix sort becomes little more than a generalization of distribution-counting sort, as in the following implementation for sorting *a[1..N]* on the w rightmost bits:

```
procedure straightradix;
  var i,j,pass: integer;
        count: array[0..M] of integer;
  begin
  for pass:=0 to (w div m)-1 do
    begin
    for j:=0 to M-1 do count[j]:=0;
    for i:=1 to N do
      count[bits(a[i],pass*m,m)]:=count[bits(a[i],pass*m,m)]+1;
    for j:=1 to M-1 do
      count[j]:=count[j-1]+count[j];
    for i:=N downto 1 do
      begin
      b[count[bits(a[i],pass*m,m)]]:=a[i];
      count[bits(a[i],pass*m,m)]:=count[bits(a[i],pass*m,m)]-1;
      end;
    for i:=1 to N do a[i]:=b[i];
    end;
  end;
```

For clarity, this procedure uses two calls on *bits* to increment and decrement *count* when one would suffice. Also, the correspondence $M = 2^m$ has been preserved in the variable names, though some versions of "pascal" can't tell the difference between m and M.

The procedure above works properly only if w is a multiple of m. Normally, this is not a particularly restrictive assumption for radix sort: it simply corresponds to dividing the keys to be sorted into an integral number of equal-size pieces.

When $m{=}w$ we have distribution counting sort; when $m{=}1$ we have straight radix sort, the right-to-left bit-by-bit radix sort described in the example above.

The implementation above moves the file from a to b during each distribution counting phase, then back to a in a simple loop. This "array copy" loop could be eliminated if desired by making two copies of the distribution counting code, one to sort from a into b, the other to sort from b into a.

Performance Characteristics of Radix Sorts

The running times of both basic radix sorts for sorting N records with b-bit keys are essentially Nb. On the one hand, one can think of this running time as being essentially the same as $N \log N$, since if the numbers are all different, b must be at least $\log N$. On the other hand, both methods usually use many fewer than Nb operations: the left-to-right method because it can stop once differences between keys have been found, and the right-to-left method because it can process many bits at once.

Property 10.1 *Radix-exchange sort uses on the average about $N \lg N$ bit comparisons.*

If the file size is a power of two and the bits are random, then we expect half of the leading bits to be 0 and half to be 1, so the recurrence $C_N = 2C_{N/2} + N$ should describe the performance, as argued for Quicksort in Chapter 9. Again, this description of the situation is not accurate, because the partition falls in the center only on the average (and because the number of bits in the keys is finite). However, in this model, the partition is much more likely to be in the center than in Quicksort, so the result stated turns out to be true. (Detailed analysis beyond the scope of this book is required to prove this.) ▪

Property 10.2 *Both radix sorts use less than Nb bit comparisons to sort N b-bit keys.*

In other words, the radix sorts are *linear* in the sense that the time taken is proportional to the number of bits of input. This follows directly from examination of the programs: no bit is examined more than once. ▪

For large random files, radix-exchange sort behaves rather like Quicksort, as shown in Figure 9.6, but straight radix sort behaves rather differently. Figure 10.5 shows the stages of straight radix sort on a random file of five-bit keys. The progressive organization of the file during the sort shows up clearly in these diagrams. For example, after the third stage (bottom left), the file consists of four intermixed sorted subfiles: the keys beginning with 00 (bottom stripe), the keys beginning with 01, etc.

Property 10.3 *Straight radix sort can sort N records with b-bit keys in b/m passes, using extra space for 2^m counters (and a buffer for rearranging the file).*

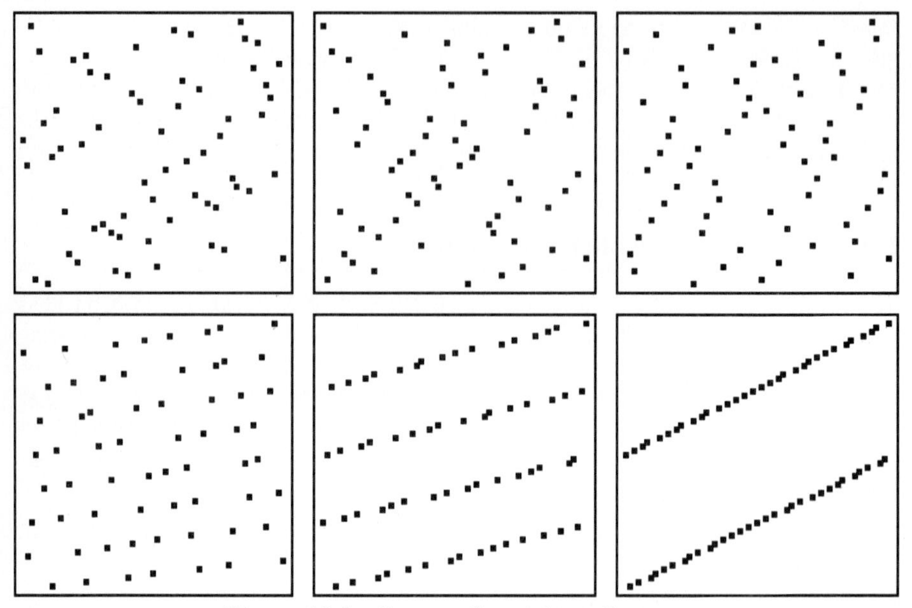

Figure 10.5 Stages of straight radix sort.

Proof of this fact is straightforward from the implementation. In particular, if we can take $m = b/4$ without using too much extra memory, we get a linear sort! The practical ramifications of this property are discussed in more detail in the next section. ■

A Linear Sort

The straight radix sort implementation given in the previous section makes b/m passes through the file. By making m large, we get a very efficient sorting method, as long as we have $M = 2^m$ words of memory available. A reasonable choice is to make m about one quarter the word-size ($b/4$), so that the radix sort is four distribution counting passes. The keys are treated as base-M numbers, and each (base-M) digit of each key is examined, but there are only four digits per key. (This corresponds directly to the architectural organization of many computers: one typical organization has 32-bit words, each consisting of four 8-bit bytes. The *bits* procedure then winds up extracting particular bytes from words in this case, which obviously can be done very efficiently on such computers.) Now, each distribution-counting pass is linear, and since there are only four of them, the entire sort is linear, certainly the best performance we could hope for in a sort.

In fact, it turns out that we can get by with only two distribution counting passes. (Even a careful reader is likely to have difficulty telling right from left by this time, so some effort may be necessary to understand this method.) We do this by taking advantage of the fact that the file will be *almost* sorted if only the leading $b/2$ bits of the b-bit keys are used. As with Quicksort, the sort can be completed efficiently by using insertion sort on the whole file afterwards. This method is obviously a trivial modification to the implementation above: to do a right-to-left sort using the leading half of the keys, we simply start the outer loop at *pass=b* **div** *(2*m)* rather than *pass=1*. Then a conventional insertion sort can be used on the nearly ordered file that results. To become convinced that a file sorted on its leading bits is quite well-ordered, the reader should examine the first few columns of Figure 10.2. For example, insertion sort run on the file already sorted on the first three bits would require only six exchanges.

Using two distribution counting passes (with m about one-fourth the word size) and then using insertion sort to finish the job will yield a sorting method that is likely to run faster than any of the others we've seen for large files whose keys are random bits. Its main disadvantage is that it requires an extra array of the same size as the array being sorted. It is possible to eliminate the extra array using linked-list techniques, but extra space proportional to N (for the links) is still required.

A linear sort is obviously desirable for many applications, but there are reasons why it is not the panacea that it might seem. First, its efficiency really does depend on the keys being random bits, randomly ordered. If this condition is not satisfied, severely degraded performance is likely. Second, it requires extra space proportional to the size of the array being sorted. Third, the "inner loop" of the program actually contains quite a few instructions, so even though it's linear, it won't be as much faster than Quicksort (say) as one might expect, except for quite large files (at which point the extra array becomes a real liability). The choice between Quicksort and radix sort is a difficult one that is likely to depend not only on features of the application, such as key, record, and file size, but also on features of the programming and machine environment that relate to the efficiency of access and use of individual bits. Again, such tradeoffs need to be studied by an expert and this type of study is likely to be worthwhile only for serious sorting applications.

□

Exercises

1. Compare the number of exchanges used by radix-exchange sort with the number used by Quicksort for the file 001, 011, 101, 110, 000, 001, 010, 111, 110, 010.
2. Why is it not as important to remove the recursion from the radix-exchange sort as it was for Quicksort?
3. Modify radix-exchange sort to skip leading bits which are identical on all keys. In what situations would this be worthwhile?
4. True or false: the running time of straight radix sort does not depend on the order of the keys in the input file. Explain your answer.
5. Which method is likely to be faster for a file of all equal keys: radix-exchange sort or straight radix sort?
6. True or false: both radix-exchange sort and straight radix sort examine all the bits of all the keys in the file. Explain your answer.
7. Aside from the extra memory requirement, what is the major disadvantage of the strategy of doing straight radix sorting on the leading bits of the keys, then cleaning up with insertion sort afterwards?
8. Exactly how much memory is required to do a four-pass straight radix sort of N b-bit keys?
9. What type of input file will make radix-exchange sort run the most slowly (for very large N)?
10. Empirically compare straight radix sort with radix-exchange sort for a random file of 1000 32-bit keys.

11

Priority Queues

☐ In many applications, records with keys must be processed in order, but not necessarily in full sorted order and not necessarily all at once. Often a set of records must be collected, then the largest processed, then perhaps more records collected, then the next largest processed, and so forth. An appropriate data structure in such an environment is one which supports the operations of inserting a new element and deleting the largest element. Such a data structure, which can be contrasted with queues (delete the oldest) and stacks (delete the newest) is called a *priority queue*. In fact, the priority queue might be thought of as a generalization of the stack and the queue (and other simple data structures), since these data structures can be implemented with priority queues, using appropriate priority assignments.

Applications of priority queues include simulation systems (where the keys might correspond to "event times" which must be processed in order), job scheduling in computer systems (where the keys might correspond to "priorities" indicating which users should be processed first), and numerical computations (where the keys might be computational errors, so the largest can be worked on first).

Later on in this book, we'll see how to use priority queues as basic building blocks for more advanced algorithms. In Chapter 22, we'll develop a file-compression algorithm using routines from this chapter, and in Chapters 31 and 33, we'll see how priority queues can serve as the basis for several fundamental graph-searching algorithms. These are but a few examples of the important role played by the priority queue as a basic tool in algorithm design.

It is useful to be somewhat more precise about how to manipulate a priority queue, since there are several operations we may need to perform on priority queues in order to maintain them and use them effectively for applications such as those mentioned above. Indeed, the main reason that priority queues are so useful is their flexibility in allowing a variety of different operations to be efficiently performed on sets of records with keys. We want to build and maintain a data structure containing records with numerical keys (*priorities*) and supporting some of the following operations:

> *Construct* a priority queue from N given items.
> *Insert* a new item.
> *Remove* the largest item.
> *Replace* the largest item with a new item (unless the new item is larger).
> *Change* the priority of an item.
> *Delete* an arbitrary specified item.
> *Join* two priority queues into one large one.

(If records can have duplicate keys, we take "largest" to mean "any record with the largest key value.")

The *replace* operation is almost equivalent to an *insert* followed by a *remove* (the difference being that *insert/remove* requires the priority queue to grow temporarily by one element); note that this is quite different from a *remove* followed by an *insert*. This is included as a separate capability because, as we will see, some implementations of priority queues can do the *replace* operation quite efficiently. Similarly, the *change* operation could be implemented as a *delete* followed by an *insert* and the *construct* could be implemented with repeated uses of the *insert* operation, but these operations can be directly implemented more efficiently for some choices of data structure. The *join* operation requires advanced data structures for efficient implementation; we'll concentrate instead on a "classical" data structure called a *heap* that allows efficient implementations of the first five operations.

The priority queue as described above is an excellent example of an *abstract data structure* as described in Chapter 3: it is very well defined in terms of the operations performed on it, independent of how the data is organized and processed in any particular implementation.

Different implementations of priority queues involve different performance characteristics for the various operations to be performed, leading to cost tradeoffs. Indeed, performance differences are really the only differences that can arise in the abstract data structure concept. First, we'll illustrate this point by discussing a few elementary data structures for implementing priority queues. Next, we'll examine a more advanced data structure and then show how the various operations can be implemented efficiently using this data structure. We'll then look at an important sorting algorithm that follows naturally from these implementations.

Elementary Implementations

One way to organize a priority queue is as an *unordered list*, simply keeping the items in an array $a[1..N]$ without paying attention to the keys. Thus *construct* is a "no-op" for this organization, but *insert* and *remove* are easily implemented as follows:

```
procedure insert(v: integer);
  begin
  N:=N+1; a[N]:=v;
  end;
function remove: integer;
  var j, max: integer;
  begin
  max:=1;
  for j:=2 to N do
    if a[j]>a[max] then max:=j;
  remove:=a[max];
  a[max]:=a[N]; N:=N-1;
  end;
```

To *insert*, we simply increment N and put the new item into $a[N]$, a constant-time operation. But *remove* requires scanning through the array to find the element with the largest key, which takes linear time (all the elements in the array must be examined), then exchanging $a[N]$ with the element with the largest key and decrementing N. The implementation of *replace* would be quite similar and is omitted.

To implement the *change* operation (change the priority of the item in $a[k]$), we could simply store the new value, and to *delete* the item in $a[k]$, we could exchange it with $a[N]$ and decrement N, as in the last line of *remove*. (Recall from Chapter 3 that such operations, which refer to individual data items, make sense only in a "pointer" or "indirect" implementation, where a reference is maintained for each item to its current place in the data structure.)

Another elementary organization to use is an *ordered list*, again using an array $a[1..N]$ but keeping the items in increasing order of their keys. Now *remove* simply involves returning $a[N]$ and decrementing N (a constant-time operation), but *insert* involves moving larger elements in the array right one position, which could take linear time.

Any priority queue algorithm can be turned into a sorting algorithm by repeatedly using *insert* to build a priority queue containing all the items to be sorted, then repeatedly using *remove* to empty the priority queue, receiving the items in reverse order. Using a priority queue represented as an unordered list in this way corresponds to selection sort; using the ordered list corresponds to insertion sort.

Linked lists can also be used for the unordered list or the ordered list rather than the array implementation given above. This doesn't change the fundamental performance characteristics for *insert*, *remove*, or *replace*, but it does make it possible to do *delete* and *join* in constant time. These implementations are omitted here because they are so similar to the basic list operations given in Chapter 3, and because implementations of similar methods for the searching problem (find a record with a given key) are given in Chapter 14.

As usual, it is wise to keep these simple implementations in mind because they often can outperform more complicated methods in many practical situations. For example, the unordered list implementation might be appropriate in an application where only a few "remove-largest" operations are performed as opposed to a large number of insertions, while an ordered list would be appropriate if the items inserted always tended to be close to the largest element in the priority queue.

Heap Data Structure

The data structure that we'll use to support the priority queue operations involves storing the records in an array in such a way that each key is guaranteed to be larger than the keys at two other specific positions. In turn, each of those keys must be larger than two more keys, and so forth. This ordering is very easy to see if we draw the array in a two-dimensional tree structure with lines down from each key to the two keys known to be smaller, as in Figure 11.1.

Recall from Chapter 4 that this structure is called a "complete binary tree": it may be constructed by placing one node (called the *root*) and then proceeding down the page and from left to right, connecting two nodes beneath each node on the previous level until N nodes have been placed. The two nodes below each node are called its *children*; the node above each node is called its *parent*. Now, we want the keys in the tree to satisfy the *heap condition*: the key in each node should be larger than (or equal to) the keys in its children (if it has any). Note that this implies in particular that the largest key is in the root.

We can represent complete binary trees sequentially within an array by simply putting the root at position 1, its children at positions 2 and 3, the nodes at the next level in positions 4, 5, 6 and 7, etc., as numbered in Figure 11.1. For example, the array representation for the tree above is shown in Figure 11.2.

This natural representation is useful because it is very easy to get from a node to its parent and children. The parent of the node in position j is in position j **div** 2, and, conversely, the two children of the node in position j are in position $2j$

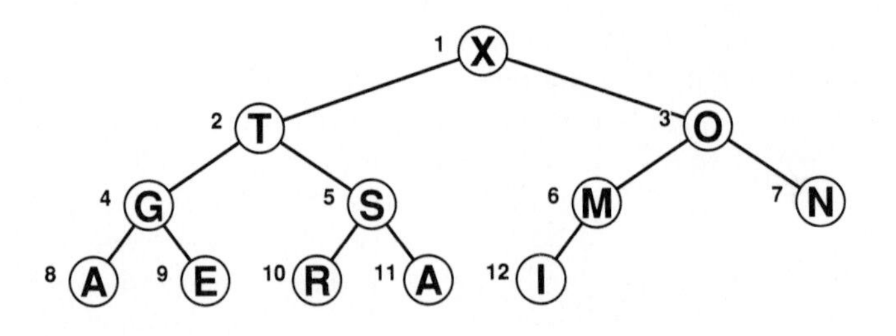

Figure 11.1 Complete tree representation of a heap.

k	1	2	3	4	5	6	7	8	9	10	11	12
$a[k]$	X	T	O	G	S	M	N	A	E	R	A	I

Figure 11.2 Array representation of a heap.

and $2j + 1$. This makes traversal of such a tree even easier than if the tree were implemented with a standard linked representation (with each element containing a pointer to its parent and children). The rigid structure of complete binary trees represented as arrays does limit their utility as data structures, but there is just enough flexibility to allow the implementation of efficient priority queue algorithms. A *heap* is a complete binary tree, represented as an array, in which every node satisfies the heap condition. In particular, the largest key is always in the first position in the array.

All of the algorithms operate along some *path* from the root to the bottom of the heap (just moving from parent to child or from child to parent). It is easy to see that, in a heap of N nodes, all paths have about $\lg N$ nodes on them. (There are about $N/2$ nodes on the bottom, $N/4$ nodes with children on the bottom, $N/8$ nodes with grandchildren on the bottom, etc. Each "generation" has about half as many nodes as the next, which implies that there can be at most $\lg N$ generations.) Thus all of the priority queue operations (except *join*) can be done in logarithmic time using heaps.

Algorithms on Heaps

The priority queue algorithms on heaps all work by first making a simple structural modification which could violate the heap condition, then traveling through the heap modifying it to ensure that the heap condition is satisfied everywhere. Some of the algorithms travel through the heap from bottom to top, others from top to bottom. In all of the algorithms, we'll assume that the records are one-word integer keys stored in an array a of some maximum size, with the current size of the heap kept in an integer N. Note that N is as much a part of the definition of the heap as the keys and records themselves.

To be able to build a heap, it is necessary first to implement the *insert* operation. Since this operation will increase the size of the heap by one, N must be incremented. Then the record to be inserted is put into $a[N]$, but this may violate the heap property. If the heap property is violated (the new node is greater than its parent), then the violation can be fixed by exchanging the new node with its parent. This may, in turn, cause a violation, and thus can be fixed in the same way. For example, if P is to be inserted in the heap above, it is first stored in $a[N]$ as the right child of M. Then, since it is greater than M, it is exchanged with M, and since it is greater than O, it is exchanged with O, and the process terminates since it is less that X. The heap shown in Figure 11.3 results.

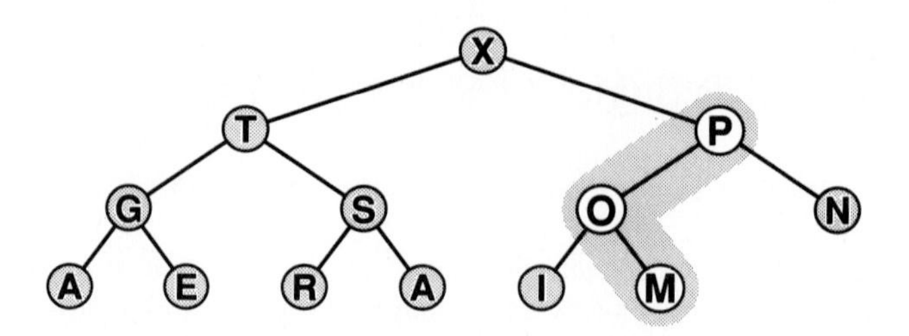

Figure 11.3 Inserting a new element (P) into a heap.

The code for this method is straightforward. In the following implementation, *insert* adds a new item to $a[N]$, then calls *upheap*(N) to fix the heap condition violation at N:

```
procedure upheap(k: integer);
  var v: integer;
  begin
  v:=a[k]; a[0]:=maxint;
  while a[k div 2]<=v do
    begin a[k]:=a[k div 2]; k:=k div 2 end;
  a[k]:=v
  end;
procedure insert(v: integer);
  begin
  N:=N+1; a[N]:=v;
  upheap(N)
  end;
```

If k **div** 2 were replaced by $k-1$ everywhere in this program, we would have in essence one step of insertion sort (implementing a priority queue with an ordered list); here, instead, we are "inserting" the new key along the path from N to the root. As with insertion sort, it is not necessary to do a full exchange within the loop, because v is always involved in the exchanges. A sentinel key must be put in $a[0]$ to stop the loop for the case that v is greater than all the keys in the heap.

The *replace* operation involves replacing the key at the root with a new key, then moving down the heap from top to bottom to restore the heap condition. For example, if the X in the heap above is to be replaced with C, the first step is to store C at the root. This violates the heap condition, but the violation can be fixed by exchanging C with T, the larger of the two children of the root. This creates a violation at the next level, which can again be fixed by exchanging C with the larger of its two children (in this case S). The process continues until the

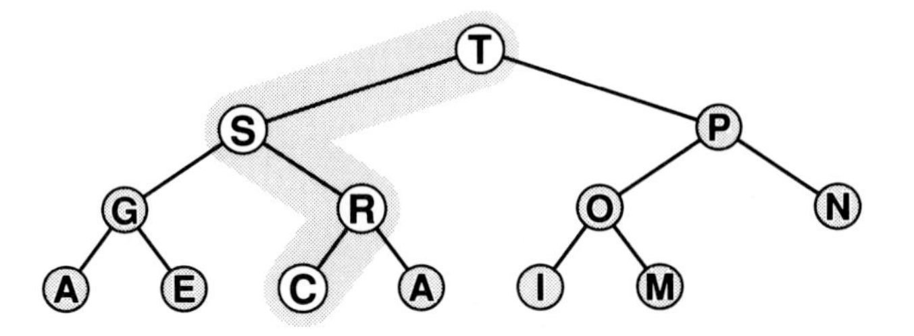

Figure 11.4 Replacing the largest key in a heap (with C).

heap condition is no longer violated at the node occupied by C. In the example, C makes it all the way to the bottom of the heap, leaving the heap depicted in Figure 11.4.

The "*remove* the largest" operation involves almost the same process. Since the heap will be one element smaller after the operation, it is necessary to decrement N, leaving no place for the element that was stored in the last position. But the largest element (which is in $a[1]$) is to be removed, so the *remove* operation amounts to a *replace*, using the element that was in $a[N]$. The heap shown in Figure 11.5 is the result of removing the T from the heap in Figure 11.4 by replacing it with the M, then moving down, promoting the larger of the two children, until reaching a node with both children smaller than M.

The implementation of both of these operations is centered around the process of fixing up a heap which satisfies the heap condition everywhere except possibly at the root. If the key at the root is too small, it must be moved down the heap without violating the heap property at any of the nodes touched. It turns out that the same operation can be used to fix up the heap after the value in any position is lowered. It may be implemented as follows:

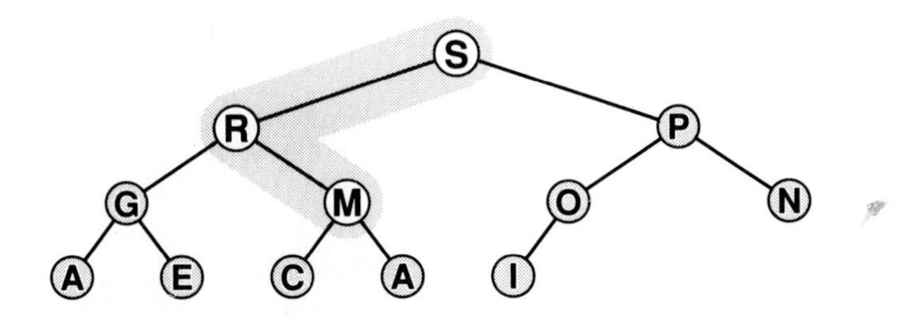

Figure 11.5 Removing the largest element in a heap.

```
procedure downheap(k: integer);
  label 0;
  var i,j,v: integer;
  begin
  v:=a[k];
  while k<=N div 2 do
    begin
    j:=k+k;
    if j<N then if a[j]<a[j+1] then j:=j+1;
    if v>=a[j] then goto 0;
    a[k]:=a[j]; k:=j;
    end;
0: a[k]:=v
  end;
```

This procedure moves down the heap, exchanging the node at position k with the larger of its two children if necessary and stopping when the node at k is larger than both children or the bottom is reached. (Note that it is possible for the node at k to have only one child: this case must be treated properly!) As above, a full exchange is not needed because v is always involved in the exchanges. The inner loop in this program is an example of a loop which really has two distinct exits: one for the case that the bottom of the heap is hit (as in the first example above), and another for the case that the heap condition is satisfied somewhere in the interior of the heap. The **goto** could be avoided, with some work, and at some expense in clarity.

Now the implementation of the *remove* operation is a direct application of this procedure:

```
function remove: integer;
  begin
  remove:=a[1];
  a[1]:=a[N]; N:=N-1;
  downheap(1);
  end;
```

The return value is set from $a[1]$ and then the element from $a[N]$ is put into $a[1]$ and the size of the heap decremented, leaving only a call to *downheap* to fix up the heap condition everywhere.

The implementation of the *replace* operation is only slightly more complicated:

```
function replace(v: integer):integer;
  begin
  a[0]:=v;
  downheap(0);
  replace:=a[0];
  end;
```

This code uses $a[0]$ in an artificial way: its children are 0 (itself) and 1, so if v is larger than the largest element in the heap, the heap is not touched; otherwise v is put into the heap and $a[1]$ is returned.

The *delete* operation for an arbitrary element from the heap and the *change* operation can also be implemented by using a simple combination of the methods above. For example, if the priority of the element at position k is raised, then *upheap*(k) can be called, and if it is lowered then *downheap*(k) does the job.

Property 11.1 *All of the basic operations insert, remove, replace, (downheap and upheap), delete, and change require less than 2* $\lg N$ *comparisons when performed on a heap of* N *elements.*

All these operations involve moving along a path between the root and the bottom of the heap, which includes no more than $\lg N$ elements for a heap of size N. The factor of two comes from *downheap*, which makes two comparisons in its inner loop; the other operations require only $\lg N$ comparisons. ■

Note carefully that the *join* operation is not included on this list. Doing this operation efficiently seems to require a much more sophisticated data structure. On the other hand, in many applications, one would expect this operation to be required much less frequently than the others.

Heapsort

An elegant and efficient sorting method can be defined from the basic operations on heaps outlined above. This method, called Heapsort, uses no extra memory and is guaranteed to sort M elements in about $M \log M$ steps no matter what the input. Unfortunately, its inner loop is quite a bit longer than the inner loop of Quicksort, and it is about twice as slow as Quicksort on the average.

The idea is simply to build a heap containing the elements to be sorted and then to remove them all in order. In this section, N will continue to be the size of the heap, so we will use M for the number of elements to be sorted. One way to sort is to implement the *construct* operation by doing M *insert* operations, as in the first two lines of the following code, then do M *remove* operations, putting the element removed into the place just vacated by the shrinking heap:

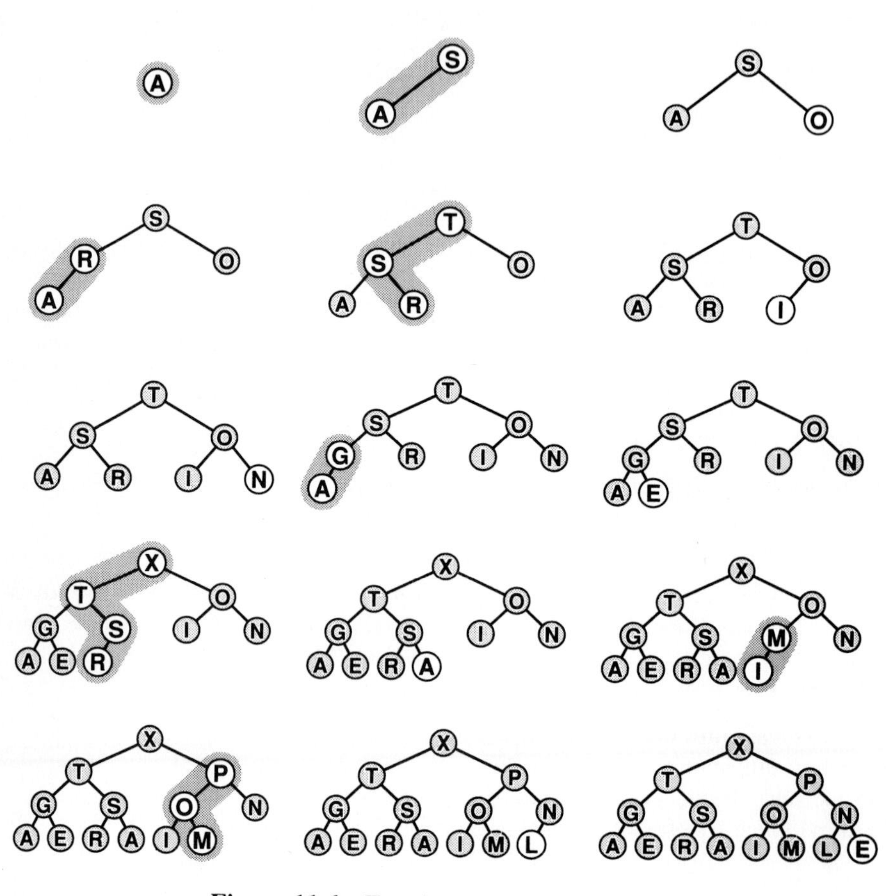

Figure 11.6 Top-down heap construction.

```
N:=0;
for k:=1 to M do insert(a[k]);
for k:=M downto 1 do a[k]:=remove;
```

This code breaks all the rules of abstract data structures by assuming a particular representation for the priority queue (during each loop, the priority queue resides in $a[1..k-1]$), but it is reasonable to do this here because we are implementing a sort, not a priority queue. The priority queue procedures are used only for descriptive purposes: in an actual implementation of the sort, we might simply use the code from the procedures to avoid unnecessary procedure calls.

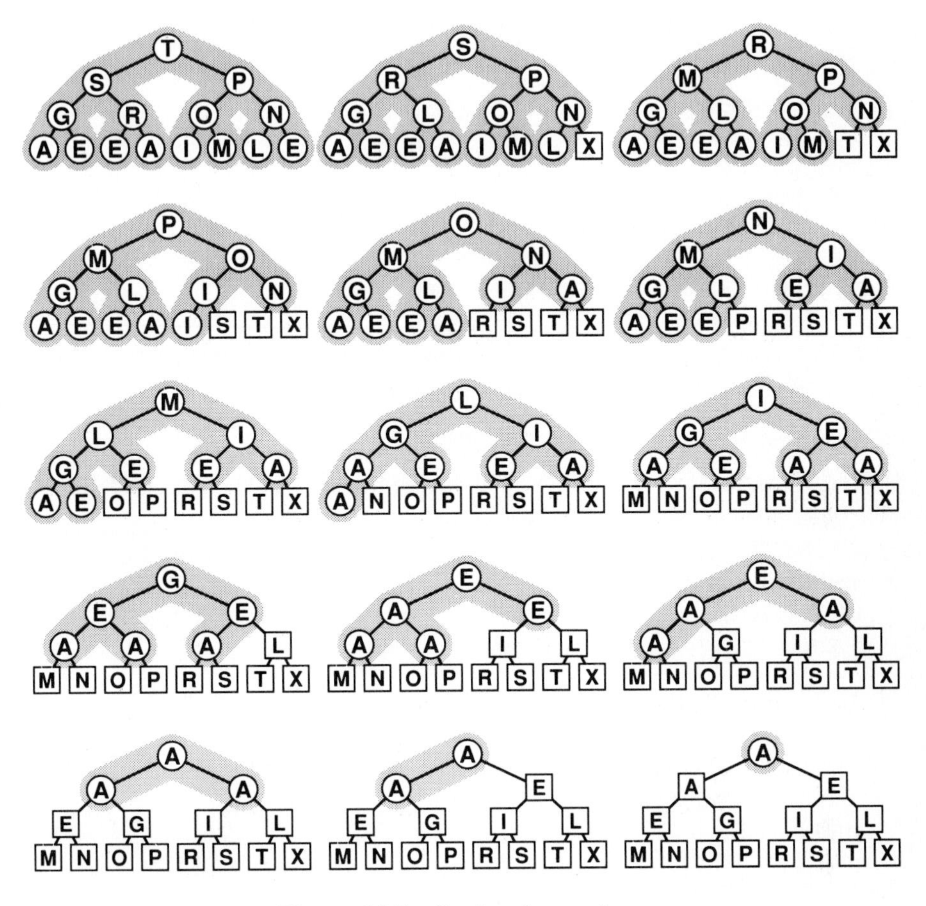

Figure 11.7 Sorting from a heap.

Figure 11.6 shows the heaps constructed as the keys A S O R T I N G E X A M P L E are inserted in that order into an initially empty heap, and Figure 11.7 shows how those keys are sorted by removing the X, then removing the T, etc.

It is actually a little better to build the heap by going backwards through it, making little heaps from the bottom up, as shown in Figure 11.8. This method views every position in the array as the root of a small heap and takes advantage of the fact that *downheap* will work as well for such small heaps as for the big heap. Working backwards through the heap, every node is the root of a heap which is heap-ordered except possibly for the root; *downheap* finishes the job. (There's no need to do anything with the heaps of size 1, so the scan starts halfway back through the array.)

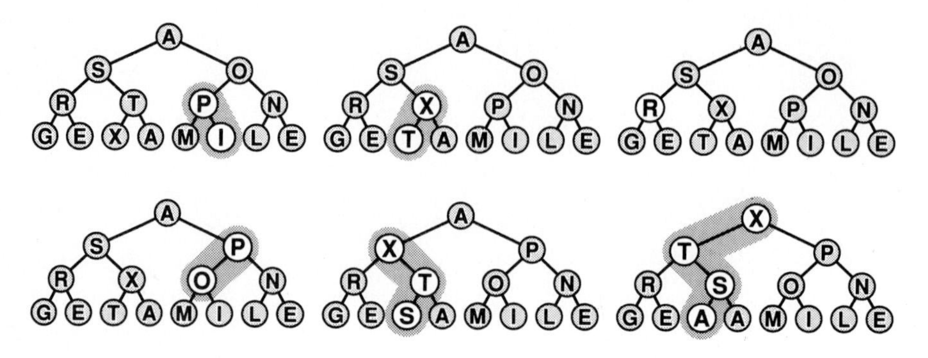

Figure 11.8 Bottom-up heap construction.

We've already noted that *remove* can be implemented by exchanging the first and last elements, decrementing N, and calling *downheap*(*1*). This leads to the following implementation of Heapsort:

```
procedure heapsort;
  var k,t: integer;
  begin
  N:=M;
  for k:=M div 2 downto 1 do downheap(k);
  repeat
    t:=a[1]; a[1]:=a[N]; a[N]:=t;
    N:=N-1; downheap(1)
  until N<=1;
  end;
```

The first two lines of this code implement *construct*(M: *integer*) to build a heap of M elements. (As just mentioned, the keys in $a[(M$ **div** $2)+1..M]$ each form heaps of one element, so they trivially satisfy the heap condition and don't need to be checked.) It is interesting to note that, though the loops in this program seem to do very different things, they can be built around the same fundamental procedure.

Figure 11.9 illustrates the data movement in Heapsort by showing the contents of each heap operated on by *downheap* for our sorting example, just after *downheap* has made the heap condition hold everywhere.

Property 11.2 *Bottom-up heap construction is linear-time.*

The reason for this property is that most of the heaps processed are small. For example, to build a heap of 127 elements, the method calls *downheap* on (64 heaps of size 1), 32 heaps of size 3, 16 heaps of size 7, 8 heaps of size 15, 4 heaps of

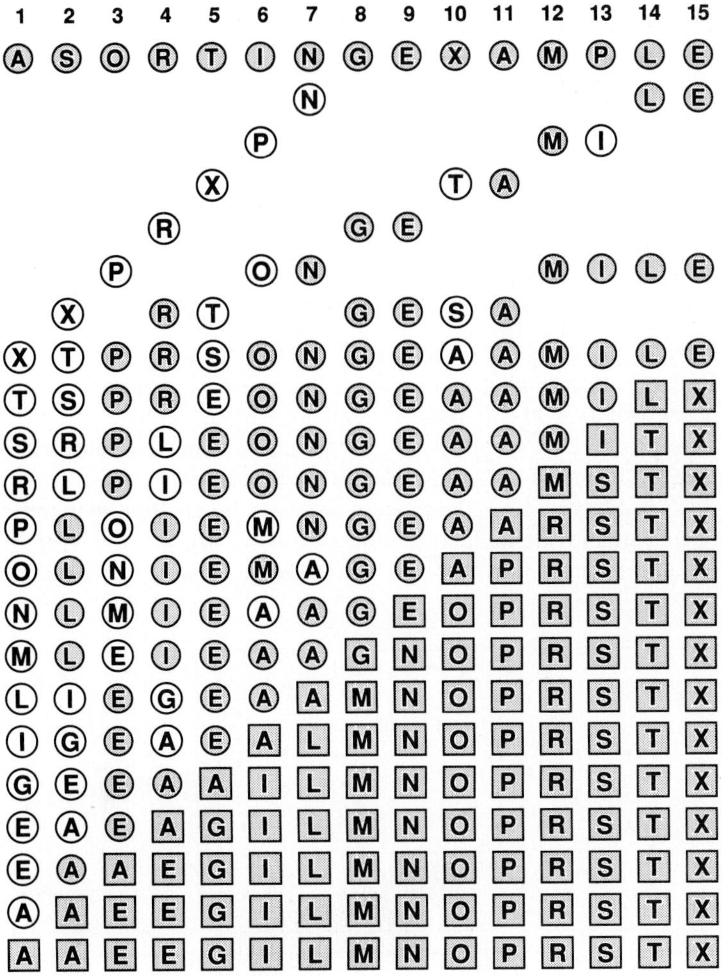

Figure 11.9 Heapsort data movement.

size 31, 2 heaps of size 63, and 1 heap of size 127, so $64 \cdot 0 + 32 \cdot 1 + 16 \cdot 2 + 8 \cdot 3 + 4 \cdot 4 + 2 \cdot 5 + 1 \cdot 6 = 120$ "promotions" (twice as many comparisons) are required in the worst case. For $M = 2^m$, an upper bound on the number of comparisons is

$$\sum_{1 \le k \le m} (k-1)2^{m-k} = 2^m - m - 1 < M,$$

and a similar proof holds when M is not a power of two. ■

This property is not of particular importance for Heapsort, since its time is still dominated by the $M \log M$ time for sorting, but it is important for other priority-queue applications, where a linear time *construct* can lead to a linear time algorithm. Note that constructing a heap with M successive *inserts* requires $M \log M$ steps in the worst case (though it turns out to be linear on the average).

Property 11.3 *Heapsort uses fewer than* $2M \lg M$ *comparisons to sort* M *elements.*

A slightly higher bound, say $3M \lg M$, is immediate from Property 11.1. The bound given here follows from more careful counting, taking Property 11.2 into account. ■

As mentioned above, Property 11.3 is the primary reason that Heapsort is of practical interest: the number of steps required to sort M elements is *guaranteed* to be proportional to $M \log M$, no matter what the input. Unlike the other methods that we've seen, there is no "worst-case" input that will make Heapsort run slower.

Figures 11.10 and 11.11 show Heapsort in operation on a larger randomly ordered file. In Figure 11.10, the process seems to be anything but sorting, since large elements are moving to the beginning of the file. But Figure 11.11 shows this structure being maintained as the file is sorted by picking off the large elements.

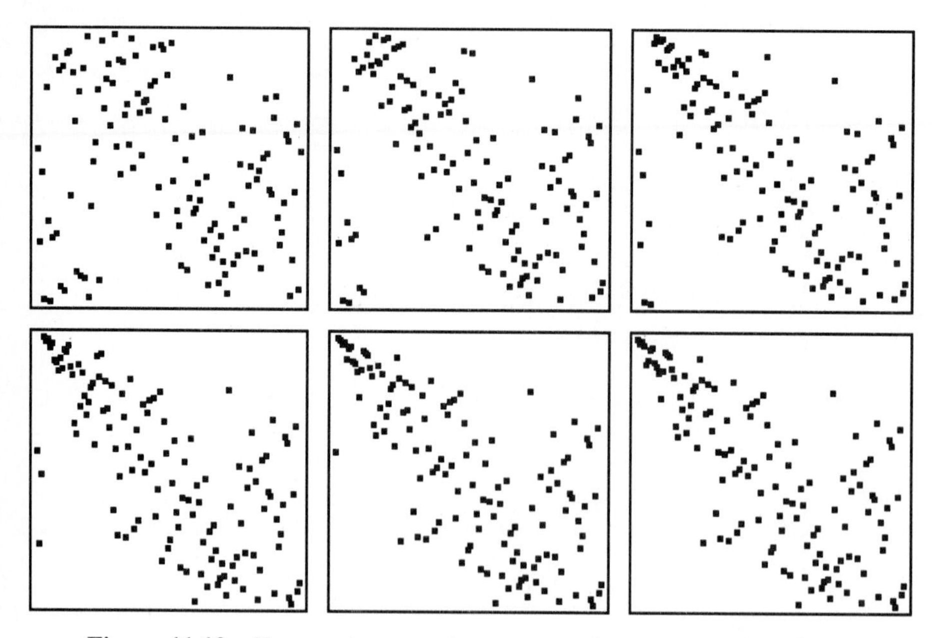

Figure 11.10 Heapsorting a random permutation: construction phase.

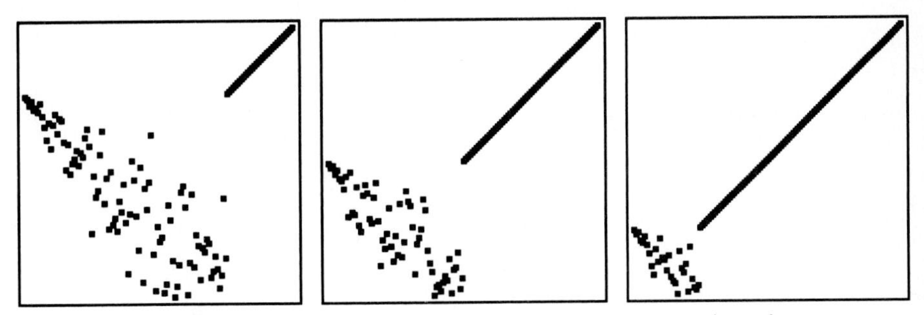

Figure 11.11 Heapsorting a random permutation: sorting phase.

Indirect Heaps

For many applications of priority queues, we don't want the records moved around at all. Instead, we want the priority queue routine not to return values but to tell us *which* of the records is the largest, etc. This is akin to the "indirect sort" or the "pointer sort" concept described in Chapter 8. Modifying the above programs to work in this way is straightforward, though sometimes confusing. It will be worthwhile to examine this in more detail here because it is so convenient to use heaps in this way.

As in Chapter 8, instead of rearranging the keys in the array a the priority queue routines will work with an array p of indices into the array a, such that $a[p[k]]$ is the record corresponding to the kth element of the heap, for k between 1 and N. (We'll continue to assume records of one-word keys: as in Chapter 8, a primary advantage of things in this way is that we can easily switch to more complicated records and keys.) Moreover, we want to maintain another array q which keeps the heap position of the kth array element. This is the mechanism that we use to allow the *change* and *delete* operations. Thus the q entry for the largest element in the array is *1*, and so on. For example, if we wished to change the value of $a[k]$ we could find its heap position in $q[k]$ and use *upheap* or *downheap*. Figure 11.12

k	1	2	3	4	5	6	7	8	9	10	11	12	13	14	15
$a[k]$	A	S	O	R	T	I	N	G	E	X	A	M	P	L	E
$p[k]$	10	5	13	4	2	3	7	8	9	1	11	12	6	14	15
$a[p[k]]$	X	T	P	R	S	O	N	G	E	A	A	M	I	L	E
$q[k]$	10	5	6	4	2	13	7	8	9	1	11	12	3	14	15

Figure 11.12 Indirect heap data structures.

gives the values in these arrays for our sample heap; note that $p[q[k]]=q[p[k]]=k$ for all k from 1 to N.

We start with $p[k]=q[k]=k$ for k from 1 to N, which indicates that no rearrangement has been done. The code for heap construction looks much the same as before:

```
procedure pqconstruct;
  var k: integer;
  begin
  N:=M;
  for k:=1 to N do
    begin p[k]:=k; q[k]:=k end;
  for k:=M div 2 downto 1 do pqdownheap(k);
  end;
```

(We'll prefix implementations of priority-queue routines based on indirect heaps with "*pq*" for identification when they are used in later chapters.)

Now, to modify *downheap* to work indirectly, we need only examine the places where it references a. Where it did a *comparison* before, it must now access a indirectly through p. Where it did a *move* before, it must now make the move in p, not a, and it must modify q accordingly. This leads to the following implementation:

```
procedure pqdownheap(k: integer);
  label 0;
  var j,v: integer;
  begin
  v:=p[k];
  while k<= N div 2 do
    begin
    j:=k+k;
    if j<N then if a[p[j]]<a[p[j+1]] then j:=j+1;
    if a[v]>=a[p[j]] then goto 0;
    p[k]:=p[j]; q[p[j]]:=k; k:=j;
    end;
0: p[k]:=v; q[v]:=k
end;
```

The other procedures given above can be modified in a similar fashion to implement "*pqinsert*," "*pqchange*," etc.

A similar indirect implementation can be developed based on maintaining p as an array of pointers to separately allocated records. In this case, a little more work is required to implement the function of q (find the heap position, given the record).

Advanced Implementations

If the *join* operation must be done efficiently, then the implementations that we have done so far are insufficient and more advanced techniques are needed. Although we don't have space here to go into the details of such methods, we can discuss some of the considerations that go into their design.

By "efficiently," we mean that a *join* should be done in about the same time as the other operations. This immediately rules out the linkless representation for heaps that we have been using, since two large heaps can be joined only by moving all the elements in at least one of them to a large array. It is easy to translate the algorithms we have been examining to use linked representations; in fact, sometimes there are other reasons for doing so (for example, it might be inconvenient to have a large contiguous array). In a direct linked representation, links would have to be kept in each node pointing to the parent and both children.

It turns out that the heap condition itself seems to be too strong to allow efficient implementation of the *join* operation. The advanced data structures designed to solve this problem all weaken either the heap or the balance condition in order to gain the flexibility needed for the *join*. These structures allow all the operations to be completed in logarithmic time.

□

Exercises

1. Draw the heap that results when the following operations are performed on an initially empty heap: *insert(10)*, *insert(5)*, *insert(2)*, *replace(4)*, *insert(6)*, *insert(8)*, *remove*, *insert(7)*, *insert(3)*.

2. Is a file in reverse sorted order a heap?

3. Give the heap constructed by successive application of *insert* on the keys E A S Y Q U E S T I O N.

4. Which positions could be occupied by the 3rd largest key in a heap of size 32? Which positions could not be occupied by the 3rd smallest key in a heap of size 32?

5. Why not use a sentinel to avoid the $j < N$ test in *downheap*?

6. Show how to obtain the functions of stacks and normal queues as special cases of priority queues.

7. What is the minimum number of keys that must be moved during a "*remove* the largest" operation in a heap? Draw a heap of size 15 for which the minimum is achieved.

8. Write a program to delete the element at position d in a heap.

9. Empirically compare bottom-up heap construction with top-down heap construction, by building heaps with 1000 random keys.

10. Give the contents of the q array after *pqconstruct* is used on the keys E A S Y Q U E S T I O N.

12

Mergesort

In Chapter 9 we studied the operation of *selection*, finding the kth smallest element in a file. We saw that selection is akin to dividing a file into two parts, the k smallest elements and the $N - k$ largest elements. In this chapter we examine a somewhat complementary process, *merging*, the operation of combining two sorted files to make one larger sorted file. As we'll see, merging is the basis for a straightforward recursive sorting algorithm.

Selection and merging are complementary operations in the sense that selection splits a file into two independent files and merging joins two independent files to make one file. The relationship between these operations also becomes evident if one tries to apply the "divide-and-conquer" paradigm to create a sorting method. The file can either be rearranged so that when two parts are sorted the whole file is sorted, or broken into two parts to be sorted and then combined to make the sorted whole file. We've already seen what happens in the first instance: that's Quicksort, which consists basically of a selection procedure followed by two recursive calls. Below, we'll look at Mergesort, Quicksort's complement in that it consists basically of two recursive calls followed by a merging procedure.

Mergesort, like Heapsort, has the advantage that it sorts a file of N elements in time proportional to $N \log N$ even in the worst case. The prime disadvantage of Mergesort is that extra space proportional to N seems to be required, unless one is willing to go to a lot of effort to overcome this handicap. The length of the inner loop is somewhere between that of Quicksort and Heapsort, so Mergesort is a candidate if speed is of the essence, especially if space is available. Moreover, Mergesort can be implemented so that it accesses the data primarily in a sequential manner (one item after the other), and this is sometimes a distinct advantage. For example, Mergesort is the method of choice for sorting a linked list, where sequential access is the only kind of access available. Similarly, as we'll see in Chapter 13, merging is the basis for sorting on sequential access devices, though the methods used in that context are somewhat different from those used for Mergesort.

Merging

In many data processing environments a large (sorted) data file is maintained to which new entries are regularly added. Typically, a number of new entries are "batched," appended to the (much larger) main file, and the whole thing is resorted. This situation is tailor-made for merging: a much better strategy is to sort the (small) batch of new entries, then merge it with the large main file. Merging has many other similar applications which make its study worthwhile. We'll also examine a sorting method based on merging.

In this chapter we'll concentrate on programs for *two-way merging*: programs which combine two sorted input files to make one sorted output file. In the next chapter, we'll look in more detail at *multiway merging*, when more than two files are involved. (The most important application of multiway merging is external sorting, the subject of that chapter.)

To begin, suppose that we have two sorted arrays $a[1..M]$ and $b[1..N]$ of integers which we wish to merge into a third array $c[1..M+N]$. The following is a direct implementation of the obvious strategy of successively choosing for c the smallest remaining element from a and b:

```
i:=1; j:=1;
a[M+1]:=maxint; b[N+1]:=maxint;
for k:=1 to M+N do
  if a[i]<b[j]
    then begin c[k]:=a[i]; i:=i+1 end
    else  begin c[k]:=b[j]; j:=j+1 end;
```

The implementation is simplified by making room in the a and b arrays for sentinel keys with values larger than all the other keys. When the $a(b)$ array is exhausted, the loop simply moves the rest of the $b(a)$ array into the c array. This method obviously uses $M+N$ comparisons. If $a[M+1]$ and $b[N+1]$ are not available for the sentinel keys, then tests to make sure that i is always less than M and j is less than N would have to be added. Another way around this difficulty is used below for the implementation of Mergesort.

Instead of using extra space proportional to the size of the merged file, it would be desirable to have an in-place method which uses $c[1..M]$ for one input and $c[M+1..M+N]$ for the other. At first, it is difficult to convince oneself that this can't be done easily, and such methods do exist, but they are so complicated that even an inplace *sort* is likely to be more efficient unless a great deal of care is exercised. We will return to this issue below.

Since extra space appears to be required for a practical implementation, we may as well consider a linked-list implementation. In fact, this method is very well suited to linked lists. A full implementation which illustrates all the conventions we'll use is given below; note that the code for the actual merge is just about as simple as the code above:

```
type link=↑node;
        node= record key: integer; next: link end;
var t,z: link;
       N: integer;
function merge(a,b: link): link;
   var c: link;
   begin
   c:=z;
   repeat
      if a↑.key<=b↑.key
         then begin c↑.next:=a; c:=a; a:=a↑.next end
         else  begin c↑.next:=b; c:=b; b:=b↑.next end
   until c↑.key=maxint;
   merge:=z↑.next; z↑.next:=z;
   end;
```

This program merges the list pointed to by a with the list pointed to by b with the help of an auxiliary pointer c. The lists are assumed to have a dummy "tail" node, as discussed in Chapter 3: all lists end with the dummy node z, which normally points to itself and also serves as a sentinel, with $z\uparrow.k{=}maxint$. During the merge, z is used to hold onto the beginning of the newly merged list (in a manner similar to the implementation of *readlist*), and c points to the end of the newly merged list (the node whose link field must be changed to add a new element to the list). After the merged list is built, the pointer to its first node is retrieved from z and z is reset to point to itself.

The key comparison in *merge* includes equality so that the merge will be stable, if the b list is considered to follow the a list. We'll see below how this stability in the merge implies stability in the sorting programs which use this merge.

Mergesort

Once we have a merging procedure, it's not difficult to use it as the basis for a recursive sorting procedure. To sort a given file, divide it in half, sort the two halves (recursively), and then merge the two halves together. The following implementation of this process sorts an array $a[l..r]$ (using an auxiliary array $b[l..r]$):

```
procedure mergesort(l,r: integer);
  var i,j,k,m: integer;
  begin
  if r-l>0 then
    begin
    m:=(r+l) div 2;
    mergesort(l,m); mergesort(m+1,r);
    for i:=m downto l do b[i]:=a[i];
    for j:=m+1 to r do b[r+m+1-j]:=a[j];
    for k:=l to r do
      if b[i]<b[j]
        then begin a[k]:=b[i]; i:=i+1 end
        else begin a[k]:=b[j]; j:=j-1 end;
    end;
  end;
```

This program manages the merge without sentinels by copying the second array into position back-to-back with the first, but in reverse order. Thus each array serves as the "sentinel" for the other: the largest element (which is in one array or the other) keeps things moving properly after the other array has been exhausted for the merge. The "inner loop" of this program is rather short (move to b, move back to a, increment i or j, increment and test k), and could be shortened even more by having two copies of the code (one for merging from a into b and one for merging b into a), though this would require going back to sentinels.

Our file of sample keys is processed as shown in Figure 12.1. Each line shows the result of a call on *merge*. First we merge A and S to get A S, then we merge O and R to get O R and this with A S to get A O R S. Later we merge I T with G N to get G I N T, and merge this with A O R S to get A G I N O R S T, etc.. Thus this method recursively builds up small sorted files into larger ones.

List Mergesort

This process involves enough data movement that a linked list representation should also be considered. The following program is a direct recursive implementation of a function which takes as input a pointer to an unsorted list and returns as its value a pointer to the sorted version of the list. The program does this by rearranging the nodes of the list: no temporary nodes or lists need be allocated. (It is convenient to pass the list length as a parameter to the recursive program; alternatively, it could be stored with the list or the program could scan the list to find its length.)

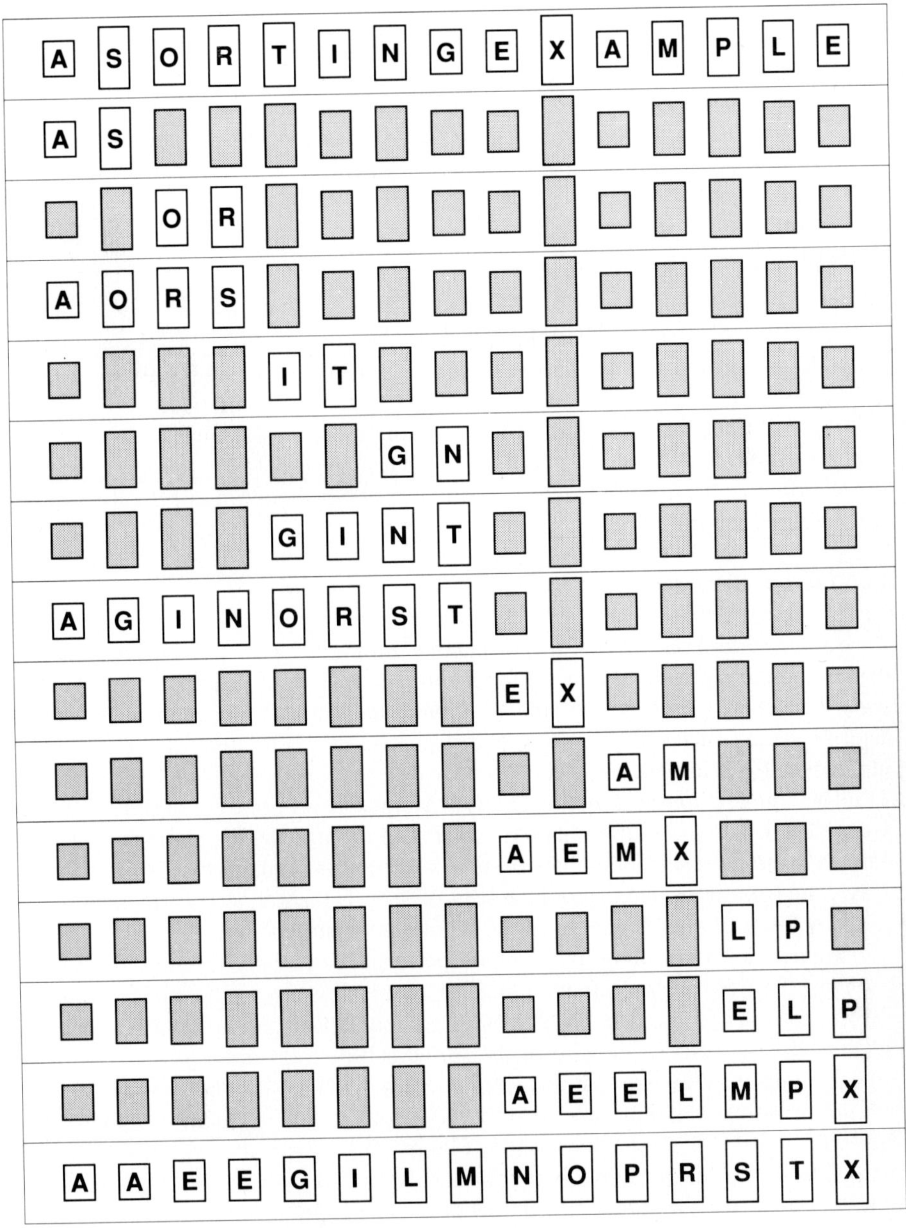

Figure 12.1 Recursive Mergesort.

```
function mergesort(c: link): link;
  var a,b: link;
  begin
  if c↑.next=z then mergesort:=c else
    begin
    a:=c; b:=c↑.next; b:=b↑.next; b:=b↑.next;
    while b<>z do begin c:=c↑.next; b:=b↑.next; b:=b↑.next end;
    b:=c↑.next; c↑.next:=z;
    mergesort:=merge(mergesort(a), mergesort(b));
    end;
  end;
```

This program sorts by splitting the list pointed to by c into two halves pointed to by a and b, sorting the two halves recursively, and then using *merge* to produce the final result. Again, this program adheres to the convention that all lists end with z: the input list must end with z (and therefore so does the b list), and the explicit instruction $c\uparrow.next:=z$ puts z at the end of the a list. This program is quite simple to understand in a recursive formulation even though it is actually a rather sophisticated algorithm.

Bottom-Up Mergesort

As discussed in Chapter 5, every recursive program has a nonrecursive analog which, though equivalent, may perform computations in a different order. Mergesort is actually a prototype of the "combine and conquer" strategy which characterizes many such computations, and it is worthwhile to study its nonrecursive implementations in detail.

The simplest nonrecursive version of Mergesort processes a slightly different set of files in a slightly different order: first scan through the list performing 1-by-1 merges to produce sorted sublists of size 2, then scan through the list performing 2-by-2 merges to produce sorted sublists of size 4, then do 4-by-4 merges to get sorted sublists of size 8, etc., until the whole list is sorted.

Figure 12.2 shows how this method performs essentially the same merges as in Figure 12.1 for our sample file (since its size is close to a power of two), but in a different order. In general, $\log N$ passes are required to sort a file of N elements, since each pass doubles the size of the sorted subfiles.

It is important to note that the actual merges made by this "bottom-up" method are *not* the same as the merges done by the recursive implementation above. Consider the sort of 95 elements shown in Figure 12.3. The last merge is a 64-by-31 merge, while in the recursive sort it would be a 47-by-48 merge. It is possible, however, to arrange things so that the sequence of merges made by the two methods is the same, though there is no particular reason to do so.

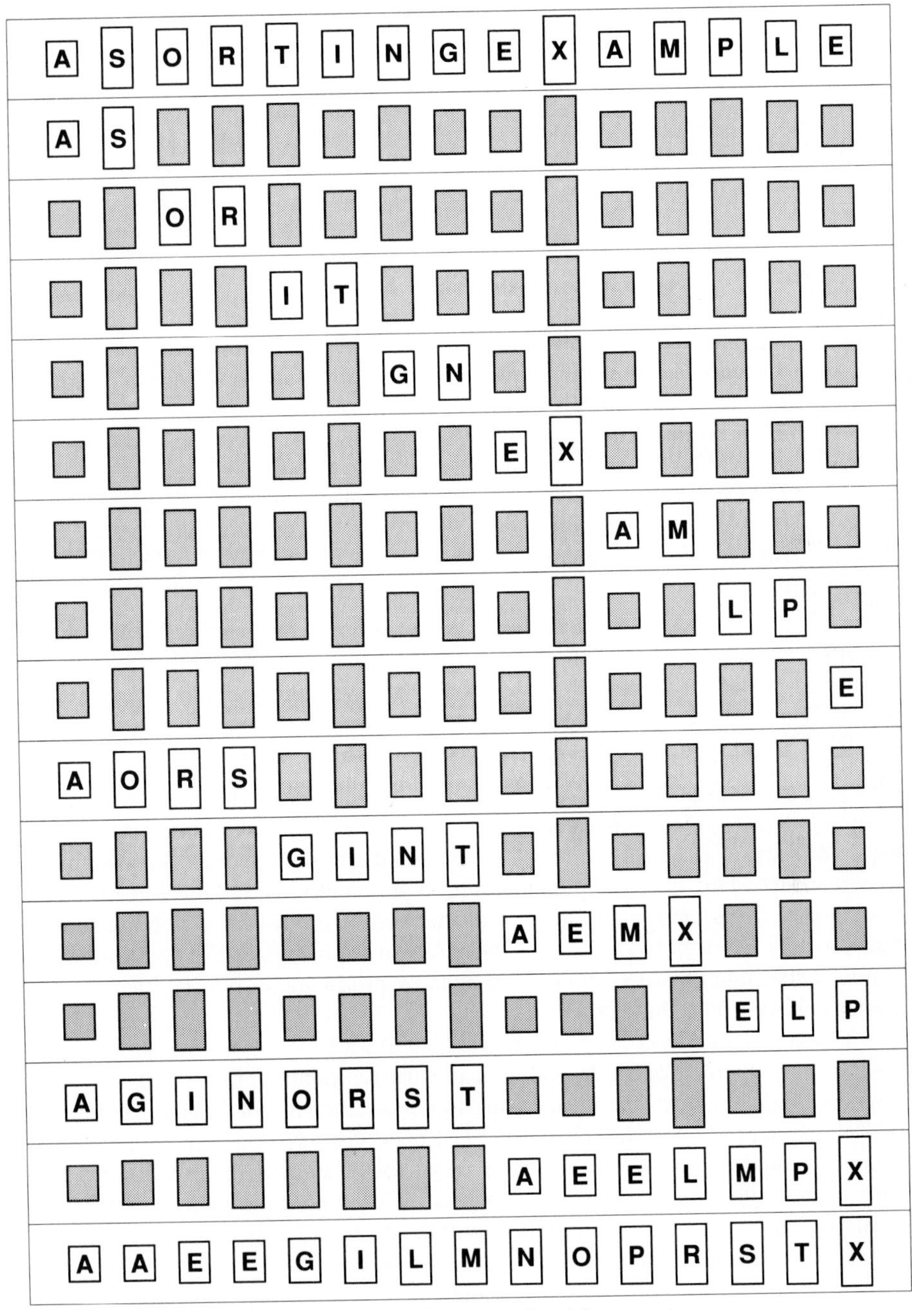

Figure 12.2 Nonrecursive Mergesort.

A detailed implementation of this bottom-up approach, using linked lists, is given below.

```
function mergesort(c: link): link;
  var a,b,head,todo,t: link;
      i,N: integer;
  begin
  N:=1; new(head); head↑.next:=c;
  repeat
    todo:=head↑.next; c:=head;
    repeat
      t:=todo;
      a:=t; for i:=1 to N-1 do t:=t↑.next;
      b:=t↑.next; t↑.next:=z;
      t:=b; for i:=1 to N-1 do t:=t↑.next;
      todo:=t↑.next; t↑.next:=z;
      c↑.next:=merge(a,b);
      for i:=1 to N+N do c:=c↑.next
    until todo=z;
    N:=N+N;
  until a=head↑.next;
  mergesort:=head↑.next
  end;
```

This program uses a "list-header" node (pointed to by *head*) whose link field points to the file being sorted. Each iteration of the outer **repeat** loop passes through the file, producing a linked list comprised of sorted subfiles twice as long as in the previous pass. This is done by maintaining two pointers, one to the part of the list not yet seen (*todo*) and one to the end of that part of the list for which the subfiles have already been merged (*c*). The inner **repeat** loop merges the two subfiles of length N starting at the node pointed to by *todo*, producing a subfile of length $N+N$ which is linked onto the *c* result list.

The actual merge is accomplished by saving a link to the first subfile to be merged in a, then skipping N nodes (using the temporary link t), linking z onto the end of a's list, then doing the same to get another list of N nodes pointed to by b (updating *todo* with the link of the last node visited), and then calling *merge*. (Then c is updated by simply chasing down to the end of the list just merged. This is a simpler (but slightly less efficient) method than the various alternatives available, such as having *merge* return pointers to both the beginning and the end or maintaining multiple pointers in each list node.)

Bottom-up Mergesort is also an interesting method to use for an array implementation; this is left as an instructive exercise for the reader.

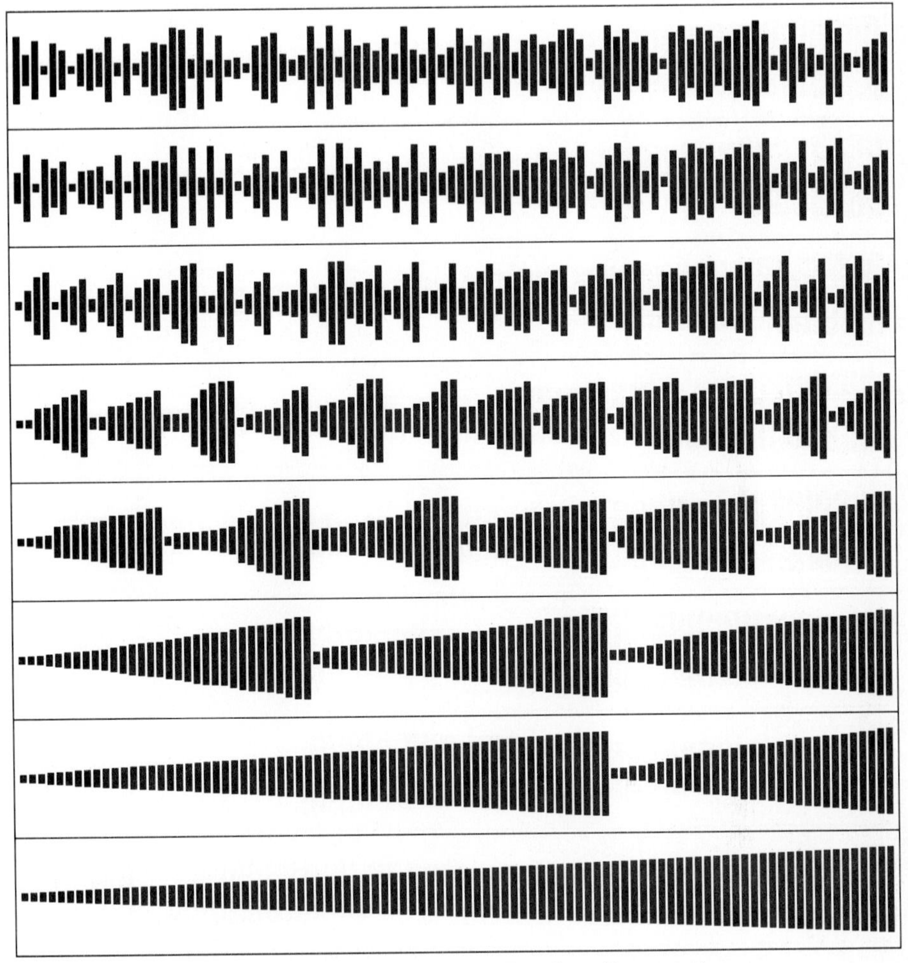

Figure 12.3 Mergesorting a Random Permutation.

Performance Characteristics

Property 12.1 *Mergesort requires about* $N \lg N$ *comparisons to sort any file of* N *elements.*

In the implementations above, each M-by-N merge will require $M+N$ comparisons (this could vary by one or two depending upon how sentinels are used). Now, for bottom-up Mergesort, $\lg N$ passes are used, each requiring about N comparisons. For the recursive version, the number of comparisons is described by the standard "divide-and-conquer" recurrence $M_N = 2M_{N/2} + N$, with $M_1 = 0$. We know from

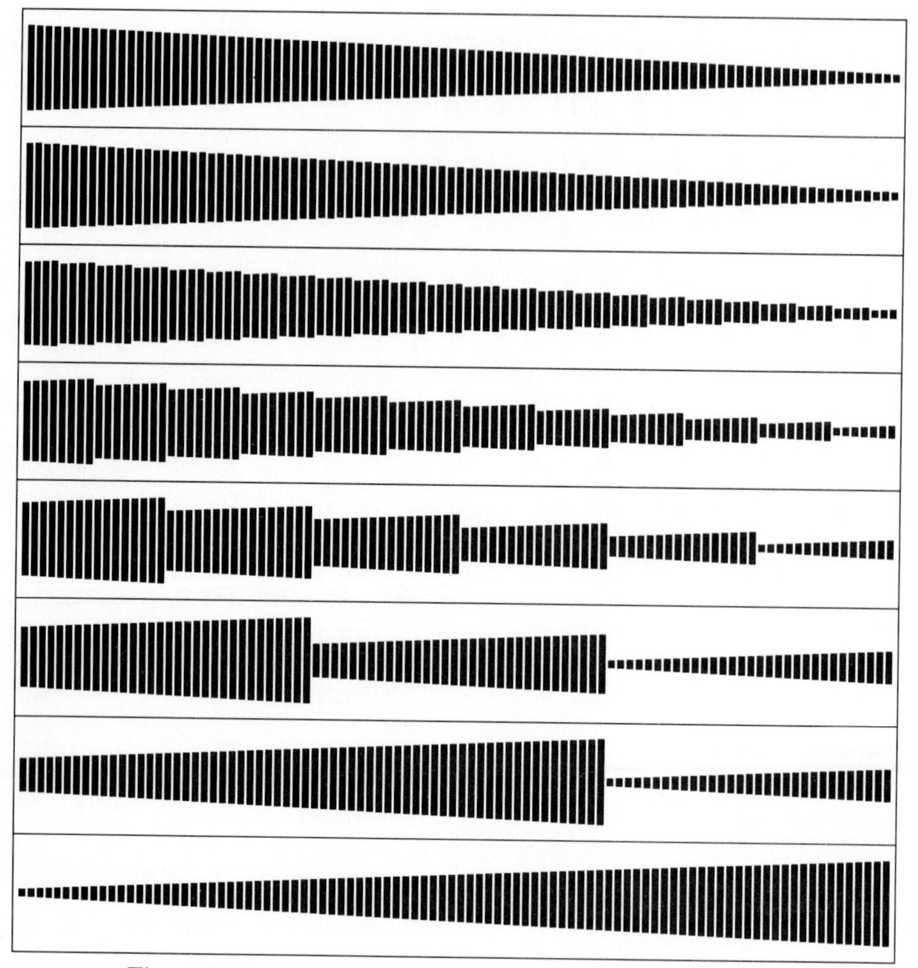

Figure 12.4 Mergesorting a Reverse-Ordered Permutation.

Chapter 6 that this has the solution $M_N \approx N \lg N$. These arguments are both precisely true if N is a power of two; it is left as an exercise to show that they hold for general N as well. Furthermore, it turns out that they also hold in the *average* case. ■

Property 12.2 *Mergesort uses extra space proportional to* N.

This is clear from the implementations, though steps can be taken to lessen the impact of this problem. Of course, if the "file" to be sorted is a linked list, the problem does not arise, since the "extra space" (for the links) is there for another purpose.

For arrays, first note that it is easy to do an M-by-N merge using extra space for the smaller of the two arrays only (see Exercise 2). This cuts the space requirement for Mergesort by half. It is actually possible to do much better and do merges in place, though this is unlikely to be worthwhile in practice. ∎

Property 12.3 *Mergesort is stable.*

Since all the implementations actually move keys only during merges, it is necessary to verify merely that the merges themselves are stable. But this is trivial to show: the relative position of equal keys is undisturbed by the merging process. ∎

Property 12.4 *Mergesort is insensitive to the initial order of its input.*

In our implementations, the input determines only the order in which elements are processed in the merges, so this statement is literally true (except for some variation depending on how the *if* statement is compiled and executed, which should be negligible). Other implementations of merging which involve an explicit test for the first file exhausted may lead to some greater variation depending on the input, but not much. The number of passes required clearly depends only on the size of the file, not its contents, and each pass certainly requires about N comparisons (actually $N - O(1)$ on the average, as explained below). But the worst case is about the same as the average case. ∎

Figure 12.4 shows bottom-up Mergesort operating on a file which is initially in reverse order. It is interesting to compare this figure with Figure 8.9, which shows Shellsort doing the same operation.

Figure 12.5 presents another view of Mergesort in operation on a random permutation, for comparison with similar views in earlier chapters. In particular, Figure 12.5 bears a striking resemblance to Figure 10.5: in this sense, Mergesort is the "transpose" of straight radix sort!

Optimized Implementations

We have already paid some attention to the inner loop of array-based Mergesort in our discussion of sentinels, where we saw that array bounds tests in the inner loop could be avoided by reversing the order of one of the arrays. This calls attention to a major inefficiency in the implementations above: the move from a to b. As we saw for straight radix sort in Chapter 10, this move can be avoided by having two copies of the code, one where we merge from a into b, another where we merge from b into a.

To accomplish these two improvements in combination, it is necessary to change things so that *merge* can output arrays in either increasing or decreasing order. In the nonrecursive version, this is accomplished by alternating between increasing and decreasing output; in the recursive version, we have four recursive routines: to merge from *a(b)* into *b(a)* with the result in decreasing or increasing order. Either of these will reduce the inner loop of Mergesort to a comparison,

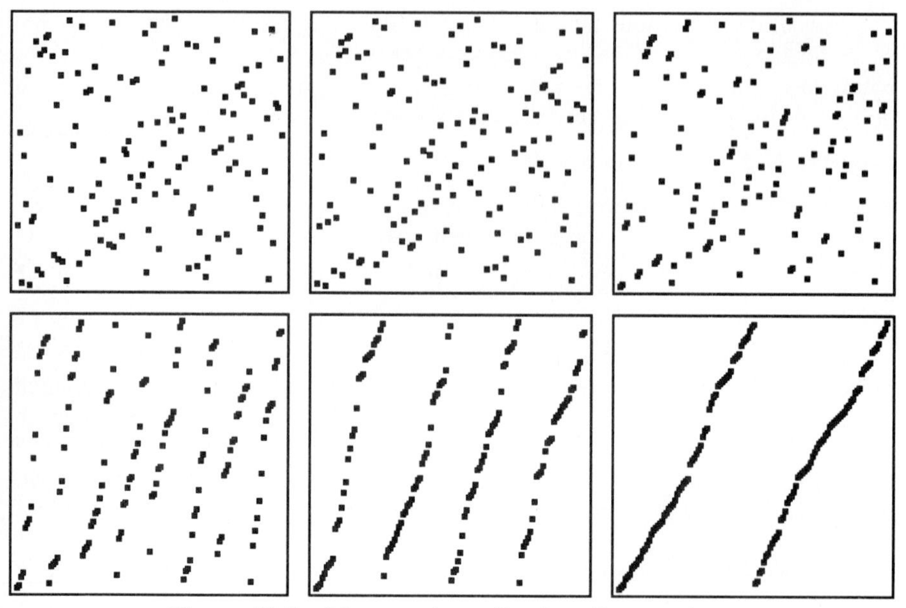

Figure 12.5 Mergesorting a Random Permutation.

a store, two pointer increments (i or j, and k), and a pointer test. This competes favorably with Quicksort's compare, increment and test, and (partial) exchange, and Quicksort's inner loop is executed $2 \ln N \approx 1.38 \lg N$ times, about 38% more often than Mergesort's.

Recursion Revisited

The programs of this chapter, together with Quicksort, are typical of implementations of divide-and-conquer algorithms. We'll see several algorithms with similar structure in later chapters, so it's worthwhile to take a more detailed look at some basic characteristics of these implementations.

Quicksort is actually a "conquer-and-divide" algorithm: in a recursive implementation, most of the work is done *before* the recursive calls. On the other hand, the recursive Mergesort has more the spirit of divide-and-conquer: first the file is divided into two parts, then each part is conquered individually. The first problem for which Mergesort does actual processing is a small one; at the finish the largest subfile is processed. Quicksort starts with actual processing on the largest subfile and finishes up with the small ones.

This difference becomes manifest in the non-recursive implementations of the two methods. Quicksort must maintain a stack, since it has to save large subproblems which are divided up in a data-dependent manner. Mergesort admits a simple nonrecursive version because the way in which it divides the file is independent of the data, so the order in which it processes subproblems can be rearranged somewhat to give a simpler program.

Another practical difference which manifests itself is that Mergesort is stable (if properly implemented), Quicksort is not (without going to a lot of extra trouble). For Mergesort, if we assume (inductively) that the subfiles have been sorted stably, then we need be sure only that the merge is done in a stable manner, which is easily arranged. But for Quicksort, no easy way of doing the partitioning in a stable manner suggests itself, so the possibility of stability is foreclosed even before the recursion comes into play.

One final note: like Quicksort or any other recursive program, Mergesort can be improved by treating small subfiles in a different way. In the recursive versions of the program, this can be implemented exactly as for Quicksort, either doing small subfiles with insertion sort on the fly or doing a cleanup pass afterwards. In the nonrecursive versions, small sorted subfiles can be built up in an initial pass using a suitable modification of insertion or selection sort. Another idea that has been suggested for Mergesort is to take advantage of "natural" ordering in the file by using a bottom-up method which merges the first two sorted runs in the file (however long they happen to be), then the next two runs, etc., repeating the process until the file is sorted. Attractive as this may seem, it doesn't stand up against the standard method that we've discussed because the cost of identifying the runs, which falls in the inner loop, more than offsets the savings achieved except for certain degenerate cases (such as a file that is already sorted).

□

Exercises

1. Implement a recursive Mergesort with a cutoff to insertion sort for subfiles with less than M elements; determine empirically the value of M for which it runs fastest on a random file of 1000 elements.

2. Empirically compare the recursive and nonrecursive Mergesorts for linked lists and $N = 1000$.

3. Implement recursive Mergesort for an array of N integers, using an auxiliary array of size less than $N/2$.

4. True or false: the running time of Mergesort does not depend on the value of the keys in the input file. Explain your answer.

5. What is the smallest number of steps Mergesort could use (to within a constant factor)?

6. Implement a bottom-up nonrecursive Mergesort that uses two arrays instead of linked lists.

7. Show the merges done when the recursive Mergesort is used to sort the keys E A S Y Q U E S T I O N.

8. Show the contents of the linked list at each iteration when the non-recursive Mergesort is used to sort the keys E A S Y Q U E S T I O N.

9. Try doing a recursive Mergesort, using arrays, using the idea of doing *3-way* rather than 2-way merges.

10. Empirically test, for random files of size 1000, the claim in the text that the idea of taking advantage of "natural" order in the file doesn't pay off.

13

External Sorting

☐ Many important sorting applications involve processing very large files, much too large to fit into the primary memory of any computer. Methods appropriate for such applications are called *external* methods, since they involve a large amount of processing external to the central processing unit (as opposed to the *internal* methods that we've looked at so far).

There are two major factors which make external algorithms quite different from those we've seen. First, the cost of accessing an item is orders of magnitude greater than any bookkeeping or calculating costs. Second, over and above with this higher cost, there are severe restrictions on access, depending on the external storage medium used: for example, items on a magnetic tape can be accessed only in a sequential manner.

The wide variety of external storage device types and costs makes the development of external sorting methods very dependent on current technology. These methods can be complicated, and many parameters affect their performance: that a clever method might go unappreciated or unused because of a simple change in the technology is a definite possibility in external sorting. For this reason, we'll concentrate in this chapter on general methods rather than on developing specific implementations.

In short, for external sorting, the "systems" aspect of the problem is certainly as important as the "algorithms" aspect. Both areas must be carefully considered if an effective external sort is to be developed. The primary costs in external sorting are for input-output. A good exercise for someone planning to implement an efficient program to sort a very large file is first to implement an efficient program to copy a large file, then (if that was too easy) implement an efficient program to reverse the order of the elements in a large file. The systems problems that arise in trying to solve these problems efficiently are similar to those that arise in external sorts. Permuting a large external file in any non-trivial way is about as difficult as sorting it, even though no key comparisons, etc. are required. In external sorting, we are concerned mainly with limiting the number of times each piece of data is moved between the external storage medium and the primary memory, and being sure that such transfers are done as efficiently as allowed by the available hardware.

External sorting methods have been developed which are suitable for the punched cards and paper tape of the past, the magnetic tapes and disks of the present, and emerging technologies such as bubble memories and videodisks. The essential differences among the various devices are the relative size and speed of available storage and the types of data access restrictions. We'll concentrate on basic methods for sorting on magnetic tape and disk because these devices are likely to remain in widespread use and illustrate the two fundamentally different modes of access that characterize many external storage systems. Often, modern computer systems have a "storage hierarchy" of several progressively slower, cheaper, and larger memories. Many of the algorithms that we will consider can be adapted to run well in such an environment, but we'll deal exclusively with "two-level" memory hierarchies consisting of main memory and disk or tape.

Sort-Merge

Most external sorting methods use the following general strategy: make a first pass through the file to be sorted, breaking it up into blocks about the size of the internal memory, and *sort* these blocks. Then *merge* the sorted blocks together by making several passes through the file, creating successively larger sorted blocks until the whole file is sorted. The data is most often accessed in a sequential manner, a property which makes this method appropriate for most external devices. Algorithms for external sorting strive to reduce the number of passes through the file and to reduce the cost of a single pass to be as close to the cost of a copy as possible.

Since most of the cost of an external sorting method is for input-output, we can get a rough measure of the cost of a sort-merge by counting the number of times each word in the file is read or written (the number of passes over all the data). For many applications, the methods that we consider involve on the order of ten or fewer such passes. Note that this implies that we're interested in methods that can eliminate even a single pass. Also, the running time of the whole external sort can be easily estimated from the running time of something like the "reverse file copy" exercise suggested above.

Balanced Multiway Merging

To begin, we'll trace through the various steps of the simplest sort-merge procedure for a small example. Suppose that we have records with the keys A S O R T I N G A N D M E R G I N G E X A M P L E on an input tape; these are to be sorted and put onto an output tape. Using a "tape" simply means that we're restricted to reading the records sequentially: the second record can't be read until the first is

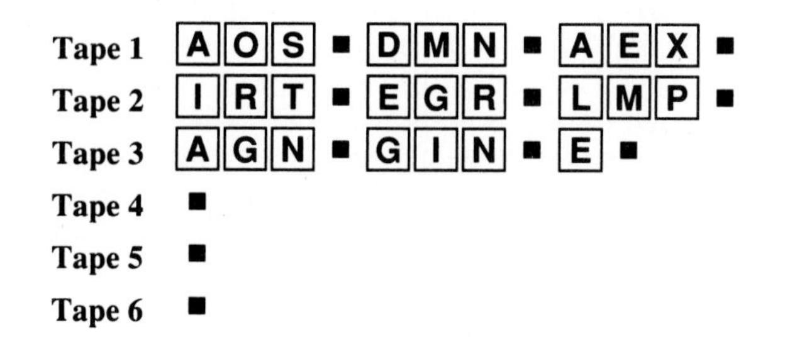

Figure 13.1 Balanced three-way merge: result of the first pass.

read, and so on. Assume further that we have only enough room for three records in our computer memory but that we have plenty of tapes available.

The first step is to read in the file three records at a time, sort them to make three-record blocks, and output the sorted blocks. Thus, first we read in A S O and output the block A O S, next we read in R T I and output the block I R T, and so forth. Now, in order for these blocks to be merged together, they must be on different tapes. If we want to do a three-way merge, then we would use three tapes, ending up after the sorting pass with the configuration shown in Figure 13.1.

Now we're ready to merge the sorted blocks of size three. We read the first record off each input tape (there's just enough room in the memory) and output the one with the smallest key. Then the next record from the same tape as the record just output is read in and, again, the record in memory with the smallest key is output. When the end of a three-word block in the input is encountered, that tape is ignored until the blocks from the other two tapes have been processed and nine records have been output. Then the process is repeated to merge the second three-word block on each tape into a nine-word block (which is output on a different tape, to get ready for the next merge). By continuing in this way, we get three long blocks configured as shown in Figure 13.2.

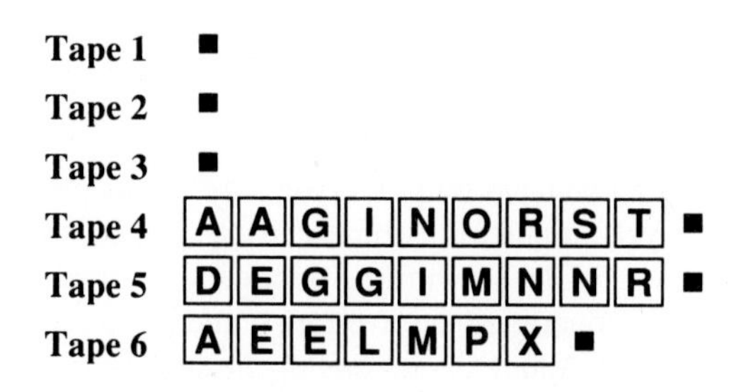

Figure 13.2 Balanced three-way merge: result of the second pass.

Now one more three-way merge completes the sort. If we had a much longer file with many blocks of size 9 on each tape, then we would finish the second pass with blocks of size 27 on tapes 1, 2, and 3, then a third pass would produce blocks of size 81 on tapes 4, 5, and 6, and so forth. We need six tapes to sort an arbitrarily large file: three for the input and three for the output of each three-way merge. (Actually, we could get by with just four tapes: the output could be put on just one tape, and then the blocks from that tape distributed to the three input tapes in between merging passes.)

This method is called the *balanced multiway merge*: it is a reasonable algorithm for external sorting and a good starting point for the implementation of an external sort. The more sophisticated algorithms below can make the sort run a little faster, but not much. (However, when execution times are measured in hours, as is not uncommon in external sorting, even a small percentage decrease in running time can be quite significant.)

Suppose that we have N words to be manipulated by the sort and an internal memory of size M. Then the "sort" pass produces about N/M sorted blocks. (This estimate assumes one-word records: for larger records, the number of sorted blocks is computed by multiplying further by the record size.) If we do P-way merges on each subsequent pass, then the number of subsequent passes is about $\log_P(N/M)$, since each pass reduces the number of sorted blocks by a factor of P.

Though small examples can help one understand the details of the algorithm, it is best to think in terms of very large files when working with external sorts. For example, the formula above says that using a four-way merge to sort a 200-million-word file on a computer with a million words of memory should take a total of about five passes. A very rough estimate of the running time can be found by multiplying by five the running time for the reverse file copy implementation suggested above.

Replacement Selection

It turns out that the details of the implementation can be developed in an elegant and efficient way using priority queues. First, we'll see that priority queues provide a natural way to implement a multiway merge. More important, it turns out that we can use priority queues for the initial sorting pass in such a way that they can produce sorted blocks much longer than could fit into internal memory.

The basic operation needed to do P-way merging is repeatedly to output the smallest of the smallest elements not yet output from each of the P blocks to be merged. That smallest element should be replaced with the next element from the block from which it came. The *replace* operation on a priority queue of size P is exactly what is needed. (Actually, the "indirect" versions of the priority queue routines as described in Chapter 11 are more appropriate for this application.) Specifically, to do a P-way merge we begin by filling up a priority queue of size P with the smallest element from each of the P inputs using the *pqinsert* procedure from Chapter 11 (appropriately modified so that the smallest element rather than the largest is at the top of the heap). Then, using the *pqreplace* procedure from Chapter 11 (modified in the same way) we output the smallest element and replace it in the priority queue with the next element from its block.

The process of merging A O S with I R T and A G N (the first merge from our example above), using a heap of size three in the merging process is shown in Figure 13.3. The "keys" in these heaps are the smallest (first) key in each node. For clarity, we show entire blocks in the nodes of the heap; of course, an actual implementation would be an indirect heap of pointers into the blocks. First, the A is output so that the O (the next key in its block) becomes the "key" of the root. This violates the heap condition, so that node is exchanged with the node containing A, G, and N. Then that A is output and replaced with the next key in its block, the G. This does not violate the heap condition, so no further change is necessary. Continuing in this way, we produce the sorted file (read the smallest key in the root node of the trees in Figure 13.3 to see the keys in the order in which they appear in the first heap position and are output). When a block is exhausted, a sentinel is put on the heap and considered to be larger than all the other keys. When the heap consists of all sentinels, the merge is completed. This way of using priority queues is sometimes called *replacement selection.*

Thus to do a P-way merge, we can use replacement selection on a priority queue of size P to find each element to be output in $\log P$ steps. This performance difference has no particular practical relevance, since a brute-force implementation can find each element to output in P steps and P is normally so small that this cost is dwarfed by the cost of actually outputting the element. The real importance of replacement selection is the way that it can be used in the first part of the sort-

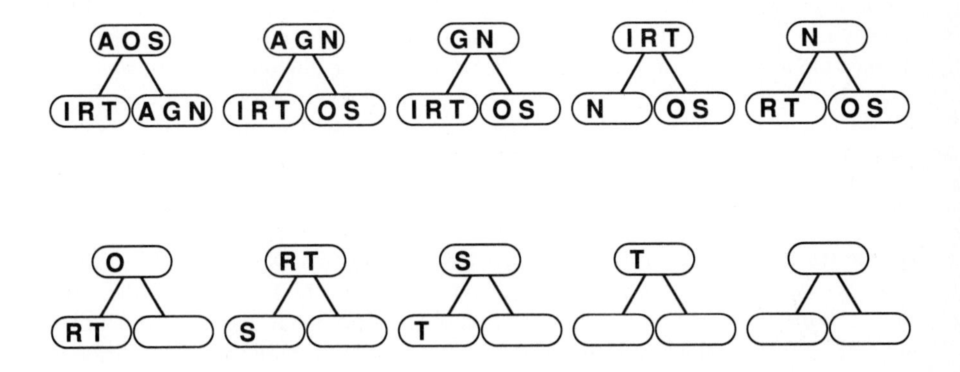

Figure 13.3 Replacement selection for merging, with heap of size three.

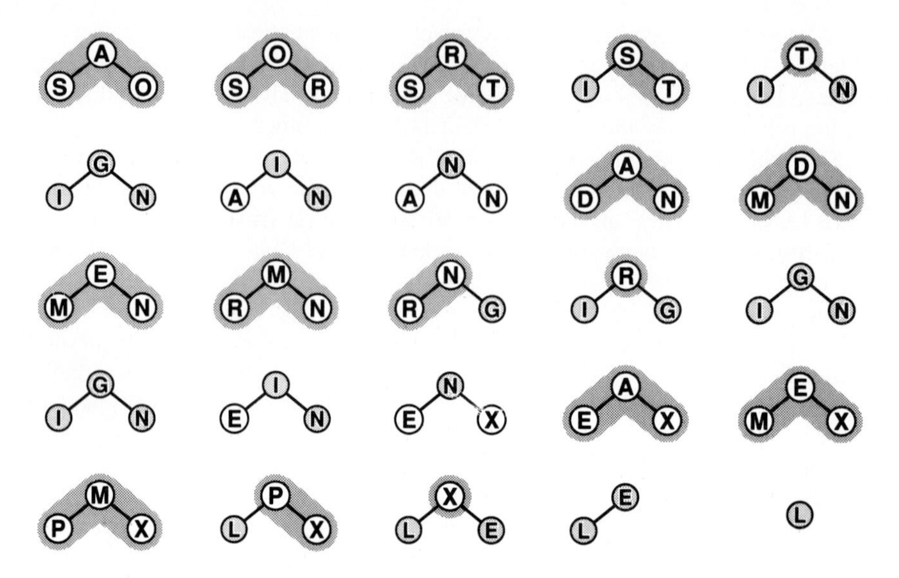

Figure 13.4 Replacement selection for producing initial runs.

merge process: to form the initial sorted blocks which provide the basis for the merging passes.

The idea is to pass the (unordered) input through a large priority queue, always writing out the smallest element on the priority queue as above, and always replacing it with the next element from the input, with one additional proviso: if the new element is smaller than the last one output, then, since it could not possibly become part of the current sorted block, it should be marked as a member of the next block and treated as greater than all elements in the current block. When a marked element makes it to the top of the priority queue, the old block is ended and a new block started. Again, this is easily implemented with *pqinsert* and *pqreplace* from Chapter 11, appropriately modified so that the smallest element is at the top of the heap and with *pqreplace* changed to treat marked elements as always greater than unmarked elements.

Our example file clearly demonstrates the value of replacement selection. With an internal memory capable of holding only three records, we can produce sorted blocks of size 5, 3, 6, 4, 5, and 2, as illustrated in Figure 13.4. As before, the order in which the keys occupy the first position in the heap is the order in which they are output. The shading indicates which keys in the heap belong to which different blocks: an element marked the same way as the element at the root belongs to the current sorted block and the others belong to the next sorted block. The heap condition (first key less than the second and third) is maintained throughout, with elements in the next sorted block considered to be greater than elements in the current sorted block. The first run ends with I N G in the heap, since these keys all arrived with larger keys at the root (so they couldn't be included in the first run), the second run ends with A N D in the heap, etc.

Property 13.1 *For random keys, the runs produced by replacement selection are about twice the size of the heap used.*

Proof of this property actually requires a rather sophisticated analysis, but it is easy to verify experimentally. ▪

The practical effect of this is to save one merging pass: rather than starting with sorted runs about the size of the internal memory and then taking a merging pass to produce runs about twice the size of the internal memory, we can start right off with runs about twice the size of the internal memory, by using replacement selection with a priority queue of size M. If there is some order in the keys, then the runs will be much, much longer. For example, if no key has more than M larger keys before it in the file, the file will be completely sorted by the replacement selection pass, and no merging will be necessary! This is the most important practical reason to use the method.

In summary, the replacement selection technique can be used for both the "sort" and the "merge" steps of a balanced multiway merge.

Property 13.2 *A file of N records can be sorted using an internal memory capable of holding M records and $P + 1$ tapes in about $1 + \log_P(N/2M)$ passes.*

As discussed above, we first use replacement selection with a priority queue of size M to produce initial runs of size about $2M$ (in a random situation) or more (if the file is partially ordered), then use replacement selection with a priority queue of size P for about $\log_P(N/2M)$ (or fewer) merge passes. ▪

Practical Considerations

To finish implementing the sorting method outlined above, it is necessary to implement the input-output functions which actually transfer data between the processor and the external devices. These functions are obviously the key to good performance for the external sort, and they just as obviously require careful consideration of some systems (as opposed to algorithm) issues. (Readers not concerned with computers at the "systems" level may wish to skim the next few paragraphs.)

A major goal in the implementation should be to overlap reading, writing, and computing as much as possible. Most large computer systems have independent processing units for controlling the large-scale input/output (I/O) devices which make this overlapping possible. The efficiency to be achieved by an external sorting method depends on the number of such devices available.

For each file being read or written, the standard systems programming technique called *double-buffering* can be used to maximize the overlap of I/O with computing. The idea is to maintain two "buffers," one for use by the main processor, one for use by the I/O device (or the processor which controls the I/O device). For input, the processor uses one buffer while the input device is filling the other. When the processor has finished using its buffer, it waits until the input device has filled its buffer, and then the buffers switch roles: the processor uses the new data in the just-filled buffer while the input device refills the buffer with the data already used by the processor. The same technique works for output, with the roles of the processor and the device reversed. Usually the I/O time is far greater than the processing time and so the effect of double-buffering is to overlap the computation time entirely; thus the buffers should be as large as possible.

A difficulty with double-buffering is that it really uses only about half the available memory space. This can lead to inefficiency if many buffers are involved, as is the case in P-way merging when P is not small. This problem can be dealt with using a technique called *forecasting*, which requires the use of only one extra buffer (not P) during the merging process. Forecasting works as follows. Certainly the best way to overlap input with computation during the replacement selection process is to overlap the input of the buffer that needs to be filled next with the processing part of the algorithm. And it is easy to determine which buffer this is: the next input buffer to be emptied is the one whose *last* item is smallest. For example, when merging A O S with I R T and A G N we know that the third buffer will be the first to empty, then the first. A simple way to overlap processing with input for multiway merging is therefore to keep one extra buffer which is filled by the input device according to this rule. When the processor encounters an empty buffer, it waits until the input buffer is filled (if it hasn't been filled already), then switches to begin using that buffer and directs the input device to begin filling the buffer just emptied according to the forecasting rule.

The most important decision to be made in the implementation of the multiway merge is the choice of the value of P, the "order" of the merge. For tape sorting, when only sequential access is allowed, this choice is easy: P must be one less than the number of tape units available, since the multiway merge uses P input tapes and one output tape. Obviously, there should be at least two input tapes, so it doesn't make sense to try to do tape sorting with less than three tapes.

For disk sorting, when access to arbitrary positions is allowed but is somewhat more expensive than sequential access, it is also reasonable to choose P to be one less than the number of disks available, to avoid the higher cost of nonsequential access that would be involved, for example, if two different input files were on the same disk. Another alternative commonly used is to pick P large enough that the sort will be complete in two merging phases: it is usually unreasonable to try to do the sort in one pass, but a two-pass sort can often be done with a reasonably small P. Since replacement selection produces about $N/2M$ runs and each merging pass divides the number of runs by P, this means that P should be chosen to be the smallest integer with $P^2 > N/2M$. For our example of sorting a 200-million-word file on a computer with a one-million-word memory, this implies that $P = 11$ would be a safe choice to ensure a two-pass sort. (The right value of P could be computed exactly after the sort phase is completed.) The best choice between these two alternatives of the lowest reasonable value of P and the highest reasonable value of P is very dependent on many systems parameters: both alternatives (and some in between) should be considered.

Polyphase Merging

One problem with balanced multiway merging for tape sorting is that it requires either an excessive number of tape units or excessive copying. For P-way merging either we must use $2P$ tapes (P for input and P for output) or we must copy almost all of the file from a single output tape to P input tapes between merging passes, which effectively doubles the number of passes to be about $2\log_P(N/2M)$. Several clever tape-sorting algorithms have been invented which eliminate virtually all of this copying by changing the way in which the small sorted blocks are merged together. The most prominent of these methods is called *polyphase merging*.

The basic idea behind polyphase merging is to distribute the sorted blocks produced by replacement selection somewhat unevenly among the available tape units (leaving one empty) and then to apply a "merge-until-empty" strategy, at which point one of the output tapes and the input tape switch roles.

For example, suppose that we have just three tapes, and we start out with the initial configuration of sorted blocks on the tapes shown at the top of Figure

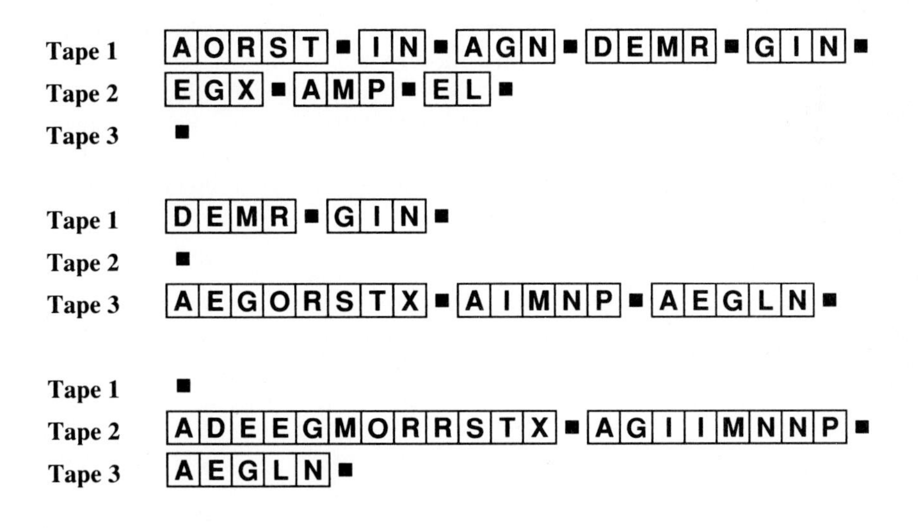

Figure 13.5 Initial stages of polyphase merging with three tapes.

13.5. (This comes from applying replacement selection to our example file with an internal memory that can only hold two records.) Tape 3 is initially empty, the output tape for the first merges. Now, after three two-way merges from tapes 1 and 2 to tape 3, the second tape becomes empty, as shown in the middle of Figure 13.5. Then, after two two-way merges from tapes 1 and 3 to tape 2, the first tape becomes empty, as shown at the bottom of Figure 13.5 The sort is completed in two more steps. First, a two-way merge from tapes 2 and 3 to tape 1 leaves one file on tape 2, one file on tape 1. Then a two-way merge from tapes 1 and 2 to tape 3 leaves the entire sorted file on tape 3.

This merge-until-empty strategy can be extended to work for an arbitrary number of tapes. Figure 13.6 shows how six tapes might be used to sort 497 initial runs. If we start out as indicated in the first column of Figure 13.6, with Tape 2 being the output tape, Tape 1 having 61 initial runs, Tape 3 having 120 initial runs, etc. as indicated in the first column of Figure 13.6, then after running a five-way "merge until empty," we have Tape 1 empty, Tape 2 with 61 (long) runs, Tape 3 with 59 runs, etc., as shown in the second column of Figure 13.6. At this point, we can rewind Tape 2 and make it an input tape, and rewind Tape 1 and make it the output tape. Continuing in this way, we eventually get the whole sorted file onto Tape 1. The merge is broken up into many *phases* which don't involve all the data, but no direct copying is involved.

The main difficulty in implementing a polyphase merge is to determine how to distribute the initial runs. It is not difficult to see how to build the table above by working backwards: take the largest number in each column, make it zero, and add it to each of the other numbers to get the previous column. This corresponds to defining the highest-order merge for the previous column which could give the present column. This technique works for any number of tapes (at least three): the numbers which arise are "generalized Fibonacci numbers" which have many interesting properties. Of course, the number of initial runs may not be known in advance, and it probably won't be exactly a generalized Fibonacci number. Thus a number of "dummy" runs must be added to make the number of initial runs exactly

Tape 1	*61*	*0*	*31*	*15*	*7*	*3*	*1*	*0*	*1*
Tape 2	*0*	*61*	*30*	*14*	*6*	*2*	*0*	*1*	*0*
Tape 3	*120*	*59*	*28*	*12*	*4*	*0*	*2*	*1*	*0*
Tape 4	*116*	*55*	*24*	*8*	*0*	*4*	*2*	*1*	*0*
Tape 5	*108*	*47*	*16*	*0*	*8*	*4*	*2*	*1*	*0*
Tape 6	*92*	*31*	*0*	*16*	*8*	*4*	*2*	*1*	*0*

Figure 13.6 Run distribution for six-tape polyphase merge.

what is needed for the table.

The analysis of polyphase merging is complicated and interesting, and yields surprising results. For example, it turns out that the very best method for distributing dummy runs among the tapes involves using extra phases and more dummy runs than would seem to be needed. The reason for this is that some runs are used in merges much more often than others.

Many other factors must be taken into consideration in implementing a most efficient tape-sorting method. A major factor which we have not considered at all is the time that it takes to rewind a tape. This subject has been studied extensively, and many fascinating methods have been defined. However, as mentioned above, the savings achievable over the simple multiway balanced merge are quite limited. Even polyphase merging is better than balanced merging only for small P, and then not substantially. For $P > 8$, balanced merging is likely to run faster than polyphase, and for smaller P the effect of polyphase is basically to save two tapes (a balanced merge with two extra tapes will run faster).

An Easier Way

Many modern computer systems provide a large *virtual memory* capability which should not be overlooked in implementing a method for sorting very large files. In a good virtual-memory system, the programmer can address a very large amount of data, leaving to the system the responsibility of making sure that the addressed data is transferred from external to internal storage when needed. This strategy relies on the fact that many programs have a relatively small "locality of reference": each reference to memory is likely to be to an area of memory that is relatively close to other recently referenced areas. This implies that transfers from external to internal storage are needed infrequently. An internal sorting method with a small locality of reference can work very well on a virtual-memory system. (For example, Quicksort has two "localities": most references are near one of the two partitioning pointers.) But check with your systems programmer before expecting to reap significant savings: a method such as radix sorting, which has no locality of reference whatsoever, would be disastrous on a virtual memory system, and even Quicksort could cause problems, depending on how well the available virtual memory system is implemented. On the other hand, the strategy of using a simple internal sorting method for sorting disk files deserves serious consideration in a good virtual-memory environment.

□

Exercises

1. Describe how you would do external *selection*: find the kth largest element in a file of N elements, where N is much too large for the file to fit in main memory.
2. Implement the replacement selection algorithm, then use it to test the claim that the runs produced are about twice the internal memory size.
3. What is the *worst* that can happen when replacement selection is used to produce initial runs in a file of N records, using a priority queue of size M with $M < N$?
4. How would you sort the contents of a disk if no other storage (except main memory) were available for use?
5. How would you sort the contents of a disk if only one tape (and main memory) were available for use?
6. Compare the four-tape and six-tape multiway balanced merge to polyphase merge with the same number of tapes, for 31 initial runs.
7. How many phases does five-tape polyphase merge use when started up with four tapes containing 26, 15, 22, and 28 runs initially?
8. Suppose the 31 initial runs in a four-tape polyphase merge are each one record long (initially distributed 0, 13, 11, 7). How many records are there in each of the files involved in the last three-way merge?
9. How should small files be handled in a Quicksort implementation to be run on a very large file in a virtual-memory environment?
10. How would you organize an external priority queue? (Specifically, design a way to support the *insert* and *remove* operations of Chapter 11, when the number of elements in the priority queue could grow to be much too large for the queue to fit in main memory.)

SOURCES for Sorting

The primary reference for this section is Volume 3 of D. E. Knuth's series, on sorting and searching. Further information on virtually every topic that we've touched upon can be found in this book. In particular, the results discussed here on performance characteristics of the various algorithms are backed up there by complete mathematical analyses.

There is a vast literature on sorting. Knuth and Rivest's 1973 bibliography contains hundreds of entries, and this doesn't include the treatment of sorting in countless books and articles on other subjects. A more up-to-date reference, with an extensive bibliography covering work to 1984, is Gonnet's book.

For Quicksort, the best reference is Hoare's original 1962 paper, which suggests all the important variants, including the use for selection discussed in Chapter 9. Many more details on the mathematical analysis and the practical effects of many of the modifications and embellishments suggested over the years may be found in this author's 1978 book.

A good example of an advanced priority queue structure is J. Vuillemin's "binomial queues" as implemented and analyzed by M. R. Brown. This data structure supports all the priority queue operations in an elegant and efficient manner. The state of the art in this type of data structure, for practical implementations, is the "pairing heap" described by Fredman, Sedgewick, Sleator, and Tarjan.

To get an impression of the myriad details of reducing algorithms like those we have discussed to general-purpose practical implementations, the reader would be advised to study the reference material for his particular computer system's sort utility. Such material necessarily deals primarily with formats of keys, records and files as well as many other details, and it is often interesting to identify how the algorithms themselves are brought into play.

M. R. Brown, "Implementation and analysis of binomial queue algorithms," *SIAM Journal of Computing*, **7**, 3 (August, 1978).

M. L. Fredman, R. Sedgewick, D. D. Sleator, and R. E. Tarjan, "The pairing heap: a new form of self-adjusting heap," *Algorithmica*, **1**, 1 (1986).

G. H. Gonnet, *Handbook of Algorithms and Data Structures*, Addison-Wesley, Reading, MA, 1984.

C. A. R. Hoare, "Quicksort," *Computer Journal*, **5**, 1 (1962).

D. E. Knuth, *The Art of Computer Programming. Volume 3: Sorting and Searching*, second printing, Addison-Wesley, Reading, MA, 1975.

R. L. Rivest and D. E. Knuth, "Bibliography 26: Computing Sorting," *Computing Reviews*, **13**, 6 (June, 1972).

R. Sedgewick, *Quicksort*, Garland, New York, 1978. (Also appeared as the author's Ph.D. dissertation, Stanford University, 1975.)

Searching Algorithms

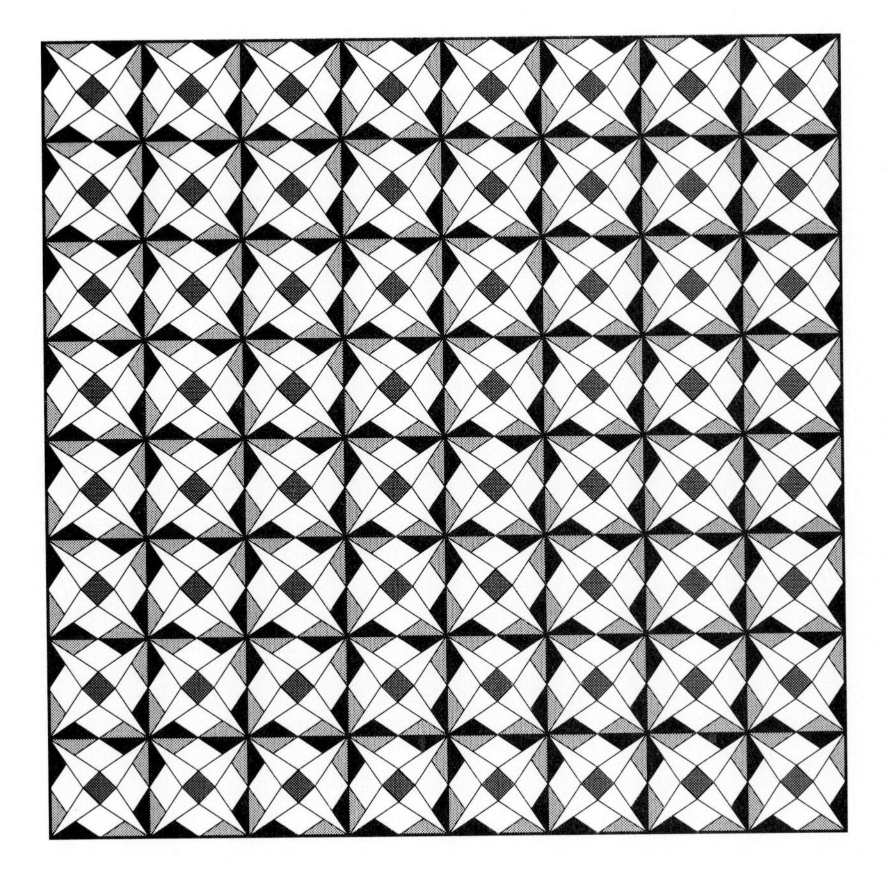

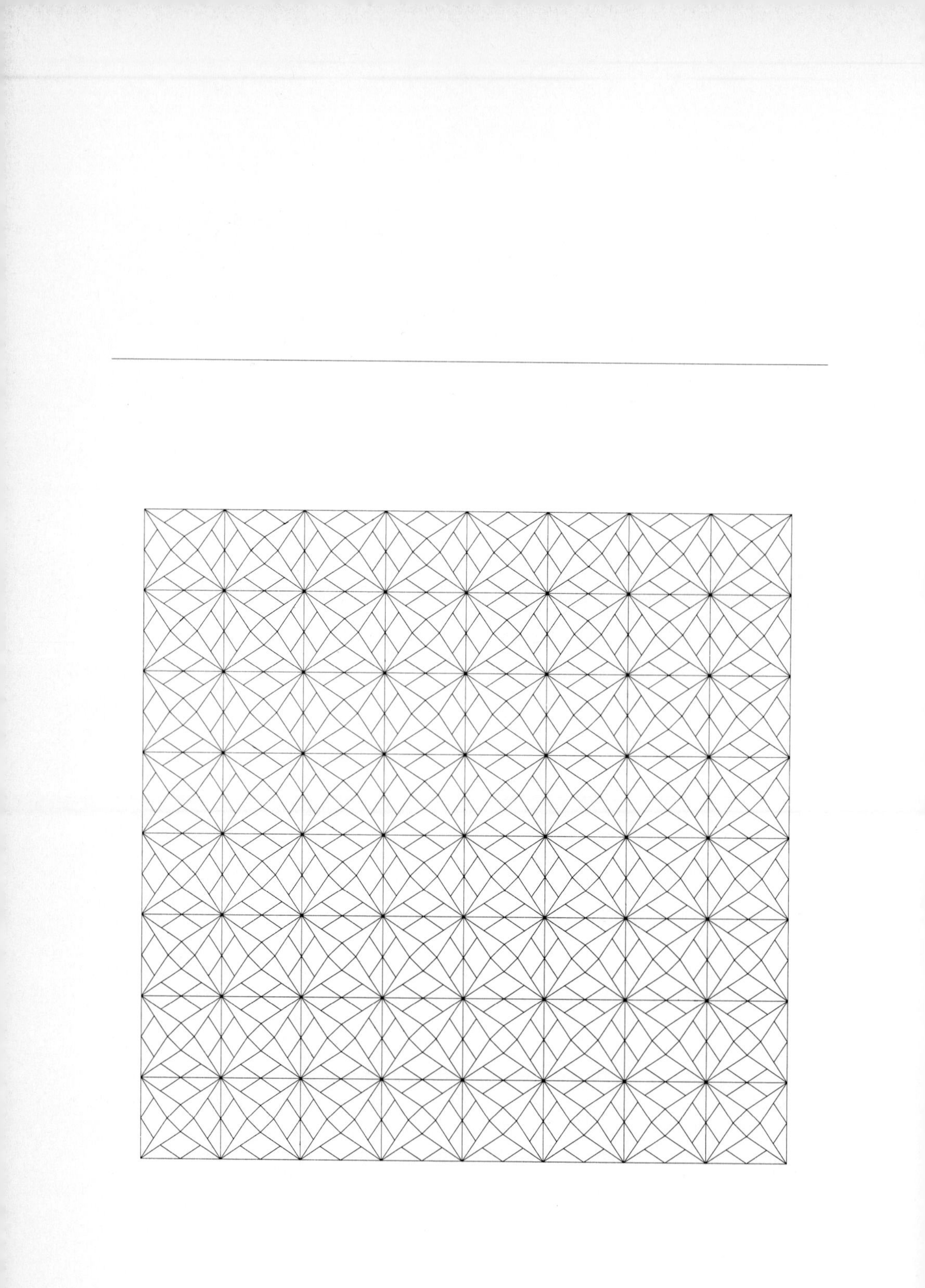

14

Elementary Searching Methods

☐ A fundamental operation intrinsic to a great many computational tasks is *searching*: retrieving some particular piece or pieces of information from a large amount of previously stored information. Normally we think of the information as divided up into *records*, each record having a *key* for use in searching. The goal of the search is to find all records with keys matching a given *search key*. The purpose of the search is usually to access information within the record (not merely the key) for processing.

Applications of searching are widespread, and involve a variety of different operations. For example, a bank needs to keep track of all its customers' account balances and to search through them to check various types of transactions. An airline reservation system has similar demands, in some ways, but most of the data is rather short-lived.

Two common terms often used to describe data structures for searching are *dictionaries* and *symbol tables*. For example, in an English language dictionary, the "keys" are the words and the "records" the entries associated with the words which contain the definition, pronunciation, and other information. One can prepare for learning and appreciating searching methods by thinking about how one would implement a system for searching in an English language dictionary. A symbol table is the dictionary for a program: the "keys" are the symbolic names used in the program, and the "records" contain information describing the object named.

In searching (as in sorting) we have programs which are in widespread and frequent use, so that it will be worthwhile to study a variety of methods in some detail. As with sorting, we'll begin by looking at some elementary methods which are very useful for small tables and in other special situations and illustrate fundamental techniques exploited by more advanced methods. We'll look at methods which store records in arrays which are either searched with key comparisons or indexed by key value, and we'll look at a fundamental method which builds structures defined by the key values.

As with priority queues, it is best to think of search algorithms as belonging to packages implementing a variety of generic operations which can be separated from particular implementations, so that alternate implementations can be substituted easily. The operations of interest include:

Initialize the data structure.
Search for a record (or records) having a given key.
Insert a new record.
Delete a specified record.
Join two dictionaries to make a large one.
Sort the dictionary; output all the records in sorted order.

As with priority queues, it is sometimes convenient to combine some of these operations. For example, a *search and insert* operation is often included for efficiency in situations where records with duplicate keys are not to be kept within the data structure. In many methods, once it has been determined that a key does not appear in the data structure, then the internal state of the search procedure contains precisely the information needed to insert a new record with the given key.

Records with duplicate keys can be handled in one of several ways, depending on the application. First, we could insist that the primary searching data structure contain only records with distinct keys. Then each "record" in this data structure might contain, for example, a link to a list of all records having that key. This is the most convenient arrangement from the point of view of the designer of searching algorithms, and it is convenient in some applications since *all* records with a given search key are returned with one *search*. The second possibility is to leave records with equal keys in the primary searching data structure and return *any* record with the given key for a *search*. This is simpler for applications that process one record at a time, where the order in which records with duplicate keys are processed is not important. It is inconvenient in terms of the algorithm design because a mechanism for retrieving all records with a given key must still be provided. A third possibility is to assume that each record has a unique identifier (apart from the key) and require that a *search* find the record with a given identifier, given the key. Or some more complicated mechanism could be used to distinguish among records with equal keys.

Each of the fundamental operations listed above has important applications, and quite a large number of basic organizations have been suggested to support efficient use of various combinations of the operations. In this and the next few chapters, we'll concentrate on implementations of the fundamental functions *search* and *insert* (and, of course, *initialize*), with some comment on *delete* and *sort* when appropriate. Like priority queues, the *join* operation normally requires advanced techniques which we won't be able to consider here.

Sequential Searching

The simplest method for searching is simply to store the records in an array, and then look through the array sequentially each time a record is sought. The following code shows an implementation of the basic functions using this simple organization and illustrates some of the conventions that we'll use in implementing searching methods.

```
type node= record key, info: integer end;
var a: array [0..maxN] of node;
    N: integer;
procedure initialize;
    begin N:=0 end;
function seqsearch(v: integer; x: integer): integer;
    begin
    a[N+1].key:=v;
    if x<=N then
        repeat x:=x+1 until v=a[x].key;
    seqsearch:=x
    end;
function seqinsert(v: integer): integer;
    begin
    N:=N+1; a[N].key:=v;
    seqinsert:=N;
    end;
```

The code above processes records that have integer keys (*key*) and "associated information" (*info*). As with sorting, it will be necessary in many applications to extend the programs to handle more complicated records and keys, but this will make no fundamental change in the algorithms. For example, *info* could be made into a pointer to an arbitrarily complicated record structure. In such a case, this field can serve as the record's unique identifier for use in distinguishing among records with equal keys.

The *seqsearch* procedure takes two arguments in this implementation: the key value being sought and an index (x) into the array. The index is included to handle the case in which several records have the same key value: by successively executing t:= *seqsearch*(v, t) starting at $t=0$, we can successively set t to the index of each record with key value v.

A sentinel record containing the key value being sought is used; this ensures that the search will always terminate and therefore involves only one completion test within the inner loop. After the inner loop has finished, testing whether the index returned is greater than N will tell whether the search found the sentinel or a key from the table. This is analogous to our use of a sentinel record containing the smallest or largest key value to simplify the coding of the inner loop of various sorting algorithms in Chapters 8–12.

Property 14.1 *Sequential search (array implementation) uses $N+1$ comparisons for an unsuccessful search (always) and about $N/2$ comparisons for a successful search (on the average).*

For unsuccessful search, this property follows directly from the code: each record must be examined to decide that a record with any particular key is absent. For successful search, if we assume that each record is equally likely to be sought, then the average number of comparisons is

$$\frac{1}{N}(1+2+\cdots+N)=\frac{N+1}{2},$$

exactly half the cost of unsuccessful search. ∎

Sequential searching obviously can be adapted in a natural way to use a linked-list representation for the records. One advantage of doing so is that it becomes easy to keep the list sorted, which leads to a quicker search:

```
type link=↑node;
        node= record key, info: integer; next: link end;
var head, t, z: link;
      i: integer;
procedure initialize;
   begin
   new(z); z↑.next:=z;
   new(head); head↑.next:=z;
   end;
function listsearch(v: integer; t: link): link;
   begin
   z↑.key:=v;
   repeat t:=t↑.next until v<=t↑.key;
   if v=t↑.key
      then listsearch:=t
      else listsearch:=z
   end;
```

With a sorted list, a search can be terminated unsuccessfully when a record with a key larger than the search key is found. Thus, on the average, only about half the records (not all) need to be examined for an unsuccessful search. The sorted order is easy to maintain because a new record can simply be inserted into the list at the point at which the unsuccessful search terminates:

```
function listinsert(v: integer; t: link): link;
  var x: link;
  begin
  z↑.key:=v;
  while t↑.next↑.key<v do t:=t↑.next;
  new(x); x↑.next:=t↑.next; t↑.next:=x;
  x↑.key:=v;
  listinsert:=x;
  end;
```

As usual with linked lists, a dummy header node *head* and a tail node z allow substantial simplification of the code. Thus, the call *listinsert*(v, *head*) will put a new node with key v into the list pointed to by the *next* field of the *head*, and *listsearch* is similar. Repeated calls on *listsearch* using the links returned will return records with duplicate keys. The tail node z is used as a sentinel in the same way as above. If *listsearch* returns z, then the search was unsuccessful.

Property 14.2 *Sequential search (sorted list implementation) uses about $N/2$ comparisons for both successful and unsuccessful search (on the average).*

For successful search, the situation is the same as before. For unsuccessful search, if we assume that the search is equally likely to be terminated by the tail node z or by each of the elements in the list (which is the case for a number of "random" search models), then the average number of comparisons is

$$\frac{1}{N+1}(1+2+\cdots+N+(N+1)) = \frac{N+2}{2},$$

slightly more than half the cost without sorting. ∎

If something is known about the relative frequency of access for various records, then substantial savings can often be realized simply by ordering the records intelligently. The "optimal" arrangement is to put the most frequently accessed record at the beginning, the second most frequently accessed record in the second position, etc. This technique can be very effective, especially if only a small set of records is frequently accessed.

If information is not available about the frequency of access, then an approximation to the optimal arrangement can be achieved with a "self-organizing" search: each time a record is accessed, move it to the beginning of the list. This method is more conveniently implemented when a linked-list implementation is used. Of course the running time for the method depends on the record access distributions, so it is difficult to predict how it will do in general. However, it is well suited to the quite common situation when most of the accesses to each record tend to be close together.

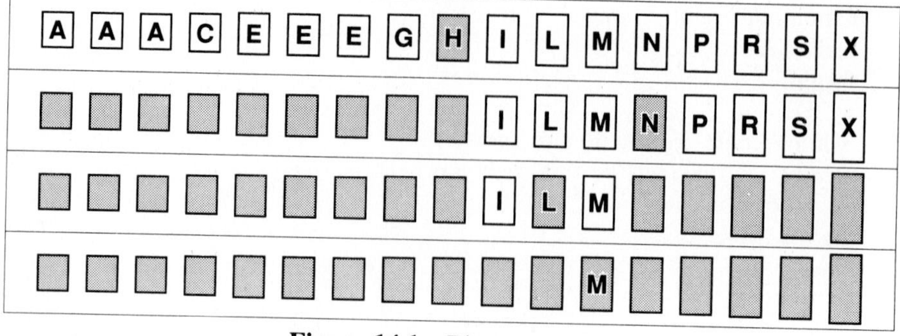

Figure 14.1 Binary search.

Binary Search

If the set of records is large, then the total search time can be significantly reduced by using a search procedure based on applying the "divide-and-conquer" paradigm: divide the set of records into two parts, determine which of the two parts the key sought belongs to, then concentrate on that part. A reasonable way to divide the sets of records into parts is to keep the records sorted, then use indices into the sorted array to delimit the part of the array being worked on. To find if a given key v is in the table, first compare it with the element at the middle position of the table. If v is smaller, then it must be in the first half of the table; if v is greater, then it must be in the second half of the table. Then apply this method recursively. (Since only one recursive call is involved, it is simpler to express the method iteratively.) This brings us directly to the following implementation, which assumes that the array a is sorted.

```
function binarysearch(v: integer): integer;
  var x, l, r: integer;
  begin
  l:=1; r:=N;
  repeat
    x:=(l+r) div 2;
    if v<a[x].key then r:=x-1 else l:=x+1
  until (v=a[x].key) or (l>r);
  if v=a[x].key
    then binarysearch:=x
    else binarysearch:=N+1
  end;
```

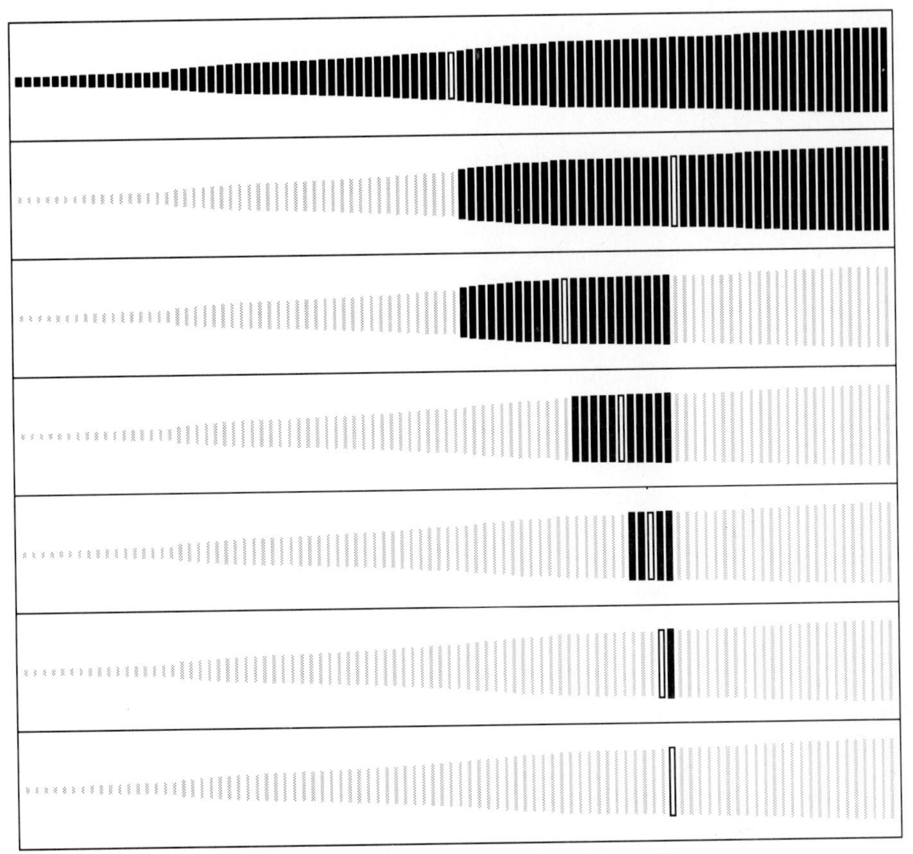

Figure 14.2 Binary search in a larger file.

Like Quicksort and radix exchange sort, this method uses the pointers l and r to delimit the subfile currently being worked on. Each time through the loop the variable x is set to point to the midpoint of the current interval, and there are three possibilities: the loop terminates successfully, or the left pointer is changed to $x+1$, or the right pointer is changed to $x-1$, depending on whether the search value v is equal to, less than, or greater than the key value of the record stored at $a[x]$.

Figure 14.1 shows the subfiles examined by this method when searching for M in a table built by inserting the keys A S E A R C H I N G E X A M P L E. The interval size is at least halved at each step, so only four comparisons are used for this search. Figure 14.2 shows a larger example, with 95 records; here only seven comparisons are required for any search.

Property 14.3 *Binary search never uses more than* $\lg N+1$ *comparisons for either successful or unsuccessful search.*

This property follows directly from the fact that the number of records is at least halved at each step: an upper bound on the number of comparisons satisfies the recurrence $C_N = C_{N/2} + 1$ with $C_1 = 1$, which implies the stated result (Formula 2 in Chapter 6). ■

It is important to note that the time required to *insert* new records is high for binary search: the array must be kept sorted, so some records must be moved to make room for any new record. For example, if a new record has a smaller key than any record in the table, then every entry must be moved over one position. A random insertion requires that $N/2$ records be moved, on the average. Thus, this method should not be used for applications involving many insertions. It is best suited for situations in which the table can be "built" ahead of time, perhaps using Shellsort or Quicksort, and then used for a large number of (very efficient) searches.

Some care must be exercised to handle properly records with equal keys for this algorithm: the index returned could fall in the middle of a block of records with key v, so loops which scan in both directions from that index must be used to pick up all the records. Of course, in this case the running time for the search is proportional to $\lg N$ plus the number of records found.

The sequence of comparisons made by the binary search algorithm is predetermined: the specific sequence used depends on the value of the key being sought and the value of N. The comparison structure can be simply described by a binary tree structure. Figure 14.3 shows the comparison structure for our example set of keys. In searching for a record with the key M for instance, it is first compared to H. Since M is greater, it is next compared to N (otherwise it would have been compared to C), then it is compared to L, and then the search terminates successfully on the fourth comparison. Below we will see algorithms that use an explicitly constructed binary tree structure to guide the search.

One improvement possible in binary search is to try to guess more precisely where the key being sought falls within the current interval of interest (rather than

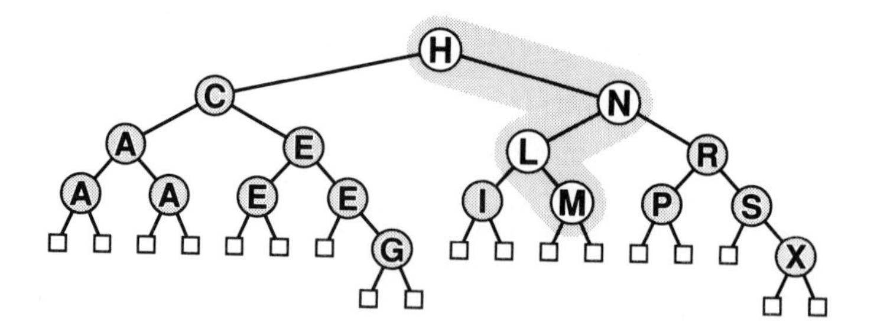

Figure 14.3 Comparison tree for binary search.

Figure 14.4 Interpolation search.

blindly using the middle element at each step). This mimics the way one looks up a number in the telephone directory, for example: if the name sought begins with B, one looks near the beginning, but if it begins with Y, one looks near the end. This method, called *interpolation search*, requires only a simple modification to the program above. In the program above, the new place to search (the midpoint of the interval) is computed with the statement $x:=(l+r)$ **div** 2, which is derived from the expression

$$x = l + \frac{1}{2}(r - l).$$

The middle of the interval is computed by adding half the size of the interval to the left endpoint. Interpolation search simply amounts to replacing $1/2$ in this formula by an estimate of where the key might be based on the values available: $1/2$ would be appropriate if v were in the middle of the interval between $a[l].key$ and $a[r].key$, but $x:=l+(v-a[l].key)*(r-l)$ **div** $(a[r].key-a[l].key)$ might be a better guess. Of course, this assumes numerical evenly distributed key values.

Suppose in our example that the ith letter in the alphabet is represented by the number i. Then, in search for M, the first table position examined would be 9, since $1+(13-1)*(17-1)/(24-1) = 9.3\ldots$ The search is completed in just three steps, as shown in Figure 14.4. Other search keys are found even more efficiently: for example the first and last elements are found in the first step. Figure 14.5 shows interpolation search on the file of 95 elements from Figure 14.2; it uses only four comparisons where binary search required seven.

Property 14.4 *Interpolation search uses fewer than* $\lg\lg N + 1$ *comparisons for both successful and unsuccessful search, in files of random keys.*

The proof of this fact is quite beyond the scope of this book. This function is a very slowly growing one which can be thought of as a constant for practical purposes: if N is one billion, $\lg\lg N < 5$. Thus, any record can be found using only a few accesses (on the average), a substantial improvement over the conventional binary search method. ∎

However, interpolation search does depend heavily on the assumption that the keys are rather well distributed over the interval: it can be badly "fooled" by

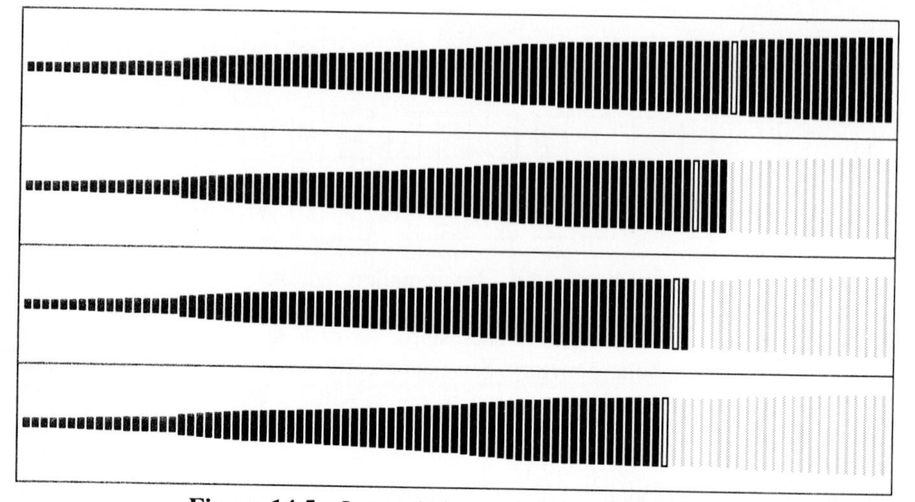

Figure 14.5 Interpolation search in a larger file.

poorly distributed keys, which do commonly arise in practice. Also, the method requires some computation: for small N, the $\lg N$ cost of straight binary search is close enough to $\lg\lg N$ that the cost of interpolating is not likely to be worthwhile. On the other hand, interpolation search certainly should be considered for large files, for applications where comparisons are particularly expensive, or for external methods where very high access costs are involved.

Binary Tree Search

Binary tree search is a simple, efficient dynamic searching method which qualifies as one of the most fundamental algorithms in computer science. It's classified here as an "elementary" method because it is so simple; but in fact it is the method of choice in many situations.

We've discussed trees at some length in Chapter 4. To review the terminology: The defining property of a *tree* is that every node is pointed to by only one other node called its *parent*. The defining property of a *binary tree* is that each node has left and right links. For searching, each node also has a record with a key value; in a *binary search tree* we insist that all records with smaller keys are in the left subtree and that all records in the right subtree have larger (or equal) key values. We'll soon see that it is quite simple to ensure that binary search trees built by successively inserting new nodes satisfy this defining property. An example of a binary search tree is shown in Figure 14.6; as usual, empty subtrees are represented by small square nodes.

A search procedure like *binarysearch* immediately suggests itself for this structure. To find a record with a given key v, first compare it against the root. If it is

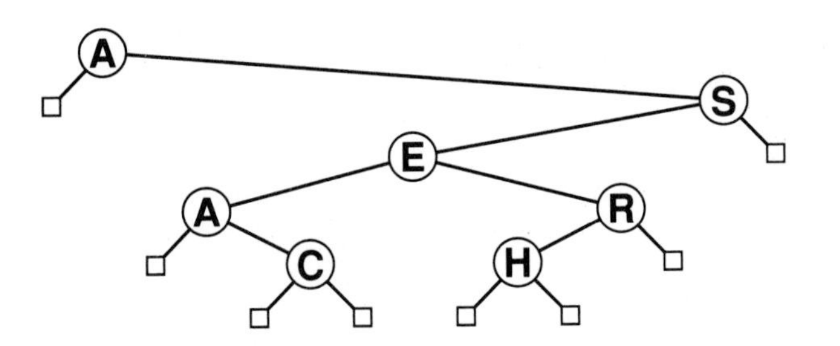

Figure 14.6 A binary search tree.

smaller, go to the left subtree; if it is equal, stop; if it is greater, go to the right subtree. Apply this method recursively. At each step, we're guaranteed that no parts of the tree other than the current subtree can contain records with key v, and, just as the size of the interval in binary search shrinks, the "current subtree" always gets smaller. The procedure stops either when a record with key v is found or, if there is no such record, when the "current subtree" becomes empty. (The words "binary," "search," and "tree" are admittedly somewhat overused at this point, and the reader should be sure to understand the difference between the *binarysearch* function given earlier in this chapter and the binary search trees described here. In binary search, we used a binary tree to describe the sequence of comparisons made by a function searching in an array; here we actually construct a data structure of records connected with links and use it for the search.)

```
type link=↑node;
        node=record key,info: integer; l,r: link end;
var t,head,z: link;
function treesearch(v: integer; x: link): link;
   begin
   z↑.key:=v;
   repeat
      if v<x↑.key then x:=x↑.l else x:=x↑.r
   until v=x↑.key;
   treesearch:=x
   end;
```

It is convenient to use a tree header node *head* whose right link points to the actual root node of the tree and whose key is smaller than all other key values (for simplicity, we use 0, assuming the keys are all positive integers). The left link of *head* is not used. The need for *head* will become more clear below when we discuss insertion.

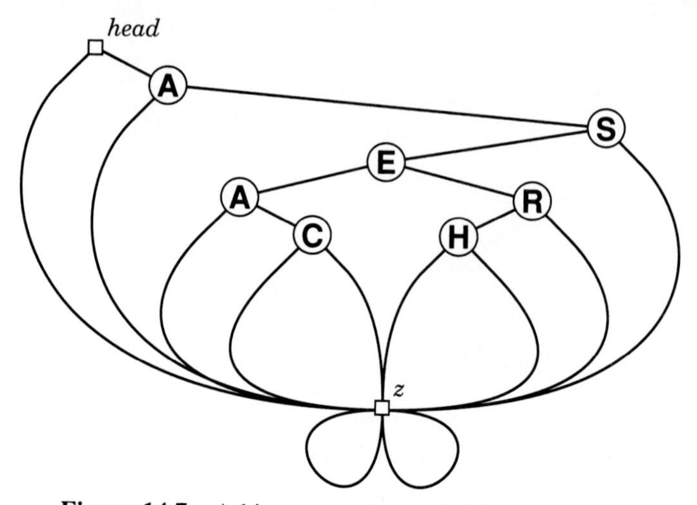

Figure 14.7 A binary search tree (with dummy nodes).

To search for a record with key v we set x:=*treesearch*(v, *head*). If a node has no left (right) subtree then its left (right) link is set to point to a "tail" node z. As in sequential search, we put the value sought in z to stop an unsuccessful search. Thus, the "current subtree" pointed to by x never becomes empty and all searches are "successful": the calling program can check whether the link returned points to z to determine whether the search was successful.

As shown above in Figure 14.6, it is convenient to think of links which point to z as pointing to imaginary *external nodes*, with all unsuccessful searches ending at external nodes. The normal nodes which contain our keys are called *internal nodes*; by introducing external nodes we can say that every internal node points to two other nodes in the tree, even though, in our implementation, all the external nodes are represented by the single node z. Figure 14.7 shows these links and the dummy nodes explicitly.

The empty tree is represented by having the right link of *head* point to z, as constructed by the following code:

```
procedure treeinitialize;
  begin
  new(z); z↑.l:=z; z↑.r:=z;
  new(head); head↑.key:=0; head↑.r:=z;
  end;
```

This initializes the links of z to point to z itself; though the programs in this chapter never access the links of z, this initialization is "safe" and convenient for the more advanced programs that we will see later.

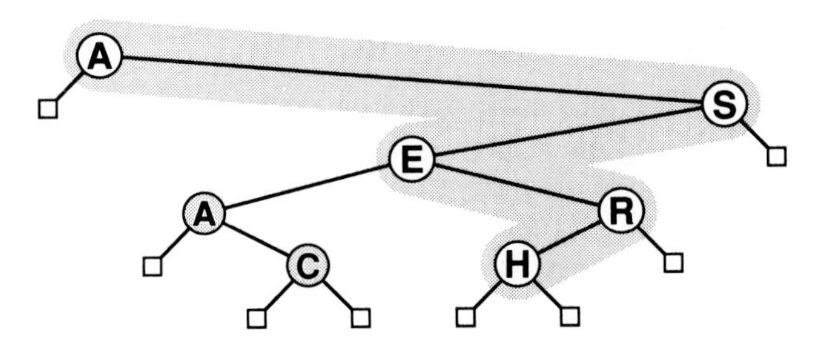

Figure 14.8 Searching (for I) in a binary search tree.

Figure 14.8 shows what happens when I is sought in our sample tree, using *treesearch*. First, it is compared against A, the key at the root. Since I is greater, it is next compared against S, the key in the right child of the node containing A. Continuing in this way, I is compared next against the E to the left of that node, then R, then H. The links in the node containing H are pointers to z so the search terminates: I is compared to itself in z and the search is unsuccessful.

To insert a node into the tree, we do an unsuccessful search for it, then attach it in place of z at the point at which the search terminated. To do the insertion, the following code keeps track of the parent p of x as it proceeds down the tree. When the bottom of the tree (x=z) is reached, p points to the node whose link must be changed to point to the new node inserted.

```
function treeinsert(v: integer; x:link): link;
  var p: link;
  begin
  repeat
    p:=x;
    if v<x↑.key then x:=x↑.l else x:=x↑.r
  until x=z;
  new(x); x↑.key:=v; x↑.l:=z; x↑.r:=z;
  if v<p↑.key then p↑.l:=x else p↑.r:=x;
  treeinsert:=x
  end;
```

A key v can be added using the call *treeinsert*(v, *head*). This function returns a link to the newly created node so that the calling routine can fill in the *info* field as appropriate.

When a new node whose key is equal to some key already in the tree is inserted, it will be inserted to the right of the node already in the tree. All records with key equal to v can be processed by successively setting t to *search*(v, t), as

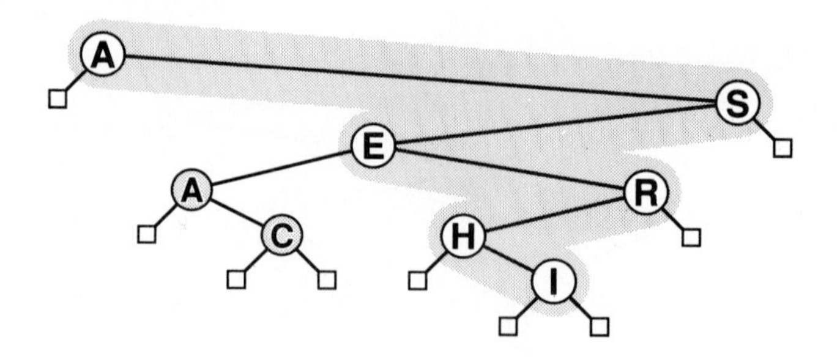

Figure 14.9 Insertion (of I) into a binary search tree.

we did for sequential searching.

The tree in Figure 14.9 results when the keys A S E A R C H I are inserted into an initially empty tree; Figure 14.10 shows the completion of our example, when N G E X A M P L E are added. The reader should pay particular attention to the position of equal keys in this tree: for example, even though the three As seem to be spread out through the tree, there are no keys "between" them.

The *sort* function comes almost for free when binary search trees are used, since a binary search tree represents a sorted file if you look at it the right way. In our figures, the keys appear in order if read from left to right on the page (ignoring their height and the links). A program has only the links to work with, but a sorting method follows directly from the defining properties of binary search trees. The following inorder traversal will do the job (see Chapter 4):

```
procedure treeprint(x: link);
  begin
  if x<>z then
    begin
    treeprint(x↑.l);
    printnode(x);
    treeprint(x↑.r)
    end
  end;
```

The call *treeprint*(*head*↑.*r*) will print out the keys of the tree in order. This defines a sorting method which is remarkably similar to Quicksort, with the node at the root of the tree playing a role similar to that of the partitioning element in Quicksort. A major difference is that the tree-sorting method must use extra memory for the links, while Quicksort sorts with only a little extra memory.

The running times of algorithms on binary search trees are quite dependent on the shapes of the trees. In the best case, the tree could be shaped like Figure

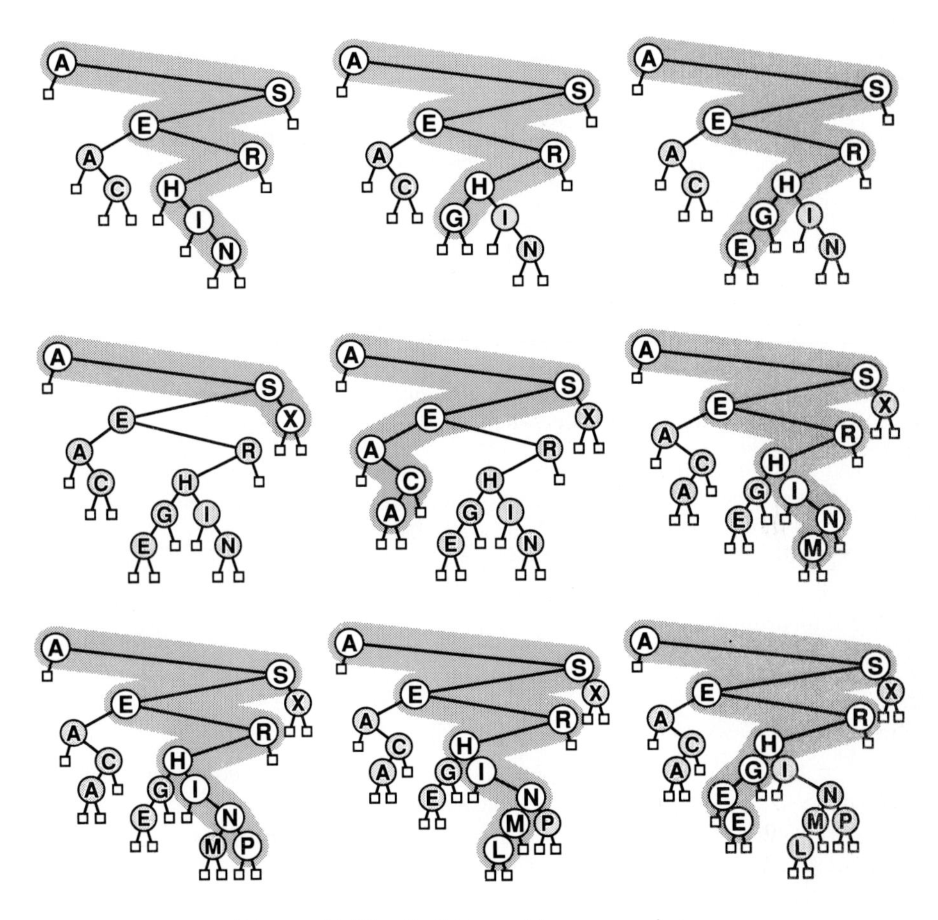

Figure 14.10 Building a binary search tree.

14.3, with about $\lg N$ nodes between the root and each external node. We might expect roughly logarithmic search times on the average, because the first element inserted becomes the root of the tree; if N keys are to be inserted at random, then this element would divide the keys in half (on the average), and this would yield logarithmic search times (using the same argument on the subtrees). Indeed, were it not for the equal keys, it could happen that the tree given above for describing the comparison structure for binary search would be built. This would be the best case of the algorithm, with guaranteed logarithmic running time for all searches. Actually, in a truly random situation, the root is equally likely to be any key so such a perfectly balanced tree is extremely rare. But if random keys are inserted, it turns out that the trees are nicely balanced.

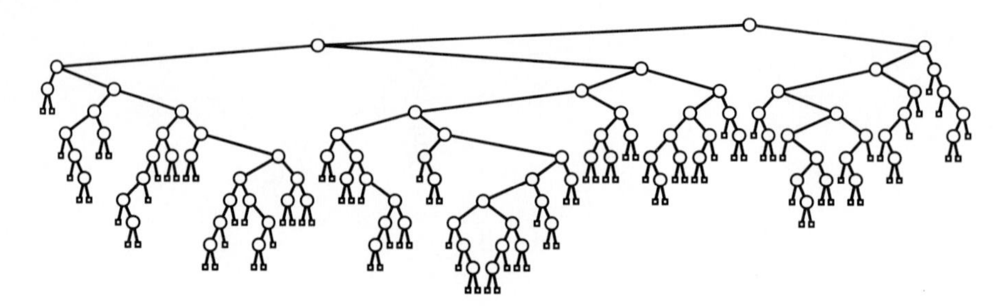

Figure 14.11 A large binary search tree.

Property 14.5 *A search or insertion in a binary search tree requires about* $2 \ln N$ *comparisons, on the average, in a tree built from* N *random keys.*

For each node in the tree, the number of comparisons used for a successful search to that node is the distance to the root. The sum of these distances for all nodes is called the *internal path length* of the tree. Dividing the internal path length by N, we get the average number of comparisons for successful search. But if C_N denotes the average internal path length of a binary search tree of N nodes, we have the recurrence

$$C_N = N - 1 + \frac{1}{N} \sum_{1 \le k \le N} (C_{k-1} + C_{N-k})$$

with $C_1 = 1$. (The $N - 1$ takes into account the fact that the root contributes 1 to the path length of each of the other $N - 1$ nodes in the tree; the rest of the expression comes from observing that the key at the root (the first inserted) is equally likely to be the kth largest, leaving random subtrees of size $k - 1$ and $N - k$.) But this is very nearly the same recurrence we solved in Chapter 9 for Quicksort, and it can easily be solved in the same way to derive the stated result. The argument for unsuccessful search is similar, though slightly more complicated. ■

Figure 14.11 shows a large binary search tree built from a random permutation of 95 elements. While it has some short paths and some long paths, it may be characterized as quite well-balanced: any search will require less than twelve comparisons, and the "average" number of comparisons to find any key in the tree is 7.00, as compared to 5.74 for binary search. (The average number of comparisons for a random unsuccessful search is one more than for successful search.) Moreover, a new key can be inserted at about the same cost, flexibility not available with binary search. However, if the keys are not randomly ordered, the algorithm can perform badly.

Property 14.6 *In the worst case, a search in a binary search tree with* N *keys can require* N *comparisons.*

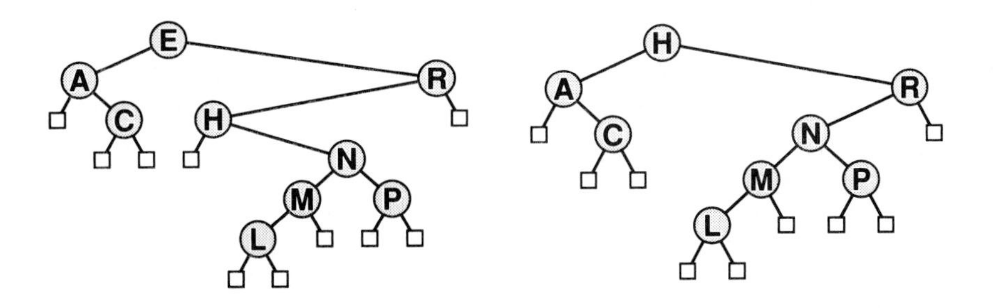

Figure 14.12 Deletion (of E) from a binary search tree.

For example, when the keys are inserted in order (or in reverse order), the binary-tree search method is no better than the sequential search method that we saw at the beginning of this chapter. Moreover, there are many other degenerate types of trees that can lead to the same worst case (for example, consider the tree formed when the keys A Z B Y C X ... are inserted in that order into an initially empty tree). In the next chapter, we'll examine a technique for eliminating this worst case and making all trees look more like the best-case tree. ■

Deletion

The implementations given above for the fundamental *search*, *insert*, and *sort* functions using binary tree structures are quite straightforward. However, binary trees also provide a good example of a recurrent theme in searching algorithms: the *delete* function is often quite cumbersome to implement.

Consider the tree shown at the left in Figure 14.12: to delete a node is easy if the node has no children, like L or P (lop it off by making the appropriate link in its parent null); or if it has just one child, like A, H, or R (move the link in the child to the appropriate parent link); or even if one of its two children has no children, like N (use that node to replace the parent); but what about nodes higher up in the tree, such as E?

Figure 14.12 shows one way to delete E: replace it with the node with the next highest key (in this case H). This node is guaranteed to have at most one child (since there are no nodes between it and the node deleted, its left link must be null), and is easily removed. To remove E from the tree on the left in Figure 14.12, then, we make the left link of R point to the right link (N) of H, copy the links from the node containing E to the node containing H, and make $head\uparrow.r$ point to H. This yields the tree on the right in the figure.

The code to cover all these cases is rather more complex than the simple routines for search and insertion, but it is worth studying carefully to prepare for the more complicated tree manipulations we will be doing in the next chapter. The following procedure deletes the node pointed to by t from the tree rooted at x. Thus, a node with key v can be deleted with the call *treedelete(treesearch(v,head),head)*.

The variable p is used to keep track of the parent of x in the tree and the variable c is used to find the successor node of the node to be deleted. After the deletion, x is the child of p. The actual *dispose* of the node pointed to by t is left to the calling routine (or perhaps the node is to be added to some other data structure).

```
procedure treedelete(t,x: link);
  var p,c: link;
  begin
  repeat
    p:=x;
    if t↑.key<x↑.key then x:=x↑.l else x:=x↑.r
  until x=t;
  if t↑.r=z then x:=x↑.l
  else  if t↑.r↑.l=z then
    begin x:=x↑.r; x↑.l:=t↑.l end
  else
    begin
    c:=x↑.r; while c↑.l↑.l<>z do c:=c↑.l;
    x:=c↑.l; c↑.l:=x↑.r;
    x↑.l:=t↑.l; x↑.r:=t↑.r
    end;
  if t↑.key<p↑.key then p↑.l:=x else p↑.r:=x;
  end;
```

The program first searches the tree in the normal way to get to the location of t in the tree. (Actually, the main purpose of this search is to set p, so that another node can be linked in after t is gone.) Next, the program checks three cases: if t has no right child, then the child of p after the deletion will be the left child of t (this would be the case for C, L, M, P, and R in Figure 14.12); if t has a right child with no left child then that right child will be the child of p after the deletion, with its left link copied from t (this would be the case for A and N in Figure 14.12); otherwise, x is set to the node with the smallest key in the subtree to the right of t; that node's right link is copied to the left link of its parent, and both of its links are set from t (this would be the case for H and E in Figure 14.12). To keep the number of cases low, this code always deletes by looking to the right, even though it might be easier in some cases to look to the left (for example, to delete H in Figure 14.12).

The approach seems asymmetric and rather *ad hoc*: for example, why not use the key immediately *before* the one to be deleted, instead of the one after? Various similar modifications have been suggested, but differences are not likely to be noticed in practical applications, though it has been shown that the algorithm above can tend to leave a tree slightly unbalanced (average height proportional to

$\sqrt{N}$) if subjected to a very large number of random delete-insert pairs.

It is actually quite typical of searching algorithms to require significantly more complicated implementations for deletion: the keys themselves tend to be integral to the structure, and removal of a key can involve complicated repairs. One alternative which is often appropriate is so-called *lazy deletion*, where a node is left in the data structure but marked as "deleted" for searching purposes. In the code above, this can be implemented by adding one further check for such nodes before stopping the search. One must make sure that large numbers of "deleted" nodes don't lead to excessive waste of time or space, but this turns out not to be an issue for many applications. Alternatively, one could periodically rebuild the entire data structure, leaving out the "deleted" nodes.

Indirect Binary Search Trees

As we saw with heaps in Chapter 11, for many applications we want a searching structure to simply help us find records, not move them around. For example, we might have an array $a[1..N]$ of records with keys and we might want the *search* routine to give us the index into that array of the record matching a certain key. Or we might want to remove the record with a given index from the searching structure, but still keep it in the array for some other use.

To adapt binary search trees to such a situation, we simply make the *info* field of the nodes the array index. Then we can eliminate the *key* field by having the search routines access the keys in the records directly, e.g. via an instruction like **if** $v< a[x\uparrow.info]$ **then**.... However, it is often better to make a copy of the key and use the code above just as given. We'll use the function name *bstinsert*(v, *info*: *integer*; x: *link*) to refer to a function just like *treeinsert*, except that it also sets the *info* field to the value given in the argument. Similarly, a function *bstdelete*(v, *info*: *integer*; x: *link*) to delete the node with key v and array index *info* from the binary search tree rooted at x will refer to an implementation of the *delete* function as described above. These functions use an extra copy of the keys (one in the array, one in the tree), but this allows the same function to be used for more than one array or, as we'll see in Chapter 27, for more than one key field in the same array. (There are other ways to achieve this: for example, a procedure could be associated with each tree which extracts keys from records.)

Another direct way to achieve "indirection" for binary search trees is simply to do away entirely with the linked implementation. That is, all links just become indices into an array $a[1..N]$ of records which contain a *key* field and l and r index fields. Then link references such as **if** $v< x\uparrow.key$ **then** $x:=x\uparrow.l$ **else** ... become array references such as **if** $v< a[x].key$ **then** $x:=a[x].l$ **else**... . No calls to *new* are used, since the tree exists within the record array: *new*(*head*) becomes *head*:=0, *new*(z) becomes $z:=N+1$, and to insert the Mth node, we would pass M, not v, to *treeinsert*, and then simply refer to $a[M].key$ instead of v and replace the line containing *new*(x) in *treeinsert* with $x:=M$.

This way of implementing binary search trees to aid in searching large arrays of records is preferred for many applications, since it avoids the extra expense of copying keys as in the previous paragraph, and it avoids the overhead of the storage-allocation mechanism implied by *new*. Its disadvantage is that unused links might waste space in the record array.

A third alternative is to use parallel arrays, as we did for linked lists in Chapter 3. The implementation of this is very much as described in the previous paragraph, except that three arrays are used, one each for the keys, left links, and right links. The advantage of this is its flexibility. Extra arrays (extra information associated with each node) can be easily added without changing the tree manipulation code at all, and when the search routine gives the index for a node it gives a way to immediately access all the arrays.

□

Exercises

1. Implement a sequential searching algorithm which averages about $N/2$ steps for both successful and unsuccessful search, keeping the records in a sorted array.
2. Give the order of the keys after records with the keys E A S Y Q U E S T I O N have been put into an initially empty table with *search and insert* using the self-organizing search heuristic.
3. Give a recursive implementation of binary search.
4. Suppose $a[i]=2i$ for $1 \leq i \leq N$. How many table positions are examined by interpolation search during the unsuccessful search for $2k - 1$?
5. Draw the binary search tree that results from inserting into an initially empty tree records with the keys E A S Y Q U E S T I O N.
6. Write a recursive program to compute the *height* of a binary tree: the longest distance from the root to an external node.
7. Suppose that we have an estimate ahead of time of how often search keys are to be accessed in a binary tree. Should the keys be inserted into the tree in increasing or decreasing order of likely frequency of access? Why?
8. Modify binary tree search so that it keeps equal keys together in the tree. (If any other nodes in the tree have the same key as any given node, then either its parent or one of its children should have an equal key.)
9. Write a nonrecursive program to print out the keys from a binary search tree in order.
10. Draw the binary search tree that results from inserting into an initially empty tree records with the keys E A S Y Q U E S T I O N, and then deleting the Q.

15

Balanced Trees

☐ The binary-tree algorithms in the previous chapter work very well for a wide variety of applications, but they do have the problem of bad worst-case performance. What's more, as with Quicksort, it's embarrassingly true that the bad worst case is one that's likely to occur in practice if the user of the algorithm is not watching for it. Files already in order, files in reverse order, files with alternating large and small keys, or files with any large segment having a simple structure can cause the binary-tree search algorithm to perform very badly.

With Quicksort, our only recourse for improving the situation was to resort to randomness: by choosing a random partitioning element, we could rely on the laws of probability to save us from the worst case. Fortunately, for binary tree searching, we can do much better: there is a general technique that will enable us to *guarantee* that this worst case will not occur. This technique, called *balancing*, has been used as the basis for several different "balanced-tree" algorithms. We'll look closely at one such algorithm and discuss briefly how it relates to some of the other methods that are used.

As will become apparent below, implementing balanced-tree algorithms is certainly a case of "easier said than done." Often, the general concept behind an algorithm is easily described, but an implementation is a morass of special and symmetric cases. The program developed in this chapter is not only an important searching method, but it also illustrates nicely the relationship between a "high-level" description and a "low-level" Pascal program to implement an algorithm.

Top-Down 2-3-4 Trees

To eliminate the worst case for binary search trees, we'll need some flexibility in the data structures that we use. To get this flexibility, let's assume that the nodes in our trees can hold more than one key. Specifically, we'll allow *3-nodes* and *4-nodes*, which can hold two and three keys respectively. A 3-node has three links coming out of it, one for all records with keys smaller than both its keys, one for

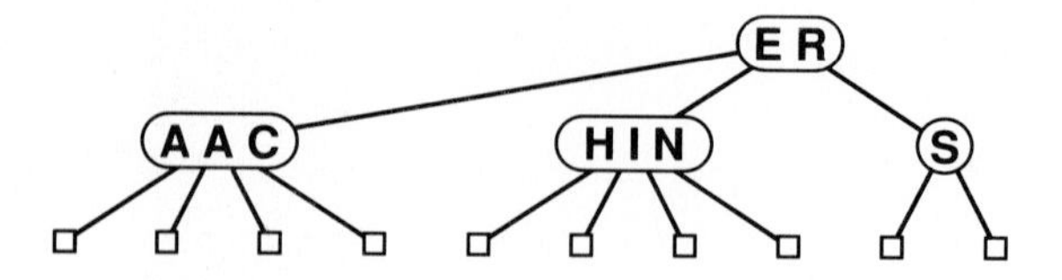

Figure 15.1 A 2-3-4 Tree.

all records with keys in between its two keys, and one for all records with keys larger than both its keys. Similarly, a 4-node has four links coming out of it, one for each of the intervals defined by its three keys. (The nodes in a standard binary search tree could thus be called *2-nodes*: one key, two links.) We'll see below some efficient ways to define and implement the basic operations on these extended nodes; for now, let's assume we can manipulate them conveniently and see how they can be put together to form trees.

For example, Figure 15.1 shows a *2-3-4 tree* which contains the keys A S E A R C H I N. It is easy to see how to search in such a tree. For example, to search for G in the tree in Figure 15.1, we would follow the middle link from the root, since G is between E and R, then terminate the unsuccessful search at the left link from the node containing H, I, and N.

To insert a new node in a 2-3-4 tree, we would like, as before, to do an unsuccessful search and then hook the node on. It is easy to see what to do if the node at which the search terminates is a 2-node: just turn it into a 3-node. For example, X could be added to the tree in Figure 15.1 by adding it (and another link) to the node containing S. Similarly, a 3-node can easily be turned into a 4-node. But what should we do if we need to insert new node into a 4-node? For example, how shall this be done if we insert G into the tree in Figure 15.1? One possibility would be to hook it on as a new leftmost child of the 4-node containing H, I, and N, but a better solution is shown in Figure 15.2: first split the 4-node into two 2-nodes and pass one of its keys up to its parent. First the 4-node containing H, I, and N is split into two 2-nodes (one containing H, the other containing N) and the "middle key" I is passed up to the 3-node containing E and R, turning it into a 4-node. Then there is room for G in the 2-node containing H.

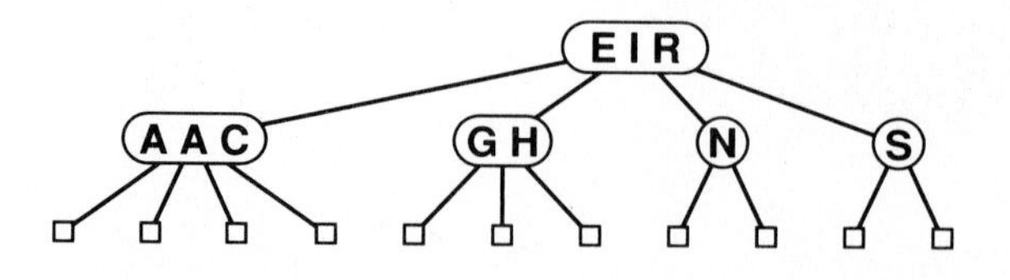

Figure 15.2 Insertion (of G) into a 2-3-4 Tree.

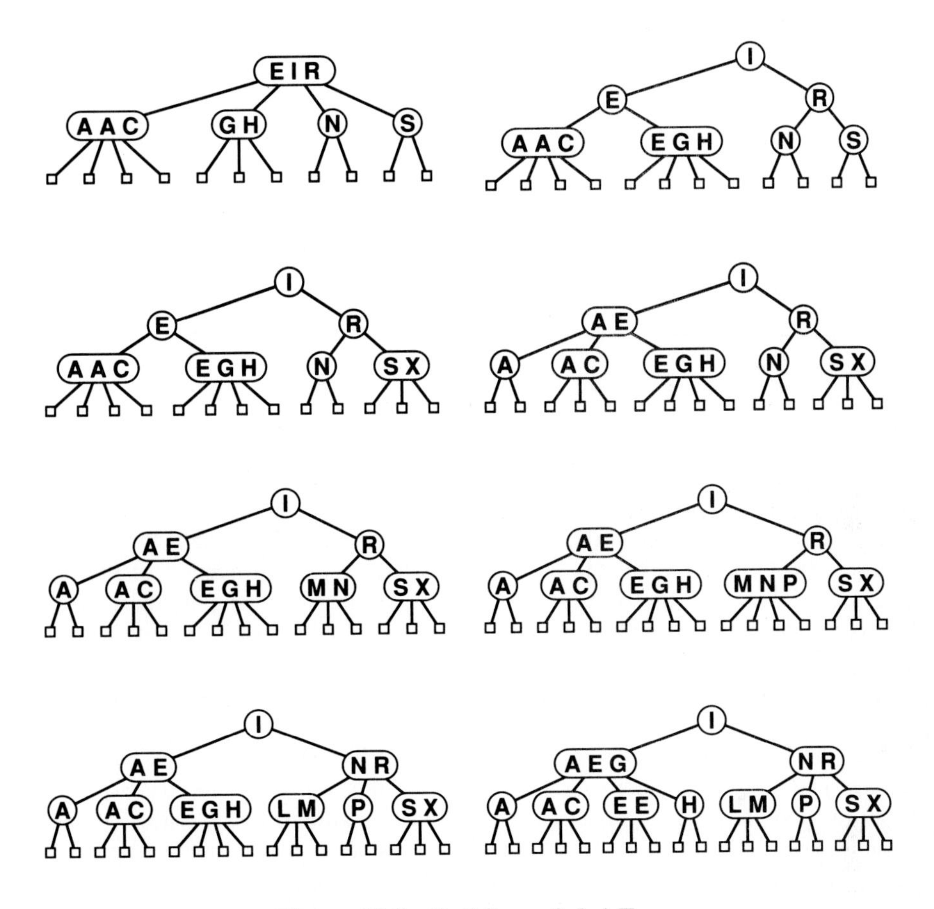

Figure 15.3 Building a 2-3-4 Tree.

But what if we need to split a 4-node whose parent is also a 4-node? One method would be to split the parent also, but we could keep having to do this all the way back up the tree. An easier way is to make sure that the parent of any node we see won't be a 4-node by splitting any 4-node we see on the way down the tree. Figure 15.3 completes the construction of a 2-3-4 tree for our full set of keys A S E A R C H I N G E X A M P L E. On the first line, we see that the root node is split during the insertion of the second E; other splits occur when the second A, the L, and the third E are inserted.

The above example shows that we can easily insert new nodes into 2-3-4 trees by doing a search and splitting 4-nodes on the way down the tree. Specifically, as shown in Figure 15.4, every time we encounter a 2-node connected to a 4-node,

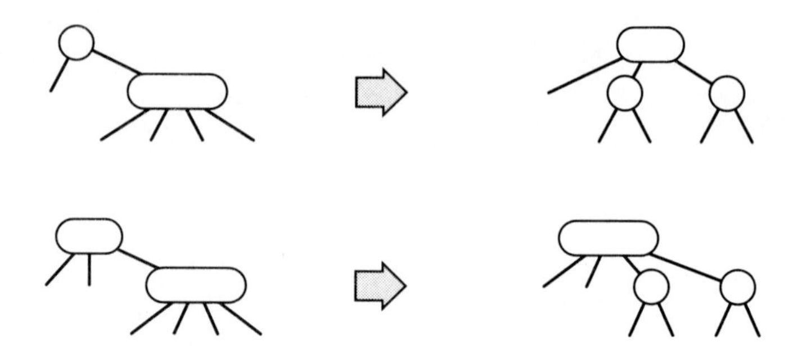

Figure 15.4 Splitting 4-nodes.

we should transform it into a 3-node connected to two 2-nodes, and every time we encounter a 3-node connected to a 4-node, we should transform it into a 4-node connected to two 2-nodes.

This "split" operation works because of the way not only the keys but also the *pointers* can be moved around. Two 2-nodes have the same number of pointers (four) as a 4-node, so the split can be executed without changing anything below the split node. And a 3-node can't be changed to a 4-node just by adding another key; another pointer is needed also (in this case, the extra pointer provided by the split). The crucial point is that these transformations are purely "local": no part of the tree need be examined or modified other than that shown in Figure 15.4. Each of the transformations passes up one of the keys from a 4-node to its parent in the tree and restructures links accordingly. Note that we needn't worry explicitly about the parent being a 4-node, since our transformations ensure that as we pass through each node in the tree, we come out on a node that is not a 4-node. In particular, when we come out the bottom of the tree, we are not on a 4-node, and we can insert the new node directly by transforming either a 2-node to a 3-node or a 3-node to a 4-node. Actually, it is convenient to treat the insertion as a split of an imaginary 4-node at the bottom which passes up the new key to be inserted. Whenever the root of the tree becomes a 4-node, we'll split it into three 2-nodes, as we did for our first node split in the example above. This (and only this) makes the tree grow one level "higher."

The algorithm sketched above gives a way to do searches and insertions in 2-3-4 trees; since the 4-nodes are split up on the way from the top down, the trees are called *top-down 2-3-4 trees*. What's interesting is that, even though we haven't been worrying about balancing at all, the resulting trees are perfectly balanced!

Property 15.1 *Searches in N-node 2-3-4 trees never visit more than* $\lg N + 1$ *nodes.*

The distance from the root to every external node is the same: the transformations

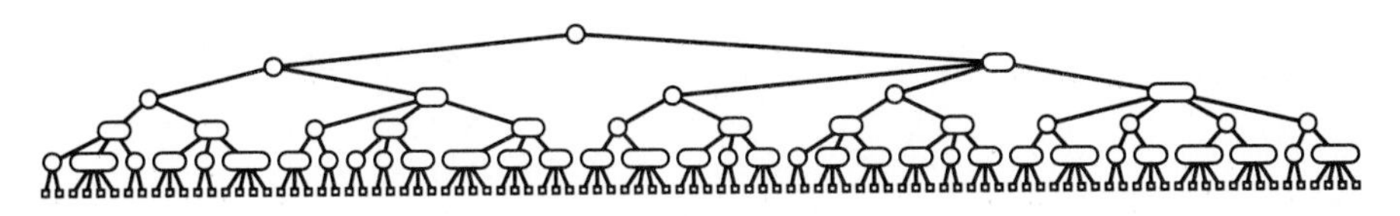

Figure 15.5 A Large 2-3-4 Tree.

that we perform have no effect on the distance from any node to the root, except when we split the root, and in this case the distance from all nodes to the root is increased by one. If all the nodes are 2-nodes, the stated result holds since the tree is like a full binary tree; if there are 3-nodes and 4-nodes, the height can only be lower. ■

Property 15.2 *Insertions into N-node 2-3-4 trees require fewer than* $\lg N + 1$ *node splits in the worst case and seem to require less than one node split on the average.*

The worst thing that can happen is that all the nodes on the path to the insertion point are 4-nodes, all of which would be split. But in a tree built from a random permutation of N elements, not only is this worst case unlikely to occur, but also few splits seem to be required on the average, because there are not many 4-nodes. Figure 15.5 shows a tree built from a random permutation of 95 elements: there are nine 4-nodes, only one of which is not on the bottom level. Analytical results on the average-case performance of 2-3-4 trees have so far eluded the experts, but empirical studies consistently show that very few splits are done. ■

The description given above is sufficient to define an algorithm for searching using 2-3-4 trees which has guaranteed good worst-case performance. However, we are only halfway towards an actual implementation. While it would be possible to write algorithms which actually perform transformations on distinct data types representing 2-, 3-, and 4-nodes, most of the things that need to be done are very inconvenient in this direct representation. (One can become convinced of this by trying to implement even the simpler of the two node transformations.) Furthermore, the overhead incurred in manipulating the more complex node structures is likely to make the algorithms slower than standard binary-tree search. The primary purpose of balancing is to provide "insurance" against a bad worst case, but it would be unfortunate to have to pay the overhead cost for that insurance on every run of the algorithm. Fortunately, as we'll see below, there is a relatively simple representation of 2-, 3-, and 4-nodes that allows the transformations to be done in a uniform way with very little overhead beyond the costs incurred by standard binary-tree search.

Red-Black Trees

Remarkably, it is possible to represent 2-3-4 trees as standard binary trees (2-nodes only) by using only one extra bit per node. The idea is to represent 3-nodes and

Figure 15.6 Red-Black Representation of 3-nodes and 4-nodes.

4-nodes as small binary trees bound together by "red" links; these contrast with the "black" links that bind the 2-3-4 tree together. The representation is simple: as shown in Figure 15.6, 4-nodes are represented as three 2-nodes connected by red links and 3-nodes are represented as two 2-nodes connected by a red link (red links are drawn as thick lines). (Either orientation is legal for a 3-node.)

Figure 15.7 shows one way to represent the final tree from Figure 15.3. If we eliminate the red links and collapse together the nodes they connect, the result is the 2-3-4 tree in Figure 15.3. The extra bit per node is used to store the color of the link pointing to that node: we'll refer to 2-3-4 trees represented in this way as *red-black trees*.

The "slant" of each 3-node is determined by the dynamics of the algorithm to be described below. There are many red-black trees corresponding to each 2-3-4 tree. It would be possible to enforce a rule that 3-nodes all slant the same way, but there is no reason to do so.

These trees have many structural properties that follow directly from the way in which they are defined. For example, there are never two red links in a row along any path from the root to an external node, and all such paths have an equal number of black links. Note that it is possible that one path (alternating black-red) be twice as long as another (all black), but that all path lengths are still proportional to $\log N$.

A striking feature of the tree above is the positioning of duplicate keys. On reflection, it is clear that any balanced tree algorithm must allow records with

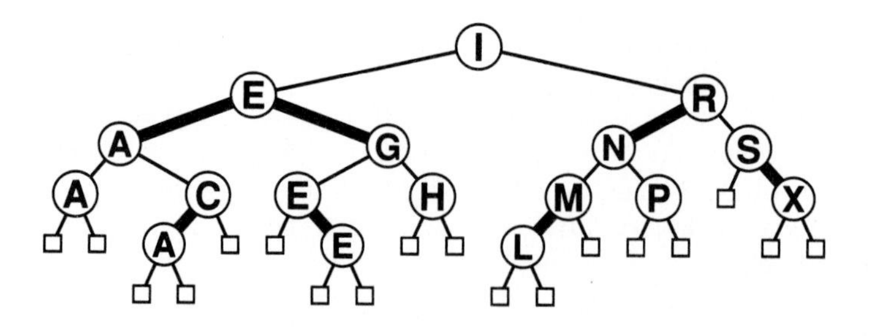

Figure 15.7 A Red-Black Tree.

keys equal to a given node to fall on both sides of that node: otherwise, severe imbalance could result from long strings of duplicate keys. This implies that we can't find all nodes with a given key by repeated calls to the searching procedure, as in the previous chapter. However, this does not present a real problem, because all nodes in the subtree rooted at a given node with the same key as that node can be found with a simple recursive procedure like the *treeprint* procedure of the previous chapter. One could alternatively require distinct keys in the data structure (with linked lists of records with duplicate keys).

One very nice property of red-black trees is that the *treesearch* procedure for standard binary tree search works without modification (except for the matter of duplicate keys discussed in the previous paragraph). We'll implement the link colors by adding a boolean field *red* to each node which is true if the link pointing to the node is red, false if it is black; the *treesearch* procedure simply never examines that field. Thus, no "overhead" is added by the balancing mechanism to the time taken by the fundamental searching procedure. Since each key is inserted just once, but may be searched for many times in a typical application, the end result is that we get improved search times (because the trees are balanced) at relatively little cost (because no work for balancing is done during the searches).

Moreover, the overhead for insertion is very small: we have to do something different only when we see 4-nodes, and there aren't many 4-nodes in the tree because we're always breaking them up. The inner loop needs only one extra test (if a node has two red children, it's a part of a 4-node), as shown in the following implementation of the insert procedure:

```
function rbtreeinsert(v: integer; x:link): link;
  var gg,g,p: link;
  begin
  p:=x; g:=x;
  repeat
    gg:=g; g:=p; p:=x;
    if v<x↑.key then x:=x↑.l else x:=x↑.r;
    if x↑.l↑.red and x↑.r↑.red then x:=split(v,gg,g,p,x);
  until x=z;
  new(x); x↑.key:=v; x↑.l:=z; x↑.r:=z;
  if v<p↑.key then p↑.l:=x else p↑.r:=x;
  rbtreeinsert:=x;
  x:=split(v,gg,g,p,x);
  end;
```

In this program, x moves down the tree as before and gg, g, and p are kept pointing to x's great-grandparent, grandparent, and parent in the tree. To see why all these links are needed, consider the addition of Y to the tree above. When the external

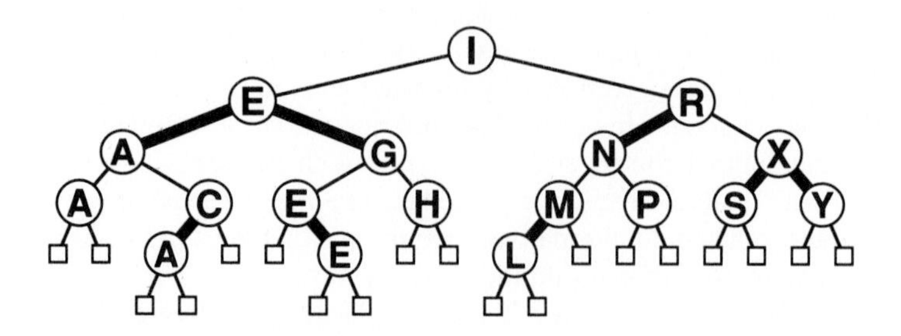

Figure 15.8 Insertion (of Y) into a Red-Black Tree.

node at the right of the 3-node containing S and X is reached, *gg* is R, *g* is S, and *p* is X. Now Y must be added to make a 4-node containing S, X, and Y, resulting in the tree shown in Figure 15.8.

We need a pointer to R (*gg*) because R's right link must be changed to point to X, not S. To see exactly how this comes about, we need to look at the operation of the *split* procedure. Let's consider the red-black representation for the two transformations we must perform: if we have a 2-node connected to a 4-node, then we should convert them into a 3-node connected to two 2-nodes; if we have a 3-node connected to a 4-node, we should convert them into a 4-node connected to two 2-nodes. When a new node is added at the bottom, it is considered to be the middle node of an imaginary 4-node (that is, think of z as being *red*, though this is never explicitly tested).

The transformation required when we encounter a 2-node connected to a 4-node is easy, and the same transformation works if we have a 3-node connected to a 4-node in the "right" way, as shown in Figure 15.9. Thus, *split* begins by marking x to be red and the children of x to be black.

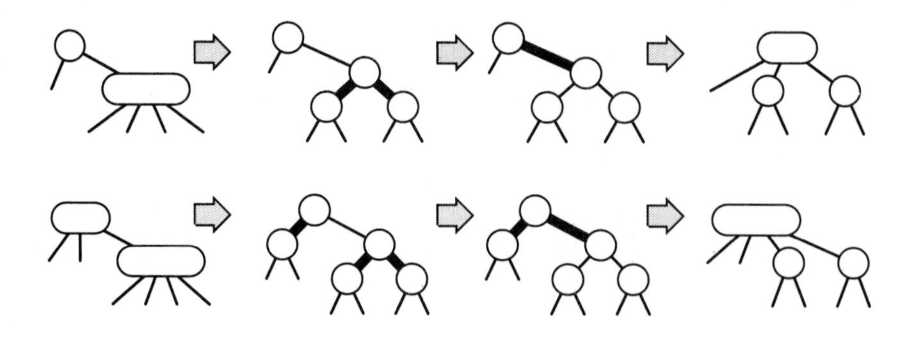

Figure 15.9 Splitting 4-nodes with a color flip.

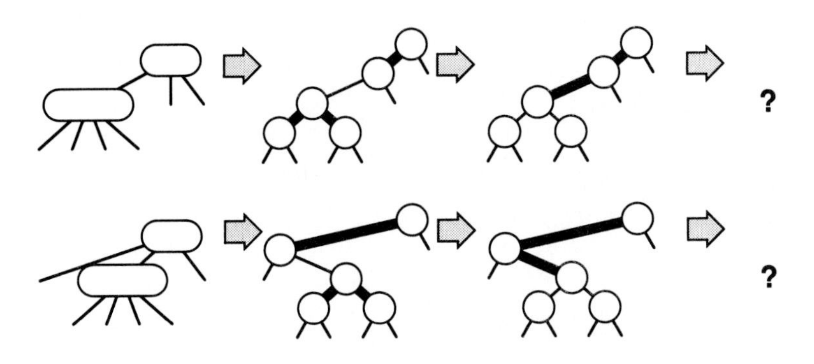

Figure 15.10 Splitting 4-nodes with a color flip: rotation needed.

This leaves the two other situations that can arise if we encounter a 3-node connected to a 4-node, as shown in Figure 15.10. (Actually, there are four situations, since the mirror images of these two can also occur for 3-nodes of the other orientation.) In these cases, splitting the 4-node has left two red links in a row, an illegal situation which must be corrected. This is easily tested for in the code: we just marked x red, so if x's parent p is also red, we must take further action. The situation is not too bad because we do have three nodes connected by red links: all we need do is transform the tree so that the red links point down from the same node.

Fortunately, there is a simple operation which achieves the desired effect. Let's begin with the easier of the two, the first (top) case from Figure 15.10, where the red links are oriented the same way. The problem is that the 3-node was oriented the wrong way: accordingly, we restructure the tree to switch the orientation of the 3-node and thus reduce this case to be the same as the second case from Figure 15.9, where the color flip of x and its children was sufficient. Restructuring the tree to reorient a 3-node involves changing three links, as shown in Figure 15.11; note that Figure 15.11 is the same as Figure 15.8, but with the 3-node containing N and R rotated. The left link of R was changed to point to P, the right link of N was changed to point to R, and the right link of I was changed to point to N. Also, note carefully that the colors of the two nodes are switched.

This *single rotation* operation is defined on any binary search tree (if we disregard operations involving the colors) and is the basis for several balanced-tree algorithms, because it preserves the essential character of the search tree and is a local modification involving only three link changes. It is important to note, however, that doing a single rotation doesn't necessarily improve the balance of the tree. In Figure 15.11, the rotation brings all the nodes to the left of N one step closer to the root, but all the nodes to the right of R are *lowered* one step: in this case the rotation makes the tree less, not more balanced. Top-down 2-3-4 trees may be viewed simply as a convenient way to identify single rotations which *are*

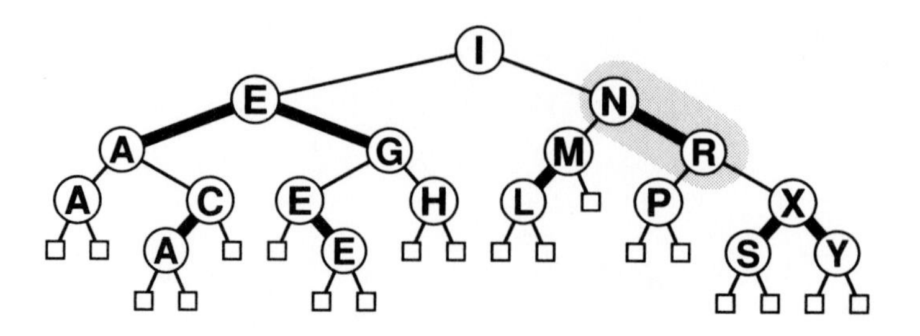

Figure 15.11 Rotating a 3-node in Figure 15.8.

likely to improve the balance.

Doing a single rotation involves modifying the structure of the tree, something that should be done with caution. As we saw when considering the deletion algorithm in Chapter 14, the code is more complicated than might seem necessary because there are a number of similar cases with left-right symmetries. For example, suppose that the links y, c, and gc point to I, R, and N respectively in Figure 15.8. Then the transformation to Figure 15.11 is effected by the link changes $c\uparrow.l{:=}gc\uparrow.r$; $gc\uparrow.r{:=}c$; $y\uparrow.r{:=}gc$. There are three other analogous cases: the 3-node could be oriented the other way, or it could be on the left side of y (oriented either way). A convenient way to handle these four different cases is to use the search key v to "rediscover" the relevant child (c) and grandchild (gc) of the node y. (We know that we'll only be reorienting a 3-node if the search took us to its bottom node.) This leads to somewhat simpler code than the alternative of remembering

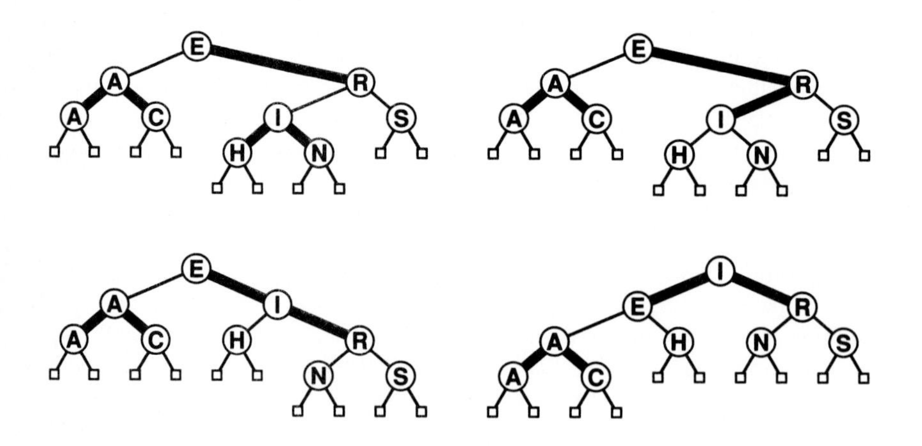

Figure 15.12 Splitting a Node in a Red-Black Tree.

during the search not only the two links corresponding to c and gc but also whether they are right or left links. We have the following function for reorienting a 3-node along the search path for v whose parent is y:

```
function rotate(v: integer; y: link): link;
  var c,gc: link;
  begin
  if v<y↑.key then c:=y↑.l else c:=y↑.r;
  if v<c↑.key
    then begin gc:=c↑.l; c↑.l:=gc↑.r; gc↑.r:=c end
    else  begin gc:=c↑.r; c↑.r:=gc↑.l; gc↑.l:=c end;
  if v<y↑.key then y↑.l:=gc else y↑.r:=gc;
  rotate:=gc
  end;
```

If y points to the root, c is the right link of y and gc is the left link of c, this makes exactly the link transformations needed to produce the tree in Figure 15.11 from Figure 15.8. The reader may wish to check the other cases. This function returns the link to the top of the 3-node, but does not do the color switch itself.

Thus, to handle the third case for *split* (see Figure 15.10), we can make g red, then set x to $rotate(v,gg)$, then make x black. This reorients the 3-node consisting of the two nodes pointed to by g and p and reduces this case to be the same as the second case, when the 3-node was oriented the right way.

Finally, to handle the case when the two red links are oriented in different directions (see Figure 15.10), we simply set p to $rotate(v,g)$. This reorients the "illegal" 3-node consisting of the two nodes pointed to by p and x. These nodes are the same color, so no color change is necessary, and we are immediately reduced to the third case. Combining this and the rotation for the third case is called a *double rotation* for obvious reasons.

Figure 15.12 shows the *split* occuring in our example when G is added. First, there is a color flip to split up the 4-node containing H, I, and N. Next, a double rotation is needed: the first part around the edge between I and R, and the second part around the edge between E and I. After these modifications, G can be inserted on the left of H, as shown in the first tree in Figure 15.13.

This completes the description of the operations to be performed by *split*. It must switch the colors of x and its children, do the bottom part of a double rotation if necessary and then do the single rotation if necessary, as follows:

```
function split(v: integer; gg,g,p,x: link): link;
  begin
  x↑.red:=true; x↑.l↑.red:=false; x↑.r↑.red:=false;
  if p↑.red then
    begin
    g↑.red:=true;
    if (v<g↑.key)<>(v<p↑.key) then p:=rotate(v,g);
    x:=rotate(v,gg);
    x↑.red:=false
    end;
  head↑.r↑.red:=false;
  split:=x
  end;
```

This procedure fixes the colors after a rotation and also restarts *x* high enough in the tree to ensure that the search doesn't get lost due to all the link changes. The long argument list is included for clarity; this procedure should more properly be declared local to *rbtreeinsert*, with access to its variables.

If the root is a 4-node then the *split* procedure makes the root red: this corresponds to transforming it, along with the dummy node above it, into a 3-node. Of course, there is no reason to do this, so a statement is included at the end of *split* to keep the root black. At the beginning of the process, it is necessary to initialize the dummy nodes carefully, as in the following code:

```
type link=↑node;
     node= record key,info: integer; l,r: link; red: boolean end;
var head,z: link;
procedure rbtreeinitialize;
  begin
  new(z); z↑.l:=z; z↑.r:=z; z↑.red:=false;
  new(head); head↑.key:=0; head↑.r:=z;
  end;
```

Assembling the code fragments above gives a very efficient, relatively simple algorithm for insertion using a binary tree structure that is guaranteed to take a logarithmic number of steps for all searches and insertions. This is one of the few searching algorithms with that property, and its use is justified whenever bad worst-case performance simply cannot be tolerated.

Figure 15.13 shows how this algorithm constructs the red-black tree for our sample set of keys. Here, at a cost of only a few rotations, we get a tree that has far better balance than the one for the same keys built in Chapter 14.

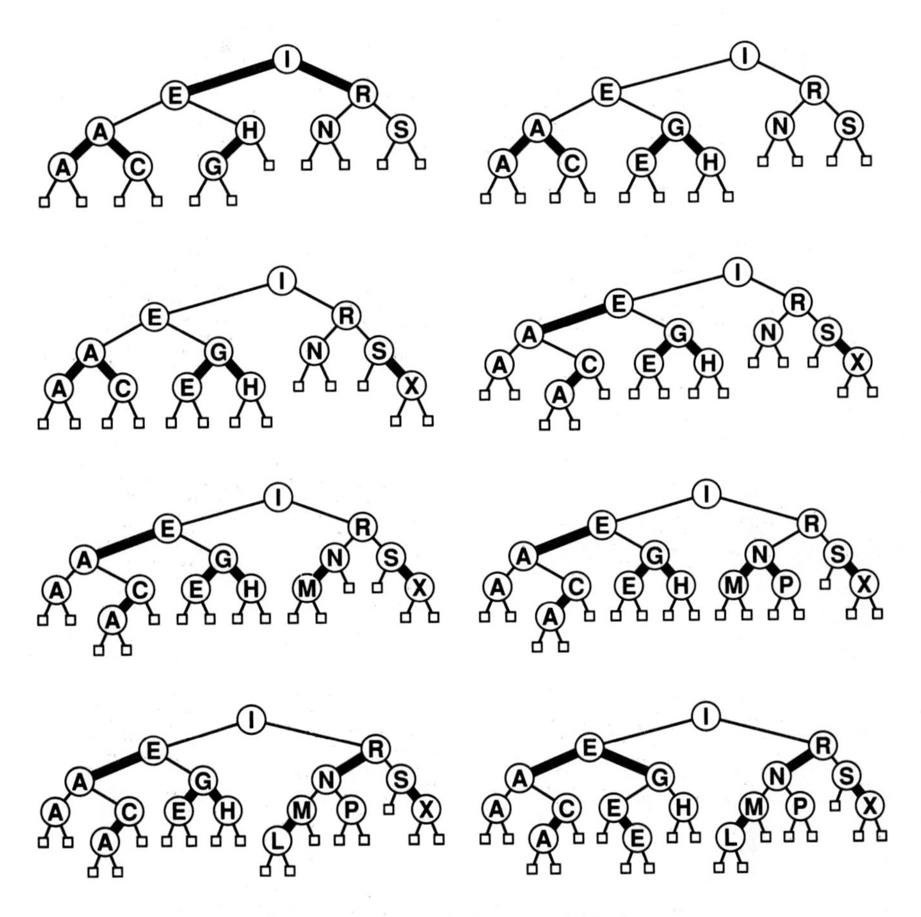

Figure 15.13 Building a red-black tree.

Property 15.3 *A search in a red-black tree with N nodes built from random keys seems to require about* $\lg N$ *comparisons, and an insertion seems to require less than one rotation, on the average.*

A precise average-case analysis of this algorithm is yet to be done, but there are convincing results from partial analyses and simulations. Figure 15.14 shows a tree built from the larger example we've using: the average number of nodes visited during a search for a random key in this tree is just 5.81, as compared to 7.00 for the tree built from same the keys in Chapter 14, and 5.74, the best possible for a perfectly balanced tree. ■

But the real significance of red-black trees is their worst-case performance, and the fact that this performance is achieved at very little cost. Figure 15.15 shows

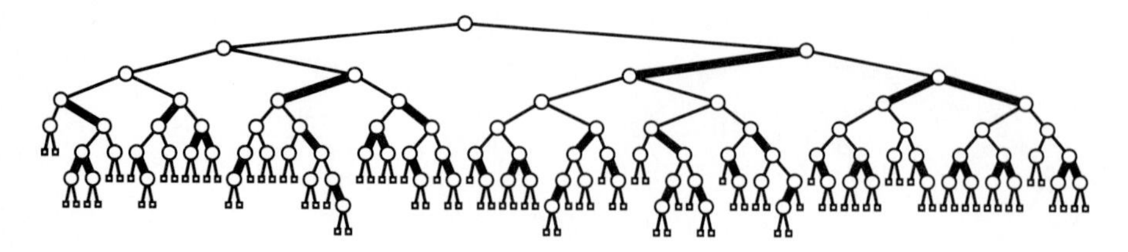

Figure 15.14 A large red-black tree.

the tree built if the numbers 1 to 95 are inserted in order into an intially empty tree; even this tree is very well-balanced. The search cost per node is just as low as if the balanced tree were constructed by the elementary algorithm, and insertion involves only one extra bit test and an occasional *split*.

Property 15.4 *A search in a red-black tree with N nodes requires fewer than* $2 \lg N + 2$ *comparisons, and an insertion requires fewer than one-quarter as many rotations as comparisons.*

Only "splits" which correspond to a 3-node connected to a 4-node in a 2-3-4 tree require a rotation in the corresponding red-black tree, so this property follows from Property 15.2. The worst case arises when the path to the insertion point consists of alternating 3- and 4-nodes. ■

To summarize: by using this method, a key in a file of, say, half a million records can be found by comparing it against only about twenty other keys. In a bad case, maybe twice as many comparisons might be needed, but no more. Furthermore, very little overhead is associated with each comparison, so a very quick search is assured.

Other Algorithms

The "top-down 2-3-4 tree" implementation using the red-black framework given in the previous section is one of several similar strategies that have been proposed for

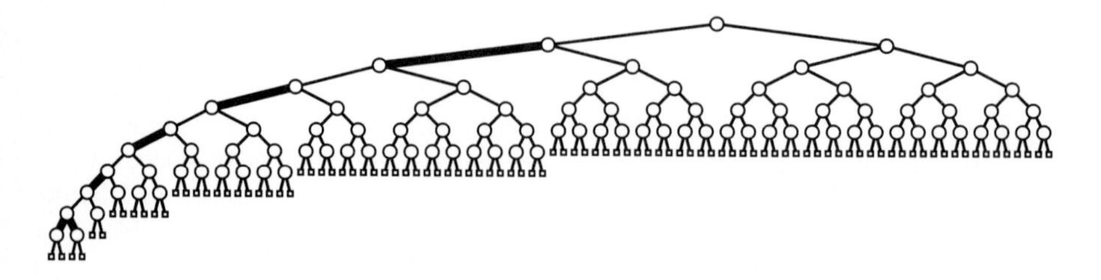

Figure 15.15 A red-black tree for a degenerate case.

implementing balanced binary trees. As we saw above, it is actually the "rotate" operations that balance the trees: we've been looking at a particular view of the trees that makes it easy to decide when to rotate. Other views of the trees lead to other algorithms, a few of which we'll mention briefly here.

The oldest and most well-known data structure for balanced trees is the *AVL tree*. These trees have the property that the heights of the two subtrees of each node differ by at most one. If this condition is violated because of an insertion, it turns out that it can be reinstated using rotations. But this requires an extra loop: the basic algorithm is to search for the value being inserted, then proceed *up* the tree along the path just traveled, adjusting the heights of nodes using rotations. Also, it is necessary to know whether each node has a height that is one less than, the same, or one greater than the height of its sibling. This requires two bits if encoded in a straightforward way, though there is a way to get by with just one bit per node, using the red-black framework.

A second well-known balanced tree structure is the *2-3 tree*, where only 2-nodes and 3-nodes are allowed. It is possible to implement *insert* using an "extra loop" involving rotations as with AVL trees, but there is not quite enough flexibility to give a convenient top-down version. Again, the red-black framework can simplify the implementation, but it is actually better to use *bottom-up 2-3-4 trees*, where we search to the bottom of the tree and insert there, then (if the bottom node was a 4-node) move back up the search path, splitting 4-nodes and inserting the middle node into the parent, until encountering a 2-node or 3-node as a parent, at which point a rotation might be involved to handle cases as in Figure 15.10. This method has the advantage of using at most one rotation per insertion, which can be an advantage in some applications. The implementation is slightly more complicated than for the top-down method given above.

In Chapter 18, we'll study the most important type of balanced tree, an extension of 2-3-4 trees called *B-trees*. These allow up to M keys per node for large M and are widely used for searching applications involving very large files.

□

Exercises

1. Draw the top-down 2-3-4 tree built when the keys E A S Y Q U E S T I O N are inserted (in that order) into an initially empty tree.
2. Draw a red-black representation of the tree from the previous question.
3. Exactly what links are modified by *split* and *rotate* when Z is inserted (after Y) into the example tree for this chapter?
4. Draw the red-black tree that results when the letters A through K are inserted in order, and describe what happens in general when keys are inserted into the trees in ascending order.
5. How many tree links actually must be changed for a double rotation, and how many are actually changed in the implementation given?
6. Generate two random 32-node red-black trees, draw them (either by hand or with a program), and compare them with the unbalanced binary search trees built with the same keys.
7. Generate ten random 1000-node red-black trees. Compute the number of rotations required to build the trees and the average distance in them from the root to an external node. Discuss the results.
8. With one bit per node for "color," we can represent 2-, 3-, and 4-nodes. How many different types of nodes could we represent if we used two bits per node for color?
9. Rotations are required in red-black trees when 3-nodes are made into 4-nodes in an "unbalanced" way. Why not eliminate rotations by allowing 4-nodes to be represented as any three nodes connected by two red links (perfectly balanced or not)?
10. Give a sequence of insertions that will construct the red-black tree shown in Figure 15.11.

16

Hashing

☐ A completely different approach to searching from the comparison-based tree structures of the previous chapter is provided by *hashing*: a method for directly referencing records in a table by doing arithmetic transformations on keys into table addresses. If we know that the keys are distinct integers from 1 to N, then we can store the record with key i in table position i, ready for immediate access with the key value. Hashing is a generalization of this trivial method for typical searching applications when we don't have such specialized knowledge about the key values.

The first step in a search using hashing is to compute a *hash function* which transforms the search key into a table address. Ideally, different keys should map to different addresses, but no hash function is perfect, and two or more different keys may hash to the same table address. The second part of a hashing search is thus a *collision-resolution* process which deals with such keys. One of the collision-resolution methods that we'll study uses linked lists, and is appropriate in highly dynamic situations where the number of search keys cannot be predicted in advance. The other two collision-resolution methods that we'll examine achieve fast search times on records stored within a fixed array.

Hashing is a good example of a *time-space tradeoff*. If there were no memory limitation, then we could do any search with only one memory access by simply using the key as a memory address. If there were no time limitation, then we could get by with only a minimum amount of memory by using a sequential search method. Hashing provides a way to use a reasonable amount of both memory and time to strike a balance between these two extremes. Efficient use of available memory and fast access to the memory are prime concerns of any hashing method.

Hashing is a "classical" computer science problem in the sense that the various algorithms have been studied in some depth and are very widely used. There is a great deal of empirical and analytic evidence to support the utility of hashing for a broad variety of applications.

Hash Functions

The first problem we must address is the computation of the hash function which transforms keys into table addresses. This is an arithmetic computation with properties similar to the random number generators that we will study in Chapter 33. What is needed is a function which transforms keys (usually integers or short character strings) into integers in the range $[0..M-1]$, where M is number of records that can fit into the amount of memory available. An ideal hash function is one which is easy to compute and approximates a "random" function: for each input, every output should be in some sense equally likely.

Since the methods that we will use are arithmetic, the first step is to transform keys into *numbers* upon which we can perform arithmetic operations. For small keys, this might involve no work at all in some programming environments, if we're allowed to use binary representations of keys as numbers (see the discussion at the beginning of Chapter 10). For longer keys, one might contemplate removing bits from character strings and packing them together in a machine word; however we'll see below a uniform way to handle keys of any length.

First, suppose that we do have a large integer which directly corresponds to our key. Perhaps the most commonly used method for hashing is to choose M to be prime and, for any key k, compute $h(k) = k \bmod M$. This is a straightforward method which is easy to compute in many environments and spreads the key values out well.

For example, suppose that our table size is 101 and we have to compute an index for the four-character key A K E Y: If the key is encoded the simple five-bit code used in Chapter 10 (where the ith letter in the alphabet is represented by the binary representation of the number i), then we may view it as the binary number

$$00001010110010111001,$$

which is equivalent to 44217 in decimal. Now, $44217 \equiv 80 \pmod{101}$, so the key A K E Y "hashes to" position 80 in the table. There are many possible keys and relatively few table positions, so many other keys hash to the same position (for example, the key B A R H also has hash address 80 in the code used above).

Why does the hash table size M have to be prime? The answer to this question depends on arithmetic properties of the mod function. In essence, we are treating the key as a base-32 number, one digit for each character in the key. We saw that our sample A K E Y corresponds to the number 44217, which also can be written as

$$1 \cdot 32^3 + 11 \cdot 32^2 + 5 \cdot 32^1 + 25 \cdot 32^0$$

since A is the first letter in the alphabet, K the eleventh letter, etc. Now, suppose that we were to make the unfortunate choice $M = 32$: because the value of $k \bmod 32$ is unaffected by adding multiples of 32, the hash function of any key is simply the value of its last character! Surely a good hash function should take all the characters of a key into account. The simplest way to ensure that it does so is to make M prime.

But the most typical situation is when the keys are not numbers and not necessary short, but rather alphanumeric strings (possibly quite long). How do we compute the hash function for something like V E R Y L O N G K E Y ? In our code, this corresponds to the 55-bit string

$$1011000101100101100101100011110111000111010110010111001,$$

or the number

$$22{\cdot}32^{10}+5{\cdot}32^{9}+18{\cdot}32^{8}+25{\cdot}32^{7}+12{\cdot}32^{6}+15{\cdot}32^{5}+14{\cdot}32^{4}+7{\cdot}32^{3}+11{\cdot}32^{2}+5{\cdot}32^{1}+25,$$

which is too large to be represented for normal arithmetic functions in most computers (and we should be able to handle much longer keys). In such a situation, it turns out that we can still compute a hash function like the one above, merely by transforming the key piece by piece. Again, we take advantage of arithmetic properties of the mod function and of a simple computing trick called *Horner's method* (see Chapter 36). This method is based on yet another way of writing the number corresponding to keys—for our example, we write the following expression:

$$(((((((((22\cdot32+5)32+18)32+25)32+12)32+15)32+14)32+7)32+11)32+5)32+25.$$

This leads to a direct arithmetic way to compute the hash function:

```
h:=key[1];
for j:=2 to keysize do
  begin
  h:=((h*32)+key[j]) mod M;
  end;
```

Here h is the computed hash value and *key*[i] is assumed to contain j if the ith character in the key is the jth letter in the alphabet. The Pascal *ord* character transfer function can be used to compute this (and a larger alphabet can be provided for, if desired, by using 64 or 128 instead of 32). Without the **mod**, this code would compute the number corresponding to the key as in the equation above, but the computation would overflow for long keys. With the **mod** present, however, it computes the hash function precisely because of the additive and multiplicative properties of the mod operation, and overflow is avoided because the **mod** always results in a value less than M. The hash address computed by this program for V E R Y L O N G K E Y with $M = 101$ is 97.

For clarity in describing the algorithms that follow, we assume that keys are integers and that the hash function is a simple **mod**. This is not meant to suggest that this type of situation is likely to be used in practice; indeed, most applications will involve data structures with longer keys which require using the hash function above. Appropriate modifications to the programs given below for the fundamental algorithms are straightforward.

Separate Chaining

The hash functions above convert keys into table addresses: we still need to decide how to handle the case when two keys hash to the same address. The most straightforward method is simply to build, for each table address, a linked list of the records whose keys hash to that address. Since the keys which hash to the same table position are kept in a linked list, they might as well be kept in order. This leads directly to a generalization of the elementary list-searching method that we discussed in Chapter 14. Rather than maintaining a single list with a single list header node *head* as discussed there, we maintain *M* lists with *M* list header nodes, initialized as follows:

```
type link=↑node;
        node=record key, info: integer; next: link end;
var heads: array[0..M] of link;
      t, z: link;
procedure initialize;
   var i: integer;
   begin
   new(z); z↑.next:=z;
   for i:=0 to M-1 do
      begin new(heads[i]); heads[i]↑.next:=z end;
   end;
```

Now the procedures from Chapter 14 can be used as is and a hash function can be used to choose among the lists. For example, *listinsert*(*v*, *heads*[*v* **mod** *M*]) can be used to add something to the table and *t*:=*listsearch*(*v*, *heads*[*v* **mod** *M*]) can be used to find the first record with key *v* and successively set *t*:=*listsearch*(*v*, *t*) until *t*=*z* to find subsequent records with key *v*.

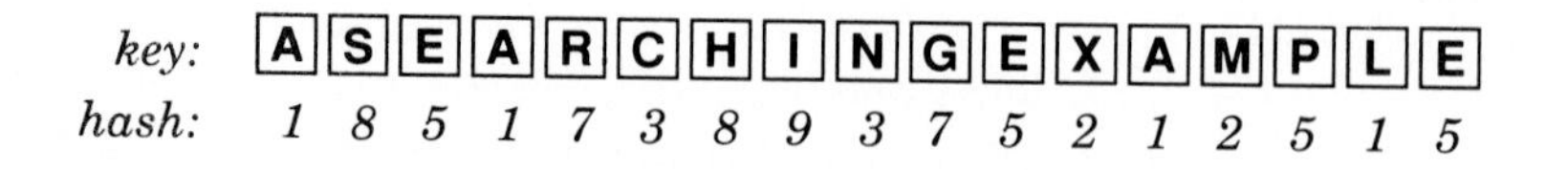

Figure 16.1 A hash function ($M = 11$).

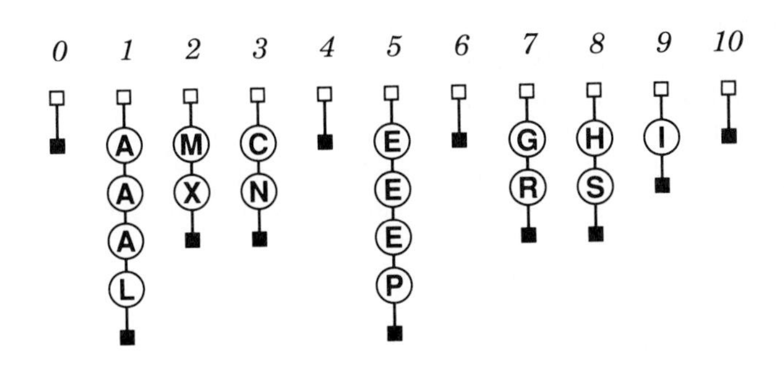

Figure 16.2 Separate chaining.

For example, if our sample keys are successively inserted into an initially empty table using the hash function in Figure 16.1, then the set of lists shown in Figure 16.2 would result. This method is traditionally called *separate chaining* because colliding records are "chained" together in separate linked lists.

Obviously, the amount of time required for a search depends on the length of the lists (and the relative positions of the keys in them). The lists could be left unordered: maintaining sorted lists is not as important for this application as it was for the elementary sequential search because the lists are quite short.

For an "unsuccessful search" (a search for a record with a key not in the table), we can assume that the hash function scrambles things enough that each of the M lists is equally likely to be searched and, as in sequential list searching, that the list searched is only traversed halfway (on the average). The average length of the list examined (not counting z) for unsuccessful search in this example is $(0+4+2+2+0+4+0+2+2+1+0)/11 \approx 1.55$. This would be the average time for an unsuccessful search were the lists unordered; by keeping them ordered we cut the time about in half. For a "successful search" (a search for one of the records in the table), we assume that each record is equally likely to be sought: seven of the keys would be found as the first list item examined, six would be found as the second item examined, etc., so the average is $(7 \cdot 1 + 6 \cdot 2 + 2 \cdot 3 + 2 \cdot 4)/17) \approx 1.94$. (This count assumes that equal keys are distinguished with a unique identifier or some other mechanism, and that the search routine is modified appropriately to be able to search for each individual key.)

Property 16.1 *Separate chaining reduces the number of comparisons for sequential search by a factor of M (on the average), using extra space for M links.*

If N, the number of keys in the table, is much larger than M then a good approximation to the average length of the lists is N/M, since each of the M hash values is "equally likely" by design of the hash function. As in Chapter 14, unsuccessful and successful searches are expected to go about halfway down some list. ∎

The implementation given above uses a hash table of links to headers of the lists containing the actual keys. One alternative to maintaining M list-header nodes is to eliminate them and make *heads* be a table of links to the first keys in the lists. This leads to some complications in the algorithm. For example, adding a new record to the beginning of a list becomes a different operation from adding a new record anywhere else in a list, because it involves modifying an entry in the table of links, not a field of a record. Yet another implementation is to put the first key within the table. Though the alternatives use less space in some situations, M is usually small enough in comparison to N that the extra convenience of using list-header nodes is probably justified.

In a separate chaining implementation, M is typically chosen relatively small so as not to use up a large area of contiguous memory. But it's probably best to choose M sufficiently large that the lists are short enough to make sequential search the most efficient method for them: "hybrid" methods (such as using binary trees instead of linked lists) are probably not worth the trouble. As a rule of thumb, one might choose M to be about one-tenth the number of keys expected be be in the table, so that the lists are expected to contain about ten keys each. One of the virtues of separate chaining is that this decision is not critical: if more keys arrive than expected, then searches will take a little longer; if fewer keys are in the table, then perhaps a little extra space was used. If memory really is a critical resource, choosing M as large as one can afford still yields a factor of M improvement in performance.

Linear Probing

If the number of elements to be put in the hash table can be estimated in advance and enough contiguous memory is available to hold all the keys with some room to spare, then it is probably not worthwhile to use any links at all in the hash table. Several methods have been devised which store N records in a table of size $M > N$, relying on empty places in the table to help with collision resolution. Such methods are called *open-addressing* hashing methods.

The simplest open-addressing method is called *linear probing*: when there is a collision (when we hash to a place in the table which is already occupied and whose key is not the same as the search key), then just *probe* the next position in the table, that is, compare the key in the record there against the search key. There are three possible outcomes of the probe: if the keys match, then the search terminates successfully; if there's no record there, then the search terminates unsuccessfully; otherwise probe the next position, continuing until either the search key or an empty table position is found. If a record containing the search key is to be inserted following an unsuccessful search, then it can simply be put into the empty table space which terminated the search. This method is easily implemented as follows:

```
procedure hashinitialize;
  var i: integer;
  begin
  for i:=0 to M do a[i].key:=maxint;
  end;
function hashinsert(v: integer): integer;
  var x: integer;
  begin
  x:=h(v);
  while a[x].key<>maxint do x:=(x+1) mod M;
  a[x].key:=v;
  hashinsert:=x;
  end;
```

Linear probing requires a special key value to signal an empty spot in the table; this program uses *maxint* for that purpose. The computation x:=(x+1) **mod** M corresponds to examining the next position (wrapping back to the beginning when the end of the table is reached). Note that this program does not check whether the table is filled to capacity. (What would happen in this case?)

For our example set of keys with $M = 19$, we might get the hash values shown in Figure 16.3. If these keys are inserted in the order given into an initially empty table, we get the sequence shown in Figure 16.4.

The implementation of *hashsearch* is similar to *hashinsert*: simply add the condition "*a*[*x*].*key*<> *v*" to the **while** loop and delete the following instruction which stores v. This leaves to the calling routine the task of checking if the search was unsuccessful, by testing whether the table position returned actually contains v (successful) or *maxint* (unsuccessful). Other conventions could be used, for example *hashsearch* could return M for unsuccessful search. For reasons that will become obvious below, open addressing is not appropriate if large numbers of records with duplicate keys are to be processed, but *hashsearch* can easily be adapted to handle equal keys in the case where each record has a unique identifier.

The table size for linear probing is greater than for separate chaining, since we must have $M > N$, but the total amount of memory space used is less, since no links are used. The average number of items that must be examined for a

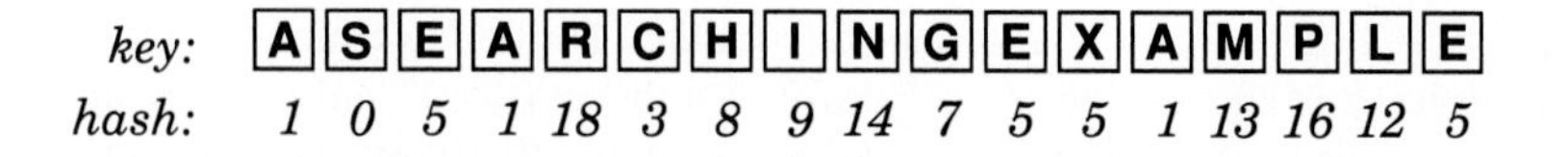

Figure 16.3 A hash function ($M = 19$).

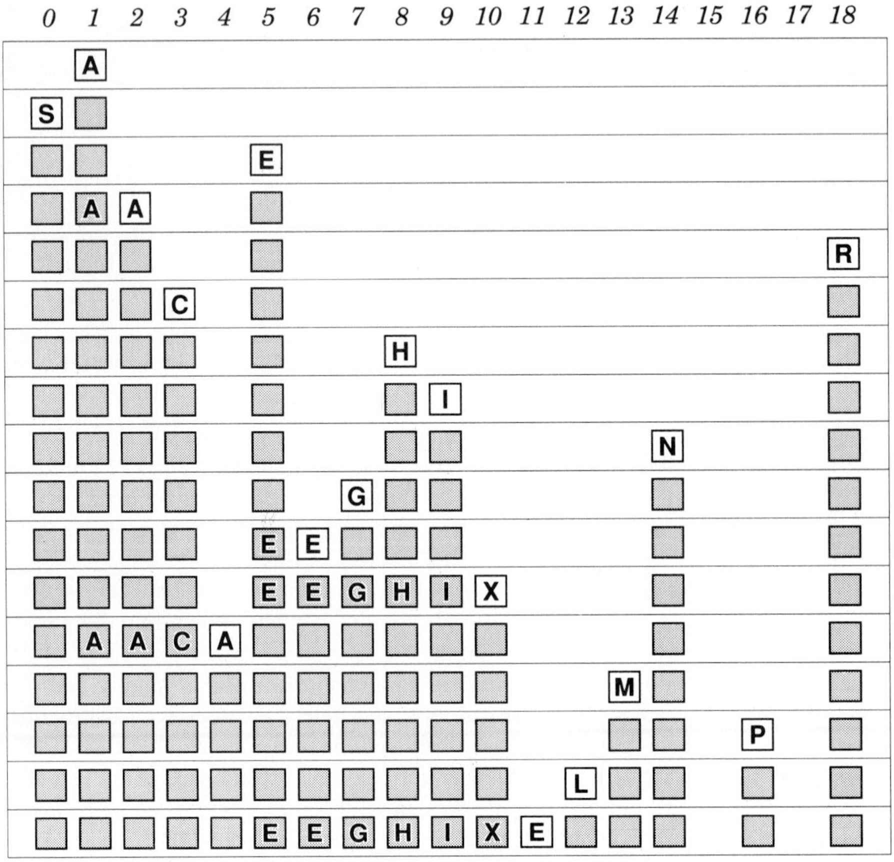

Figure 16.4 Linear probing.

successful search for this example is $33/17 \approx 1.94$.

Property 16.2 *Linear probing uses less than five probes, on the average, for a hash table which is less than two-thirds full.*

The exact formula for the average number of probes required, in terms of the "load factor" of the hash table $\alpha = N/M$, is $1/2 + 1/2(1-\alpha)^2$ for an unsuccessful search and $1/2 + 1/2(1-\alpha)$ for a successful search. Thus, if we take $\alpha = 2/3$, we get five probes for an average unsuccessful search, as stated, and two for an average successful search. Unsuccessful search is always the more expensive of the two: a successful search will require less than five probes until the table is about 90% full. As the table fills up (as α approaches 1) these numbers get very large; this should not be allowed to happen in practice, as we discuss further below. ■

Figure 16.5 Linear probing in a larger table.

Double Hashing

Linear probing (and indeed any hashing method) works because it guarantees that, when searching for a particular key, we look at every key that hashes to the same table address (in particular, the key itself if it's in the table). Unfortunately, in linear probing, other keys are also examined, especially when the table begins to fill up: in the example above, the search for X involved looking at G, H, and I, none of which had the same hash value. What's worse, insertion of a key with one hash value can drastically increase the search times for keys with other hash values: in the example, an insertion at position 17 would cause greatly increased search times for position 16. This phenomenon, called *clustering*, can make linear probing run very slowly for nearly full tables. Figure 16.5 shows clusters forming in a larger example.

Fortunately, there is an easy way to virtually eliminate the clustering problem: *double hashing*. The basic strategy is the same; the only difference is that, instead of examining each successive entry following a collided position, we use a second hash function to get a fixed increment to use for the "probe" sequence. This is easily implemented by inserting u:=$h2(v)$ at the beginning of the procedure and changing x:=$(x+1)$ **mod** M to x:=$(x+u)$ **mod** M within the **while** loop.

The second hash function $h2$ must be chosen with some care, since otherwise the program may not work at all. First, we obviously don't want to have $u = 0$, since that would lead to an infinite loop on collision. Second, it is important that M and u be relatively prime here, since otherwise some of the probe sequences could be very short (consider the case $M = 2u$). This is easily enforced by making M prime and $u < M$. Third, the second hash function should be "different from" the first, since otherwise a slightly more complicated clustering can occur. A function

key:	A	S	E	A	R	C	H	I	N	G	E	X	A	M	P	L	E
hash 1:	1	0	5	1	18	3	8	9	14	7	5	5	1	13	16	12	5
hash 2:	7	3	3	7	6	5	8	7	2	1	3	8	7	3	8	4	3

Figure 16.6 Double hash function ($M = 19$).

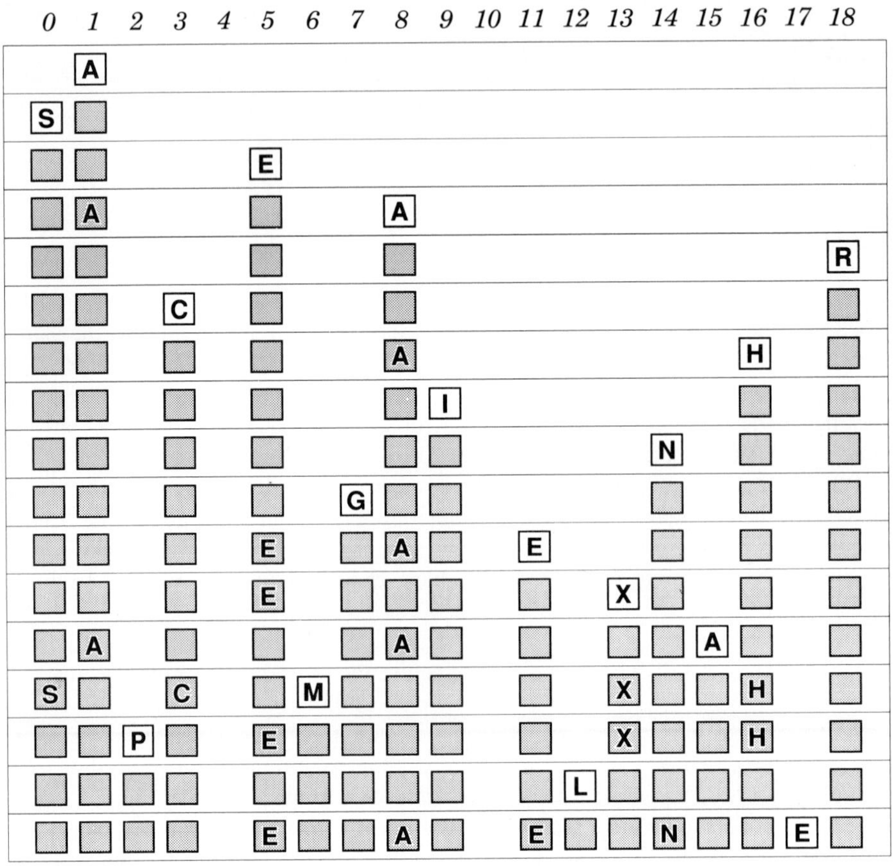

Figure 16.7 Double hashing.

such as $h_2(k) = M - 2 - k \bmod (M - 2)$ will produce a good range of "second" hash values, but this is perhaps going too far, since, especially for long keys, the cost of computing the second hash function essentially doubles the cost of the search, only to save a few probes by eliminating clustering. In practice, a much simpler second hash function will suffice, such as $h_2(k) = 8 - (k \bmod 8)$. This function uses only the last three bits of k; it might be appropriate to use a few more for a large table, though the effect, even if noticable, is not likely to be significant in practice.

For our sample keys, these functions produce the hash values shown in Figure 16.6. Figure 16.7 shows the table produced by successively inserting our sample keys into an initially empty table using double hashing with these values.

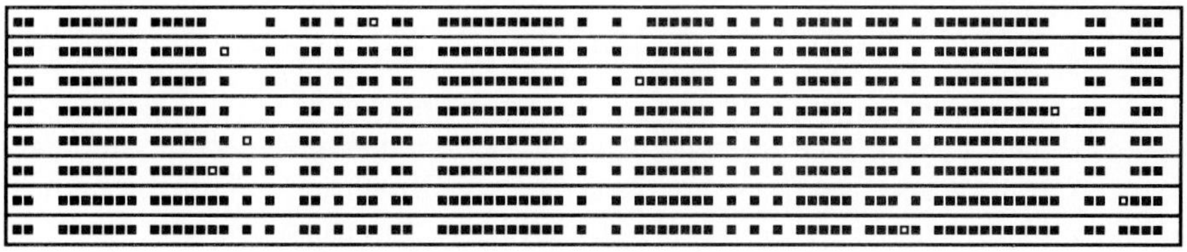

Figure 16.8 Double hashing in a larger table.

The average number of items examined for successful search is slightly larger than with linear probing for this example: $35/17 \approx 2.05$. But in a sparser table, there is far less clustering, as shown in Figure 16.8. For this example, there are twice as many clusters as for linear probing (Figure 16.5), or, equivalently, the average cluster is about half as long.

Property 16.3 *Double hashing uses fewer probes, on the average, than linear probing.*

The actual formula for the average number of probes made for double hashing with an "independent" double-hash function is $1/(1-\alpha)$ for unsuccessful search and $-\ln(1-\alpha)/\alpha$ for successful search. (These formulas are the result of deep mathematical anaylsis, and haven't even been verified for large α.) The simpler easy-to-compute second hash function recommended above won't behave quite this well, but it will be rather close, especially if enough bits are used to make the range of values possible close to M. Practically, this means that a smaller table can be used to get the same search times with double hashing as with linear probing for a given application: the average number of probes is less than five for an unsuccessful search if the table is less than 80% full, and for a successful search if the table is less than 99% full. ■

Open addressing methods can be inconvenient in a dynamic situation when an unpredictable number of insertions and deletions may have to be processed. First, how big should the table be? Some estimate must be made of how many insertions are expected but performance degrades drastically as the table starts to get full. A common solution to this problem is to rehash everything into a larger table on a (very) infrequent basis. Second, a word of caution is necessary about deletion: a record can't simply be removed from a table built with linear probing or double hashing. The reason is that later insertions into the table might have skipped over that record, and searches for those records will terminate at the hole left by the deleted record. A way to solve this problem is to have another special key which can serve as a placeholder for searches but can be identified and remembered as an empty position for insertions. Note that neither table size nor deletion are a particular problem with separate chaining.

Perspective

The methods discussed above have been analyzed completely and it is possible to compare their performance in some detail. The formulas given above are summarized from detailed analyses described by D. E. Knuth in his book on sorting and searching. The formulas indicate how badly performance degrades for open addressing as α gets close to 1. For large M and N, with a table about 90% full, linear probing will take about 50 probes for an unsuccessful search, compared to 10 for double hashing. But in practice, one should never let a hash table get to be 90% full! For small load factors, only a few probes are required; if small load factors can't be arranged, hashing shouldn't be used.

Comparing linear probing and double hashing against separate chaining is more complicated, because more memory is available in the open-addressing methods (since there are no links). The value of α used should be modified to take this into account, based on the relative size of keys and links. This means that it is not normally justifiable to choose separate chaining over double hashing on the basis of performance.

The choice of the very best hashing method for a particular application can be very difficult. However, the very best method is rarely needed for a given situation, and the various methods do have similar performance characteristics as long as the memory resource is not being severely strained. Generally, the best course of action is to use the simple separate chaining method to reduce search times drastically when the number of records to be processed is not known in advance (and a good storage allocator is available) and to use double hashing to search a set of keys whose size can be roughly predicted ahead of time.

Many other hashing methods have been developed which have application in some special situations. Although we can't go into details, we'll briefly consider two examples to illustrate the nature of specially adapted hashing methods. These and many other methods are fully described in the books by Knuth and Gonnet.

The first, called *ordered hashing*, exploits ordering within an open addressing table. In standard linear probing, we stop the search when we find an empty table position or a record with a key equal to the search key; in ordered hashing, we stop the search when we find a record with a key greater than or equal to the search key (the table must be cleverly constructed to make this work). This method turns out to reduce the time for unsuccessful search to approximately that for successful search. (This is the same kind of improvement that comes in separate chaining.) This method is useful for applications where unsuccessful searching is frequently used. For example, a text-processing system might have an algorithm for hyphenating words that works well for most words but not for bizarre cases (such as "bizarre"). The situation could be handled by looking up all words in a relatively small *exception dictionary* of words which must be handled in a special way, with most searches likely to be unsuccessful.

Similarly, there are methods for moving some records around during unsuccessful search to make successful searching more efficient. In fact, R. P. Brent developed a method for which the average time for a successful search can be bounded by a constant, giving a very useful method for applications involving frequent successful searching in very large tables such as dictionaries.

These are only two examples of a large number of algorithmic improvements which have been suggested for hashing. Many of these improvements are interesting and have important applications. However, our usual cautions must be raised against premature use of advanced methods except by experts with serious searching applications, because separate chaining and double hashing are simple, efficient, and quite acceptable for most applications.

Hashing is preferred to the binary tree structures of the previous two chapters for many applications because it is somewhat simpler and can provide very fast (constant) searching times, if space is available for a large enough table. Binary tree structures have the advantages that they are dynamic (no advance information on the number of insertions is needed), they can provide guaranteed worst-case performance (everything could hash to the same place even in the best hashing method), and they support a wider range of operations (most important, the *sort* function). When these factors are not important, hashing is certainly the searching method of choice.

□

Exercises

1. Describe how you might implement a hash function by making use of a good random number generator. Would it make sense to implement a random number generator by making use of a hash function?

2. How long could it take in the worst case to insert N keys into an initially empty table, using separate chaining with unordered lists? Answer the same question for sorted lists.

3. Give the contents of the hash table that results when the keys E A S Y Q U E S T I O N are inserted in that order into an initially empty table of size 13 using linear probing. (Use $h_1(k) = k \bmod 13$ for the hash function for the kth letter of the alphabet.)

4. Give the contents of the hash table that results when the keys E A S Y Q U E S T I O N are inserted in that order into an initially empty table of size 13 using double hashing. (Use $h_1(k)$ from the previous question, $h_2(k) = 1+(k \bmod 11)$ for the second hash function.)

5. About how many probes are involved when double hashing is used to build a table consisting of N equal keys?

6. Which hashing method would you use for an application in which many equal keys are likely to be present?

7. Suppose that the number of items to be put into a hash table is known in advance. Under what conditions will separate chaining be preferable to double hashing?

8. Suppose a programmer has a bug in his double-hashing code so that one of the hash functions always returns the same value (not 0). Describe what happens in each situation (when the first one is wrong and when the second one is wrong).

9. What hash function should be used if it is known in advance that the key values fall into a relatively small range?

10. Criticize the following algorithm for deletion from a hash table built with linear probing. Scan right from the element to be deleted (wrapping as necessary) to find an empty position, then scan left to find an element with the same hash value. Then replace the element to be deleted with that element, leaving its table position empty.

17

Radix Searching

☐ Several searching methods proceed by examining the search keys one bit at a time, rather than using full comparisons between keys at each step. These methods, called *radix-searching methods*, work with the bits of the keys themselves, as opposed to the transformed version of the keys used in hashing. As with radix-sorting methods (see Chapter 10), these methods can be useful when the bits of the search keys are easily accessible and the values of the search keys are well distributed.

The principal advantages of radix-searching methods are that they provide reasonable worst-case performance without the complication of balanced trees; they provide an easy way to handle variable-length keys; some allow some savings in space by storing part of the key within the search structure; and they can provide very fast access to data, competitive with both binary search trees and hashing. The disadvantages are that biased data can lead to degenerate trees with bad performance (and data comprised of characters is biased) and that some of the methods can make very inefficient use of space. Also, as with radix sorting, these methods are designed to take advantage of particular characteristics of a computer's architecture: since they use digital properties of the keys, it's difficult or impossible to do efficient implementations in languages such as Pascal.

We'll examine a series of methods, each one correcting a problem inherent in the previous one, culminating in an important method which is quite useful for searching applications where very long keys are involved. In addition, we'll see the analogue to the "linear-time sort" of Chapter 10, a "constant-time" search based on the same principle.

Digital Search Trees

The simplest radix-search method is digital tree searching: the algorithm is precisely the same as that for binary tree searching, except that we branch in the tree not according to the result of the comparison between the keys, but according to the key's bits. At the first level the leading bit is used, at the second level the second leading bit, and so on until an external node is encountered. The code for this is virtually the same as the code for binary tree search. The only difference is that the key comparisons are replaced by calls on the *bits* function that we used in radix sorting. (Recall from Chapter 10 that $bits(x,k,j)$ is the j bits which appear k from the right; it can be efficiently implemented in machine language by shifting right k bits, then setting to 0 all but the rightmost j bits.)

```
function digitalsearch(v: integer; x: link): link;
  var b: integer;
  begin
  z↑.key:=v; b:=maxb;
  repeat
    if bits(v,b,1)=0 then x:=x↑.l else x:=x↑.r;
    b:=b-1;
  until v=x↑.key;
  digitalsearch:=x
  end;
```

The data structures for this program are the same as those that we used for elementary binary search trees. The constant *maxb* is the number of bits in the keys to be sorted. The program assumes that the first bit in each key (the (*maxb*+1)st from the right) is 0 (perhaps the key is the result of a call to *bits* with a third argument of *maxb*), so that searching is done by setting x:=*digitalsearch*(v, *head*), where *head* is a link to a tree-header node with 0 key and a left link pointing to the search tree. Thus the initialization procedure for this program is the same as for binary tree search, except that we begin with *head*↑.*l*:=*z* instead of *head*↑.*r*:=*z*.

A	0 0 0 0 1
S	1 0 0 1 1
E	0 0 1 0 1
R	1 0 0 1 0
C	0 0 0 1 1
H	0 1 0 0 0
I	0 1 0 0 1
N	0 1 1 1 0
G	0 0 1 1 1
X	1 1 0 0 0
M	0 1 1 0 1
P	1 0 0 0 0
L	0 1 1 0 0

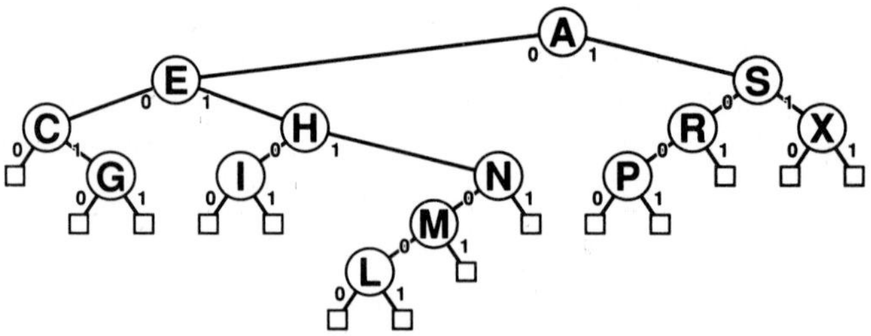

Figure 17.1 A digital search tree.

We saw in Chapter 10 that equal keys are anathema in radix sorting; the same is true in radix searching, not in this particular algorithm, but in the ones that we'll be examining later. Thus we'll assume in this chapter that all the keys to appear in the data structure are distinct: if necessary, a linked list could be maintained for each key value of the records whose keys have that value. As in previous chapters, we'll assume that the ith letter of the alphabet is represented by the five-bit binary representation of i. The sample keys to be used in this chapter are given in Figure 17.1. To be consistent with *bits*, we consider the bits as numbered 0–4, from right to left. Thus bit 0 is A's only nonzero bit and bit 4 is P's only nonzero bit.

The insert procedure for digital search trees also derives directly from the corresponding procedure for binary search trees:

```
function digitalinsert(v: integer; x: link): link;
  var p: link; b: integer;
  begin
  b:=maxb;
  repeat
    p:=x;
    if bits(v,b,1)=0 then x:=x↑.l else x:=x↑.r;
    b:=b-1;
  until x=z;
  new(x); x↑.key:=v; x↑.l:=z; x↑.r:=z;
  if bits(v,b+1,1)=0 then p↑.l:=x else p↑.r:=x;
  digitalinsert:=x
  end;
```

The tree built by this program when our sample keys are inserted into an initially empty tree is shown in Figure 17.1. Figure 17.2 shows what happens when a new key Z=11010 is added to the tree in Figure 17.1. We go right twice because the leading two bits of Z are 1 and then we go left, where we hit the external node at the left of X, where Z is inserted.

The worst case for trees built with digital searching is much better than for binary search trees, if the number of keys is large and the keys are not long. The length of the longest path in a digital search tree is the length of the longest match in the leading bits between any two keys in the tree, and this is likely to be relatively small for many applications (for example, if the keys are comprised of random bits).

Property 17.1 *A search or insertion in a digital search tree requires about* $\lg N$ *comparisons on the average and* b *comparisons in the worst case in a tree built from* N *random* b*-bit keys.*

It is obvious that no path will ever be any longer than the number of bits in the

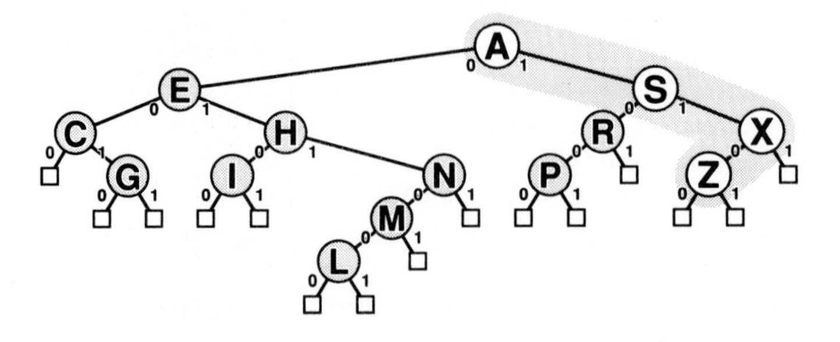

Figure 17.2 Insertion (of Z) into a digital search tree.

keys: for example, a digital search tree built from eight-character keys with, say, six bits per character will have no path longer than 48, even if there are hundreds of thousands of keys. The result that digital search trees are nearly perfectly balanced on the average requires analysis beyond the scope of this book, though it validates the simple intuitive notion that the "next" bit of a random keys should be equally likely to begin with a 0 bit as a 1 bit, so half should fall on either side of any node. Figure 17.3 shows a digital search tree made from 95 random 7-bit keys—this tree is quite well-balanced. ■

Thus, digital search trees provide an attractive alternative to standard binary search trees, *provided* that bit extraction is as easy to do as key comparison (which is not really the case in Pascal).

Radix Search Tries

It is quite often the case that search keys are very long, perhaps consisting of twenty characters or more. In such a situation, the cost of comparing a search key for equality with a key from the data structure can be a dominant cost which cannot be neglected. Digital tree searching uses such a comparison at each tree node; in this section we'll see that it is possible in most cases to get by with only one comparison per search.

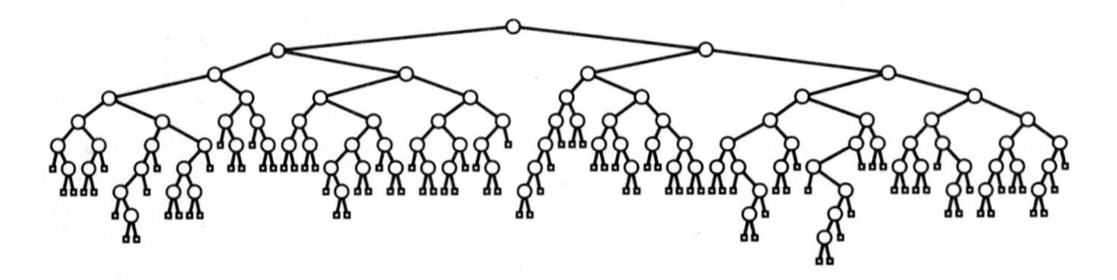

Figure 17.3 A large digital search tree.

The idea is to not store keys in tree nodes at all, but rather to put all the keys in external nodes of the tree. That is, instead of using z for external nodes of the structure, we put nodes which contain the search keys. Thus, we have two types of nodes: internal nodes, which just contain links to other nodes, and external nodes, which contain keys and no links. (Fredkin named this method "trie" because it is useful for re*trie*val; in conversation this word is usually pronounced "try-ee" or just "try" for obvious reasons.) To search for a key in such a structure, we just branch according to its bits, as above, but we don't compare it to anything until we get to an external node. Each key in the tree is stored in an external node on the path described by the leading bit pattern of the key and each search key winds up at one external node, so one full key comparison completes the search.

Figure 17.4 shows the (binary) radix search trie for the keys A S E R C. For example, to reach E, we go left, left, right from the root, since the first three bits of E are 001; but none of the keys in the trie begin with the bits 101, because an external node is encountered if one goes right, left, right. Before thinking about insertion, the reader should ponder the rather surprising property that the trie structure is independent of the order in which the keys are inserted: there is a unique trie for any given set of distinct keys.

As usual, after an unsuccessful search, we can insert the key sought by replacing the external node which terminated the search, *provided* it doesn't contain a key. This is the case when H is inserted into the trie of Figure 17.4, as shown in the first trie of Figure 17.5. If the external node which terminates the search does contain a key, then it must be replaced by an internal node which will have the key sought and the key which terminated the search in external nodes below it. Unfortunately, if these keys agree in more bit positions, it is necessary to add some external nodes which correspond to no keys in the tree (or put another way, some internal nodes with an empty external node as a child). This happens when I is inserted, as shown in the second trie of Figure 17.5. The rest of Figure 17.5 shows the completion of our example as the keys N G X M P L are added.

Implementing this method in Pascal is actually relatively complicated because

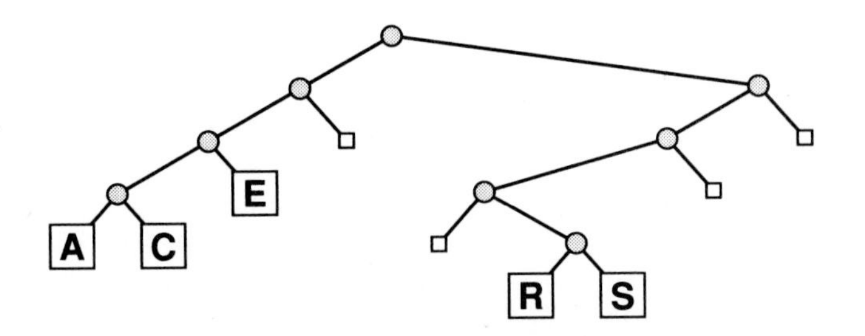

Figure 17.4 A radix search trie.

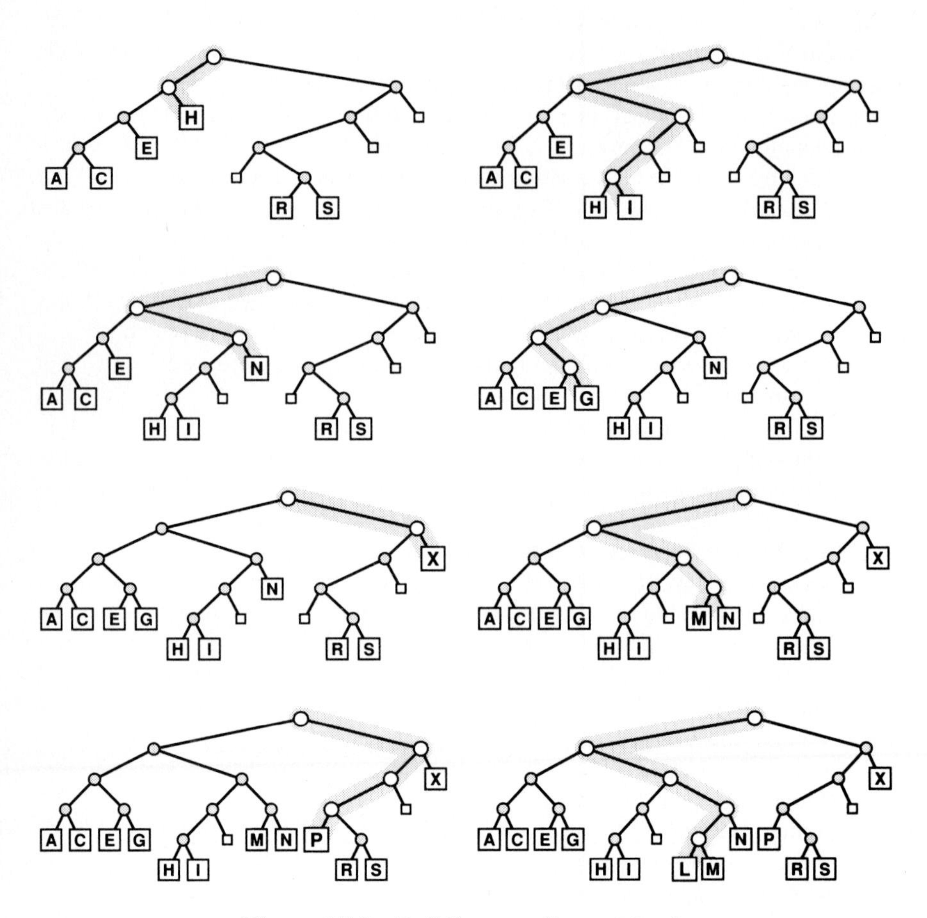

Figure 17.5 Building a radix search trie.

of the necessity to maintain two types of nodes, both of which could be pointed to by links in internal nodes. This is an example of an algorithm for which a low-level implementation might be simpler than a high-level implementation. We'll omit the code for this because we'll see an improvement below which avoids this problem.

The left subtree of a binary radix search trie has all the keys which have 0 for the leading bit; the right subtree has all the keys which have 1 for the leading bit. This leads to an immediate correspondence with radix sorting: binary trie searching partitions the file in exactly the same way as radix exchange sorting. (Compare the trie above with Figure 10.1, the partitioning diagram for radix exchange sorting, after noting that the keys are slightly different.) This correspondence is analogous to that between binary tree searching and Quicksort.

Property 17.2 *A search or insertion in a radix search trie requires about* $\lg N$ *bit comparisons for an average search and* b *bit comparisons in the worst case in a tree built from* N *random* b*-bit keys.*

As above, the worst-case result comes directly from the algorithm and the average-case result requires mathematical analysis beyond the scope of this book, though it validates the rather simple intuitive notion that each bit examined should be as likely to be a 0 bit as a 1 bit, so about half the keys should fall on each side of any trie node. ▪

An annoying feature of radix tries, and one which distinguishes them from the other types of search trees we've seen, is the "one-way" branching required for keys with a large number of bits in common. For example, keys which differ only in the last bit require a path whose length is equal to the key length, no matter how many keys there are in the tree. The number of internal nodes can be somewhat larger than the number of keys.

Property 17.3 *A radix search trie built from* N *random* b*-bit keys has about* $N/\ln 2 \approx 1.44N$ *nodes on the average.*

Again, proof of this result is quite beyond the scope of this book, though it is easily verified empirically. Figure 17.6 shows a trie built from 95 random 10-bit keys which has 131 nodes. ▪

The height of tries is still limited by the number of bits in the keys, but we would like to consider the possibility of processing records with very long keys (say 1000 bits or more) which perhaps have some uniformity, as might arise in encoded character data. One way to shorten the paths in the trees is to use many more than two links per node (though this exacerbates the "space" problem of using too many nodes); another way is to "collapse" paths containing one-way branches into single links. We'll discuss these methods in the next two sections.

Multiway Radix Searching

For radix sorting, we found that we could get a significant improvement in speed

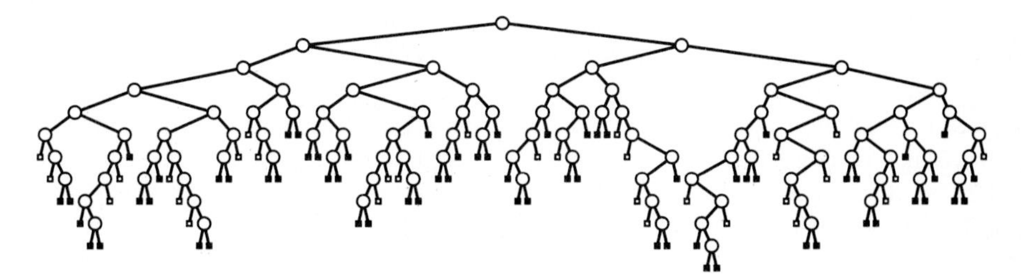

Figure 17.6 A large radix search trie.

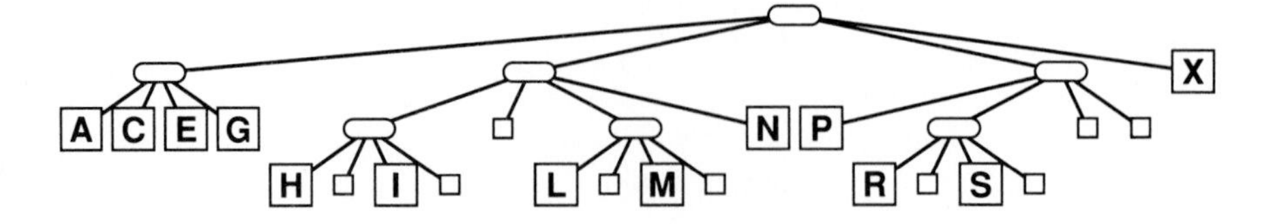

Figure 17.7 A 4-way radix trie.

by considering more than one bit at a time. The same is true for radix searching: by examining m bits at a time, we can speed up the search by a factor of 2^m. However, there's a catch which makes it necessary to be more careful in applying this idea than was necessary for radix sorting. The problem is that considering m bits at a time corresponds to using tree nodes with $M = 2^m$ links, which can lead to a considerable amount of wasted space for unused links.

For example, if $M = 4$ the trie shown in Figure 17.7 is formed for our sample keys. To search in this trie, consider the bits in the key two bits at a time: if the first two bits are 00, then take the left link at the first node; if they are 01 take the second link; if they are 10 take the third link; and if they are 11 take the right link. Then branch on the next level according to the third and fourth bits, etc. For example, to search for T=10100 in the trie in Figure 17.7, take the third link from the root, and then the third link from the third child of the root to access an external node, so the search is unsuccessful. To insert T, that node could be replaced by a new node containing T (and four external links).

Note that there is some wasted space in this tree because of the large number of unused external links. As M gets larger, this effect gets worse: it turns out that the number of links used is about $MN/\ln M$ for random keys. On the other hand, this is a very efficient searching method: the running time is about $\log_M N$. A reasonable compromise can be struck between the time efficiency of multiway tries and the space efficiency of other methods by using a "hybrid" method with a large value of M at the top (say the first two levels) and a small value of M (or some elementary method) at the bottom. Again, efficient implementations of such methods can be quite complicated, however, because of multiple node types.

For example, a two-level 32-way tree divides the keys into 1024 categories, each accessible in two steps down the tree. This would be quite useful for files of thousands of keys, because there are likely to be (only) a few keys per category. On the other hand, a smaller M would be appropriate for files of hundreds of keys, because otherwise most categories would be empty and too much space would be wasted, and a larger M would be appropriate for files with millions of keys, because otherwise most categories would have too many keys and too much time would be wasted.

It is amusing to note that "hybrid" searching corresponds quite closely to the way humans search for things, for example, names in a telephone book. The first step is a multiway decision ("Let's see, it starts with 'A' "), followed perhaps by some two-way decisions ("It's before 'Andrews', but after 'Aitken'") followed by sequential search (" 'Algonquin' ... 'Algren' ... No, 'Algorithms' isn't listed!"). Of course, computers are likely to be somewhat better than humans at multiway search, so two levels are appropriate. Also, 26-way branching (with even more levels) is a quite reasonable alternative to consider for keys which are composed simply of letters (for example, in a dictionary).

In the next chapter, we'll see a systematic way to adapt the structure to take advantage of multiway radix searching for arbitrary file sizes.

Patricia

The radix trie searching method as outlined above has two annoying flaws: the "one-way branching" leads to the creation of extra nodes in the tree, and there are two different types of nodes in the tree, which complicates the code somewhat (especially the insertion code). D. R. Morrison discovered a way to avoid both of these problems in a method which he named *Patricia* ("Practical Algorithm To Retrieve Information Coded In Alphanumeric"). The algorithm given below is not in precisely the same form as presented by Morrison, because he was interested in "string searching" applications of the type that we'll see in Chapter 19. In the present context, Patricia allows searching for N arbitrarily long keys in a tree with just N nodes, but requires only one full key comparison per search.

One-way branching is avoided by a simple device: each node contains the index of the bit to be tested to decide which path to take out of that node. External nodes are avoided by replacing links to external nodes with links that point upwards in the tree, back to our normal type of tree node with a key and two links. But in Patricia, the keys in the nodes are not used on the way down the tree to control the search; they are merely stored there for reference when the bottom of the tree is reached. To see how Patricia works, we'll first look at how it operates on a typical tree and then examine how the tree is constructed in the first place. The Patricia tree shown in Figure 17.8 is constructed when our example keys are successively inserted.

To search in this tree, we start at the root and proceed down the tree, using the bit index in each node to tell us which bit to examine in the search key—we go right if that bit is 1, left if it is 0. The keys in the nodes are not examined at all on the way down the tree. Eventually, an upwards link is encountered: each upward link points to the unique key in the tree that has the bits that would cause a search to take that link. For example, S is the only key in the tree that matches the bit pattern 10*11. Thus if the key at the node pointed to by the first upward link encountered is equal to the search key, then the search is successful; otherwise it is unsuccessful. For tries, all searches terminate at external nodes, whereupon one full key comparison is done to determine whether or not the search

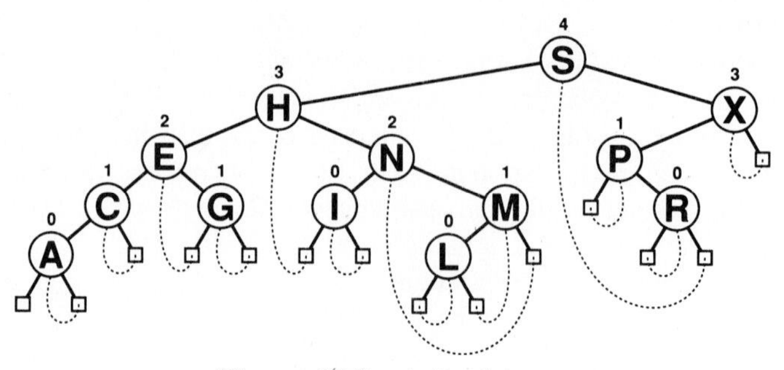

Figure 17.8 A Patricia tree.

was successful; for Patricia all searches terminate at upwards links, whereupon one full key comparison is done to determine whether or not the search was successful. Furthermore, it's easy to test whether a link points up, because the bit indices in the nodes (by definition) decrease as we travel down the tree. This leads to the following search code for Patricia, which is as simple as the code for radix tree or trie searching:

```
type link=↑node;
       node= record key,info,b: integer; l,r: link end;
var head,z: link;
function patriciasearch(v: integer; x: link): link;
  var p: link;
  begin
  repeat
    p:=x;
    if bits(v,x↑.b,1)=0 then x:=x↑.l else x:=x↑.r;
  until p↑.b<=x↑.b;
  patriciasearch:=x
  end;
```

This function returns a link to the unique node which could contain the record with key v. The calling routine then can test whether the search was successful or not. Thus to search for Z=11010 in the above tree we go right and then up at the right link of X. The key there is not Z, so the search is unsuccessful.

Figure 17.9 shows the result of inserting Z=11010 into the Patricia tree of Figure 17.8. As described above, the search for Z ends at the node containing X=11000. By the defining property of the tree, X is the only key in the tree for which a search would terminate at that node. If Z is inserted, there would be two such nodes, so the upward link that was followed into the node containing X must

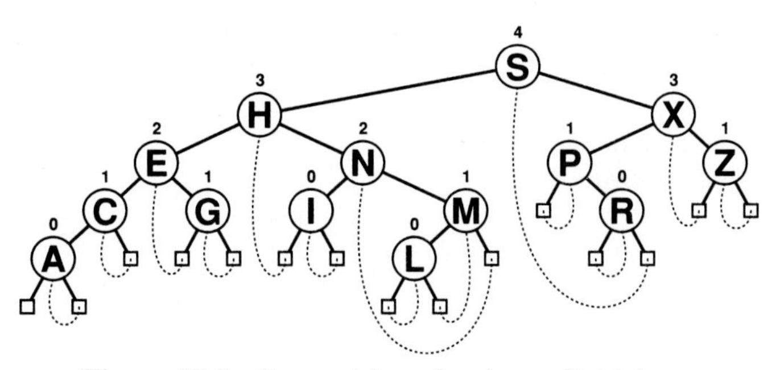

Figure 17.9 External insertion into a Patricia tree.

be made to point to a new node containing Z, with a bit index corresponding to the leftmost point where X and Z differ, and with two upward links: one pointing to X and the other pointing to Z. This corresponds precisely to replacing the external node containing X with a new internal node with X and Z as children in radix trie insertion, with one-way branching eliminated by including the bit index.

Inserting T=10100 illustrates a more complicated case, as shown in Figure 17.10. The search for T ends at P=10000, indicating that P is the only key in the tree with the pattern 10∗0∗. Now, T and P differ at bit 2, a position that was skipped during the search. The requirement that the bit indices decrease as we go down the tree dictates that T be inserted between X and P, with an upward self-pointer corresponding to its own bit 2. Note carefully that the fact that bit 2 was skipped before the insertion of T implies that P and R have the same bit-2 value.

These examples illustrate the only two cases that arise in insertion for Patricia. The following implementation gives the details:

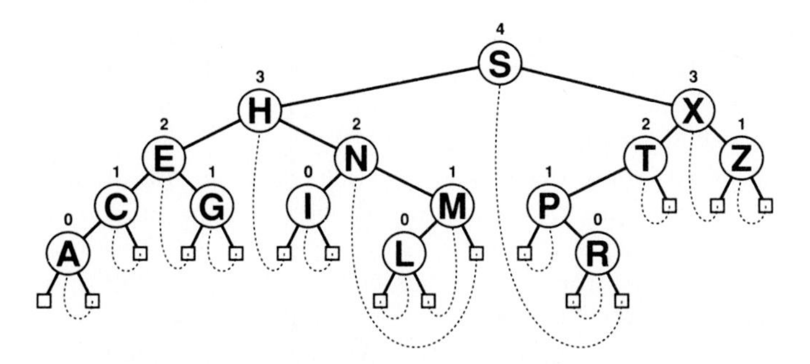

Figure 17.10 Internal insertion into a Patricia tree.

```
function patriciainsert(v: integer; x: link): link;
  label 0;
  var t,p: link; i: integer;
  begin
  t:=patriciasearch(v,x);
  if v=t↑.key then goto 0;
  i:=maxb;
  while bits(v,i,1)=bits(t↑.key,i,1) do i:=i-1;
  repeat
    p:=x;
    if bits(v,x↑.b,1)=0 then x:=x↑.l else x:=x↑.r;
  until (x↑.b<=i) or (p↑.b<=x↑.b);
  new(t); t↑.key:=v; t↑.b:=i;
  if bits(v,t↑.b,1)=0
    then begin t↑.l:=t; t↑.r:=x end
    else  begin t↑.r:=t; t↑.l:=x end;
  if bits(v,p↑.b,1)=0 then p↑.l:=t else p↑.r:=t;
0: patriciainsert:=t
  end;
```

(This code assumes that *head* is initialized with key field of 0, a bit index of *maxb* and both links self-pointers.) First, we do a search to find the key which must be distinguished from v. The conditions $x\uparrow.b<=i$ and $p\uparrow.b<=x\uparrow.b$ characterize the situations shown in Figures 17.10 and 17.9, respectively. Then we determine the leftmost bit position at which they differ, travel down the tree to that point, and insert a new node containing v at that point.

Patricia is the quintessential radix searching method: it manages to identify the bits which distinguish the search keys and build them into a data structure (with no surplus nodes) that quickly leads from any search key to the only key in the data structure that could be equal. Clearly, the same technique as used in Patricia can be used in binary radix trie searching to eliminate one-way branching, but this only exacerbates the multiple-node-type problem. Figure 17.11 shows the Patricia tree for the same keys used to build the trie of Figure 17.6—this tree not only has 44% less nodes, but it is quite well-balanced.

Unlike standard binary tree search, the radix methods are insensitive to the order in which keys are inserted; they depend only upon the structure of the keys themselves. For Patricia the placement of the upwards links depend on the order of insertion, but the tree structure depends only on the bits in the keys, as in the other methods. Thus, even Patricia would have trouble with a set of keys like 001, 0001, 00001, 000001, etc., but for normal key sets, the tree should be relatively well-balanced so the number of bit inspections, even for very long keys, will be roughly proportional to $\lg N$ when there are N nodes in the tree.

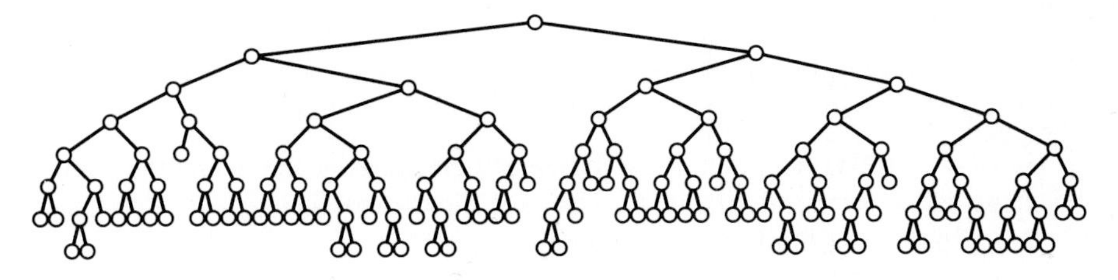

Figure 17.11 A large Patricia tree.

Property 17.4 *A Patricia trie built from N random b-bit keys has N nodes and requires* $\lg N$ *bit comparisons for an average search.*

As for the other methods of this chapter, the analysis of the average case is rather difficult: it turns out that Patricia involves one less comparison, on the average, than does a standard trie. ■

The most useful feature of radix trie searching is that it can be done efficiently with keys of varying length. In all of the other searching methods we have seen the length of the key is "built into" the searching procedure in some way, so that the running time is dependent on the length as well as the number of the keys. The specific savings available depends on the method of bit access used. For example, suppose we have a computer which can efficiently access 8-bit "bytes" of data, and we have to search among hundreds of 1000-bit keys. Then Patricia would require accessing only about 9 or 10 bytes of the search key for the search, plus one 125-byte equality comparison, while hashing would require accessing all 125 bytes of the search key to compute the hash function plus a few equality comparisons, and comparison-based methods require several long comparisons. This effect makes Patricia (or radix trie searching with one-way branching removed) the search method of choice when very long keys are involved.

□

Exercises

1. Draw the digital search tree that results when the keys E A S Y Q U E S T I O N are inserted in that order into an initially empty tree.
2. Generate a 1000 node digital search tree and compare its height and the number of nodes at each level against a standard binary search tree and a red-black tree (Chapter 15) built from the same keys.
3. Find a set of 12 keys that make a particularly badly balanced digital search trie.
4. Draw the radix search trie that results when the keys E A S Y Q U E S T I O N are inserted in that order into an initially empty tree.
5. A problem with 26-way multiway radix search tries is that some letters of the alphabet are very infrequently used. Suggest a way to fix this problem.
6. Describe how you would delete an element from a multiway radix search tree.
7. Draw the Patricia tree that results when the keys E A S Y Q U E S T I O N are inserted in that order into an initially empty tree.
8. Find a set of 12 keys that make a particularly badly balanced Patricia tree.
9. Write a program that prints out all keys in a Patricia tree having the same initial t bits as a given search key.
10. For which of the radix methods is it reasonable to write a program which prints out the keys in sorted order? Which of the methods are not amenable to this operation?

18

External Searching

☐ Searching algorithms appropriate for accessing items from very large files are of immense practical importance. Searching is the fundamental operation on large data files, and certainly consumes a very significant fraction of the resources used in many computer installations.

We'll be concerned mainly with methods for searching on large disk files, since disk searching is of the most practical interest. With sequential devices such as tapes, searching quickly degenerates to the trivially slow method: to search a tape for an item, one can't do much better than to mount the tape and read it until the item is found. Remarkably, the methods that we'll study can find an item from a disk as large as a billion words with only two or three disk accesses.

As with external sorting, the "systems" aspect of using complex I/O hardware is a primary factor in the performance of external searching methods but one we won't be able to study in detail. However, unlike sorting, where the external methods are really quite different from the internal methods, we'll see that external searching methods are logical extensions of the internal methods that we've studied.

Searching is a fundamental operation for disk devices. Files are typically organized to take advantage of particular device characteristics to make access of information as efficient as possible. As we did with sorting, we'll work with a rather simple and imprecise model of "disk" devices in order to explain the principal characteristics of the fundamental methods. Determining the best external searching method for a particular application is extremely complicated and very dependent on characteristics of the hardware (and systems software), and so it is quite beyond the scope of this book. However, we can suggest some general approaches to use.

For many applications we would frequently like to change, add, delete or (most important) quickly access small bits of information inside very, very large files. In this chapter, we'll examine some methods for such dynamic situations which offer the same kinds of advantages over the straightforward methods that binary search trees and hashing offer over binary search and sequential search.

A very large collection of information to be processed using a computer is called a *database*. A great deal of study has gone into methods of building, maintaining and using databases. However, large databases have very high inertia: once a very large database has been built around a particular searching strategy, it can be very expensive to rebuild it around another. For this reason, the older, static methods are in widespread use and likely to remain so, though the newer, dynamic methods are beginning to be used for new databases.

Database applications systems typically support much more complicated operations than a simple search for an item based on a single key. Searches are often based on criteria involving more than one key and are expected to return a large number of records. In later chapters we'll see some examples of algorithms which are appropriate for some search requests of this type, but general search requests are sufficiently complicated that it is typical to do a sequential search over the entire database, testing each record to see if it meets the criteria.

The methods that we will discuss are of practical importance in the implementation of large file systems in which every file has a unique identifier and the purpose of the file system is to support efficient access, insertion and deletion based on that identifier. Our model will consider the disk storage as divided up into *pages*, contiguous blocks of information that can be efficiently accessed by the disk hardware. Each page will hold many records; our task is to organize the records within the pages in such a way that any record can be accessed by reading only a few pages. We assume that the I/O time required to read a page completely dominates the processing time required to do any computing involving that page. As mentioned above, this model is oversimplified in many ways, but it retains enough of the characteristics of actual external storage devices to allow us to consider some of the fundamental methods used.

Indexed Sequential Access

Sequential disk searching is the natural extension of the elementary sequential searching methods considered in Chapter 14: the records are stored in increasing order of their keys, and searches are done simply by reading in the records one after the other until one containing a key greater than or equal to the search key is found. For example, if our search keys come from E X T E R N A L S E A R C H I N G E X A M P L E and we have disks capable of holding three pages of four records each, then we have the configuration shown in Figure 18.1. (As for external sorting, we must consider very small examples to understand the algorithms, and think about very large examples to appreciate their performance.) Obviously, pure sequential searching is unattractive because, for example, searching for W in Figure 18.1 would require reading all the pages.

To vastly improve the speed of a search, we can keep, for each disk, an "index" of which keys belong to which pages on that disk, as in Figure 18.2. The first page of each disk is its index: the small letters indicate that only the key value is

Figure 18.1 Sequential access.

stored, not the full record and small numbers are page indices (0 means the first page on the disk, 1 the next page, etc.). In the index, each page number appears below the value of the last key on the previous page. (The blank is a sentinel key, smaller than all the others, and the "+" means "look on the next disk".) Thus, for example, the index for disk 2 says that its first page contains records with keys between E and I inclusive and its second page contains records with keys between I and N inclusive. It is normally possible to fit many more keys and page indices on an index page than records on a "data" page; in fact, the index for a whole disk should require only a few pages.

To further expedite the search, these indices may be coupled with a "master index" which tells which keys are on which disk. For our example, the master index would say that disk 1 contains keys less than or equal to E, disk 2 contains keys less than or equal to N (but not less than E), and disk 3 contains keys less than or equal to X (but not less than N). The master index is likely to be small enough that it can be kept in memory, so that most records can be found by accessing only two pages, one for the index on the appropriate disk and one for the page containing the appropriate record. For example, a search for W would involve first reading the index page from disk 3, then reading the second data page from disk 3 which is the only one that could contain W. Searches for keys which appear in the index require reading three pages: the index plus the two pages flanking the key value in the index. If no duplicate keys are in the file, then the extra page access can be avoided. On the other hand, if there are many equal keys in the file, several page accesses might be called for (records with equal keys might fill several pages).

Because it combines a sequential key organization with indexed access, this organization is called *indexed sequential access*. It is the method of choice for applications in which changes to the database are likely to be made infrequently.

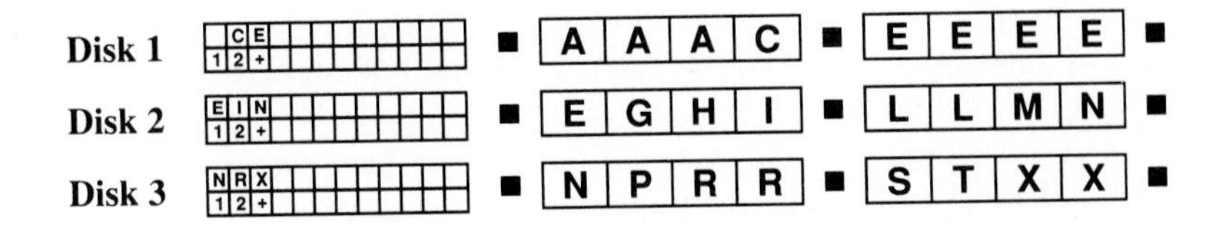

Figure 18.2 Indexed sequential access.

The disadvantage of using indexed sequential access is that it is very inflexible. For example, adding B to the configuration above requires that virtually the whole database be rebuilt, with new positions for many of the keys and new values for the indices.

Property 18.1 *A search in an indexed sequential file requires only a constant number of disk accesses, but an insertion can involve rearranging the entire file.*

Actually, the "constant" involved here depends on the number of disks and on the relative size of records, indices and pages. For example, a large file of one-word keys certainly couldn't be stored on just one disk in such a way as to allow searching with a constant number of accesses. Or, to take another absurd example at the other extreme, a large number of very small disks each capable of holding only one record might also be hard to search. ■

B-Trees

A better way to handle searching in a dynamic situation is to use balanced trees. In order to reduce the number of (relatively expensive) disk accesses, it is reasonable to allow a large number of keys per node so that the nodes have a large branching factor. Such trees were named B-trees by R. Bayer and E. McCreight, who were the first to consider the use of multiway balanced trees for external searching. (Many people reserve the term "B-tree" to describe the exact data structure built by the algorithm suggested by Bayer and McCreight; we'll use it as a generic term to mean "external balanced trees.")

The top-down algorithm that we used for 2-3-4 trees (see Chapter 15) extends readily to handle more keys per node: assume that there are anywhere from 1 to $M - 1$ keys per node (and so anywhere from 2 to M links per node). Searching proceeds in a way analogous to 2-3-4 trees: to move from one node to the next, first find the proper interval for the search key in the current node and then exit through the corresponding link to get to the next node. Continue in this way until an external node is reached, then insert the new key into the last internal node reached. As with top-down 2-3-4 trees, it is necessary to "split" nodes that are "full" on the way down the tree: any time we see a k-node attached to an M-node, we replace it by a $(k + 1)$-node attached to two $(M/2)$-nodes (for even splits, we assume that M is even). This guarantees that when the bottom is reached there is room to insert the new node.

The B-tree constructed for $M = 4$ and our sample keys is shown in Figure 18.3. This tree has 13 nodes, each corresponding to a disk page. Each node must contain links as well as records. The choice $M = 4$, even though it leaves us with familiar 2-3-4 trees, is meant to emphasize this point: earlier we could fit four records per page, now only three will fit, to leave room for the links. The actual

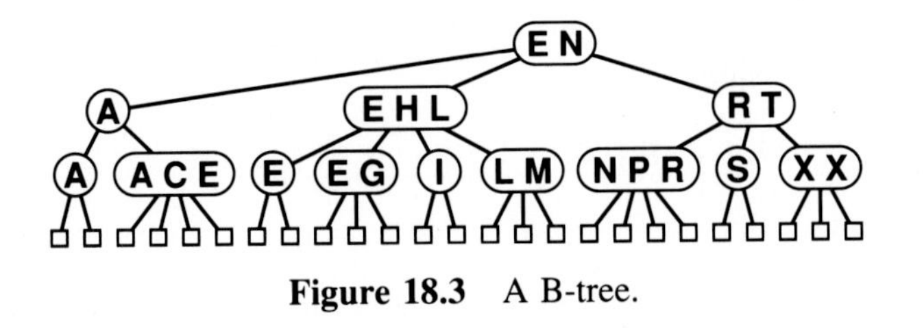

Figure 18.3 A B-tree.

amount of space used depends on the relative size of records and links. We'll see below a method which avoids this mixing of records and links.

Just as we kept the master index for indexed sequential search in memory, it's reasonable to keep the root node of the B-tree in memory. For the B-tree in Figure 18.3, this might indicate that the root of the subtree containing records with keys less than or equal to E is on page 0 of disk 1, the root of the subtree with keys less than or equal to N (but not less than E) is on page 1 of disk 1, and the root of the subtree with keys greater than or equal to N is on page 2 of disk 1. The other nodes for our example might be stored as shown in Figure 18.4.

Nodes are assigned to disk pages in this example simply by proceeding down the tree, working from right to left at each level, assigning nodes to disk 1, then disk 2, etc. We avoid storing null links by keeping track of when the bottom level is reached: in this case, all nodes on disks 2, 3, and 4 have all null links (which need not be stored). In an actual application, other considerations come into play. For example, it might be better to avoid having all searches going through disk 1 by assigning first to page 0 of all the disks, etc. In fact, more sophisticated strategies are needed because of the dynamics of the tree construction (consider the difficulty of implementing a *split* routine that respects either of the above strategies).

Property 18.2 *A search or an insertion in a B-tree of order M with N records is guaranteed to require fewer than $\log_{M/2} N$ disk accesses, a constant number for practical purposes (as long as M is not small).*

This property follows from the observation that all the nodes in the "interior" of

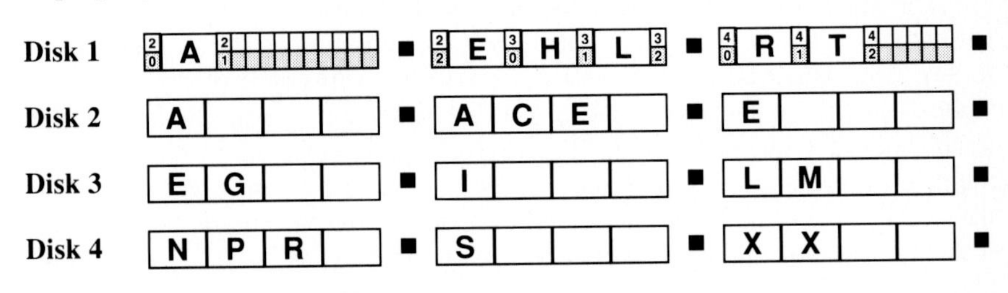

Figure 18.4 B-tree access.

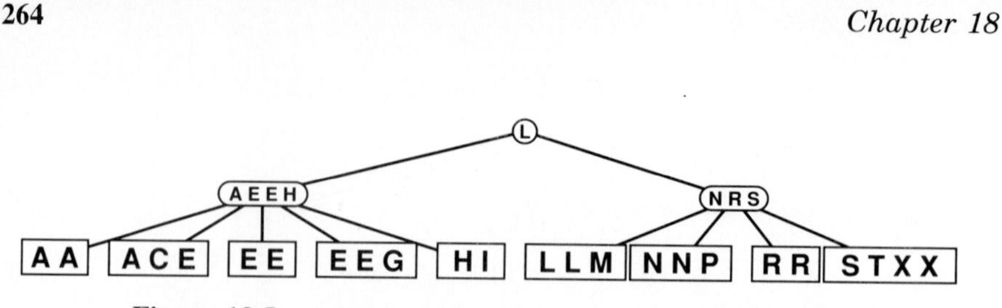

Figure 18.5 A B-tree with records only at external nodes.

the B-tree (nodes which are not the root and not leaves) have between $M/2$ and M keys, since they are formed from a split of a full node with M keys, and can only grow in size (when a lower node is split). In the worst case, these nodes form a complete tree of degree $M/2$, which leads immediately to the stated bound. ■

Property 18.3 *A B-tree of order M constructed from N random records may be expected to have about 1.44N/M nodes.*

Proof of this fact is beyond the scope of this book, but note that the amount of space wasted ranges up to about N, in the worst case when all of the nodes are about half full. ■

In the above example, we were forced to choose $M = 4$ because of the need to save room for links in the nodes. But we ended up *not* using links in most of the nodes, since most of the nodes in a B-tree are external and most of the links are null. Furthermore, a much larger value of M can be used at the higher levels of the tree if we store just keys (not full records) in the interior nodes, as in indexed sequential access. To see how to take advantage of these observations in our example, suppose that we can fit up to seven keys and eight links on a page, so that we can use $M = 8$ for the interior nodes and $M = 5$ for the bottom-level nodes (*not* $M = 4$ because no space for links need be reserved at the bottom). A bottom node splits when a fifth record is added to it (into one node with two records and one with three records); the split ends by "inserting" the key of the middle record into the node above, where there is room because the tree above has operated as a normal B-tree for $M = 8$ (on stored keys, not records). This leads to the tree shown in Figure 18.5.

The effect for a typical application is likely to be much more dramatic since the branching factor of the tree is increased by roughly the ratio of the record size to key size, which is likely to be large. Also, with this type of organization, the "index" (which contains keys and links) can be separated from the actual records, as in indexed sequential search. Figure 18.6 shows how the tree in Figure 18.5 might be stored: the root node is on page 0 of disk 1 (there is room for it since the tree in Figure 18.5 has one less node than the tree in Figure 18.3), though in most applications it probably would be kept in memory, as above. Other comments above regarding node placement on the disks also apply here.

Now we have two values of M, one for the interior nodes which determines the branching factor of the tree (M_I) and one for the bottom-level nodes which

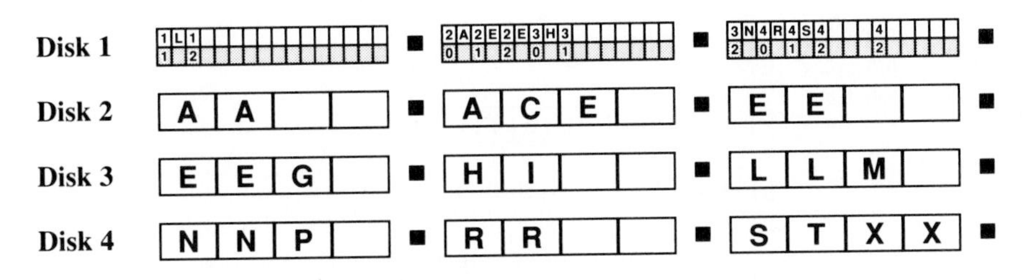

Figure 18.6 B-tree access with records only at external nodes.

determines the allocation of records to pages (M_B). To minimize the number of disk accesses, we want to make both M_I and M_B as large as possible, even at the expense of some extra computation. On the other hand, we don't want to make M_I huge, because then most tree nodes would be largely empty and space would be wasted, and we don't want to make M_B huge, because this would reduce to sequential search of the bottom-level nodes. Usually, it is best to relate both M_I and M_B to the page size. The obvious choice for M_B is the number of records that can fit on a page (plus one): the goal of the search is to find the page containing the record sought. If M_I is taken as the number of keys that can fit on two to four pages, then the B-tree is likely to be only three levels deep, even for very large files (a three-level tree with $M_I = 2048$ can handle up to 1024^3, or over a billion, entries). But recall that the root node of the tree, which is accessed for every operation on the tree, is kept in memory, so that only two disk accesses are required to find any element in the file.

As briefly mentioned at the end of Chapter 15, a more complicated "bottom-up" insertion method is commonly used for B-trees (though the distinction between top-down and bottom up methods loses importance for three-level trees). Technically, the trees described here should be referred to as "top-down" B-trees to distinguish them from those commonly discussed in the literature. Many other variations have been described, some of which are quite important for external searching. For example, when a node becomes full, splitting (and the resultant half-empty nodes) can be forestalled by dumping some of the contents of the node into its "sibling" node (if it's not too full). This leads to better space utilization within the nodes, which is likely to be a major concern in a large-scale disk searching application.

Extendible Hashing

An alternative to B-trees which extends digital searching algorithms to apply to external searching was developed in 1978 by R. Fagin, J. Nievergelt, N. Pippenger, and R. Strong. This method, called *extendible hashing*, involves two disk accesses for each search in typical applications while at the same time allowing efficient insertion. As with B-trees, our records are stored on pages which are split into two pieces when they fill up; as with indexed sequential access, we maintain an index

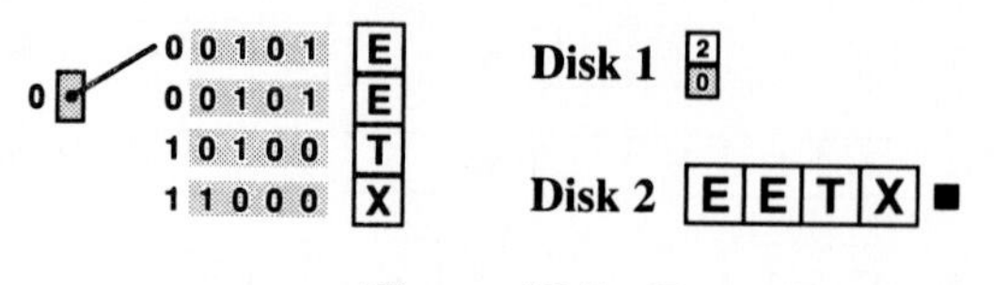

Figure 18.7 Extendible hashing: first page.

which we access to find the page containing the records which match our search key. Extendible hashing combines these approaches by using digital properties of the search keys.

To see how extendible hashing works, we'll consider how it handles successive insertions of keys from E X T E R N A L S E A R C H I N G E X A M P L E, using pages with a capacity of up to four records. We start with an "index" with just one entry, a pointer to the page which is to hold the records. The first four records fit on the page, leaving the trivial structure shown in Figure 18.7.

The directory on disk 1 says that all records are on page 0 of disk 2, where they are kept in sorted order of their keys. For reference, we also give the binary value of the keys, using our standard encoding of the five-bit binary representation of i for the ith letter of the alphabet. Now the page is full and must be split in order to add the key R=10010. The strategy is simple: put records with keys that begin with 0 on one page and records with keys that begin with 1 on another page. This necessitates doubling the size of the directory and moving half the keys from page 0 of disk 2 to a new page, leaving the structure shown in Figure 18.8.

Now N=01110 and A=00001 can be added, but this again fills the first page, as shown in Figure 18.9. Another split is needed before L=01100 can be added. To insert L=01100, then, we proceed in the same way as for the first split, by splitting the first page into two pieces, one for keys that begin with 00 and one for keys that begin with 01. What's not immediately clear is what to do with the directory. One alternative would be to simply add another entry, one pointer to each page. This is unattractive because it essentially reduces to indexed sequential search (albeit a radix version): the directory has to be scanned sequentially to find

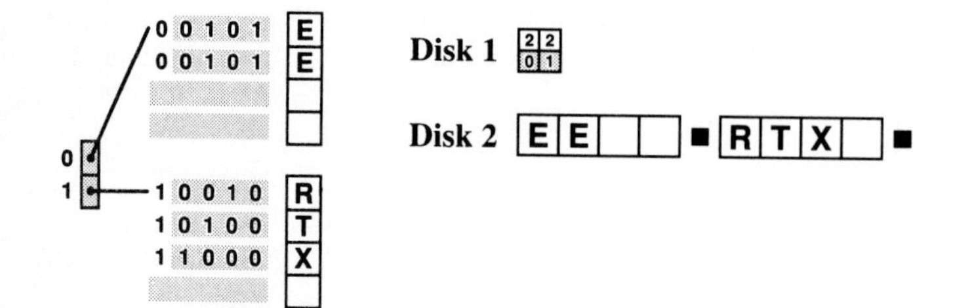

Figure 18.8 Extendible hashing: directory split.

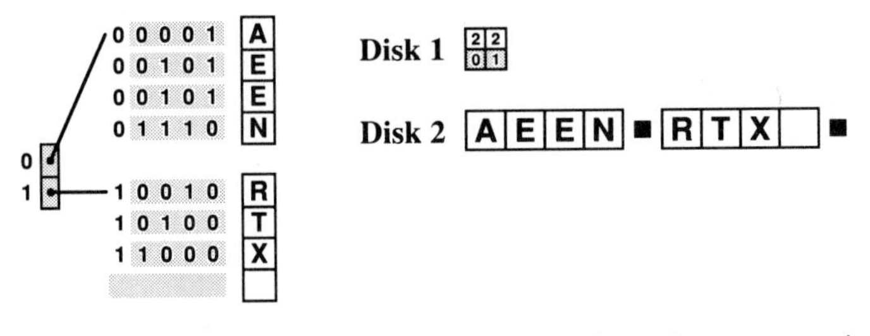

Figure 18.9 Extendible hashing: first page again full.

the proper page during a search. Alternatively, we can just double the size of the directory again, giving the structure shown in Figure 18.10. A new page (page 2 on disk 2) contains the keys that begin with 01 (L and N), the page which split (page 0 on disk 2) now contains the keys that begin with 00 (A, E, and E), and the page containing keys that begin with 1 (R, T, and X) is unaffected, though now there are two pointers to it, one to indicate that keys starting with 10 are stored there, the other to indicate that keys starting with 11 are stored there. Now we can access any record by using the first two bits of its key to access directly the directory entry containing the address of the page containing the record.

Keeping the records in sorted order within the pages may seem like a brute-force simplification, but recall our basic assumptions that we do disk I/O in page units and that processing time is negligible compared to the time to input or output a page. Thus, keeping the records in sorted order of their keys is not a real expense: to add a record to a page, we must read the page into memory, modify it, and write it back out. The extra time required to maintain sorted order is not likely to be noticeable in the typical case when the pages are not large.

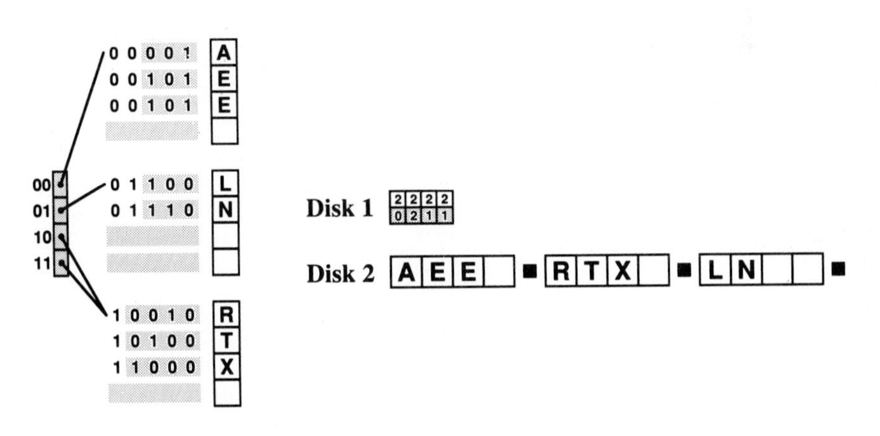

Figure 18.10 Extendible hashing: second split.

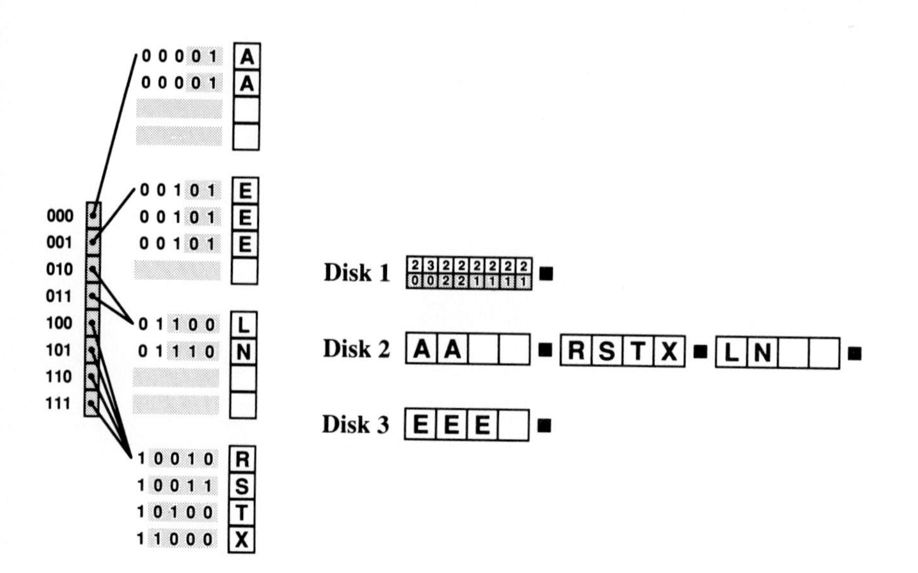

Figure 18.11 Extendible hashing: third split.

Continuing a little further, we can add S=10011 and E=00101 before another split is necessary to add A=00001. This split also requires doubling the directory, producing the structure shown in Figure 18.11. The process of doubling the directory is simple: just read the old directory, then make the new one by writing out each entry of the old one twice. This makes a space for the pointer to the new page just created by the split.

In general, the structure built by extendible hashing consists of a *directory* of 2^d words (one for each d-bit pattern) and a set of *leaf pages* which contain all records with keys beginning with a specific bit pattern (of less than or equal to d bits). A search entails using the leading d bits of the key to index into the directory, which contains pointers to leaf pages. Then the referenced leaf page is accessed and searched (using any strategy) for the proper record. A leaf page can be pointed to by more than one directory entry: to be precise, if a leaf page contains all the records with keys beginning with a specific k bits (those not shaded in the figures), then it will have 2^{d-k} directory entries pointing to it. In Figure 18.11, we have $d = 3$, and page 1 of disk 2 contains all the records with keys that begin with a 1 bit, so there are four directory entries pointing to it.

In our example so far, each page split has required a directory split, but in normal circumstances, the directory may be expected to split only rarely. This is the essence of the algorithm: the extra pointers in the directory allow the structure to accommodate dynamic growth gracefully. For example, when R is inserted into the structure in Figure 18.11, page 1 on disk 2 must be split to accommodate the five keys that begin with a 1, but the directory does not need to grow, as

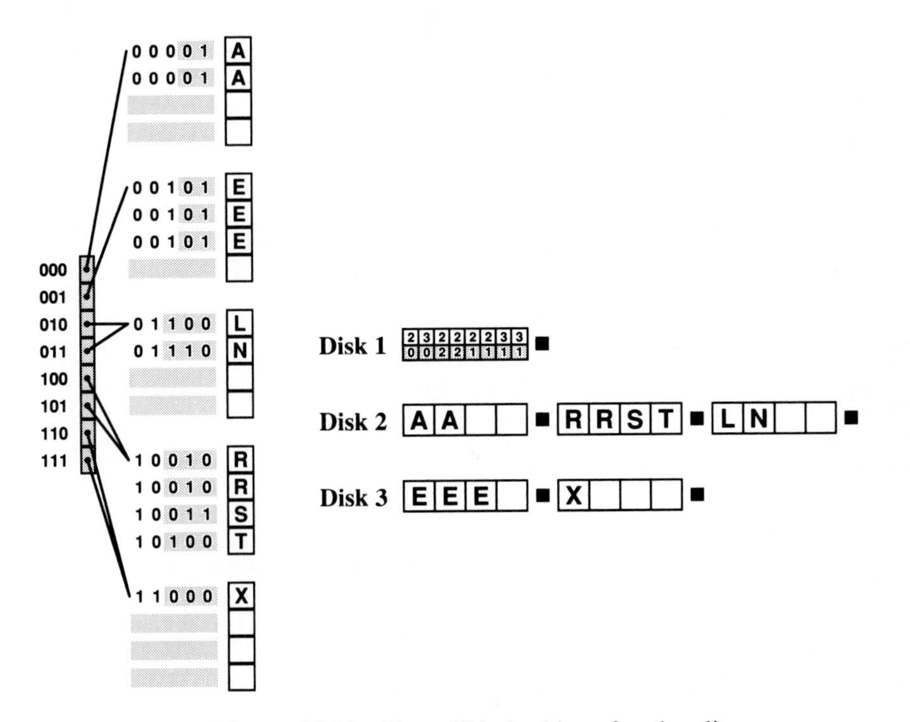

Figure 18.12 Extendible hashing: fourth split.

Figure 18.12 shows. The only change to the directory is that the last two pointers are changed to point to page 1 on disk 3, the new page created in the split to accommodate all keys in the data structure that begin with 11 (the X).

The directory contains only pointers to pages. These are likely to be smaller than keys or records, so more directory entries will fit on each page. For our example, we'll assume that we can fit twice as many directory entries as records on a page, though this ratio is likely to be much higher in practice. When the directory spans more than one page, we keep a "root node" in memory which tells where the directory pages are, using the same indexing scheme. For example, if the directory spans two pages, the root node might indicate that the directory for all the records with keys beginning with 0 is on page 0 of disk 1, and the directory for all keys beginning with 1 is on page 1 of disk 1. For our example, this split after we insert C, H, I, N, G, and E. Continuing, after we insert X, A, M, P, and L, we get the disk storage structure shown in Figure 18.13. (For clarity, we have reserved disk 1 for the directory, though in practice it might be mixed in with the other pages, page 0 of each disk might be reserved, or some other strategy used.)

Thus, insertion into an extendible hashing structure can involve one of three operations, after the leaf page which could contain the search key is accessed. If there's room in the leaf page, the new record is simply inserted there; otherwise

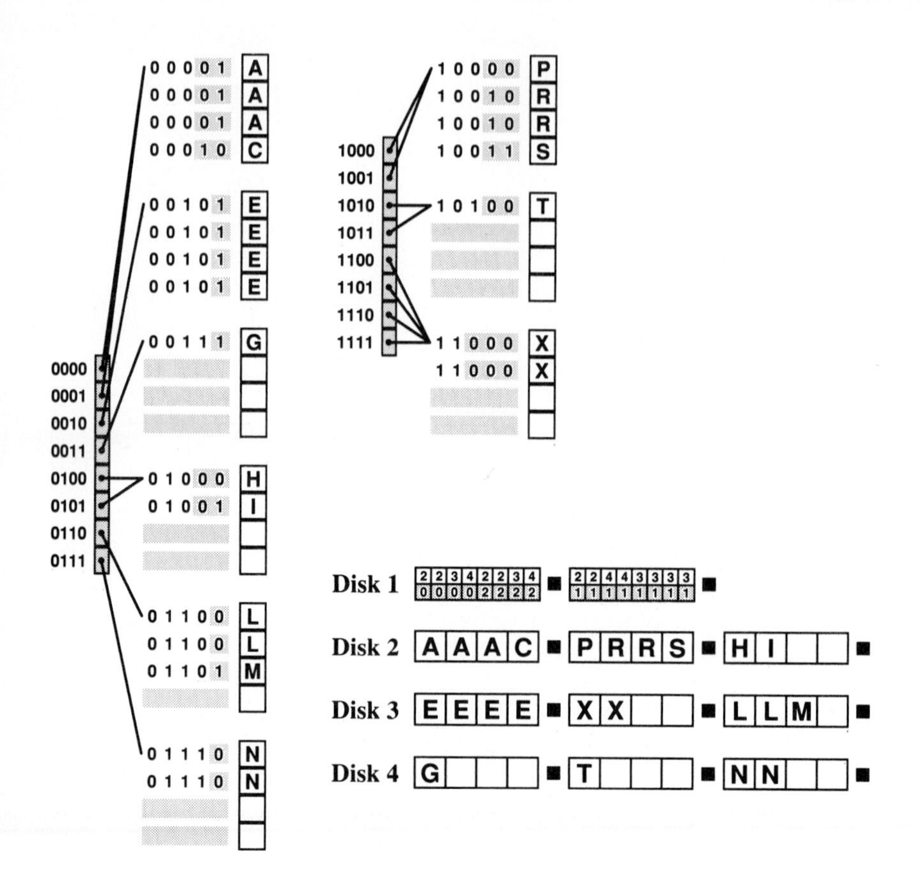

Figure 18.13 Extendible hashing access.

the leaf page is split in two (half the records are moved to a new page). If the directory has more than one entry pointing to that leaf page, then the directory entries can be split as the page is. If not, the size of the directory must be doubled.

As described so far, this algorithm is very susceptible to a bad input-key distribution: the value of d is the largest number of bits required to separate the keys into sets small enough to fit on leaf pages, and thus if a large number of keys agree in a large number of leading bits, the directory could become unacceptably large. For actual large-scale applications, this problem can be headed off by *hashing* the keys to make the leading bits (pseudo-)random. To search for a record, we hash its key to get a bit sequence which we use to access the directory; the directory tells us which page to search for a record with the same key. From a hashing standpoint, we can think of the algorithm as splitting nodes to take care of hash-value collisions: hence the name "extendible hashing." This method presents a very attractive alternative to B-trees and indexed sequential access because it always uses exactly two disk accesses for each search (like indexed sequential access), while still retaining the capability for efficient insertion (like B-trees) without wasting very much space.

Property 18.4 *With pages that can hold M records, extendible hashing may be expected to require about $1.44(N/M)$ pages for a file of N records. The directory may be expected to have about $N^{1+1/M}/M$ entries.*

This analysis is a complicated extension of the analysis of tries referred to in the previous chapter. When M is large, the amount of space wasted is roughly the same as for B-trees, but for small M the directory may get too large. ■

Even with hashing, extraordinary steps must be taken if large numbers of equal keys are present. They can make the directory artificially large; and the algorithm breaks down entirely if there are more equal keys than fit in one leaf page. (This actually occurs in our example, since we have five E's.) If many equal keys are present then we could (for example) assume distinct keys in the data structure and put pointers to linked lists of records containing equal keys in the leaf pages. To see the complication involved, consider what would happen if the last E were to be inserted into the structure in Figure 18.13.

A less disastrous situation to handle is that the insertion of one new key can cause the directory to split more than once. This occurs when one more bit is not enough to distinguish the keys on an overfull page. For example, if two keys with the value D=00100 were to be inserted into the extendible hashing structure of Figure 18.12, then two directory splits would be required because five bits are needed to distinguish D from E (the fourth bit doesn't help). This is simple to cope with in an implementation, but must not be overlooked.

Virtual Memory

The "easier way" discussed at the end of Chapter 13 for external sorting applies directly and trivially to the searching problem. A virtual memory is actually nothing more than a general-purpose external searching method: given an address (key), return the information associated with that address. However, direct use of the virtual memory is *not* recommended as an easy searching application. As mentioned in Chapter 13, virtual memories perform best when most accesses are relatively close to previous accesses. Sorting algorithms can be adapted to this, but the very nature of searching is that requests are for information from arbitrary parts of the database.

□

Exercises

1. Give the contents of the B-tree that results when the keys E A S Y Q U E S T I O N are inserted in that order into an initially empty tree, with $M = 5$.

2. Give the contents of the B-tree that results when the keys E A S Y Q U E S T I O N are inserted in that order into an initially empty tree, with $M = 6$. Use the variant of the method in which all the records are kept in external nodes.

3. Draw the B-tree that is built when sixteen equal keys are inserted into an initially empty tree, with $M = 5$.

4. Suppose that one page from the database is destroyed. Describe how you would handle this event for each of the B-tree structures described in the text.

5. Give the contents of the extendible hashing table that results when the keys E A S Y Q U E S T I O N are inserted in that order into an initially empty table, with a page capacity of four records. (Following the example in the text, don't hash, but use the five-bit binary representation of i as the key for the ith letter.)

6. Give a sequence of as few distinct keys as possible which make an extendible hashing directory grow from an intially empty table to size 16, from an initially empty table, with a page capacity of three records.

7. Outline a method for *deleting* an item from an extendible hashing table.

8. Why are "top-down" B-trees better than "bottom-up" B-trees for concurrent access to data? (For example, suppose two programs are trying to insert a new node at the same time.)

9. Implement *search* and *insert* for *internal* searching using the extendible hashing method.

10. Discuss how the program of the previous exercise compares with double hashing and radix-trie searching for internal searching applications.

SOURCES for Searching

The primary references for this section are Knuth's Volume 3, Gonnet's book, and Mehlhorn's book. Most of the algorithms we've studied are treated in great detail in these books, with mathematical analyses and suggestions for practical applications. Classical methods are covered by Knuth; the more recent methods are described by Gonnet and Mehlhorn, with further references to the literature. These three sources describe nearly all the "beyond the scope of this book" analyses referred to in this section.

The material in Chapter 15 comes from Guibas and Sedgewick's 1978 paper, which shows how to fit many classical balanced-tree algorithms into the "red-black" framework and gives several other implementations. There is actually quite a large literature on balanced trees: the persistent reader might begin with that paper. Mehlhorn's book has detailed proofs of properties of red-black trees and similar structures, and references to more recent work. Comer's 1979 survey discusses B-trees from a more practical point of view.

The extendible hashing algorithm presented in Chapter 18 comes from Fagin, Nievergelt, Pippenger and Strong's 1979 paper. This paper is a must for anyone wishing further information on external searching methods: it ties together material from our Chapters 16 and 17 to bring out the algorithm in Chapter 18. The paper also contains a detailed analysis and a discussion of practical ramifications.

Many practical applications of the methods discussed here, especially Chapter 18, arise within the context of database systems. The study of databases is a broad, maturing field, but basic search algorithms continue to play a fundamental part in most systems. An introduction to this field is given in Ullman's 1982 book.

D. Comer, "The ubiquitous B-tree," *Computing Surveys*, **11** (1979).

R. Fagin, J. Nievergelt, N. Pippenger and H. R. Strong, "Extendible hashing—a fast access method for dynamic files," *ACM Transactions on Database Systems*, **4**, 3 (September, 1979).

G. H. Gonnet, *Handbook of Algorithms and Data Structures*, Addison-Wesley, Reading, MA, 1984.

L. Guibas and R. Sedgewick, "A dichromatic framework for balanced trees," in *19th Annual Symposium on Foundations of Computer Science*, IEEE, 1978. Also in *A Decade of Progress 1970–1980*, Xerox PARC, Palo Alto, CA.

D. E. Knuth, *The Art of Computer Programming. Volume 3: Sorting and Searching*, Addison-Wesley, Reading, MA, 1975.

K. Mehlhorn, *Data Structures and Algorithms 1: Sorting and Searching*, Springer-Verlag, Berlin, 1984.

J. D. Ullman, *Principles of Database Systems*, Computer Science Press, Rockville, MD, 1982.

String Processing

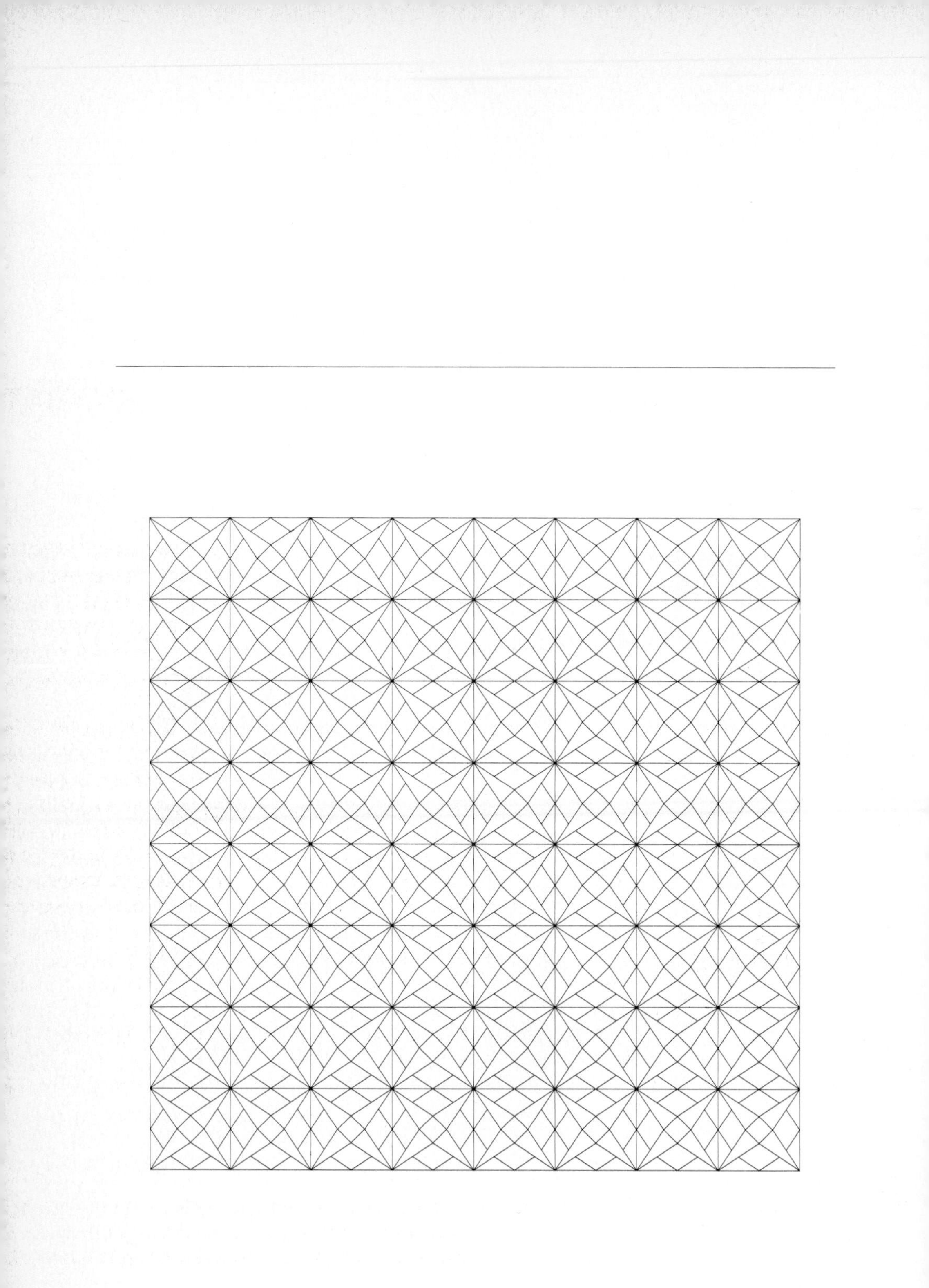

19

String Searching

☐ Data to be processed often does not decompose logically into independent records with small identifiable pieces. This type of data is characterized only by the fact that it can be written down as a *string*: a linear (typically very long) sequence of characters.

Strings are obviously central in word-processing systems, which provide a variety of capabilities for the manipulation of text. Such systems process *text strings*, which might be loosely defined as sequences of letters, numbers, and special characters. These objects can be quite large (for example, this book contains over a million characters), and efficient algorithms play an important role in manipulating them.

Another type of string is the *binary string*, a simple sequence of 0 and 1 values. This is in a sense merely a special type of text string, but it is worth making the distinction both because different algorithms are appropriate and also because binary strings arise naturally in many applications. For example, some computer graphics systems represent pictures as binary strings. (This book was printed on such a system: the present page was represented at one time as a binary string consisting of millions of bits.)

In one sense, text strings are quite different objects from binary strings, since they are made up of characters from a large alphabet. In another sense, though, the two types of strings are equivalent, since each text character can be represented by (say) eight binary bits and a binary string can be viewed as a text string by treating eight-bit chunks as characters. We'll see that the size of the alphabet from which the characters are taken to form a string is an important factor in the design of string-processing algorithms.

A fundamental operation on strings is *pattern matching*: given a *text* string of length N and a *pattern* of length M, find an occurrence of the pattern within the text. (We use the term "text" even when referring to a sequence of 0-1 values or some other special type of string.) Most algorithms for this problem can easily be extended to find *all* occurrences of the pattern in the text, since they scan through the text sequentially and can be restarted at the point directly after the beginning of a match to find the next match.

The pattern-matching problem can be characterized as a searching problem with the pattern as the key, but the searching algorithms we have studied do not apply directly because the pattern can be long and because it "lines up" with the text in an unknown way. It is an interesting problem: several very different (and surprising) algorithms have only recently been discovered which not only provide a spectrum of useful practical methods but also illustrate some fundamental algorithm design techniques.

A Short History

The algorithms we'll be examining have an interesting history: we'll summarize it here to help place the various methods into perspective.

There is an obvious brute-force algorithm for string processing which is in widespread use. While it has a worst-case running time proportional to MN, the strings which arise in many applications lead to a running time which is virtually always proportional to $M + N$. Furthermore, it is well suited to good architectural features on most computer systems, so an optimized version provides a "standard" which is difficult to beat with a clever algorithm.

In 1970, S. A. Cook proved a theoretical result about a particular type of abstract machine which implied that an algorithm exists solving the pattern-matching problem in time proportional to $M + N$ in the worst case. D. E. Knuth and V. R. Pratt laboriously followed through the construction Cook used to prove his theorem (which was not intended at all to be practical) and got an algorithm which they were then able to refine into a relatively simple practical algorithm. This seemed a rare and satisfying example of a theoretical result with immediate (and unexpected) practical applicability. But it turned out that J. H. Morris had discovered virtually the same algorithm as a solution to an annoying practical problem confronting him when implementing a text editor (he didn't want ever to "back up" in the text string). However, the fact that the same algorithm arose from two such different approaches lends it credibility as a fundamental solution to the problem.

Knuth, Morris, and Pratt didn't get around to publishing their algorithm until 1976, and in the meantime R. S. Boyer and J. S. Moore (and, independently, R. W. Gosper) discovered an algorithm which is much faster in many applications, since it often examines only a fraction of the characters in the text string. Many text editors use this algorithm to achieve a noticeable decrease in response time for string searches.

Both the Knuth-Morris-Pratt and the Boyer-Moore algorithms require some complicated preprocessing on the pattern that is difficult to understand and has limited the extent to which they are used. (In fact, the story goes that an unknown systems programmer found Morris's algorithm too difficult to understand and replaced it with a brute-force implementation.)

In 1980, R. M. Karp and M. O. Rabin observed that the problem is not as different from the standard searching problem as it had seemed, and came up with an algorithm almost as simple as the brute-force algorithm which virtually always runs in time proportional to $M+N$. Furthermore, their algorithm extends easily to two-dimensional patterns and text, which makes it more useful than the others for picture processing.

This story illustrates that the search for a "better algorithm" is still very often justified; indeed, one suspects that there are still more developments on the horizon even for this problem.

Brute-Force Algorithm

The obvious method for pattern matching that immediately comes to mind is just to check, for each possible position in the text at which the pattern could match, whether it does in fact match. The following program searches in this way for the first occurrence of a pattern $p[1..M]$ in a text string $a[1..N]$:

```
function brutesearch: integer;
  var i,j: integer;
  begin
  i:=1; j:=1;
  repeat
    if a[i]=p[j]
      then begin i:=i+1; j:=j+1 end
      else  begin i:=i-j+2; j:=1 end;
  until (j>M) or (i>N);
  if j>M then brutesearch:=i-M else brutesearch:=i
  end;
```

The program keeps one pointer (i) into the text and another pointer (j) into the pattern. As long as they point to matching characters, both pointers are incremented. If the end of the pattern is reached ($j>M$), then a match has been found. If i and j point to mismatching characters, then j is reset to point to the beginning of the pattern and i is reset to correspond to moving the pattern to the right one position for matching against the text. If the end of the text is reached ($i>N$), then there is no match. If the pattern does not occur in the text, the value $N+1$ is returned.

In a text-editing application, the inner loop of this program is seldom iterated and the running time is very nearly proportional to the number of text characters examined. For example, suppose that we are looking for the pattern STING in the

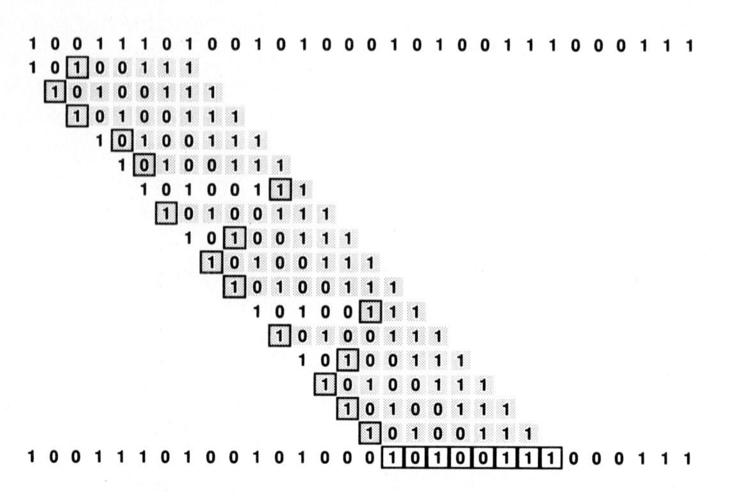

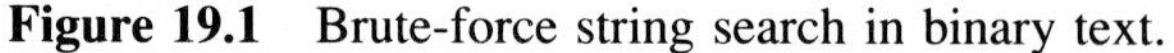
Figure 19.1 Brute-force string search in binary text.

text string

A STRING SEARCHING EXAMPLE CONSISTING OF ...

Then the statement *j:=j+1* is executed only four times (once for each S, but twice for the first ST) before the actual match is encountered.

On the other hand, brute-force searching can be very slow for some patterns, for example if the text is binary (two-character), as might occur in picture processing and systems programming applications. Figure 19.1 shows what happens when the algorithm is used to search for the pattern 10100111 in a long binary text string. Each line (except the last, which shows the match) consists of zero or more characters which match the pattern followed by one mismatch. These are the “false starts” that occur when trying to find the pattern; an obvious goal in algorithm design is to try to limit the number and length of these.

Property 19.1 *Brute-force string searching can require about NM character comparisons.*

The worst case is when both pattern and text are all 0’s followed by a 1. Then for each of the $N - M + 1$ possible match positions, all the characters in the pattern are checked against the text, for a total cost of $M(N - M + 1)$. Normally M is very small compared to N, so the total is about NM. ■

Such degenerate strings are not likely in English (or Pascal) text, but they may well occur when binary texts are being processed, so we seek better algorithms.

Knuth-Morris-Pratt Algorithm

The basic idea behind the algorithm discovered by Knuth, Morris, and Pratt is this:

when a mismatch is detected, our "false start" consists of characters that we know in advance (since they're in the pattern). Somehow we should be able to take advantage of this information instead of backing up the i pointer over all those known characters.

For a simple example of this, suppose that the first character in the pattern doesn't appear again in the pattern (say the pattern is 10000000). Then, suppose we have a false start j characters long at some position in the text. When the mismatch is detected, we know, by dint of the fact that j characters have matched, that we needn't "back up" the text pointer i, since none of the previous $j-1$ characters in the text can match the first character in the pattern. This change could be implemented by replacing *i:=i–j+2* in the program above by *i:=i+1*. The practical effect of this change is limited because such a specialized pattern is not particularly likely to occur, but the idea is worth thinking about and the Knuth-Morris-Pratt algorithm is a generalization of it. Surprisingly, it is always possible to arrange things so that the i pointer is never decremented.

Fully skipping past the pattern on detecting a mismatch as described in the previous paragraph won't work when the pattern could match itself at the point of the mismatch. For example, when searching for 10100111 in 1010100111 we first detect the mismatch at the fifth character, but we had better back up to the third character to continue the search, since otherwise we would miss the match. But we can figure out ahead of time exactly what to do, because it depends only on the pattern, as shown in Figure 19.2.

The array *next*[*1..M*] will be used to determine how far to back up when a mismatch is detected. Imagine that we slide a copy of the first $j-1$ characters of the pattern over itself, from left to right, starting with the first character of the

```
j   next[j]

2     1       10100111
               10100111

3     1       10100111
                10100111

4     2       10100111
                10100111

5     3       10100111
                10100111

6     1       10100111
                   10100111

7     2       10100111
                   10100111

8     2       10100111
                    10100111
```

Figure 19.2 Restart positions for Knuth-Morris-Pratt search.

copy over the second character of the pattern and stopping when all overlapping characters match (or there are none). These overlapping characters define the next possible place the pattern could match, if a mismatch is detected at $p[j]$. The distance to back up in the pattern ($next[j]$) is exactly one plus the number of the overlapping characters. Specifically, for $j>1$, the value of $next[j]$ is the maximum $k<j$ for which the first $k-1$ characters of the pattern match the last $k-1$ characters of the first $j-1$ characters of the pattern. As we'll soon see, it is convenient to define $next[1]$ to be 0.

This *next* array immediately gives a way to limit (in fact, as we'll see, eliminate) the "backup" of the text pointer i, as discussed above. When i and j point to mismatching characters (testing for a pattern match beginning at position $i-j+1$ in the text string), then the next possible position for a pattern match is beginning at position $i-next[j]+1$. But by definition of the *next* table, the first $next[j]-1$ characters at that position match the first $next[j]-1$ characters of the pattern, so there's no need to back up the i pointer that far: we can simply leave the i pointer unchanged and set the j pointer to $next[j]$, as in the following program:

```
function kmpsearch: integer;
  var i,j: integer;
  begin
  i:=1; j:=1; initnext;
  repeat
    if (j=0) or (a[i]=p[j])
      then begin i:=i+1; j:=j+1 end
      else  begin j:=next[j] end;
  until (j>M) or (i>N);
  if j>M then kmpsearch:=i-M else kmpsearch:=i;
  end;
```

When $j=1$ and $a[i]$ does not match the pattern, there is no overlap, so we want to increment i and set j to the beginning of the pattern. This is achieved by defining $next[1]$ to be 0, which results in j being set to 0; then i is incremented and j set to 1 next time through the loop. (For this trick to work, the pattern array must be declared to start at 0, since otherwise standard Pascal will complain about 'subscript out of range' when $j=0$ even though it doesn't really have to access $p[0]$ to determine the truth of the **or**.) Functionally, this program is the same as *brutesearch*, but it is likely to run faster for patterns which are highly self-repetitive.

It remains to compute the *next* table. The program for this is short but tricky: it is basically the same program as above except that it is used to match the pattern against itself.

```
procedure initnext;
  var i,j: integer;
  begin
  i:=1; j:=0; next[1]:=0;
  repeat
    if (j=0) or (p[i]=p[j])
      then begin i:=i+1; j:=j+1; next[i]:=j end
      else  begin j:=next[j] end;
  until i>=M;
  end;
```

Just after i and j are incremented, it has been determined that the first $j-1$ characters of the pattern match the characters in positions $p[i-j-1..i-1]$, the last $j-1$ characters in the first $i-1$ characters of the pattern. And this is the largest j with this property, since otherwise a "possible match" of the pattern with itself would have been missed. Thus, j is exactly the value to be assigned to $next[i]$.

An interesting way to view this algorithm is to consider the pattern as fixed, so that the *next* table can be "wired in" to the program. For example, the following program is exactly equivalent to the program above for the pattern that we've been considering, but it's likely to be much more efficient.

```
   i:=0;
0: i:=i+1;
1: if a[i]<>'1' then goto 0; i:=i+1;
2: if a[i]<>'0' then goto 1; i:=i+1;
3: if a[i]<>'1' then goto 1; i:=i+1;
4: if a[i]<>'0' then goto 2; i:=i+1;
5: if a[i]<>'0' then goto 3; i:=i+1;
6: if a[i]<>'1' then goto 1; i:=i+1;
7: if a[i]<>'1' then goto 2; i:=i+1;
8: if a[i]<>'1' then goto 2; i:=i+1;
   search:=i-8;
```

The **goto** labels in this program correspond precisely to the *next* table. In fact, the *initnext* program above which computes the *next* table can easily be modified to output this program! To avoid checking whether $i>N$ each time i is incremented, we assume that the pattern itself is stored at the end of the text as a sentinel, in $a[N+1..N+M]$. (This optimization could also be applied to the standard implementation.) This is a simple example of a "string-searching compiler": given a pattern, we can produce a very efficient program to scan for that pattern in an arbitrarily long text string. We'll see generalizations of this concept in the next two chapters.

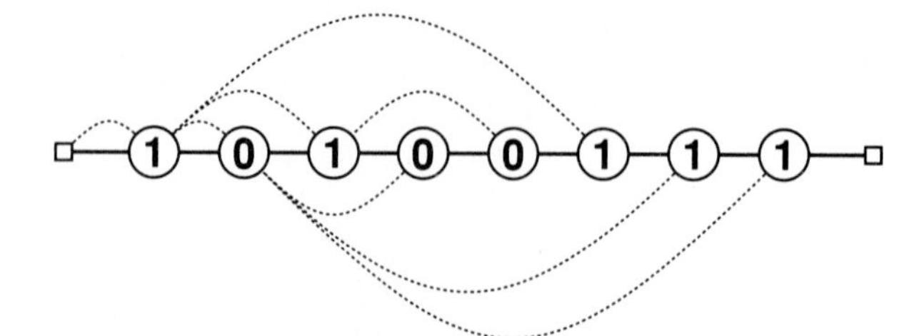

Figure 19.3 Finite state machine for the Knuth-Morris-Pratt algorithm.

The program above uses just a few very basic operations to solve the string searching problem. This means that it can easily be described in terms of a very simple machine model called a *finite-state machine*. Figure 19.3 shows the finite-state machine for the program above.

The machine consists of *states* (indicated by circled numbers) and *transitions* (indicated by lines). Each state has two transitions leaving it: a *match* transition (solid line, going right) and a *non-match* transition (dotted line, going left). The states are where the machine executes instructions; the transitions are the **goto** instructions. When in the state labeled "x," the machine can perform only one instruction: "if the current character is x then scan past it and take the match transition, otherwise take the non-match transition." To "scan past" a character means to take the next character in the string as the "current character"; the machine scans past characters as it matches them. There are two exceptions to this: the first state always takes a match transition and scans to the next character (essentially this corresponds to scanning for the first occurrence of the first character in the pattern), and the last state is a "halt" state indicating that a match has been found. In the next chapter we'll see how to use a similar (but more powerful) machine to help develop a much more powerful pattern-matching algorithm.

The alert reader may have noticed that there's still some room for improvement in this algorithm, because it doesn't take into account the character which caused the mismatch. For example, suppose that our text begins with 1011 and we are searching for our sample pattern 10100111. After matching 101, we find a mismatch on the fourth character; at this point the *next* table says to check the second character of the pattern against the fourth character of the text, since, on the basis of the 101 match, the first character of the pattern may line up with the third character of the text (but we don't have to compare these because we know that they're both 1's). However, we could not have a match here: from the mismatch, we know that the next character in the text is not 0, as required by the pattern. Another way to see this is to look at the version of the program with the next table "wired in": at label 4 we go to 2 if $a[i]$ is not 0, but at label 2 we go to 1 if $a[i]$ is not 0. Why not just go to 1 directly? Figure 19.4 shows the improved version of the finite-state machine for our example.

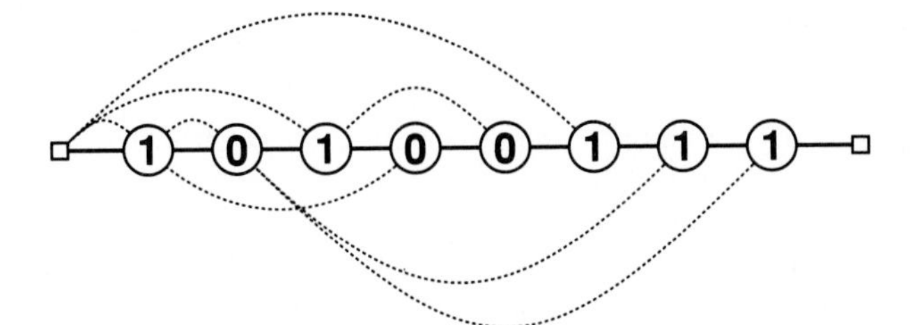

Figure 19.4 Knuth-Morris-Pratt finite-state machine (improved).

Fortunately, it is easy to put this change into the algorithm. We need only replace the statement *next*[*i*]:=*j* in the *initnext* program by

```
if p[j]<>p[i] then next[i]:=j else next[i]:=next[j];
```

Since we are proceeding from left to right, the value of *next* that is needed has already been computed, so we just use it.

Property 19.2 *Knuth-Morris-Pratt string searching never uses more than* $M+N$ *character comparisons.*

This property is illustrated in Figure 19.5, and it is also obvious from the code: we either increment *j* or reset it from the *next* table at most once for each *i*. ■

Figure 19.5 shows that this method certainly uses far fewer comparisons than the brute-force method for our binary example. However, the Knuth-Morris-Pratt algorithm is not likely to be significantly faster than the brute-force method in many actual applications, because few applications involve searching for highly self-repetitive patterns in highly self-repetitive text. However, the method does have a major practical advantage: it proceeds sequentially through the input and never "backs up" in the input. This makes it convenient for use on a large file being read in from some external device. (Algorithms requiring backup need some complicated buffering in this situation.)

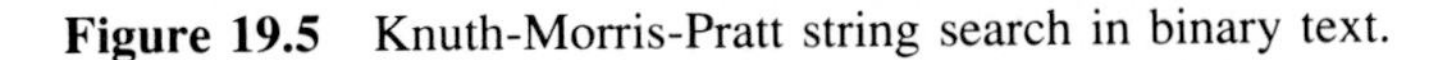

Figure 19.5 Knuth-Morris-Pratt string search in binary text.

Boyer-Moore Algorithm

If "backing up" is not difficult, then a significantly faster string-searching method can be developed by scanning the pattern from right to left when trying to match it against the text. When searching for our sample pattern 10100111, if we find matches on the eighth, seventh, and sixth character but not on the fifth, then we can immediately slide the pattern seven positions to the right, and check the fifteenth character next, because our partial match found 111, which might appear elsewhere in the pattern. Of course, the pattern at the end does appear elsewhere in general, so we need a *next* table as above.

A right-to-left version of the *next* table for the pattern 10110101 is shown in Figure 19.6: in this case *next*[*j*] is the number of character positions by which the pattern can be shifted to the right given that a mismatch in a right-to-left scan occurred on the *j*th character from the right in the pattern. This is found as before, by sliding a copy of the pattern over the last $j-1$ characters of itself from left to right, starting with the next-to-last character of the copy lined up with the last character of the pattern and stopping when all overlapping characters match (also taking into account the character which caused the mismatch). For example, *next*[3]=7 because, if there is a match of the last two characters and then a mismatch in a right-to-left scan, then 001 must have been encountered in the text; this doesn't appear in the pattern, except possibly if the 1 lines up with the first character in the pattern, so we can slide 7 positions to the right.

This leads directly to a program which is quite similar to the above implementation of the Knuth-Morris-Pratt method. We won't explore this in more detail because there is a quite different way to skip over characters with right-to-left

j	*next*[*j*]	
2	4	10110101 10110101
3	7	10110101 10110101
4	2	10110101 10110101
5	5	10110101 10110101
6	5	10110101 10110101
7	5	10110101 10110101
8	5	10110101 10110101

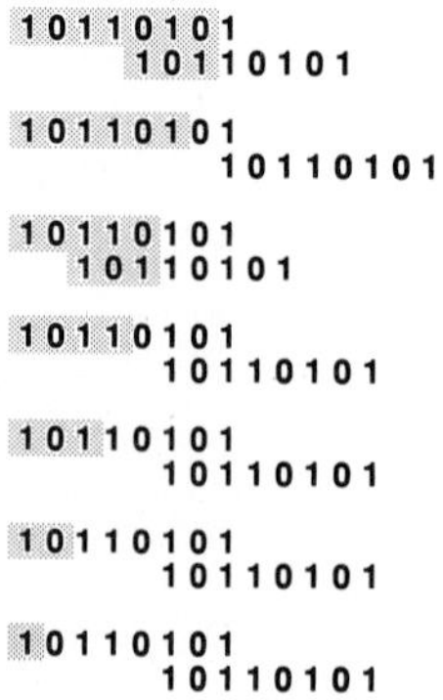

Figure 19.6 Restart Positions for Boyer-Moore Search.

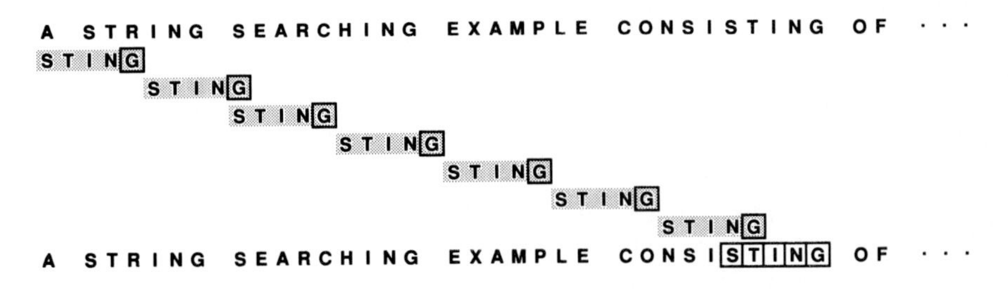

Figure 19.7 Boyer-Moore string search using the mismatched character heuristic.

pattern scanning which is much better in many cases.

The idea is to decide what to do next on the basis of the character that caused the mismatch in the *text* as well as the pattern. The preprocessing step is to decide, for each possible character that could occur in the text, what we would do if that character were to cause the mismatch. The simplest realization of this leads immediately to a quite useful program.

Figure 19.7 shows this method on our first sample text. Proceeding from right to left to match the pattern, we first check the G in the pattern against the R (the fifth character) in the text. Not only do these not match, but also we can notice that R appears *nowhere* in the pattern, so we might as well slide it all the way past the R. The next comparison is of the G in the pattern against the fifth character following the R (the S in SEARCHING). This time, we can slide the pattern to the right until its S matches the S in the text. Then the G in the pattern is compared against the C in SEARCHING, which doesn't appear in the pattern, so the pattern can be slid five more places to the right. After three more five-character skips, we arrive at the T in CONSISTING, at which point we align the pattern so that its T matches the T in the text and find the full match. This method brings us right to the match position at a cost of examining only seven characters in the text (and five more to verify the match)!

This "mismatched-character" algorithm is quite easy to implement. It simply improves a brute-force right-to-left pattern scan to initialize an array *skip* that tells, for each character in the alphabet, how far to skip if that character appears in the text and causes a mismatch during the string search. There must be an entry in *skip* for each character that possibly could occur in the text: for simplicity, we assume that we have a **function** *index*(*c*: *char*): *integer*; that returns 0 for blanks and i for the ith letter of the alphabet; we also assume a **procedure** *initskip* which initializes the *skip* array to M for characters not in the pattern and then, for j from 1 to M, sets *skip*[*index*(*p*[*j*])] to $M-j$. Then the implementation is straightforward:

```
function mischarsearch: integer;
  var i,j: integer;
  begin
  i:=M; j:=M; initskip;
  repeat
    if a[i]=p[j]
      then begin i:=i-1; j:=j-1 end
      else
        begin
        if M-j+1>skip[index(a[i])]
          then i:=i+M-j+1
          else i:=i+skip[index(a[i])];
        j:=M;
        end;
  until (j<1) or (i>N);
  mischarsearch:=i+1
  end;
```

The statement $i:=i+M-j+1$ resets i to the next position in the text string (as the pattern moves from left right across it); then $j:=M$ resets the pattern pointer to prepare for a right-to-left character-by-character match. The next statement moves the pattern even further across the text, if warranted. For the pattern STING, the *skip* entry for G would be 0, the entry for N would be 1, the entry for I would be 2, the entry for T would be 3, the entry for S would be 4, and the entries for all other letters would be 5. Thus, for example, when an S is encountered during a right-to-left search, the i pointer is incremented by 4 so that the end of the pattern is aligned four positions to the right of the S (and consequently the S in the pattern lines up with the S in the text). If there were more than one S in the pattern, we would want to use the rightmost one for this calculation: hence the *skip* array is built by scanning from left to right.

Boyer and Moore suggested combining the two methods we have outlined for right-to-left pattern scanning, choosing the larger of the two skips called for.

Property 19.3 *Boyer-Moore string searching never uses more than $M+N$ character comparisons, and uses about N/M steps if the alphabet is not small and the pattern is not long.*

The algorithm is linear in the worst case in the same way as the Knuth-Morris Pratt method (the implementation given above, which only does one of the two Boyer-Moore heuristics, is not linear). The "average-case" result N/M can be proved for various random string models, but these tend to be unrealistic, so we shall skip the details. In many practical situations it is true that all but a few of the alphabet characters appear nowhere in the pattern, so each comparison leads to

M characters being skipped, and this gives the stated result. ■

The mismatched character algorithm obviously won't help much for binary strings, because there are only two possibilities for characters which cause the mismatch (and these are both likely to be in the pattern). However, the bits can be grouped together to make "characters" which can be used exactly as above. If we take b bits at a time, then we need a *skip* table with 2^b entries. The value of b should be chosen small enough so that this table is not too large, but large enough that most b-bit sections of the text are not likely to be in the pattern. Specifically, there are $M - b + 1$ different b-bit sections in the pattern (one starting at each bit position from 1 through $M - b + 1$), so we want $M - b + 1$ to be significantly less than 2^b. For example, if we take b to be about $\lg(4M)$, then the *skip* table will be more than three-quarters filled with M entries. Also b must be less than $M/2$, since otherwise we could miss the pattern entirely if it were split between two b-bit text sections.

Rabin-Karp Algorithm

A brute-force approach to string searching which we didn't examine above would be to exploit a large memory by treating each possible M-character section of the text as a key in a standard hash table. But it is not necessary to keep a whole hash table, since the problem is set up so that only one key is being sought; all we need to do is compute the hash function for each of the possible M-character sections of the text and check if it is equal to the hash function of the pattern. The problem with this method is that it seems at first to be just as hard to compute the hash function for M characters from the text as it is merely to check to see if they're equal to the pattern. Rabin and Karp found an easy way to get around this difficulty for the hash function we used in Chapter 16: $h(k) = k \bmod q$ where q (the table size) is a large prime. In this case, nothing is stored in the hash table, so q can be taken to be very large.

The method is based on computing the hash function for position i in the text given its value for position $i - 1$, and follows quite directly from the mathematical formulation. Let's assume that we translate our M characters to numbers by packing them together in a computer word, which we then treat as an integer. This corresponds to writing the characters as numbers in a base-d number system, where d is the number of possible characters. The number corresponding to $a[i..i+M-1]$ is thus

$$x = a[i]d^{M-1} + a[i+1]d^{M-2} + \cdots + a[i+M-1]$$

and we can assume that we know the value of $h(x) = x \bmod q$. But shifting one position right in the text simply corresponds to replacing x by

$$(x - a[i]d^{M-1})d + a[i+M].$$

A fundamental property of the mod operation is that we can perform it at any time during these operations and still get the same answer. Put another way, if we take the remainder when divided by q after each arithmetic operation (to keep the numbers that we're dealing with small), then we get the same answer as if we were to perform all of the arithmetic operations, then take the remainder when divided by q.

This leads to the very simple pattern-matching algorithm implemented below. The program assumes the same *index* function as above, but *d=32* is used for efficiency (the multiplications might be implemented as shifts).

```
function rksearch: integer;
  const q=33554393; d=32;
  var h1,h2,dM,i: integer;
  begin
  dM:=1; for i:=1 to M-1 do dM:=(d*dM) mod q;
  h1:=0; for i:=1 to M do h1:=(h1*d+index(p[i])) mod q;
  h2:=0; for i:=1 to M do h2:=(h2*d+index(a[i])) mod q;
  i:=1;
  while (h1<>h2) and (i<=N-M) do
    begin
    h2:=(h2+d*q-index(a[i])*dM) mod q;
    h2:=(h2*d+index(a[i+M])) mod q;
    i:=i+1;
    end;
  rksearch:=i;
  end;
```

The program first computes a hash value *h1* for the pattern, then a hash value *h2* for the first M characters of the text. (It also computes the value of d^{M-1} mod q in the variable *dM*.) Then it proceeds through the text string, using the technique above to compute the hash function for the M characters starting at position i for each i and comparing each new hash value to *h1*. The prime q is chosen to be as large as possible, but small enough that $(d+1)*q$ doesn't cause overflow: this requires fewer **mod** operations than if we used the largest representable prime. (An extra $d*q$ is added during the *h2* calculation to make sure that everything stays positive so that the **mod** operation works as it should.)

Property 19.4 *Rabin-Karp pattern matching is extremely likely to be linear.*

This algorithm obviously takes time proportional to $N+M$, but note that it really only finds a position in the text which has the same hash value as the pattern. To be sure, we really should do a direct comparison of that text with the pattern. However, the use of the very large value of q, made possible by the **mod** computations and by the fact that we needn't keep the actual hash table around, makes it extremely unlikely that a collision will occur. Theoretically, this algorithm could still take $O(NM)$ steps in the (unbelievably) worst case, but in practice it can be relied upon to take about $N + M$ steps. ■

Multiple Searches

The algorithms we've been discussing are all oriented towards a specific string-searching problem: find an occurrence of a given pattern in a given text string. If the same text string is to be the object of many pattern searches, then it will be worthwhile to do some processing on the string to make subsequent searches efficient.

If there are a large number of searches, the string-searching problem can be viewed as a special case of the general searching problem that we studied in the previous section. We simply treat the text string as N overlapping "keys," the ith key defined to be $a[1..N]$, the entire text string starting at position i. Of course, we manipulate not the keys themselves but pointers to them: when we need to compare keys i and j we do character-by-character compares starting at positions i and j in the text string. (If we use a "sentinel" character larger than all other characters at the end, then one of the keys is always greater than the other.) Then the hashing, binary tree, and other algorithms in the previous section can be used directly. First, an entire structure is built up from the text string, and then efficient searches can be performed for particular patterns.

Many details need to be worked out in applying searching algorithms to string searching in this way; our intent is to point this out as a viable option for some string-searching applications. Different methods will be appropriate in different situations. For example, if the searches will always be for patterns of the same length, a hash table constructed with a single scan, as in the Rabin-Karp method, will yield constant search times on the average. On the other hand, if the patterns are to be of varying length, then one of the tree-based methods might be appropriate. (Patricia is especially adaptable to such an application.)

Other variations in the problem can make it significantly more difficult and lead to drastically different methods, as we'll discover in the next two chapters.

□

Exercises

1. Implement a brute-force pattern-matching algorithm that scans the pattern from right to left.
2. Give the *next* table for the Knuth-Morris-Pratt algorithm for the pattern AAAA-AAAA.
3. Give the *next* table for the Knuth-Morris-Pratt algorithm for the pattern ABRA-CADABRA.
4. Draw a finite state machine which can search for the pattern ABRACADABRA.
5. How would you search a text file for a string of 50 consecutive blanks?
6. Give the right-to-left *skip* table for the right-left scan for the pattern ABRA-CADABRA.
7. Construct an example for which the right-to-left pattern scan with only the mismatch heuristic performs badly.
8. How would you modify the Rabin-Karp algorithm to search for a given pattern with the additional proviso that the middle character is a "wild card" (any text character at all can match it)?
9. Implement a version of the Rabin-Karp algorithm to search for patterns in two-dimensional text. Assume both pattern and text are rectangles of characters.
10. Write programs to generate a random 1000-bit text string, then find all occurrences of the last k bits elsewhere in the string, for $k = 5, 10, 15$. (Different methods may be appropriate for different values of k.)

20

Pattern Matching

☐ It is often desirable to do string searching with somewhat less than complete information about the pattern to be found. For example, users of a text editor may wish to specify only part of a pattern, or to specify a pattern which could match a few different words, or to specify that any number of occurrences of some specific characters should be ignored. In this chapter we'll consider how *pattern matching* of this type can be done efficiently.

The algorithms in the previous chapter have a rather fundamental dependence on complete specification of the pattern, so we have to consider different methods. The basic mechanisms we will consider make possible a very powerful string-searching facility which can match complicated M-character patterns in N-character text strings in time proportional to MN^2 in the worst case, and much faster for typical applications.

First, we have to develop a way to describe the patterns: a "language" that can be used to specify, in a rigorous way, the kinds of partial-string-searching problems suggested above. This language will involve more powerful primitive operations than the simple "check if the ith character of the text string matches the jth character of the pattern" operation used in the previous chapter. In this chapter, we consider three basic operations in terms of an imaginary type of machine that can search for patterns in a text string. Our pattern-matching algorithm will be a way to simulate the operation of this type of machine. In the next chapter, we'll see how to translate from the pattern specification which the user employs to describe his string-searching task to the machine specification which the algorithm employs to actually carry out the search.

As we'll see, the solution we develop to the pattern-matching problem is intimately related to fundamental processes in computer science. For example, the method we will use in our program to perform the string-searching task implied by a given pattern description is akin to the method used by the Pascal system to perform the computational task implied by a given Pascal program.

Describing Patterns

We'll consider pattern descriptions made up of symbols tied together with the following three fundamental operations.

(*i*) *Concatenation.* This is the operation used in the last chapter. If two characters are adjacent in the pattern, then there is a match if and only if the same two characters are adjacent in the text. For example, AB means A followed by B.

(*ii*) *Or.* This is the operation that allows us to specify alternatives in the pattern. If we have an *or* between two characters, then there is a match if and only if either of the characters occurs in the text. We'll denote this operation by using the symbol + and use parentheses to combine it with concatenation in arbitrarily complicated ways. For example, A+B means "either A or B"; C(AC+B)D means "either CACD or CBD"; and (A+C)((B+C)D) means "either ABD or CBD or ACD or CCD."

(*iii*) *Closure.* This operation allows parts of the pattern to be repeated arbitrarily. If we have the closure of a symbol, then there is a match if and only if the symbol occurs any number of times (including 0). Closure will be denoted by placing a * after the character or parenthesized group to be repeated. For example, AB* matches strings consisting of an A followed by any number of B's, while (AB)* matches strings consisting of alternating A's and B's.

A string of symbols built up using these three operations is called a *regular expression.* Each regular expression describes many specific text patterns. Our goal is to develop an algorithm that determines if any of the pattern described by a given regular expression occur in a given text string.

We'll concentrate on concatenation, *or*, and closure in order to show the basic principles in developing a regular-expression pattern-matching algorithm. Various additions are commonly made in actual systems for convenience. For example, −A might mean "match any character *except* A." This *not* operation is the same as an *or* involving all the characters except A but is much easier to use. Similarly, "?" might mean "match any letter." Again, this is obviously much more compact than a large *or*. Other examples of additional symbols that make specification of large patterns easier are symbols to match the beginning or end of a line, any letter or any number, etc.

These operations can be remarkably descriptive. For example, the pattern description $? * (ie + ei)? *$ matches all words which have *ie* or *ei* in them (and so are likely to be misspelled!); $(1 + 01) * (0 + 1)$ describes all strings of 0's and 1's which do not have two consecutive 0's. Obviously there are many different pattern descriptions which describe the same strings: we must try to specify succinct pattern descriptions just as we try to write efficient algorithms.

The pattern-matching algorithm we'll examine may be viewed as a generalization of the brute force left-to-right string searching method (the first method looked at in Chapter 19). The algorithm looks for the leftmost substring in the text string which matches the pattern description by scanning the text string from left to right, testing at each position whether there is a substring beginning at that position which matches the pattern description.

Pattern Matching Machines

Recall that we can view the Knuth-Morris-Pratt algorithm as a finite-state machine constructed from the search pattern which scans the text. The method we will use for regular-expression pattern matching is a generalization of this.

The finite-state machine for the Knuth-Morris-Pratt algorithm changes from state to state by looking at a character from the text string and then changing to one state if there's a match, to another if not. A mismatch at any point means that the pattern cannot occur in the text starting at that point. The algorithm itself can be thought of as a simulation of the machine. The characteristic of the machine that makes it easy to simulate is that it is *deterministic*: each state transition is completely determined by the next input character.

To handle regular expressions, it will be necessary to consider a more powerful abstract machine. Because of the *or* operation, the machine can't determine whether or not the pattern could occur at a given point by examining just one character; in fact, because of closure, it can't even determine how many characters might need to be examined before a mismatch is discovered. The most natural way to overcome these problems is to endow the machine with the power of *nondeterminism*: when faced with more than one way to try to match the pattern, the machine should "guess" the right one! This operation seems impossible to allow, but we will see that it is easy to write a program to simulate the actions of such a machine.

Figure 20.1 shows a nondeterministic finite-state machine that could be used to search for the pattern description (A*B+AC)D in a text string. (The states are numbered, in a way that will become clear below.) Like the deterministic machine of the previous chapter, the machine can travel from a state labeled with a character

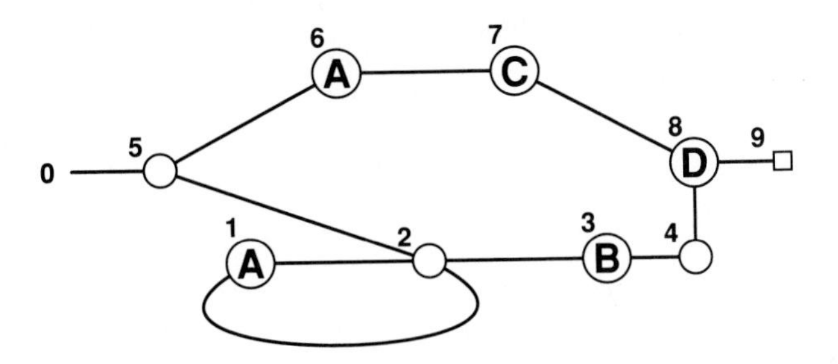

Figure 20.1 A nondeterministic pattern recognition machine for (A*B+AC)D.

to the state "pointed to" by that state by matching (and scanning past) that character in the text string. What makes the machine nondeterministic is that there are some states (called *null* states) which not only are not labeled, but also can "point to" two different successor states. (Some null states, such as state 4 in the diagram, are "no-op" states with one exit that don't affect the operation of the machine but facilitiate the implementation of the program which constructs the machine, as we'll see. State 9 is a null state with no exits, which stops the machine.) When in such a state, the machine can go to *either* successor state regardless of the input (without scanning past anything). The machine has the power to guess which transition will lead to a match for the given text string (if any will). Note that there are no "non-match" transitions as in the previous chapter: the machine fails to find a match only if there is no way even to guess a sequence of transitions leading to a match.

The machine has a unique *initial* state (indicated by the unattached line at the left) and a unique *final* state (the small square at the right). When started out in the initial state, the machine should be able to "recognize" any string described by the pattern by reading characters and changing state according to its rules, ending

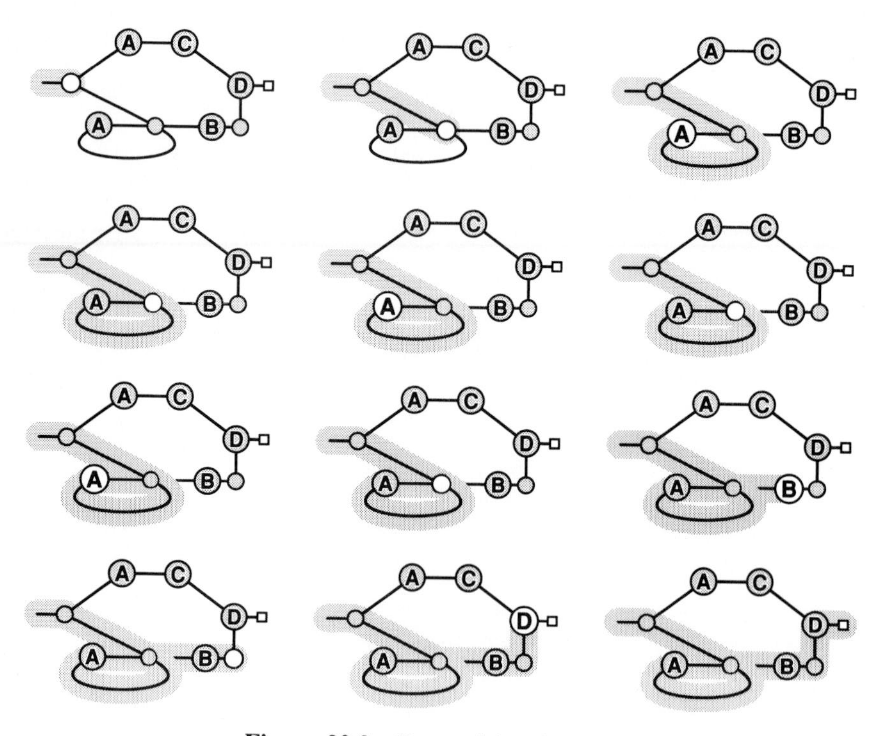

Figure 20.2 Recognizing AAABD.

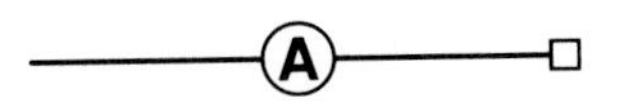

Figure 20.3 Two-state machine to recognize a character.

up in the "final state." Because the machine has the power of nondeterminism, it can guess the sequence of state changes that can lead to the solution. (But when we try to simulate the machine on a standard computer, we'll have to try all the possibilities.) For example, to determine if its pattern description (A*B+AC)D can occur in the text string

CDAABCAAABDDACDAAC

the machine would immediately report failure when started on the first or second character; it would work some to report failure on the next two characters; it would immediately report failure on the fifth or sixth characters; and it would guess the sequence of state transitions shown in Figure 20.2 to recognize AAABD if started on the seventh character.

We can construct the machine for a given regular expression by building partial machines for parts of the expression and defining the ways in which two partial machines can be composed into a larger machine for each of the three operations: concatenation, *or*, and closure.

We start with the trivial machine to recognize a particular character. It's convenient to write this as a two-state machine, with an initial state (which also recognizes the character) and a final state, as shown in Figure 20.3.

Now to build the machine for the concatenation of two expressions from the machines for the individual expressions, we simply merge the final state of the first with the initial state of the second, as shown in Figure 20.4.

Similarly, the machine for the *or* operation is built by adding a new null state pointing to the two initial states and making one final state point to the other, which becomes the final state of the combined machine, as shown in Figure 20.5.

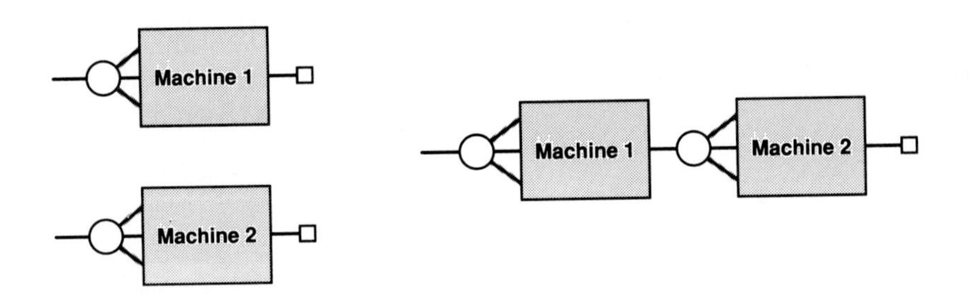

Figure 20.4 State machine construction: concatenation.

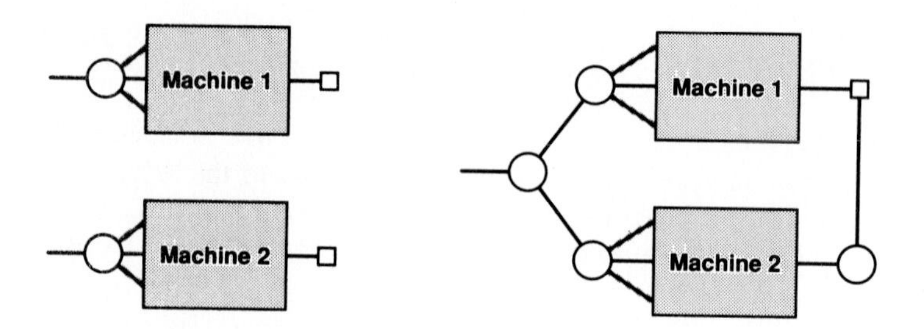

Figure 20.5 State machine construction: *or*.

Finally, the machine for the closure operation is built by making the final state the initial state and having it point back to the old initial state and a new final state, as shown in Figure 20.6.

A machine can be built that corresponds to any regular expression by successively applying these rules. The states for the example machine above are numbered in order of creation as the machine is built by scanning the pattern from left to right, so the construction of the machine from the rules above can be easily traced. Note that we have a two-state trivial machine for each letter in the regular expression and that each + and * causes one state to be created (concatenation causes one to be deleted), so the number of states is certainly less than twice the number of characters in the regular expression.

Representing the Machine

Our nondeterministic machines will all be constructed using only the three composition rules outlined above, and we can take advantage of their simple structure to manipulate them in a straightforward way. For example, no more than two lines leave any state. In fact, there are only two types of states: those labeled by a character from the input alphabet (with one line leaving) and unlabeled (null) states (with two or fewer lines leaving). This means that the machine can be represented with only a few pieces of information per node. Since we will often want to access states just by number, the most suitable organization for the machine is an array representation. We'll use the three parallel arrays *ch*, *next1*, and *next2*, indexed

Figure 20.6 State machine construction: closure.

state	*0*	*1*	*2*	*3*	*4*	*5*	*6*	*7*	*8*	*9*
ch[state]		*A*		*B*			*A*	*C*	*D*	
next1[state]	5	2	3	4	8	6	7	8	9	0
next2[state]	5	2	1	4	8	2	7	8	9	0

Figure 20.7 Array representation for machine of Figure 20.1.

by *state* to represent and access the machine. It would be possible to get by with two-thirds this amount of space, since each state really uses only two meaningful pieces of information, but we'll forgo this improvement for the sake of clarity and also because pattern descriptions are not likely to be particularly long.

The machine above can be represented as in Figure 20.7. The entries indexed by *state* can be interpreted as instructions to the nondeterministic machine of the form "If you are in *state* and you see *ch*[*state*] then scan the character and go to state *next1*[*state*] (or *next2*[*state*])." State 9 is the final state in this example, and State 0 is a pseudo-initial state whose *Next* entries are the number of the actual initial state. (Note the special representation used for null states with 0 or 1 exits.)

We've seen how to build up machines from regular expression pattern descriptions and how such machines might be represented as arrays. However, writing a program to do the translation from a regular expression to the corresponding nondeterministic machine representation is quite another matter. In fact, even writing a program to determine if a given regular expression is legal is challenging for the uninitiated. In the next chapter, we'll study this operation, called *parsing*, in much more detail. For the moment, we'll assume that this translation has been done, so that we have available the *ch*, *next1*, and *next2* arrays representing a particular nondeterministic machine that corresponds to the regular expression pattern description of interest.

Simulating the Machine

The last step in the development of a general regular-expression pattern-matching algorithm is to write a program that somehow simulates the operation of a nondeterministic pattern-matching machine. The idea of writing a program that can "guess" the right answer seems ridiculous. However, in this case it turns out that we can keep track of *all possible* matches in a systematic way, so that we do eventually encounter the correct one.

One possibility would be to develop a recursive program that mimics the nondeterministic machine (but tries all possibilities rather than guessing the right one). Instead of using this approach, we'll look at a nonrecursive implementation that exposes the basic operating principles of the method by keeping the states under consideration in a rather peculiar data structure called a *deque*.

The idea is to keep track of all states that could possibly be encountered while the machine is "looking at" the current input character. Each of these states is processed in turn: null states lead to two (or fewer) states, states for characters which do not match the current input are eliminated, and states for characters which do match the current input lead to new states for use when the machine is looking at the *next* input character. Thus, we want to maintain a list of all the states that the nondeterministic machine could possibly be in at a particular point in the text. The problem is to design an appropriate data structure for this list.

Processing null states seems to require a *stack*, since we are essentially postponing one of two things to be done, just as in recursion removal (so the new state should be put at the *beginning* of the current list, lest it get postponed indefinitely). Processing the other states seems to require a *queue*, since we don't want to examine states for the next input character until we've finished with the current character (so the new state should be put at the *end* of the current list). Rather than choosing between these two data structures, we'll use both! Deques ("double-ended queues") combine the features of stacks and queues: a deque is a list to which items can be added at either end. (Actually, we use an "output-restricted deque," since we always remove items from the beginning, not the end—that would be "dealing from the bottom of the deck.")

A crucial property of the machine is that it have no "loops" consisting of just null states, since otherwise it could decide nondeterministically to loop forever. It turns out that this implies that the number of states on the deque at any time is less than the number of characters in the pattern description.

The program given below uses a deque to simulate the actions of a nondeterministic pattern-matching machine as described above. While examining a particular character in the input, the nondeterministic machine can be in any one of several possible states: the program keeps track of these in a deque, using procedures *push*, *put*, and *pop*, like those in Chapter 3. Either an array representation (as in the queue implementation in Chapter 3) or a linked representation (as in the stack implementation in Chapter 3) could be used; the implementation is omitted.

The main loop of the program removes a state from the deque and performs the action required. If a character is to be matched, the input is checked for the required character: if it is found, the state transition is effected by putting the new state at the *end* of the deque (so that all states involving the current character are processed before those involving the next one). If the state is null, the two possible states to be simulated are put at the *beginning* of the deque. The states involving the current input character are kept separately from those involving the next by a marker $scan{=}-1$ in the deque: when *scan* is encountered, the pointer into the input string is advanced. The loop terminates when the end of the input is reached (no match found), state 0 is reached (legal match found). or only one item, the *scan* marker, is left on the deque (no match found). This leads directly to the following implementation:

```
function match(j: integer): integer;
  const scan=-1;
  var state,n1,n2: integer;
  begin
  dequeinit; put(scan);
  match:=j-1; state:=next1[0];
  repeat
    if state=scan then
      begin j:=j+1; put(scan) end
    else if ch[state]=a[j] then
      put(next1[state])
    else if ch[state]=' ' then
      begin
      n1:=next1[state]; n2:=next2[state];
      push(n1); if n1<>n2 then push(n2)
      end;
    state:=pop;
  until (j>N) or (state=0) or (dequeempty);
  if state=0 then match:=j-1;
  end;
```

This function takes as its argument the position j in the text string a at which it should start trying to match. It returns the index of the last character in the match found (if any; otherwise it returns $j-1$).

Figure 20.8 shows the contents of the deque each time a state is removed when our sample machine is run with the text string AAABD. This diagram assumes an array representation, as used for queues in Chapter 3: a plus sign is used to represent *scan*. Each time the *scan* marker reaches the front of the deque (on the bottom in the diagram), the j pointer is incremented to the next character in the text. Thus, we start with state 5 while scanning the first character in the text (the first A). First state 5 leads to states 2 and 6, then state 2 leads to states 1 and 3, all of which need to scan the same character and are on the beginning of the deque. Then state 1 leads to state 2, but at the end of the deque (for the next input character). State 3 leads to another state only while scanning a B, so it is ignored while an A is being scanned. When the "*scan*" sentinel finally reaches the front of the deque, we see that the machine could be in either state 2 or state 7 after scanning an A. Then the program tries states 2, 1, 3, and 7 while "looking at" the second A, to discover, the second time *scan* reaches the front of the deque, that state 2 is the only possibility after scanning AA. Now, while looking at the third A, the only possibilities are states 2, 1, and 3 (the AC possibility is now precluded). These three states are tried again, to lead eventually to state 4 after scanning AAAB. Continuing, the program goes to state 8, scans the D and ends up the final state. A match has been

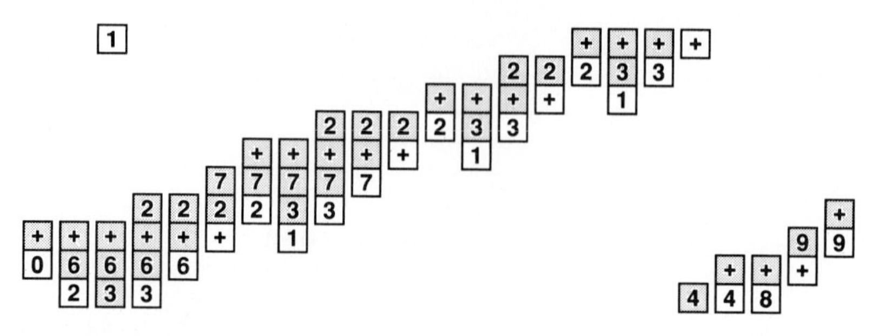

Figure 20.8 Contents of deque during recognition of AAABD.

found, but, more important, all transitions consistent with the text string have been considered.

Property 20.1 *Simulating the operation of an M-state machine to look for patterns in a text string of N characters involves NM state transitions in the worst case.*

The running time of this program obviously depends very heavily on the pattern being matched. However, for each of the N input characters, it processes at most M states of the machine, so the worst-case running time is proportional to MN *for each* starting position in the text. The grand total for determining whether any portion of the text string is described by the pattern, is $O(MN^2)$. ■

Not all nondeterministic machines can be simulated so efficiently, as discussed in more detail in Chapter 40, but the use of a simple hypothetical pattern-matching machine in this application leads to a quite reasonable algorithm for a quite difficult problem. However, to complete the algorithm, we need a program which translates arbitrary regular expressions into "machines" for interpretation by the above code. In the next chapter, we'll look at the implementation of such a program in the context of a more general discussion of compilers and parsing techniques.

□

Exercises

1. Give a regular expression for recognizing all occurrences of four or fewer consecutive 1's in a binary string.

2. Draw the nondeterministic pattern-matching machine for the pattern description (A+B)*+C.

3. Give the state transitions your machine from the previous exercise would make to recognize ABBAC.

4. Explain how you would modify the nondeterministic machine to handle the *not* function.

5. Explain how you would modify the nondeterministic machine to handle "don't-care" characters.

6. How many different patterns can be described by a regular expression with M *or* operators and no closure operators?

7. Modify *match* to handle regular expressions with the *not* function and "don't-care" characters.

8. Show how to construct a pattern description of length M and a text string of length N for which the running time of *match* is as large as possible.

9. Why must the deque in *match* contain only one "*scan*" sentinel?

10. Show the contents of the deque each time a state is removed when *match* is used to simulate the example machine in the text with the text string ACD.

21

Parsing

☐ Several fundamental algorithms have been developed to recognize legal computer programs and to decompose them into a form suitable for further processing. This operation, called *parsing*, has application beyond computer science, since it is directly related to the study of the structure of language in general. For example, parsing plays an important role in systems which try to "understand" natural (human) languages and in systems for translating from one language to another. One particular case of interest is translating from a "high-level" computer language like Pascal (suitable for human use) to a "low-level" assembly or machine language (suitable for machine execution). A program for doing such a translation is called a *compiler*. Actually, we've already touched upon a parsing method, in Chapter 4 when we built a tree representing an arithmetic expression.

Two general approaches are used for parsing. *Top-down* methods look for a legal program by first looking for parts of a legal program, then looking for parts of parts, etc. until the pieces are small enough to match the input directly. *Bottom-up* methods put pieces of the input together in a structured way making bigger and bigger pieces until a legal program is constructed. In general, top-down methods are recursive, bottom-up methods are iterative; top-down methods are thought to be easier to implement, bottom-up methods are thought to be more efficient. The method in Chapter 4 was bottom-up; in this chapter we study a top-down method in detail.

A full treatment of the issues involved in parser and compiler construction is clearly beyond the scope of this book. However, by building a simple "compiler" to complete the pattern-matching algorithm of the previous chapter, we will be able to consider some of the fundamental concepts involved. First we'll construct a top-down parser for a simple language for describing regular expressions. Then we'll modify the parser to make a program which translates regular expressions into pattern-matching machines for use by the *match* procedure of the previous chapter.

Our intent in this chapter is to give some feeling for the basic principles of parsing and compiling while at the same time developing a useful pattern-matching algorithm. Certainly we cannot treat the issues involved at the level of depth they deserve. The reader should note that subtle difficulties are likely to arise in applying the same approach to similar problems, and that compiler construction is a quite well-developed field with a variety of advanced methods available for serious applications.

Context-Free Grammars

Before we can write a program to determine whether a program written in a given language is legal, we need a description of exactly what constitutes a legal program. This description is called a *grammar*: to appreciate the terminology, think of the language as English and read "sentence" for "program" in the previous sentence (except for the first occurrence!). Programming languages are often described by a particular type of grammar called a *context-free grammar.* For example, the context-free grammar defining the set of all legal regular expressions (as described in the previous chapter) is given below.

$$\begin{aligned} \langle \mathit{expression} \rangle &::= \langle \mathit{term} \rangle \mid \langle \mathit{term} \rangle + \langle \mathit{expression} \rangle \\ \langle \mathit{term} \rangle &::= \langle \mathit{factor} \rangle \mid \langle \mathit{factor} \rangle \langle \mathit{term} \rangle \\ \langle \mathit{factor} \rangle &::= (\langle \mathit{expression} \rangle) \mid v \mid (\langle \mathit{expression} \rangle)* \mid v* \end{aligned}$$

This grammar describes regular expressions like those that we used in the last chapter, such as (1+01)*(0+1) or (A*B+AC)D. Each line in the grammar is called a *production* or *replacement rule.* The productions consist of *terminal* symbols (,), + and $*$ which are the symbols used in the language being described ("v," a special symbol, stands for any letter or digit); *nonterminal* symbols ⟨*expression*⟩, ⟨*term*⟩, and ⟨*factor*⟩ which are internal to the grammar; and *metasymbols* ::= and | which are used to describe the meaning of the productions. The ::= symbol, which may be read "*is a*," defines the left-hand side of the production in terms of the right-hand side; and the | symbol, which may be read as "*or*," indicates alternative choices. The various productions, though expressed in this concise symbolic notation, correspond in a simple way to an intuitive description of the grammar. For example, the second production in the example grammar might be read "a ⟨*term*⟩ is a ⟨*factor*⟩ or a ⟨*factor*⟩ followed by a ⟨*term*⟩." One nonterminal symbol, in this case ⟨*expression*⟩, is *distinguished* in the sense that a string of terminal symbols is in the language described by the grammar if and only if there is some way to use the productions to *derive* that string from the distinguished nonterminal by replacing (in any number of steps) a nonterminal symbol by any of the *or* clauses on the right-hand side of a production for that nonterminal symbol.

One natural way to describe the result of this derivation process is a *parse tree*: a diagram of the complete grammatical structure of the string being parsed. For example, the parse tree in Figure 21.1 shows that the string (A*B+AC)D is in

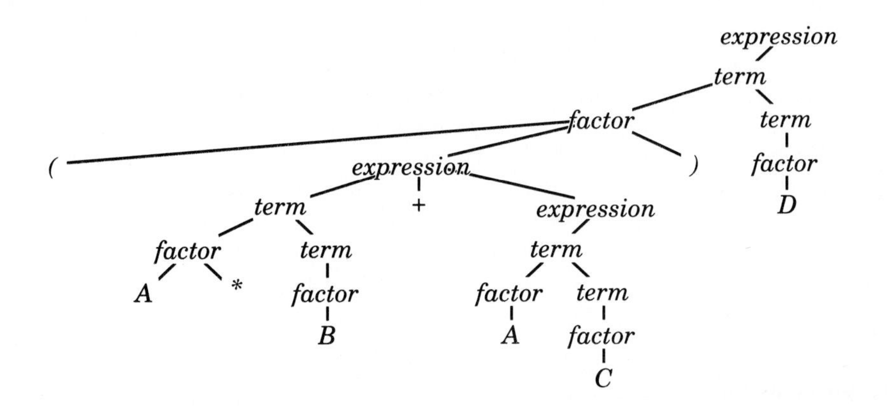

Figure 21.1 Parse tree for (A*B+AC)D.

the language described by the above grammar. Parse trees like this are sometimes used for English to break down a sentence into subject, verb, object, etc.

The main function of a parser is to accept strings which can be so derived and reject those that cannot, by attempting to construct a parse tree for any given string. That is, the parser can *recognize* whether a string is in the language described by the grammar by determining whether or not there exists a parse tree for the string. Top-down parsers do this by building the tree starting with the distinguished nonterminal at the top and working down towards the string to be recognized at the bottom; bottom-up parsers do this by starting with the string at the bottom and working backwards up towards the distinguished nonterminal at the top. As we'll see, if the meanings of the strings being recognized also imply further processing, then the parser can convert them into an internal representation which can facilitate such processing.

Another example of a context-free grammar may be found in the appendix of the *Pascal User Manual and Report*: it describes legal Pascal programs. The principles considered in this section for recognizing and using legal expressions apply directly to the complex job of compiling and executing Pascal programs. For example, the following grammar describes a very small subset of Pascal, arithmetic expressions involving addition and multiplication:

$$\langle expression \rangle ::= \langle term \rangle \mid \langle term \rangle + \langle expression \rangle$$
$$\langle term \rangle ::= \langle factor \rangle \mid \langle factor \rangle * \langle term \rangle$$
$$\langle factor \rangle ::= (\langle expression \rangle) \mid v$$

These rules describe in a formal way what we were able to take for granted in Chapter 4: they are the rules specifying what constitutes "legal" arithmetic expressions. Again, v is a special symbol which stands for any letter, but in this grammar the letters are likely to represent variables with numeric values. Examples of legal strings for this grammar are A+(B*C) and A*(((B+C)*(D*E))+F). We've already seen a parse tree for the latter, in Chapter 4, but that tree does not correspond to the grammar above—for example, parentheses are not explicitly included.

As we have defined things, some strings are perfectly legal both as arithmetic expressions and as regular expressions. For example, A*(B+C) might mean "add B to C and multiply the result by A" or "take any number of A's followed by either B or C." This points out the obvious fact that checking whether a string is legally formed is one thing, but understanding what it means is quite another. We'll return to this issue after we've seen how to parse a string to check whether or not it is described by some grammar.

Each regular expression is itself an example of a context-free grammar: any language which can be described by a regular expression can also be described by a context-free grammar. The converse is not true: for example, the concept of "balancing" parentheses can't be captured with regular expressions. Other types of grammars can describe languages which context-free grammars cannot. For example, *context-sensitive* grammars are the same as those above except that the left-hand sides of productions need not be single nonterminals. The differences between classes of languages and a hierarchy of grammars for describing them have been very carefully worked out and form a beautiful theory which lies at the heart of computer science.

Top-Down Parsing

One parsing method uses recursion to recognize strings from the language described exactly as specified by the grammar. Put simply, the grammar is such a complete specification of the language that it can be turned directly into a program!

Each production corresponds to a procedure with the name of the nonterminal on the left-hand side. Nonterminals on the right-hand side of the input correspond to (possibly recursive) procedure calls; terminals correspond to scanning the input string. For example, the following procedure is part of a top-down parser for our regular expression grammar:

```
procedure expression;
  begin
  term;
  if p[j]='+' then
    begin j:=j+1; expression end
  end;
```

An array *p* contains the regular expression being parsed and an index *j* points to the character currently begin examined. To parse a given regular expression, we put it in $p[1..M]$ (with a sentinel character in $p[M+1]$ which is not used in the grammar), set j to 1, and call *expression*. If this results in j being set to $M+1$, then the regular expression is in the language described by the grammar. If not, we'll see below how various error conditions are handled.

The first thing that *expression* does is call *term*, which has a slightly more complicated implementation:

```
procedure term;
   begin
   factor;
   if (p[j]='(') or letter(p[j]) then term
   end;
```

A direct translation from the grammar would simply have *term* call *factor* and then *term*. This obviously won't work because it leaves no way to exit from *term*: this program would go into an infinite recursive loop if called. (Such loops have particularly unpleasant effects in many systems.) The implementation above gets around this by first checking the input to decide whether *term* should be called. The first thing that *term* does is call *factor*, which is the only one of the procedures that could detect a mismatch in the input. From the grammar, we know that when *factor* is called, the current input character must be either a "(" or an input letter (represented by v). This process of checking the next character without incrementing j to decide what to do is called *lookahead*. For some grammars, this is not necessary; for others, even more lookahead is required.

Now, the implementation of *factor* follows directly from the grammar. If the input character being scanned is not a "(" or an input letter, a procedure *error* is called to handle the error condition:

```
procedure factor;
   begin
   if p[j]='(' then
      begin
      j:=j+1;
      expression;
      if p[j]=')' then j:=j+1 else error
      end
   else if letter(p[j]) then j:=j+1 else error;
   if p[j]='*' then j:=j+1
   end;
```

Another error condition occurs when a ")" is missing.

```
expression
  term
    factor
      (
      expression
        term
          factor   A   *
          term
            factor   B
        +
        expression
          term
            factor   A
            term
              factor   C
      )
    term
      factor   D
```

Figure 21.1 Parsing (A*B+AC)D.

The *expression*, *term*, and *factor* procedures are obviously recursive; in fact, they are so intertwined that they can't be compiled in Pascal without using the **forward** construct to get around the rule against using a procedure without declaring it first.

The parse tree for a given string gives the recursive call structure during parsing. Figure 21.2 traces through the operation of the above three procedures when p contains (A*B+AC)D and *expression* is called with j=*1*. Except for the plus sign, all the "scanning" is done in *factor*. For readability, the characters procedure *factor* scans, except for the parentheses, are put on the same line as the *factor* call.

The reader is encouraged to relate this process to the grammar and the tree in Figure 21.1. This process corresponds to traversing the tree in preorder, though the correspondence is not exact because our lookahead strategy essentially amounted to changing the grammar. Since we start at the top of the tree and work down, the origin of the "top-down" name is obvious. Such parsers are also often called *recursive-descent parsers* because they move down the parse tree recursively.

The top-down approach won't work for all possible context-free grammars. For example, with the production $\langle expression \rangle ::= v \mid \langle expression \rangle + \langle term \rangle$, if we were to follow the mechanical translation into Pascal as above, we would get the undesirable result:

```
procedure badexpression;
  begin
  if letter(p[j]) then j:=j+1 else
    begin
    badexpression;
    if p[j]<>'+' then error else
      begin j:=j+1; term end
    end
  end;
```

If this procedure were called with $p[j]$ a nonletter (as in our example, for $j=1$) it would go into an infinite recursive loop. Avoiding such loops is a principal difficulty in the implementation of recursive-descent parsers. For *term*, we used lookahead to avoid such a loop; in this case the proper way to get around the problem is to switch the grammar to say ⟨*term*⟩+⟨*expression*⟩. The occurrence of a nonterminal as the first item on the right-hand side of a replacement rule for itself is called *left recursion.* Actually, the problem is more subtle, because the left recursion can arise indirectly, for example with the productions ⟨*expression*⟩ ::= ⟨*term*⟩ and ⟨*term*⟩ ::= *v* | ⟨*expression*⟩ + ⟨*term*⟩. Recursive-descent parsers won't work for such grammars: they have to be transformed to equivalent grammars without left recursion, or some other parsing method must to be used. In general, there is an intimate and very widely studied connection between parsers and the grammars they recognize, and the choice of a parsing technique is often dictated by the characteristics of the grammar to be parsed.

Bottom-Up Parsing

Though there are several recursive calls in the programs above, it is an instructive exercise to remove the recursion systematically. Recall from Chapter 5 that each procedure call can be replaced by a stack push and each procedure return by a stack pop, mimicking what the Pascal system does to implement recursion. Also, recall that one reason to do this is that many calls which seem recursive are not truly recursive. When a procedure call is the last action of a procedure, then a simple **goto** can be used. This turns *expression* and *term* into simple loops that can be merged and combined with *factor* to produce a single procedure with one true recursive call (the call to *expression* within *factor*).

This view leads directly to a quite simple way to check whether regular expressions are legal. Once all the procedure calls are removed, we see that each terminal symbol is simply scanned as it is encountered. The only real processing done is to check whether there is a right parenthesis to match each left parenthesis, whether each "+" is followed by either a letter or a "(", and whether each "*" follows by either a letter or a ")". That is, checking whether a regular expression is legal is essentially equivalent to checking for balanced parentheses. This can be simply implemented by keeping a counter, initialized to 0, that is incremented when a left parenthesis is encountered and decremented when a right parenthesis is encountered. If the counter is zero at the end of the expression, and the "+" and "*" symbols in the expression meet the requirements just mentioned, then the expression was legal.

Of course, there is more to parsing than simply checking whether the input string is legal: the main goal is to build the parse tree (even if in an implicit way, as in the top-down parser) for other processing. It turns out to be possible to do this with programs with the same essential structure as the parenthesis checker described in the previous paragraph. One type of parser which works in this way is the so-called *shift-reduce parser*. The idea is to maintain a pushdown stack which holds terminal and nonterminal symbols. Each step in the parse is either a *shift* step, in which the next input character is simply pushed onto the stack, or a *reduce* step, in which the top characters on the stack are matched to the right-hand side of some production in the grammar and "reduced to" (replaced by) the nonterminal on the left side of that production. (The main difficulty in building a shift-reduce parser is deciding when to shift and when to reduce. This can be a complicated decision, depending on the grammar.) Eventually all the input characters get shifted onto the stack, and eventually the stack gets reduced to a single nonterminal symbol. The programs in Chapters 3 and 4 for constructing a parse tree from an infix expression by first converting the expression to postfix comprise a simple example of such a parser.

Bottom-up parsing is generally considered the method of choice for actual programming languages, and there is an extensive literature on developing parsers for large grammars of the type needed to describe a programming language. Our brief description only skims the surface of the issues involved.

Compilers

A *compiler* may be thought of as a program which translates from one language to another. For example, a Pascal compiler translates programs from the Pascal language into the machine language of some particular computer. We'll illustrate one way to do this by continuing with our regular-expression pattern-matching example; now, however, we wish to translate from the language of regular expressions to a "language" for pattern-matching machines, the *ch*, *next1*, and *next2* arrays of the *match* program of the previous chapter.

The translation process is essentially "one-to-one": for each character in the pattern (with the exception of parentheses) we want to produce a state for the pattern-matching machine (an entry in each of the arrays). The trick is to keep track of the information necessary to fill in the *next1* and *next2* arrays. To do so, we'll convert each of the procedures in our recursive-descent parser into functions which create pattern-matching machines. Each function will add new states as necessary onto the end of the *ch*, *next1*, and *next2* arrays, and return the index of the initial state of the machine created (the final state will always be the last entry in the arrays). For example, the function given below for the ⟨*expression*⟩ production creates the *or* states for the pattern-matching machine.

```
function expression: integer;
  var t1,t2: integer;
  begin
  t1:=term; expression:=t1;
  if p[j]='+' then
    begin
    j:=j+1; state:=state+1;
    t2:=state; expression:=t2; state:=state+1;
    setstate(t2,' ',expression,t1);
    setstate(t2-1,' ',state,state);
    end;
  end;
```

This function uses a procedure *setstate* which simply sets the *ch*, *next1*, and *next2* array entries indexed by the first argument to the values given in the second, third, and fourth arguments, respectively. The index *state* keeps track of the "current" state in the machine being built: Each time a new state is created, *state* is incremented. Thus, the state indices for the machine corresponding to a particular procedure call range between the value of *state* on entry and the value of *state* on exit. The final state index is the value of *state* on exit. (We don't actually "create" the final state by incrementing *state* before exiting, since this makes it easy to "merge" the final state with later initial states, as we'll see below.)

With this convention, it is easy to check (beware of the recursive call!) that the above program implements the rule for composing two machines with the *or* operation as diagrammed in the previous chapter. First the machine for the first part of the expression is built (recursively), then two new null states are added and the second part of the expression built. The first null state (with index *t2−1*) is the final state of the machine of the first part of the expression which is made into a "no-op" state to skip to the final state for the machine for the second part of the expression, as required. The second null state (with index *t2*) is the initial state, so its index is the return value for *expression* and its *next1* and *next2* entries are made to point to the initial states of the two expressions. Note carefully that these are constructed in the opposite order to what one might expect, because the value of *state* for the no-op state is not known until the recursive call to *expression* has been made.

The function for ⟨*term*⟩ first builds the machine for a ⟨*factor*⟩ and then, if necessary, merges the final state of that machine with the initial state of the machine for another ⟨*term*⟩. This is easier done than said, since *state* is the final state index of the call to *factor*. A call to *term* without incrementing *state* does the trick:

```
function term;
  var t: integer;
  begin
  term:=factor;
  if (p[j]='(') or letter(p[j]) then t:=term
  end;
```

(We have no use for the initial state index returned by the second call to *term*, but Pascal requires us to put it somewhere, so we throw it away in a temporary variable *t*.)

The function for ⟨*factor*⟩ uses similar techniques to handle its three cases: a parenthesis calls for a recursive call on *expression*; a *v* calls for simple concatenation of a new state; and a $*$ calls for operations similar to those in *expression*, according to the closure diagram from the previous section:

```
function factor;
  var t1,t2: integer;
  begin
  t1:=state;
  if p[j]='(' then
    begin
    j:=j+1; t2:=expression;
    if p[j]=')' then j:=j+1 else error
    end
  else  if letter(p[j]) then
    begin
    setstate(state,p[j],state+1,state+1);
    t2:=state; j:=j+1; state:=state+1
    end
  else  error;
  if p[j]<>'*' then factor:=t2 else
    begin
    setstate(state,' ',state+1,t2);
    factor:=state; next1[t1-1]:=state;
    j:=j+1; state:=state+1;
    end;
  end;
```

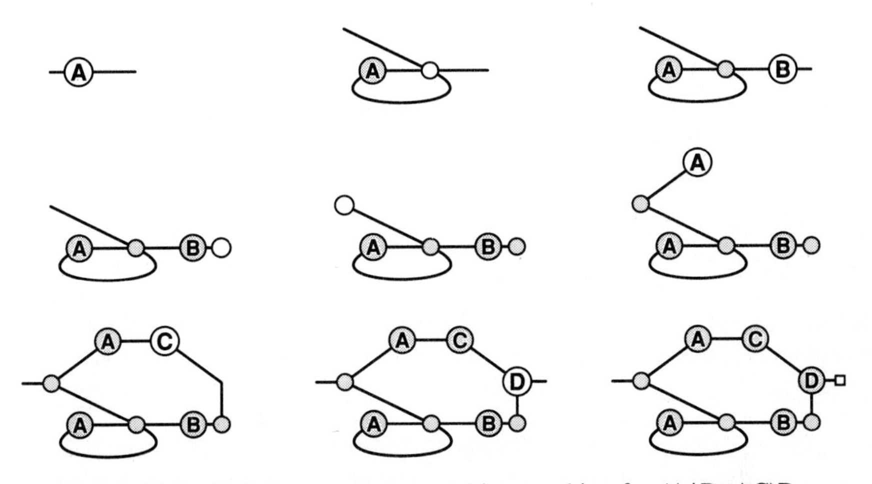

Figure 21.3 Building a pattern matching machine for (A*B+AC)D.

Figure 21.3 shows how the states are constructed for the pattern (A*B+AC)D, our example from the previous chapter. First, state 1 is constructed for the A. Then, state 2 is constructed for the closure operand and state 3 is attached for the B. Next, the "+" is encountered and states 4 and 5 are built by *expression*, but their fields can't be filled in until after a recursive call to *expression*, and this eventually results in the construction of states 6 and 7. Finally, the concatenation of the D is handled with state 8, leaving state 9 as the final state.

The final step in the development of a general regular-expression pattern-matching algorithm is to put these procedures together with the *match* procedure:

```
procedure matchall;
  begin
  j:=1; state:=1;
  next1[0]:=expression;
  setstate(state,' ',0,0);
  for i:=1 to N-1 do
      if match(i)>=i then writeln(i);
  end;
```

This program prints out all character positions in a text string $a[1..N]$ where a pattern $p[1..M]$ leads to a match.

Compiler-Compilers

The program for general regular-expression pattern-matching we have developed in this and the previous chapter is efficient and quite useful. A version of this program with a few added amenities (for handling "don't-care" characters, etc.) is likely to be among the most heavily used utilities on many computer systems.

It is interesting (some might say confusing) to reflect on this algorithm from a more philosophical point of view. In this chapter, we have considered parsers for unraveling the structure of regular expressions, based on a formal description of regular expressions using a context-free grammar. Put another way, we used the context-free grammar to specify a particular "pattern": a sequence of characters with legally balanced parentheses. The parser then checks to see if the pattern occurs in the input (but considers a match legal only if it covers the entire input string). Thus parsers, which check that an input string is in the set of strings defined by some context-free grammar, and pattern matchers, which check that an input string is in the set of strings defined by some regular expression, are essentially performing the same function! The principal difference is that context-free grammars are capable of describing a much wider class of strings. For example, regular expressions cannot describe the set of all regular expressions.

Another difference in the programs is that the context-free grammar is "built into" the parser, while the *match* procedure is "table-driven": the same program works for all regular expressions, once they have been translated into the proper format. It turns out to be possible to build parsers which are table-driven in the same way, so that the same program can be used to parse all languages which can be described by context-free grammars. A *parser generator* is a program which takes a grammar as input and produces as output a parser for the language described by that grammar. This can be carried one step further: one can build compilers that are table-driven in terms of both the input and the output languages. A *compiler-compiler* is a program which takes two grammars (and a specification of the relationships between them) as input and produces a compiler which translates strings from one language to the other as output.

Parser generators and compiler-compilers are available for general use in many computing environments, and are quite useful tools which can be used to produce efficient and reliable parsers and compilers with a relatively small amount of effort. On the other hand, top-down recursive-descent parsers of the type considered here are quite serviceable for the simple grammars which arise in many applications. Thus, as with many of the algorithms we have considered, we have a straightforward method appropriate for applications where a great deal of implementation effort might not be justified, and several advanced methods that can lead to significant performance improvements for large-scale applications. As stated above, we've only scratched the surface of this extensively researched field.

□

Exercises

1. How does the recursive-descent parser find an error in a regular expression such as (A+B)*BC+ which is incomplete?

2. Give the parse tree for the regular expression ((A+B)+(C+D)*)*.

3. Extend the arithmetic expression grammar to include exponentiation, **div** and **mod**.

4. Give a context-free grammar to describe all strings with no more than two consecutive 1's.

5. How many procedure calls are used by the recursive-descent parser to recognize a regular expression in terms of the number of concatenation, *or*, and closure operations and the number of parentheses?

6. Give the *ch*, *next1* and *next2* arrays that result from building the pattern-matching machine for the pattern ((A+B)+(C+D)*)*.

7. Modify the regular expression grammar to handle the "not" function and "don't-care" characters.

8. Build a general regular-expression pattern matcher based on the improved grammar in your answer to the previous question.

9. Remove the recursion from the recursive-descent compiler and simplify the resulting code as much as possible. Compare the running time of the nonrecursive and recursive methods.

10. Write a compiler for simple arithmetic expressions described by the grammar in the text. It should produce a list of "instructions" for a machine capable of three operations: *push* the value of a variable onto a stack; *add* the top two values on the stack, removing them from the stack, then putting the result there; and *multiply* the top two values on the stack, in the same way.

22

File Compression

For the most part, the algorithms we have studied have been designed primarily to use as little *time* as possible and only secondarily to conserve *space*. In this section, we'll examine some algorithms with the opposite orientation: methods designed primarily to reduce space consumption without using up too much time. Ironically, the techniques we'll examine to save space are "coding" methods from information theory that were developed to minimize the amount of information necessary in communications systems and were thus originally intended to save time (not space).

In general, most computer files have a great deal of redundancy. The methods we will examine save space by exploiting the fact that most files have a relatively low "information content." File compression techniques are often used for text files (in which certain characters appear much more often than others), "raster" files for encoding pictures (which can have large homogeneous areas), and files for the digital representation of sound and other analog signals (which can have large repeated patterns).

We'll look at an elementary algorithm for the problem (that is still quite useful) and an advanced "optimal" method. The amount of space saved by these methods varies depending on characteristics of the file. Savings of 20% to 50% are typical for text files, and savings of 50% to 90% might be achieved for binary files. For some types of files, for example files consisting of random bits, little can be gained. In fact, it is interesting to note that any general-purpose compression method must make some files longer (otherwise we could apply the method continually to produce an arbitrarily small file).

On the one hand, one might argue that file-compression techniques are less important than they once were because the cost of computer storage devices has dropped dramatically and far more storage is available to the typical user than in the past. On the other, it can be argued that file compression techniques are more important than ever because, since so much storage is in use, the savings they make possible are greater. Compression techniques are also appropriate for storage devices which allow extremely high-speed access and are by nature relatively expensive (and therefore small).

Run-Length Encoding

The simplest type of redundancy in a file is long runs of repeated characters. For example, consider the following string:

AAAABBBAABBBBBCCCCCCCCDABCBAAABBBBCCCD

This string can be encoded more compactly by replacing each repeated string of characters by a single instance of the repeated character along with a count of the number of times it is repeated. We would like to say that this string consists of 4 A's followed by 3 B's followed by 2 A's followed by 5 B's, etc. Compressing a string in this way is called *run-length encoding*. When long runs are involved, the savings can be dramatic. There are several ways to proceed with this idea, depending on characteristics of the application. (Do the runs tend to be relatively long? How many bits are used to encode the characters being encoded?) We'll look at one particular method, then discuss other options.

If we know that our string contains just letters, then we can encode counts simply by interspersing digits with the letters. Thus our string might be encoded as follows:

4A3BAA5B8CDABCB3A4B3CD

Here "4A" means "four A's," and so forth. Note that is is not worthwhile to encode runs of length one or two, since two characters are needed for the encoding.

For binary files, a refined version of this method is typically used to yield dramatic savings. The idea is simply to store the run lengths, taking advantage of the fact that the runs alternate between 0 and 1 to avoid storing the 0's and 1's themselves. This assumes that there are few short runs (we save bits on a run only if the length of the run is more than the number of bits needed to represent itself in binary), but no run-length encoding method will work very well unless most of the runs are long.

Figure 22.1 is a "raster" representation of the letter "q" lying on its side; this is representative of the type of information that might be processed by a text-formatting system (such as the one used to print this book); at the right is a list of numbers that might be used to store the letter in a compressed form. That is, the first line consists of 28 0's followed by 14 1's followed by 9 more 0's, etc. The 63 counts in this table plus the number of bits per line (51) contain sufficient information to reconstruct the bit array (in particular, note that no "end-of-line" indicator is needed). If six bits are used to represent each count, then the entire file is represented with 384 bits, a substantial savings over the 975 bits required to store it explicitly.

```
000000000000000000000000000011111111111111000000000   28 14  9
000000000000000000000000001111111111111111110000000   26 18  7
000000000000000000000001111111111111111111111110000   23 24  4
000000000000000000000011111111111111111111111111000   22 26  3
000000000000000000001111111111111111111111111111110   20 30  1
000000000000000000011111110000000000000000001111111   19  7 18 7
000000000000000000011111000000000000000000000011111   19  5 22 5
000000000000000000011100000000000000000000000000111   19  3 26 3
000000000000000000011100000000000000000000000000111   19  3 26 3
000000000000000000011100000000000000000000000000111   19  3 26 3
000000000000000000011100000000000000000000000000111   19  3 26 3
000000000000000000001111000000000000000000000001110   20  4 23 3 1
000000000000000000000011100000000000000000000111000   22  3 20 3 3
011111111111111111111111111111111111111111111111111   1  50
011111111111111111111111111111111111111111111111111   1  50
011111111111111111111111111111111111111111111111111   1  50
011111111111111111111111111111111111111111111111111   1  50
011111111111111111111111111111111111111111111111111   1  50
011000000000000000000000000000000000000000000000011   1   2 46 2
```

Figure 22.1 A typical bitmap, with information for run-length encoding.

Run-length encoding requires separate representations for the file and its encoded version, so that it can't work for all files. This can be quite inconvenient: for example, the character-file-compression method suggested above won't work for character strings that contain digits. If other characters are used to encode the counts, it won't work for strings that contain those characters. To illustrate a way to encode any string from a fixed alphabet of characters using only characters from that alphabet, we'll assume that we have only the 26 letters of the alphabet (and spaces) to work with.

How can we make some letters represent digits and others represent parts of the string to be encoded? One solution is to use a character that is likely to appear only rarely in the text as a so-called *escape character*. Each appearance of that character signals that the next two letters form a (count, character) pair, with counts represented by having the ith letter of the alphabet represent the number i. Thus our example string would be represented as follows with Q as the escape character:

QDABBBAAQEBQHCDABCBAAAQDBCCCD

The combination of the escape character, the count, and the one copy of the repeated character is called an *escape sequence*. Note that it's not worthwhile to encode runs less than four characters long, since at least three characters are required to encode any run.

But what if the escape character itself happens to occur in the input? We can't afford simply to ignore this possibility, because it is difficult to ensure that any particular character can't occur. (For example, someone might try to encode a string that has already been encoded.) One solution to this problem is to use an escape sequence with a count of zero to represent the escape character. Thus, in our example, the space character could represent zero, and the escape sequence "Q⟨space⟩" would represent any occurrence of Q in the input. It is interesting to note that files containing Q are the only ones made longer by this compression method. If a file that has already been compressed is compressed again, it grows by at least a number of characters equal to the number of escape sequences used.

Very long runs can be encoded with multiple escape sequences. For example, a run of 51 A's would be encoded as QZAQYA using the conventions above. If many very long runs are expected, it would be worthwhile to reserve more than one character to encode the counts.

In practice, it is advisable to make both the compression and expansion programs somewhat sensitive to errors. This can be done by including a small amount of redundancy in the compressed file so that the expansion program can be tolerant of an accidental minor change to the file between compression and expansion. For example, it probably is worthwhile to put "end-of-line" characters in the compressed version of the letter "q" above, so that the expansion program can resynchronize itself in case of an error.

Run-length encoding is not particularly effective for text files because the only character likely to be repeated is the blank, and there are simpler ways to encode repeated blanks. (It was used to great advantage in the past to compress text files created by reading in punched-card decks, which necessarily contained many blanks.) In modern systems, repeated strings of blanks are never entered, never stored: repeated strings of blanks at the beginning of lines are encoded as "tabs" and blanks at the ends of lines are obviated by the use of "end-of-line" indicators. A run-length encoding implementation like the one above (but modified to handle all representable characters) saves only about 4% when used on the text file for this chapter (and this savings all comes from the letter "q" example!).

Variable-Length Encoding

In this section we'll examine a file-compression technique that can save a substantial amount of space in text files (and many other kinds of files). The idea is to abandon the way in which text files are usually stored: instead of using the usual seven or eight bits for each character, only a few bits are used for characters which appear often and more bits for those which appear rarely.

It will be convenient to examine how the code is used on a small example before considering how it is created. Suppose we wish to encode the string

"ABRACADABRA." Encoding it in our standard compact binary code with the five-bit binary representation of i representing the ith letter of the alphabet (0 for blank) gives the following bit sequence:

0000100010100100000100011000010010000001000101001000001

To "decode" this message, simply read off five bits at a time and convert according to the binary encoding defined above. In this standard code the D, which appears only once, requires the same number of bits as the A, which appears five times. With a variable-length code, we can achieve economy in space by encoding frequently used characters with as few bits as possible so that the total number of bits used for the message is minimized.

We might try to assign the shortest bit strings to the most commonly used letters, encoding A with 0, B with 1, R with 01, C with 10, and D with 11, so ABRACADABRA would be encoded as

0 1 01 0 10 0 11 0 1 01 0

This uses only 15 bits compared to the 55 above, but it's not really a code because it depends on the blanks to delimit the characters. Without the blanks, the string 010101001101010 could be decoded as RRRARBRRA or as several other strings. Still, the count of 15 bits plus 10 delimiters is rather more compact than the standard code, primarily because *no* bits are used to encode letters not appearing in the message. To be fair, we also need to count the bits in the code itself, since the message can't be decoded without it, and the code does depend on the message (other messages will have different frequencies of letter usage). We will consider this issue later; for the moment we're interested in seeing how compact we can make the message.

First, delimiters aren't needed if no character code is the prefix of another. For example, if we encode A with 11, B with 00, C with 010, D with 10, and R with 011, there is only one way to decode the 25-bit string

1100011110101110110001111

One easy way to represent the code is with a trie (see Chapter 17). In fact, any trie with M external nodes can be used to encode any message with M different characters. For example, Figure 22.2 shows two codes which could be used for ABRACADABRA. The code for each character is determined by the path from the root to that character, with 0 for "go left" and 1 for "go right", as usual in a trie. Thus, the trie at the left corresponds to the code given above; the trie at the right corresponds to a code that produces the string

01101001111011100110100

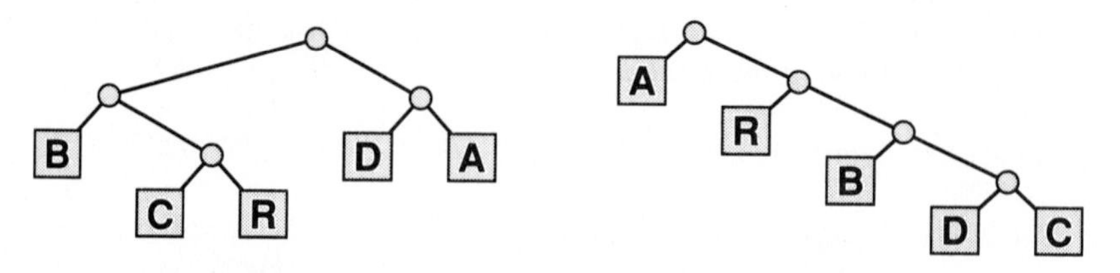

Figure 22.2 Two encoding tries for A, B, C, D, and R.

which is two bits shorter. The trie representation guarantees that no character code is the prefix of another, so the string is uniquely decodable from the trie. Starting at the root, proceed down the trie according to the bits of the message: each time an external node is encountered, output the character at that node and restart at the root.

But which trie is the best one to use? It turns out that there is an elegant way to compute a trie which leads to a bit string of minimal length for any given message. The general method for finding the code was discovered by D. Huffman in 1952 and is called *Huffman encoding*. (The implementation we'll examine uses some more modern algorithmic technology.)

Building the Huffman Code

The first step in building the Huffman code is to count the frequency of each character within the message to be encoded. The following code fills an array *count*[*0..26*] with the frequency counts for a message in a character array *a*[*1..M*]. (This program uses the *index* procedure described in Chapter 19 to keep the frequency count for the ith letter of the alphabet in *count*[i], with *count*[*0*] used for blanks.)

```
for i:=0 to 26 do count[i]:=0;
for i:=1 to M do
  count[index(a[i])]:=count[index(a[i])]+1;
```

For example, suppose we wish to encode the string "A SIMPLE STRING TO BE ENCODED USING A MINIMAL NUMBER OF BITS." The count table produced is shown in Figure 22.3: there are eleven blanks, three A's, three B's, etc.

The next step is to build the coding trie from the bottom up according to the frequencies. In building the trie, we'll view it as a binary tree with frequencies stored in the nodes: after it has been built we'll view it as a trie for coding, as above. First a tree node is created for each nonzero frequency, as shown on the left in the first row of Figure 22.4 (the order in which the nodes appear is determined by the

		A	B	C	D	E	F	G	H	I	J	K	L	M	N	O	P	Q	R	S	T	U	V	W	X	Y	Z
k	0	1	2	3	4	5	6	7	8	9	10	11	12	13	14	15	16	17	18	19	20	21	22	23	24	25	26
count[k]	11	3	3	1	2	5	1	2	0	6	0	0	2	4	5	3	1	0	2	4	3	2	0	0	0	0	0

Figure 22.3 Frequency counts for A SIMPLE STRING TO BE ENCODED ....

dynamics of the algorithm described below, but is not particularly relevant to the current discussion). Then the two nodes with the smallest frequencies are found, and a new node is created with those two nodes as children and with frequency value the sum of the values of the children. This is shown on the right in the first row of Figure 22.4. (It doesn't matter which nodes are used if there are more than two with the smallest frequency.) Then the two nodes with smallest frequency in that forest are found, and a new node created in the same way, as shown in the second row of Figure 22.4, on the left. Continuing in this way, we build up larger and larger subtrees and at the same time reduce the number of trees in the forest by one at each step (remove two, add one). Ultimately, all the nodes are combined together into a single tree.

Note that nodes with low frequencies end up far down in the tree and nodes with high frequencies end up near the root of the tree. The number labeling the external (square) nodes in this tree is a frequency count, while the number labeling each internal (round) node is the sum of the labels of its two children.

Now, the Huffman code is derived simply by replacing the frequencies at the bottom nodes with the associated letters and then viewing the tree as an encoding trie, with "left" corresponding to a code bit of 0 and "right" corresponding to a code bit of 1, exactly as above. The trie for our example is shown in Figure 22.5. The code for N is 000, the code for I is 001, the code for C is 110100, etc. The small number above each node in this tree is the index into the *count* array where the frequency is stored, for reference when examining the program that constructs the tree below. Thus, for example, *count*[33] is 11, the sum of the frequency counts for N and I (see also Figure 22.4), etc.

Clearly, letters with high frequencies are nearer the root of the tree and are encoded with fewer bits, so this is a good code, but why is this the *best* code?

Property 22.1 *The length of the encoded message is equal to the weighted external path length of the Huffman frequency tree.*

The "weighted external path length" of a tree is the sum over all external nodes of the "weight" (associated frequency count) times the distance to the root. Clearly, this is one way to compute the length of the message: it is equivalent to the sum over all letters of the number of occurrences times the number of bits per occurrence. ■

Property 22.2 *No tree with the same frequencies in external nodes has lower*

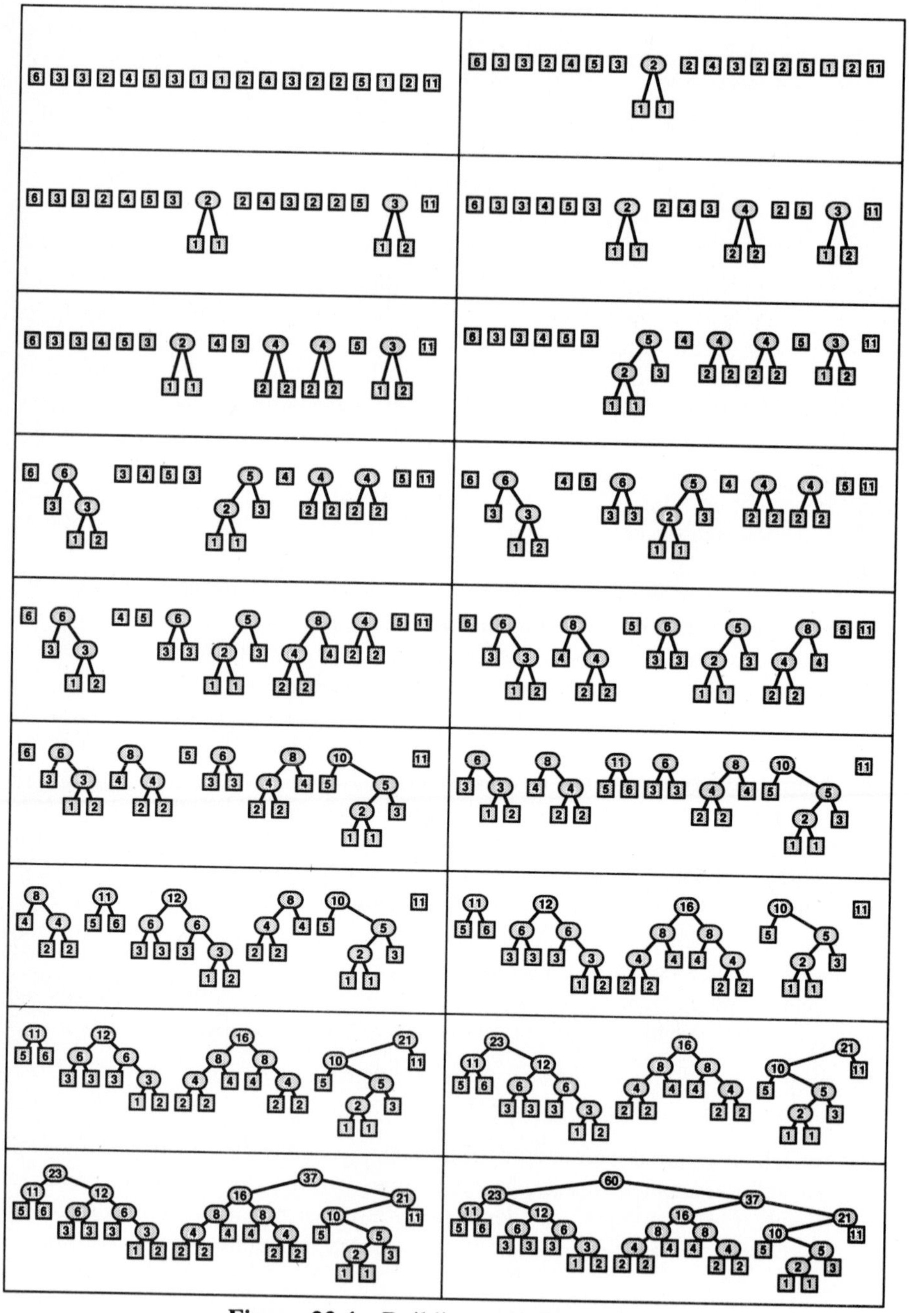

Figure 22.4 Building a Huffman tree.

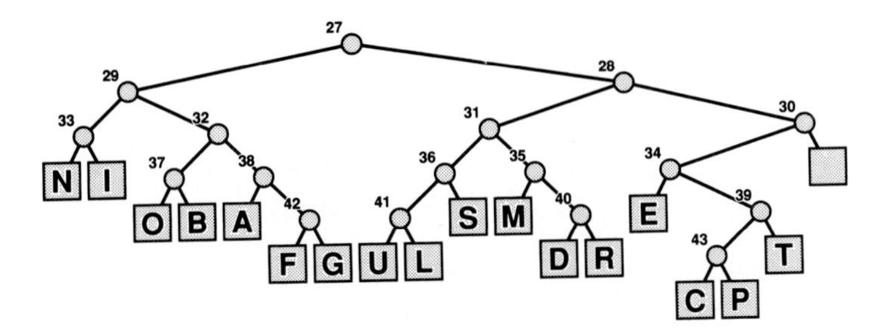

Figure 22.5 Huffman coding trie for A SIMPLE STRING TO BE ENCODED ....

weighted external path length than the Huffman tree.

Any tree can be reconstructed by the same process that we use to construct the Huffman tree, but not necessarily picking the two nodes of smallest weight at each step. It can be proven by induction that no strategy can do better than that of picking the two smallest weights first. ■

The description above gives a general outline of how to compute the Huffman encoding, in terms of algorithmic operations that we've studied. As usual, the step from such a description to an actual implementation is rather instructive, so we consider the details of the implementation next.

Implementation

The construction of the tree of frequencies involves the general process of removing the smallest from a set of unordered elements, so we'll use the *pqdownheap* procedure from Chapter 11 to build and maintain an indirect heap on the frequency values. Since we're interested in small values first, we'll assume that the sense of the inequalities in *pqdownheap* has been reversed. One advantage of using indirection is that it is easy to ignore zero frequency counts. Figure 22.6 shows the heap constructed for our example: Specifically, this heap is built by first initializing the *heap* array to point to the non-zero frequency counts and then using the *pqdownheap* procedure from Chapter 11, as follows:

```
N:=0;
for i:=0 to 26 do
  if count[i]<>0 then
    begin N:=N+1; heap[N]:=i end;
for k:=N downto 1 do pqdownheap(k);
```

k	1	2	3	4	5	6	7	8	9	10	11	12	13	14	15	16	17	18
heap[k]	3	7	16	21	12	15	6	20	9	4	13	14	5	2	18	19	1	0
count[heap[k]]	1	2	1	2	2	3	1	3	6	2	4	5	5	3	2	4	3	11

Figure 22.6 Initial heap (indirect) for Huffman tree construction.

As mentioned above, this assumes that the sense of the inequalities in the implementation of *pqdownheap* has been reversed.

Now, using this procedure to construct the tree as above is straightforward: we take the two smallest elements off the heap, add them and put the result back into the heap. At each step we create one new count, and decrease the size of the heap by one. This process creates $N-1$ new counts, one for each of the internal nodes of the tree being created, as in the following code:

```
repeat
  t:=heap[1]; heap[1]:=heap[N]; N:=N-1;
  pqdownheap(1);
  count[26+N]:=count[heap[1]]+count[t];
  dad[t]:=26+N; dad[heap[1]]:=-26-N;
  heap[1]:=26+N; pqdownheap(1);
until N=1;
dad[26+N]:=0;
```

The first two lines of this loop are actually *pqremove*; the size of the heap is decreased by one. Then a new internal node is "created" with index $26+N$ and given a value equal to the sum of the value at the root and value just removed. Then this node is put at the root, which raises its priority, necessitating another call on *pqdownheap* to restore order in the heap. The tree itself is represented with an array of "parent" links: *dad*[*t*] is the index of the parent of the node whose weight is in *count*[*t*]. The sign of *dad*[*t*] indicates whether the node is a left or right child of its parent. For example, in the tree above we have *count*[*30*]=*21*, *dad*[*30*]=−*28*, and *count*[*28*]=*37* (indicating that the node of weight 21 has index

k	27	28	29	30	31	32	33	34	35	36	37	38	39	40	41	42	43
count[k]	60	37	23	21	16	12	11	10	8	8	6	6	5	4	4	3	2
dad[k]	0	-27	27	-28	28	-29	29	30	-31	31	32	-32	-34	-35	36	-38	39

Figure 22.7 Parent link representation of Huffman tree (internal nodes).

30 and it is the right child of a parent with index 28 and weight 37). Figure 22.7 gives the *dad* array for the internal nodes of the tree of Figure 22.5.

The following program fragment constructs the actual Huffman code, as represented by the trie in Figure 22.5, from the representation of the coding tree computed during the sifting process. The code is represented by two arrays: *code*[*k*] gives the binary representation of the *k*th letter and *len*[*k*] gives the number of bits from *code*[*k*] to use in the code. For example, I is the 9th letter and has code 001, so *code*[*9*]=*1* and *len*[*9*]=*3*.

```
for k:=0 to 26 do
  if count[k]=0 then
    begin code[k]:=0; len[k]:=0 end
  else
    begin
    i:=0; j:=1; t:=dad[k]; x:=0;
    repeat
      if t<0 then begin x:=x+j; t:=-t end;
      t:=dad[t]; j:=j+j; i:=i+1
    until t=0;
    code[k]:=x; len[k]:=i;
    end;
```

Finally, we can use these computed representations of the code to encode the message:

```
for j:=1 to M do
  for i:=len[index(a[j])] downto 1 do
    write(bits(code[index(a[j])],i-1,1):1);
```

This program uses the *bits* procedure from Chapters 10 and 17 to access single bits. Our sample message is encoded in only 236 bits versus the 300 used for the straightforward encoding, a 21% savings:

01101111001001101011010110001110011110011101110111001000011111111110
11010011101011100111110000011010001001011011001011011110000100100 10
0001111111011011110100010000011010011010001111000100001010010111001
0111111010001110111010100111011100 1

Now, as mentioned above, the tree must be saved or sent along with the message in order to decode it. Fortunately, this does not present any real difficulty. It is actually necessary to store only the *code* array, because the radix search trie which results from inserting the entries from that array into an initially empty tree is the decoding tree.

Thus the storage savings quoted above is not entirely accurate, because the message can't be decoded without the trie and we must take into account the cost of storing the trie (i.e., the *code* array) along with the message. Huffman encoding is therefore only effective for long files where the savings in the message is enough to offset the cost, or in situations where the coding trie can be precomputed and used for a large number of messages. For example, a trie based on the frequencies of occurrence of letters in the English language could be used for text documents. For that matter, a trie based on the frequency of occurrence of characters in Pascal programs could be used for encoding programs (for example, ";" is likely to be near the top of such a trie). A Huffman encoding algorithm saves about 23% when run on the text for this chapter.

As before, for truly random files, even this clever encoding scheme won't work because each character will occur approximately the same number of times, which will lead to a fully balanced coding tree and an equal number of bits per letter in the code.

□

Exercises

1. Implement compression and expansion procedures for the run-length encoding method for a fixed alphabet described in the text, using Q as the escape character.

2. Could "QQ" occur somewhere in a file compressed by the method described in the text? Could "QQQ" occur?

3. Implement compression and expansion procedures for the binary file encoding method described in the text.

4. The letter "q" given in the text can be processed as a sequence of five-bit characters. Discuss the pros and cons of doing so in order to use a character-based run-length encoding method.

5. Show the construction process when the method used in the text is used to build the Huffman coding tree for the string "ABRACADABRA." How many bits does the encoded message require?

6. What is the Huffman code for a binary file? Give an example showing the maximum number of bits that could be used in a Huffman code for a N-character ternary (three-valued) file.

7. Suppose that the frequencies of the occurrence of all the characters to be encoded are different. Is the Huffman encoding tree unique?

8. Huffman coding could be extended in a straightforward way to encode in two-bit characters (using 4-way trees). What would be the main advantage and the main disadvantage of doing so?

9. What would be the result of breaking up a Huffman-encoded string into five-bit characters and Huffman-encoding that string?

10. Implement a procedure to *decode* a Huffman-encoded string, given the *code* and *len* arrays.

23

Cryptology

☐ In the previous chapter we looked at methods for encoding strings of characters to save space. Of course, there is another very important reason to encode strings of characters: to keep them secret.

Cryptology, the study of systems for secret communications, consists of two complementary fields of study: *cryptography*, the design of secret communications systems, and *cryptanalysis*, the study of ways to compromise secret communications systems. Cryptology primarily has been applied in military and diplomatic communications systems, but other significant applications are becoming apparent. Two principal examples are computer file systems (where each user would prefer to keep his files private) and electronic funds transfer systems (where very large amounts of money are involved). A computer user wants to keep his computer files just as private as papers in his file cabinet, and a bank wants electronic funds transfer to be just as secure as funds transfer by armored car.

Except for military applications, we assume that cryptographers are "good guys" and cryptanalysts are "bad guys": our goal is to protect our computer files and our bank accounts from criminals. If this point of view seems somewhat unfriendly, it must be noted (without being over-philosophical) that by using cryptography one is assuming the existence of unfriendliness! Of course, "good guys" must know something about cryptanalysis, since the very best way to be sure a system is secure is to try to compromise it yourself. (Also, there are several documented instances of wars being brought to an end, and many lives saved, through successes in cryptanalysis.)

Cryptology has many close connections with computer science and algorithms, especially the arithmetic and string-processing algorithms that we have studied. Indeed, the art (science?) of cryptology has an intimate relationship with computers and computer science that is only beginning to be fully understood. Like algorithms, cryptosystems have been around far longer than computers. Secrecy system design and algorithm design have a common heritage, and the same people are attracted to both.

It is not clear which branch of cryptology has been affected most by the availability of computers. Cryptographers now have available much more powerful encryption machines than before, but they also now have more room to make a mistake. Cryptanalysts have much more powerful tools for breaking codes than ever before, but the codes to be broken are more complicated. Cryptanalysis can place a huge strain on computational resources; not only was it among the first applications areas for computers, but it still remains a principal applications area for modern supercomputers.

More recently, the widespread use of computers has led to the emergence of a variety of important new applications for cryptology, as mentioned above. New cryptographic methods have recently been developed appropriate for such applications, and these have led to the discovery of a fundamental relationship between cryptology and an important area of theoretical computer science that we'll examine briefly in Chapter 45.

In this chapter, we'll look at some of the basic characteristics of cryptographic algorithms. We'll refrain from delving into detailed implementations: cryptography is certainly a field that should be left to experts. While it's not difficult to "keep people honest" by encrypting things with a simple cryptographic algorithm, it is dangerous to rely upon a method implemented by a non-expert.

Rules of the Game

The elements that go into providing a means for secure communications between two individuals are collectively called a *cryptosystem*. The canonical structure of a typical cryptosystem is diagrammed in Figure 23.1.

The *sender* sends a message (called the *plaintext*) to the *receiver* by transforming the plaintext into a secret form suitable for transmission (called the *ciphertext*) using a cryptographic algorithm (the *encryption method*) and some *key* parameters. To read the message, the receiver must have a matching cryptographic algorithm (the *decryption method*) and the same key parameters, which will transform the ciphertext back into the plaintext, the message. It is usually assumed that the

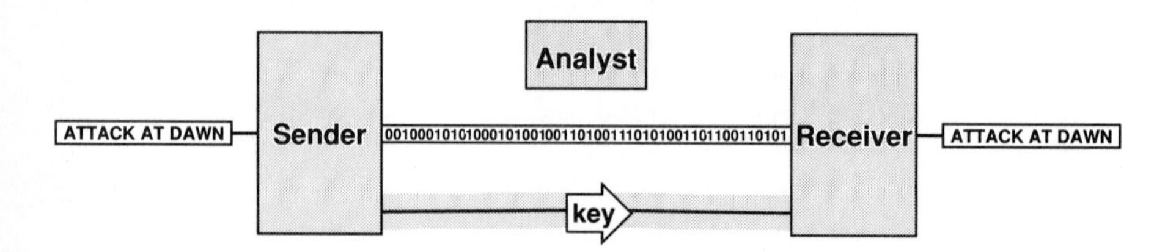

Figure 23.1 A Typical Cryptosystem.

ciphertext is sent over insecure communications lines and is available to the cryptanalyst. It is also usually assumed that the encryption and decryption methods are known to the cryptanalyst: his aim is to recover the plaintext from the ciphertext without knowing the key parameters. Note that the whole system depends on some separate prior method of communication between the sender and receiver to agree on the key parameters. As a rule, the more key parameters, the more secure the cryptosystem is but the more inconvenient it is to use. This situation is akin to that for more conventional security systems: a combination safe is more secure with more numbers on the combination lock, but it is harder to remember the combination. The parallel with conventional systems also serves as a reminder that any security system is only as reliable as the people who have the key.

It is important to remember that economic questions play a central role in cryptosystems. There is an economic motivation to build simple encryption and decryption devices (since many may need to be provided and complicated devices cost more). Also, there is an economic motivation to reduce the amount of key information to be distributed (since a very secure and expensive method of communications must be used). Balanced against the cost of implementing cryptographic algorithms and distributing key information is the amount of money the cryptanalyst would be willing to pay to break the system. For most applications, it is the cryptographer's aim to develop a low-cost system with the property that it would cost the cryptanalyst much more to read messages than he would be willing to pay. For a few applications, a "provably secure" cryptosystem may be required: one for which it can be guaranteed that the cryptanalyst can never read messages no matter how much he is willing to spend. (The very high stakes in some applications of cryptology naturally imply that very large amounts of money are used for cryptanalysis.) In algorithm design, we try to keep track of costs to help us choose the best algorithms; in cryptology, costs play a central role in the design process.

Simple Methods

Among the simplest (and among the oldest) methods for encryption is the *Caesar cipher*: if a letter in the plaintext is the Nth letter in the alphabet, replace it by the $(N+K)$th letter in the alphabet where K is some fixed integer (Caesar used $K = 3$). The table below shows how a message is encrypted using this method with $K = 1$:

Plaintext: ATTACK AT DAWN
Ciphertext: BUUBDLABUAEBXO

This method is weak because the cryptanalyst has only to guess the value of K: by trying each of the 26 choices, he can be sure that he will read the message.

A far better method is to use a general table to define the substitution to be made: for each letter in the plaintext, the table tells which letter to put in the ciphertext. For example, if the table gives the correspondence

ABCDEFGHIJKLMNOPQRSTUVWXYZ
THE QUICKBROWNFXJMPDVRLAZYG

then the message is encrypted as follows:

Plaintext: ATTACK AT DAWN
Ciphertext: HVVH OTHVTQHAF

This is much more powerful than the simple Caesar cipher because the cryptanalyst would have to try many more (about $27! > 10^{28}$) tables to be sure of reading the message. However, "simple substitution" ciphers like this are easy to break because of letter frequencies inherent in the language. For example, since E is the most frequent letter in English text, the cryptanalyst could get a good start on reading the message by looking for the most frequent letter in the ciphertext and replacing it by E. While this might not be the right choice, it is certainly better than trying all 26 letters blindly. The situation gets even better (for the cryptanalyst) when two-letter combinations ("digrams") are taken into account: certain digrams (such as QJ) never occur in English text while others (such as ER) are very common. By examining frequencies of letters and combinations of letters, a cryptanalyst can break a simple substitution cipher very easily.

One way to make this type of attack more difficult is to use more than one table. A simple example of this is an extension of the Caesar cipher called the *Vigenere cipher*: a small repeated key is used to determine the value of K for each letter. At each step, the key letter index is added to the plaintext letter index to determine the ciphertext letter index. Our sample plaintext, with the key ABC, is encrypted as follows:

Key: ABCABCABCABCAB
Plaintext: ATTACK AT DAWN
Ciphertext: BVWBENACWAFDXP

For example, the last letter of the ciphertext is P, the 16th letter of the alphabet, because the corresponding plaintext letter is N (the 14th letter), and the corresponding key letter is B (the 2nd letter).

The Vigenere cipher can obviously be made more complicated by using different general tables for each letter of the plaintext (rather than simple offsets). Also, it is obvious that the longer the key, the better. In fact, if the key is as long as the plaintext, we have the *Vernam cipher*, more commonly called the *one-time pad*. This is the only provably secure cryptosystem known, and it is reportedly used for the Washington-Moscow hotline and other vital applications. Since each key letter is used only once, the cryptanalyst can do no better than try every possible key letter for every message position, an obviously hopeless situation since this is as difficult as trying all possible messages. However, using each key letter only once obviously leads to a severe key distribution problem, and the one-time pad is only useful for relatively short messages which are to be sent infrequently.

If the message and key are encoded in binary, a more common scheme for position-by-position encryption is to use the "exclusive-or" function: to encrypt the plaintext, "exclusive-or" it (bit by bit) with the key. An attractive feature of this method is that decryption is the same operation as encryption: the ciphertext is the exclusive-or of the plaintext and the key, but doing another exclusive-or of the ciphertext and the key returns the plaintext. Notice that the exclusive-or of the ciphertext and the plaintext is the key. This seems surprising at first, but actually many cryptographic systems have the property that the cryptanalyst can discover the key if he knows the plaintext.

Encryption/Decryption Machines

Many cryptographic applications (for example, voice systems for military communications) involve the transmission of large amounts of data, and this makes the one-time pad infeasible. What is needed is an approximation to the one-time pad in which a large amount of "pseudo-key" can be generated from a small amount of true key to be distributed.

The usual setup in such situations is as follows: an encryption machine is fed some *cryptovariables* (true key) by the sender, which it uses to generate a long stream of key bits (pseudo-key). The exclusive-or of these bits and the plaintext forms the ciphertext. The receiver, having a similar machine and the same cryptovariables, uses them to generate the same key stream to exclusive-or against the ciphertext and to retrieve the plaintext.

Key generation in this context is very much like hashing and random-number generation, and the methods discussed in Chapter 16 and 35 are appropriate for key generation. In fact, some of the mechanisms discussed in Chapter 35 were first developed for use in encryption/decryption machines such as those described here. However, key generators have to be somewhat more complicated than random number generators, because there are ways to attack simple machines. The problem is that it might be easy for the cryptanalyst to get some plaintext (for example, silence in a voice system), and therefore some key. If the cryptanalyst knows enough about the machine, then the key might provide enough clues to allow the values of all the cryptovariables at some point in time to be derived—then the operation of the machine can be simulated and all the key calculated from that point on.

Cryptographers have several ways to avoid such problems. One way is to make part of the architecture of the machine itself a cryptovariable. It is usually assumed that the cryptanalyst knows everything about the structure of the machine (maybe one was stolen) except the cryptovariables, but if some of the cryptovariables are used to "configure" the machine, it may be difficult to find their values. Another method commonly used to confuse the cryptanalyst is the *product cipher*, where two different machines are combined to produce a complicated key stream (or to drive each other). Another method is *nonlinear substitution*; here the translation between plaintext and ciphertext is done in large chunks, not bit by bit. The general problem with such complex methods is that they can be too complicated for even the cryptographer to understand, and there always is the possibility that things may degenerate badly for some choices of the cryptovariables.

Public-Key Cryptosystems

In commercial applications such as electronic funds transfer and (real) computer mail, the *key distribution problem* is even more onerous than in the traditional applications of cryptography. The prospect of providing long keys that must be changed often to every citizen, while still maintaining both security and cost-effectiveness, certainly inhibits the development of such systems. Methods have recently been developed, however, which promise to eliminate the key distribution problem completely. Such systems, called *public-key cryptosystems*, are likely to come into widespread use in the near future. One of the most prominent of these systems is based on some of the arithmetic algorithms we have been studying, so we will take a close look at how it works.

The idea in public-key cryptosystems is to use a "phone book" of encryption keys. Everyone's encryption key (denoted by P) is public knowledge: a person's key could be listed, for example, next to his number in the telephone book. Everyone also has a secret key used for decryption; this secret key (denoted by S) is not known to anyone else. To transmit a message M, the sender looks up the receiver's public key, uses it to encrypt the message, and then transmits the message. We'll denote the encrypted message (ciphertext) by $C = P(M)$. The receiver uses his private decryption key to decrypt and read the message. For this system to work, at least the following conditions must be satisfied:

(*i*) $S(P(M)) = M$ for every message M.
(*ii*) All (S, P) pairs are distinct.
(*iii*) Deriving S from P is as hard as reading M.
(*iv*) Both S and P are easy to compute.

The first of these is a fundamental cryptographic property, the second two provide the security, and the fourth makes the system feasible for use.

This general scheme was outlined by W. Diffie and M. Hellman in 1976, but they had no method which satisfied all of these conditions. Such a method was discovered soon afterwards by R. Rivest, A. Shamir, and L. Adleman. Their scheme, which has come to be known as the *RSA public-key cryptosystem*, is based on arithmetic algorithms performed on very large integers. The encryption key P is the integer pair (N, p) and the decryption key S is the integer pair (N, s), where s is kept secret. These numbers are intended to be very large (typically, N might be 200 digits and p and s might be 100 digits). The encryption and decryption methods are then simple: first the message is broken up into numbers less than N (for example, by taking $\lg N$ bits at a time from the binary string corresponding to the character encoding of the message). Then these numbers are independently raised to a power modulo N: to *encrypt* a (piece of a) message M, compute $C = P(M) = M^p \bmod N$, and to *decrypt* a ciphertext C, compute $M = S(C) = C^s \bmod N$. In Chapter 36 we'll study how to perform this computation—while computing with 200-digit numbers can be cumbersome, the fact that we need only the remainder after dividing by N means that we can keep the numbers from getting large, despite the fact that M^p and C^s themselves are impossibly large numbers.

Property 23.1 *In the RSA cryptosystem, a message can be encrypted in linear time.*

For long messages, the length of the numbers used for keys may be viewed constant—an implementation detail. Similarly, raising a number to a power is done in constant time, since the numbers are not allowed to get longer than a "constant" length. It is true that this argument hides many implementation considerations related to computing with long numbers; the costs of these operations are in fact an inhibiting factor in broadening the applicability of the method. ■

Condition (*iv*) above is therefore satisfied, and condition (*ii*) can be easily enforced. We still must make sure that the cryptovariables N, p, and s can be chosen so as to satisfy conditions (*i*) and (*iii*). To be convinced of these requires an exposition of number theory which is beyond the scope of this book, but we can outline the main ideas. First, it is necessary to generate three large (approximately 100-digit) "random" prime numbers: the largest will be s and we'll call the other two x and y. Then N is chosen to be the product of x and y, and p is chosen so that $ps \bmod (x-1)(y-1) = 1$. It is possible to prove that, with N, p, and s chosen in this way, we have $M^{ps} \bmod N = M$ for all messages M.

For example, with our standard encoding, the message ATTACK AT DAWN might correspond to the 28-digit number

$$0120200103110001200004012314$$

since A is the first letter (01) in the alphabet, T is the twentieth letter (20), etc. Now, to keep the example small, we start with some 2-digit primes (rather than 100-digit, as required): take $x = 47$, $y = 79$, and $s = 97$. These values lead to $N = 3713$ (the product of x and y) and $p = 37$ (the unique integer which gives a remainder of 1 when multiplied by 97 and divided by 3588). Now, to encode the message, we break it up into 4-digit chunks and raise to the pth power (modulo N). This gives the encoded version

$$1404293235360001328422802235.$$

That is, $0120^{37} \equiv 1404,\ 2001^{37} \equiv 2932,\ 0311^{37} \equiv 3536 \pmod{3713}$, etc. The decoding process is the same, using s rather than p. Thus, we get the original message back because $1404^{97} \equiv 0120,\ 2932^{97} \equiv 2001 \pmod{3713}$, etc.

The most important part of the calculations involved is the encoding of the message, as discussed in Property 23.1 above. But there's no cryptosystem at all if it is not possible to compute the key variables. Though this involves both sophisticated number theory and relatively sophisticated programs for manipulating large numbers, the time to compute the keys is likely to be less than the square of their length (and not proportional to their magnitude, which would be unacceptable).

Property 23.2 *The keys for the RSA cryptosystem can be created without excessive computation.*

Again, some methods based on number theory beyond the scope of this book are required; it turns out that each large prime can be generated by first generating a large random number, then testing successive numbers starting at that point until a prime is found. One simple method performs a calculation on a random number that, with probability $1/2$, will "prove" that the number to be tested is not prime. (A number which is not prime will survive 20 applications of this test less than one time out of a million, 30 applications less than 1 time out of a billion.) The last step is to compute p: it turns out that a variant of Euclid's algorithm (see Chapter 1) is just what is needed. ■

Recall that the decryption key s (and the factors x and y of N) are to be kept secret, and that the success of the method depends on the cryptanalyst not being able to find the value of s, given N and p. Now, for our small example, it is easy to discover that $3713 = 47 * 79$, but if N is a 200-digit number, one has little hope of finding its factors. That is, s seems to be difficult to compute from knowledge of p (and N), though no one has been able to *prove* that to be the case. Apparently, finding p from s requires knowledge of x and y, and apparently it is necessary to factor N to calculate x and y. But factoring N is thought to be very difficult: the best factoring algorithms known would take millions of years to factor a 200-digit number, using current technology.

An attractive feature of the RSA system is that the complicated computations involving N, p, and s are performed only once for each user who subscribes to the system, while the much more frequent operations of encryption and decryption involve only breaking up the message and applying the simple exponentiation procedure. This computational simplicity, combined with all the convenience features provided by public-key cryptosystems, make this system quite attractive for secure communications, especially on computer systems and networks.

The RSA method has its drawbacks: the exponentiation procedure is actually expensive by cryptographic standards, and, worse, there is the lingering possibility that it might be possible to read messages encrypted using the method. This is true of many cryptosystems: a cryptographic method must withstand serious cryptanalytic attacks before it can be used with confidence.

Several other methods have been suggested for implementing public-key cryptosystems. Some of the most interesting are linked to an important class of problems which are generally thought to be very hard (though this is not known for sure), which we'll discuss in Chapter 45. These cryptosystems have the interesting property that a successful attack could provide insight on how to solve some well-known difficult unsolved problems (as with factoring for the RSA method). This link between cryptology and fundamental topics in computer science research, along with the potential for widespread use of public-key cryptography, have made this an active area of current research.

□

Exercises

1. Decrypt the following message, which was encrypted with a Vigenere cipher using the pattern CAB (repeated as necessary) for the key (on a 27-letter alphabet, with blank preceding A): DOBHBUAASXFZWJQQ

2. What table should be used to *decrypt* messages that have been encrypted using the table substitution method?

3. Suppose that a Vigenere cipher with a two-character key is used to encrypt a relatively long message. Write a program to infer the key, based on the assumption that the frequency of occurrence of each character in odd positions should be roughly equal to the frequency of occurrence of each character in the even positions.

4. Write matching encryption and decryption procedures that use the "exclusive or" operation between a binary version of the message a binary stream from one of the linear congruential random number generators of Chapter 3.

5. Write a program to break the method given in the previous exercise, assuming that the first 10 characters of the message are known to be blanks.

6. Could one encrypt plaintext by "and"ing it (bit by bit) with the key? Explain why or why not.

7. True or false? Public-key cryptography makes it convenient to send the same message to several different users. Discuss your answer.

8. What is P(S(M)) for the RSA method for public-key cryptography?

9. RSA encoding might involve computing M^n, where M might be a k-digit number represented in an array of k integers, say. About how many operations would be required for this computation?

10. Implement encryption/decryption procedures for the RSA method (assume that s, p and N are all given and represented in arrays of integers of size 25).

SOURCES for String Processing

The best sources for further information on many of the topics covered in the chapters in this section are the original references. Knuth, Morris, and Pratt's 1977 paper, Boyer and Moore's 1977 paper and Karp and Rabin's 1981 paper form the basis for much of the material in Chapter 19. The 1968 paper by Thompson is the basis for the regular-expression pattern matcher of Chapters 20–21. Huffman's 1952 paper, though it predates many of the algorithmic considerations here, still makes interesting reading. Rivest, Shamir, and Adleman describe fully the implementation and applications of their public-key cryptosystem in their 1978 paper.

The book by Standish is a good general reference for many of the topics covered in these chapters, especially Chapters 19, 22, and 23. That book also addresses some representations and basic practical algorithms not covered here.

Parsing and compiling are viewed by many to be the heart of computer science: we've investigated one connection to algorithms, but their relationship with programming languages, theoretical computer science and other areas is certainly more important. Many algorithmic issues have also been studied in great detail. The standard reference on the subject is the book by Aho, Sethi, and Ullman.

Obviously, the public literature on cryptography is rather sparse. However, much background information on the subject may be found in the books by Kahn and Konheim.

A. V. Aho, R. Sethi, and J. D. Ullman, *Compilers: Principles, Techniques, Tools*, Addison-Wesley, Reading, MA, 1986.

R. S. Boyer and J. S. Moore, "A fast string searching algorithm," *Communications of the ACM*, **20**, 10 (October, 1977).

D. A. Huffman, "A method for the construction of minimum-redundancy codes," *Proceedings of the IRE*, **40** (1952).

D. Kahn, *The Codebreakers*, Macmillan, New York, 1967.

R. M. Karp and M. O. Rabin, "Efficient Randomized Pattern-Matching Algorithms," Technical Report TR–31–81, Aiken Comput. Lab., Harvard U., Cambridge, MA, 1981.

D. E. Knuth, J. H. Morris, and V. R. Pratt, "Fast pattern matching in strings," *SIAM Journal on Computing*, **6**, 2 (June, 1977).

A. G. Konheim, *Cryptography: A Primer*, John Wiley & Sons, New York, 1981.

R. L. Rivest, A. Shamir and L. Adleman, "A method for obtaining digital signatures and public-key cryptosystems," *Communications of the ACM*, **21**, 2 (February, 1978).

T. A. Standish, *Data Structure Techniques*, Addison-Wesley, Reading, MA, 1980.

K. Thompson, "Regular expression search algorithm," *Communications of the ACM*, **11**, 6 (June, 1968).

Geometric Algorithms

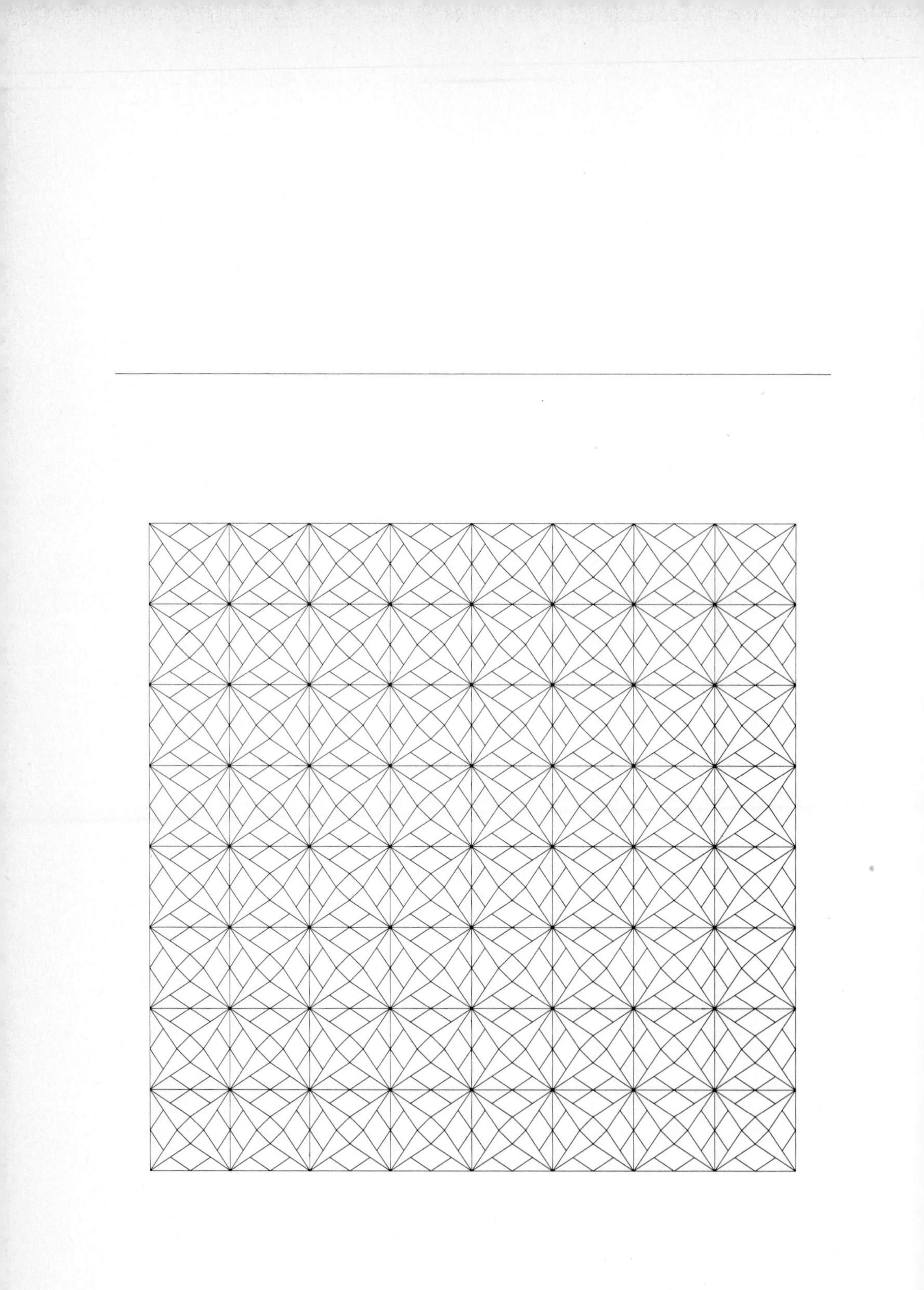

24

Elementary Geometric Methods

☐ Computers are being used more and more to solve large-scale problems that are inherently geometric. Geometric objects such as points, lines and polygons are the basis of a broad variety of important applications and give rise to an interesting set of problems and algorithms.

Geometric algorithms are important in design and analysis systems modeling physical objects ranging from buildings and automobiles to very large-scale integrated circuits. A designer working with a physical object has a geometric intuition that is difficult to support in a computer representation. Many other applications directly involve processing geometric data. For example, a political "gerrymandering" scheme to divide a district up into areas of equal population (and that satisfy other criteria such as putting most of the members of the other party in one area) is a sophisticated geometric algorithm. Other applications abound in mathematics and statistics, fields in which many types of problems can be naturally set in a geometric representation.

Most of the algorithms we've studied have involved text and numbers, which are represented and processed naturally in most programming environments. Indeed, the primitive operations required are implemented in the hardware of most computer systems. We'll see that the situation is different for geometric problems: even the most elementary operations on points and lines can be computationally challenging.

Geometric problems are easy to visualize, but that can be a liability. Many problems that can be solved instantly by a person looking at a piece of paper (example: is a given point inside a given polygon?) require non-trivial computer programs. For more complicated problems, as in many other applications, the method of solution appropriate for computer implementation may well be quite different from the method of solution appropriate for a person.

One might suspect that geometric algorithms would have a long history because of the constructive nature of ancient geometry and because useful applications are so widespread, but actually much of the work in the field has been quite recent. Nonetheless, the work of ancient mathematicians is often useful in the development of algorithms for modern computers. The field of geometric algorithms is interesting to study because of its strong historical context, because new fundamental algorithms are still being developed, and because many important large-scale applications require these algorithms.

Points, Lines, and Polygons

Most of the programs we'll study operate on simple geometric objects defined in a two-dimensional space, though we will consider a few algorithms for higher dimensions. The fundamental object is a *point*, which we consider to be a pair of integers—the "coordinates" of the point in the usual Cartesian system. A *line* is a pair of points, which we assume are connected together by a straight line segment. A *polygon* is a list of points: we assume that successive points are connected by lines and that the first point is connected to the last to make a closed figure.

To work with these geometric objects, we need to decide how to represent them. Usually we use an array representation for polygons, though a linked list or some other representation can be used when appropriate. Most of our programs will use the straightforward representations

```
type point = record x,y: integer end;
     line = record p1,p2: point end;
var polygon: array[0..Nmax] of point;
```

Note that points are restricted to have integer coordinates. A *real* representation could also be used. Using integer coordinates leads to slightly simpler and more efficient algorithms, and is not as severe a restriction as it might seem. As mentioned in Chapter 2, working with integers when possible can be a very significant timesaver in many computing environments, because integer calculations are typically much more efficient than floating-point calculations. Thus, when we can get by with dealing only with integers without introducing much extra complication, we will do so.

More complicated geometric objects will be represented in terms of these basic components. For example, polygons will be represented as arrays of *points*. Note that using arrays of *lines* would result in each point on the polygon being included twice (though that still might be the natural representation for some algorithms). Also, it is useful in some applications to include extra information associated with each point or line; we can do this by adding an *info* field to the records.

We'll use the sets of points shown in Figure 24.1 to illustrate the operation of several geometric algorithms. The sixteen points on the left are labeled with single letters for reference in explaining the examples, and have the integer coordinates

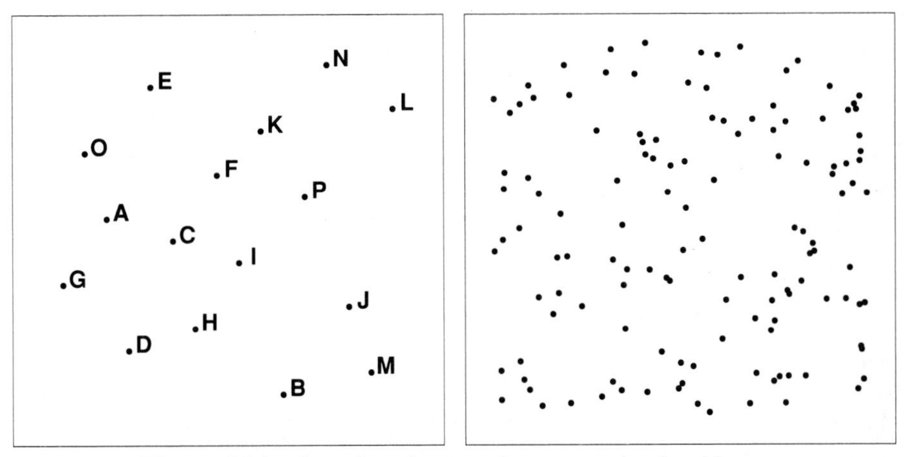

Figure 24.1 Sample point sets for geometric algorithms.

shown in Figure 24.2. (The labels we use are assigned in the order in which the points are assumed to appear in the input.) The programs usually have no reason to refer to points "by name"; they are simply stored in an array and are referred to by index. The order in which the points appear in the array may be important in some of the programs: indeed, it is the goal of some geometric algorithms to "sort" the points into some particular order. On the right in Figure 24.1 are 128 points, randomly generated with integer coordinates between 0 and 1000.

A typical program maintains an **array** $p[1..N]$ of points and simply reads in N pairs of integers, assigning the first pair to the x and y coordinates of $p[1]$, the second pair to $p[2]$, etc. When p represents a polygon, it is sometimes convenient to maintain "sentinel" values $p[0]=p[N]$ and $p[N+1]=p[1]$.

Line Segment Intersection

As our first elementary geometric problem, we'll consider determining whether or not two given line segments intersect. Figure 24.3 illustrates some of the situations that can arise. In the first case, the line segments intersect. In the second, the endpoint of one segment is on the other segment: We'll consider this an intersection

	A	B	C	D	E	F	G	H	I	J	K	L	M	N	O	P
x	3	11	6	4	5	8	1	7	9	14	10	16	15	13	3	12
y	9	1	8	3	15	11	6	4	7	5	13	14	2	16	12	10

Figure 24.2 Coordinates of points in small sample set (on the left in Figure 24.1).

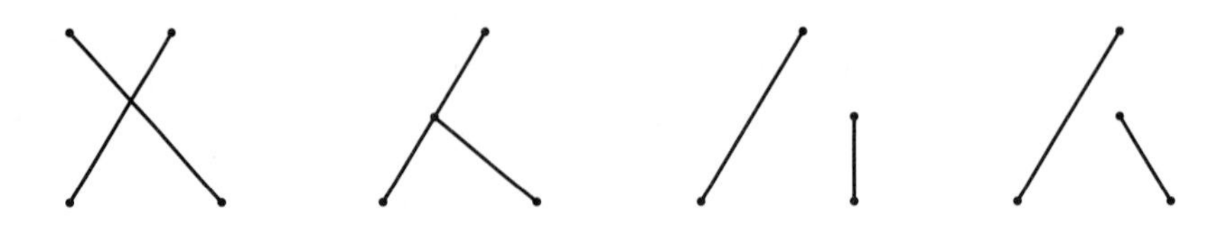

Figure 24.3 Testing whether line segments intersect: four cases.

by assuming the segments to be "closed" (endpoints are part of the segments); thus, line segments having a common endpoint intersect. In both the last two cases in Figure 24.3, the segments do not intersect, but the cases differ when we consider the intersection point of the lines defined by the segments. In the fourth case this intersection point falls on one of the segments; in the third it does not. Or, the lines could be parallel (a special case of this that frequently turns up is when one or both of the segments is a single point).

The straightforward way to solve this problem is to find the intersection point of the lines defined by the line segments and then check whether this intersection point falls between the endpoints of both of the segments. Another easy method is based on a tool that we'll find useful later, so we'll consider it in more detail. Given three points, we want to know whether, in traveling from the first to the second to the third, we turn counterclockwise or clockwise. For example, for points A, B, and C in Figure 24.1 the answer is yes, but for points A, B, and D the answer is no. This function is straightforward to compute from the equations for the lines as follows:

```
function ccw(p0,p1,p2: point): integer;
  var dx1,dx2,dy1,dy2: integer;
  begin
  dx1:=p1.x-p0.x; dy1:=p1.y-p0.y;
  dx2:=p2.x-p0.x; dy2:=p2.y-p0.y;
  if dx1*dy2>dy1*dx2 then ccw:=1;
  if dx1*dy2<dy1*dx2 then ccw:=-1;
  if dx1*dy2=dy1*dx2 then
    begin
    if (dx1*dx2<0) or (dy1*dy2<0) then ccw:=-1 else
    if (dx1*dx1+dy1*dy1)>=(dx2*dx2+dy2*dy2) then ccw:=0 else ccw:=1;
    end;
  end;
```

To understand how the program works, first suppose that all of the quantities $dx1$, *dx2*, *dy1*, and *dy2* are positive. Then note that the slope of the line connecting *p0* and *p1* is *dy1/dx1* and the slope of the line connecting *p0* and *p2* is *dy2/dx2*. Now, if the slope of the second line is greater than the slope of the first, a "left" (counterclockwise) turn is required in the journey from *p0* to *p1* to *p2*; if less, a "right" (clockwise) turn is required. Comparing slopes in the program is slightly inconvenient because the lines could be vertical (*dx1* or *dx2* could be 0): we multiply by *dx1*dx2* to avoid this. It turns out that the slopes need not be positive for this test to work properly; if, however, the slopes are the same (the three points are collinear), one can envision a variety of ways to define *ccw*. Our choice is to make the function three-valued: rather than **true** and **false** we use *1* and *-1*, reserving the value *0* for the case where *p2* is *on* the line segment *between p0* and *p1*. If the points are collinear, and *p0* is between *p2* and *p1*, we take *ccw* to be *-1*; if *p2* is between *p0* and *p1*, we take *ccw* to be 0; and if *p1* is between *p0* and *p2*, we take *ccw* to be 1. We'll see that this convention simplifies the coding for functions that use *ccw* in this and the next chapter.

This immediately gives an implementation of the *intersect* function. If both endpoints of each line are on different "sides" (have different *ccw* values) of the other, then the lines must intersect:

```
function intersect(l1,l2: line): boolean;
  begin
  intersect:=((ccw(l1.p1,l1.p2,l2.p1)*ccw(l1.p1,l1.p2,l2.p2))<=0)
            and ((ccw(l2.p1,l2.p2,l1.p1)*ccw(l2.p1,l2.p2,l1.p2))<=0);
  end;
```

This solution seems to involve a fair amount of computation for such a simple problem. The reader is encouraged to try to find a simpler solution, but should be warned to be sure that the solution works on all cases. For example, if all four points are collinear, there are six different cases (not counting situations where points coincide), only four of which are intersections. Special cases like this are the bane of geometric algorithms: they cannot be avoided, but we can lessen their impact with primitives like *ccw*.

If many lines are involved, the situation becomes much more complicated. In Chapter 27, we'll see a sophisticated algorithm for determining whether any two of a set of *N* lines intersect.

Simple Closed Path

To get the flavor of problems dealing with sets of points, let's consider the problem of finding a path through a set of *N* given points that doesn't intersect itself, visits all the points, and returns to the point at which it started. Such a path is called a *simple closed path*. One can imagine many applications for this: the points might represent homes and the path the route that a mailman might take to get to each of the homes without crossing his path. Or we might simply want a reasonable way to draw the points using a mechanical plotter. This problem is elementary because it asks only for any closed path connecting the points. The problem of finding the best such path, called the *traveling salesman problem*, is much, much more difficult, and we'll look at it in some detail in the last few chapters of this book. In the next chapter, we'll consider a related but much easier problem: finding the shortest path that surrounds a set of N given points. In Chapter 31, we'll see how to find the best way to "connect" a set of points.

An easy way to solve the elementary problem at hand is the following. Pick one of the points to serve as an "anchor." Then compute the angle made by drawing a line from each of the points in the set to the anchor and then out in the positive horizontal direction (this is part of the polar coordinate of each point with the anchor point as origin). Next, sort the points according to that angle. Finally, connect adjacent points. The result is a simple closed path connecting the points, as shown in Figure 24.4 for the points in Figure 24.1. In the small set of points, B is used as the anchor: if the points are visited in the order

B M J L N P K F I E C O A H G D B

then a simple closed polygon will be traced out.

If dx and dy are the distances along the x and y axes from the anchor point to some other point, then the angle needed in this algorithm is $\tan^{-1} dy/dx$. Although the arctangent is a built-in function in Pascal (and some other programming environments), it is likely to be slow and leads to at least two annoying extra conditions to compute: whether dx is zero and which quadrant the point is in. Since the angle

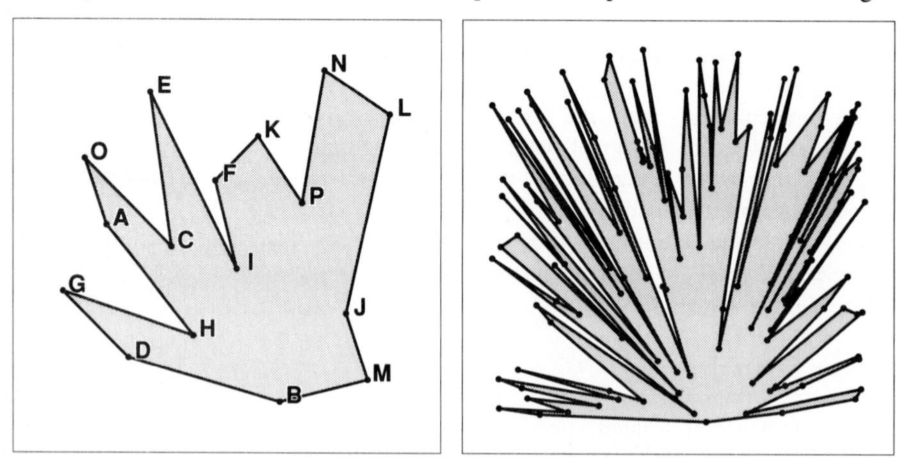

Figure 24.4 Simple closed paths.

is used only for the sort in this algorithm, it makes sense to use a function that is much easier to compute but has the same ordering properties as the arctangent (so that when we sort, we get the same result). A good candidate for such a function is simply $dy/(dy + dx)$. Testing for exceptional conditions is still necessary, but simpler. The following program returns a number between 0 and 360 that is *not* the angle made by *p1* and *p2* with the horizontal but which has the same order properties as that angle.

```
function theta(p1,p2: point): real;
  var dx,dy,ax,ay: integer;
      t: real;
  begin
  dx:=p2.x-p1.x; ax:=abs(dx);
  dy:=p2.y-p1.y; ay:=abs(dy);
  if (dx=0) and (dy=0) then t:=0
       else t:=dy/(ax+ay);
  if dx<0 then t:= 2-t
   else if dy<0 then t:=4+t;
  theta:=t*90.0;
  end;
```

In some programming environments it may not be worthwhile to use such programs instead of standard trigonometric functions; in others it may lead to significant savings. (In some cases it may be worthwhile to change *theta* to have an integer value, to avoid using *real* numbers entirely.)

Inclusion in a Polygon

The next problem we'll consider is a natural one: given a point and polygon represented as an array of points, determine whether the point is inside or outside the polygon. A straightforward solution to this problem immediately suggests itself: draw a long line segment from the point in any direction (long enough so that its other endpoint is guaranteed to be outside the polygon) and count the number of lines from the polygon that it crosses. If the number is odd, the point must be inside; if it is even, the point is outside. This is easily seen by tracing what happens as we come in from the endpoint on the outside: after the first line, we are inside, after the second we are outside, etc. If we do this an even number of times, the point at which we end up (the original point) must be outside.

The situation is not quite so simple, however, because some intersections might occur right at the vertices of the input polygon. Figure 24.5 shows some of the situations that must to be handled. The first is a straightforward "outside" case; the second is a straightforward "inside" case; in the third, the test line leaves the

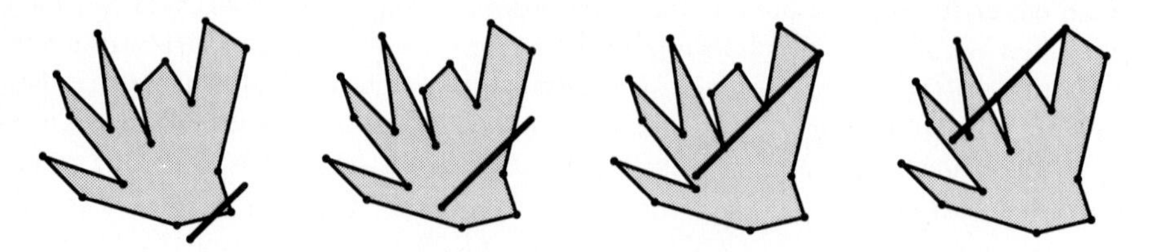

Figure 24.5 Cases to be handled by a point-in-polygon algorithm.

polygon at a vertex (after touching two other vertices; and in the fourth, the test line coincides with an edge of the polygon before leaving. In some cases where the test line intersects a vertex it should count as one intersection with the polygon; in other cases it should count as none (or two). The reader may be amused to try to find a simple test to distinguish these cases before reading further.

The need to handle cases where polygon vertices fall on the test lines forces us to do more than just count the line segments in the polygon intersecting the test line. Essentially, we want to travel around the polygon, incrementing an intersection counter whenever we go from one side of the test line to another. One way to implement this is to simply ignore points that fall on the test line, as in the following program:

```
function inside(t: point): boolean;
  var count, i, j: integer;
      lt, lp: line;
  begin
  count:=0; j:=0;
  p[0]:=p[N]; p[N+1]:=p[1];
  lt.p1:=t; lt.p2:=t; lt.p2.x:=maxint;
  for i:=1 to N do
    begin
    lp.p1:=p[i]; lp.p2:=p[i];
    if not intersect(lp, lt) then
      begin
      lp.p2:=p[j]; j:=i;
      if intersect(lp, lt) then count:=count+1;
      end;
    end;
  inside:=((count mod 2)=1);
  end;
```

This program uses a horizontal test line for ease of calculation (imagine the diagrams in Figure 24.5 as rotated 45 degrees). The variable j is maintained as the index of the last point on the polygon known not to lie on the test line. The program assumes that *p[1]* is the point with the smallest x coordinate among all the points with the smallest y coordinate, so that if *p[1]* is on the test line, then *p[0]* cannot be. The same polygon can be represented by N different *p* arrays, but as this illustrates it is sometimes convenient to fix a standard rule for *p[1]*. (For example, this same rule is useful for *p[1]* as the "anchor" for the procedure suggested above for computing a simple closed polygon.) If the next point on the polygon that is not on the test line is on the same side of the test line as the jth point, then we need not increment the intersection counter (*count*); otherwise we have an intersection. The reader may wish to check that this algorithm works properly for the cases in Figure 24.5.

If the polygon has only three or four sides, as is true in many applications, then such a complex program is not called for: a simpler procedure based on calls to *ccw* will be adequate. Another important special case is the *convex polygon*, to be studied in the next chapter, which has the property that no test line can have more than two intersections with the polygon. In this case, a procedure like binary search can be used to determine in $O(\log N)$ steps whether or not a point is inside.

Perspective

From the few examples given, it should be clear that it is easy to underestimate the difficulty of solving a particular geometric problem with a computer. There are many other elementary geometric computations that we have not treated at all. For example, a program to compute the area of a polygon makes an interesting exercise. However, the problems we've looked at have provided some basic tools that will be useful in later sections for solving the more difficult problems.

Some of the algorithms we'll study involve building geometric structures from a given set of points. The "simple closed polygon" is an elementary example of this. We will need to decide upon appropriate representations for such structures, develop algorithms to build them, and investigate their use in particular applications. As usual, these considerations are intertwined. For example, the algorithm used in the *inside* procedure in this chapter depends in an essential way on the representation of the simple closed polygon as an ordered set of points (rather than as an unordered set of lines).

Many of the algorithms we'll study involve *geometric search*: we want to know which points from a given set are close to a given point, or which points fall in a given rectangle, or which points are closest to one another. Many of the algorithms appropriate for such search problems are closely related to the search algorithms studied in Chapters 14–17. The parallels will be quite evident.

Few geometric algorithms have been analyzed to the point that precise statements can be made about their relative performance characteristics. As we've already seen, the running time of a geometric algorithm can depend on many things.

The distribution of the points themselves, the order in which they appear in the input, and whether trigonometric functions are used can all significantly affect the running time of geometric algorithms. As usual in such situations, however, we do have empirical evidence that suggests good algorithms for particular applications. Also, many of the algorithms are derived from complexity studies and are designed for good worst-case performance.

□

Exercises

1. Give the value of *ccw* for the three cases when two of the points are identical (and the third is different), and for the case when all three points are identical.
2. Give a quick algorithm for determining whether two line segments are parallel, without using any divisions.
3. Give a quick algorithm for determining whether four line segments form a square, without using any divisions.
4. Given an array of *lines*, how would you test to see whether they form a simple closed polygon?
5. Draw the simple closed polygons that result from using A, C, and D in Figure 24.1 as "anchors" in the method described in the text.
6. Suppose that we use an arbitrary point for the "anchor" in the method described in the text for computing a simple closed polygon. Give conditions which such a point must satisfy for the method to work.
7. What does the *intersect* function return when called with two copies of the same line segment?
8. Does *inside* call a vertex of the polygon inside or outside?
9. What is the maximum value achievable by *count* when *inside* is executed on a polygon with N vertices? Give an example supporting your answer.
10. Write an efficient program for determining whether a given point is inside a given quadrilateral.

25

Finding the Convex Hull

Often, when we have a large number of points to process, we're interested in the boundaries of the point set. People looking at a diagram of a set of points plotted in the plane, have little trouble distinguishing those on the "inside" of the point set from those lying on the edge. This distinction is a fundamental property of point sets; in this chapter we'll see how it can be precisely characterized by looking at algorithms for separating out the "natural boundary" points.

The mathematical way to describe the natural boundary of a point set depends on a geometric property called *convexity*. This is a simple concept that the reader may have encountered before: a *convex polygon* has the property that any line connecting any two points inside the polygon must itself lie entirely inside the polygon. For example, the "simple closed polygon" that we computed in the previous chapter is decidedly nonconvex; on the other hand, any triangle or rectangle is convex.

Now, the mathematical name for the natural boundary of a point set is the *convex hull*. The convex hull of a set of points in the plane is defined to be the smallest convex polygon containing them all. Equivalently, the convex hull is the shortest path surrounding the points. An obvious property of the convex hull that is easy to prove is that the vertices of the convex polygon defining the hull are points from the original point set. Given N points, some of them form a convex polygon within which all the others are contained. The problem is to find those points. Many algorithms have been developed to find the convex hull; in this chapter we'll examine some of the important ones.

Figure 25.1 shows our sample sets of points for Figure 24.1 and their convex hulls. There are 8 points on the hull of the small set and 15 points on the hull of the large set. In general, the convex hull can contain as few as three points (if the three points form a large triangle containing all the others) or as many as all the points (if they fall on a convex polygon, then the points comprise their own convex hull). The number of points on the convex hull of a "random" point set

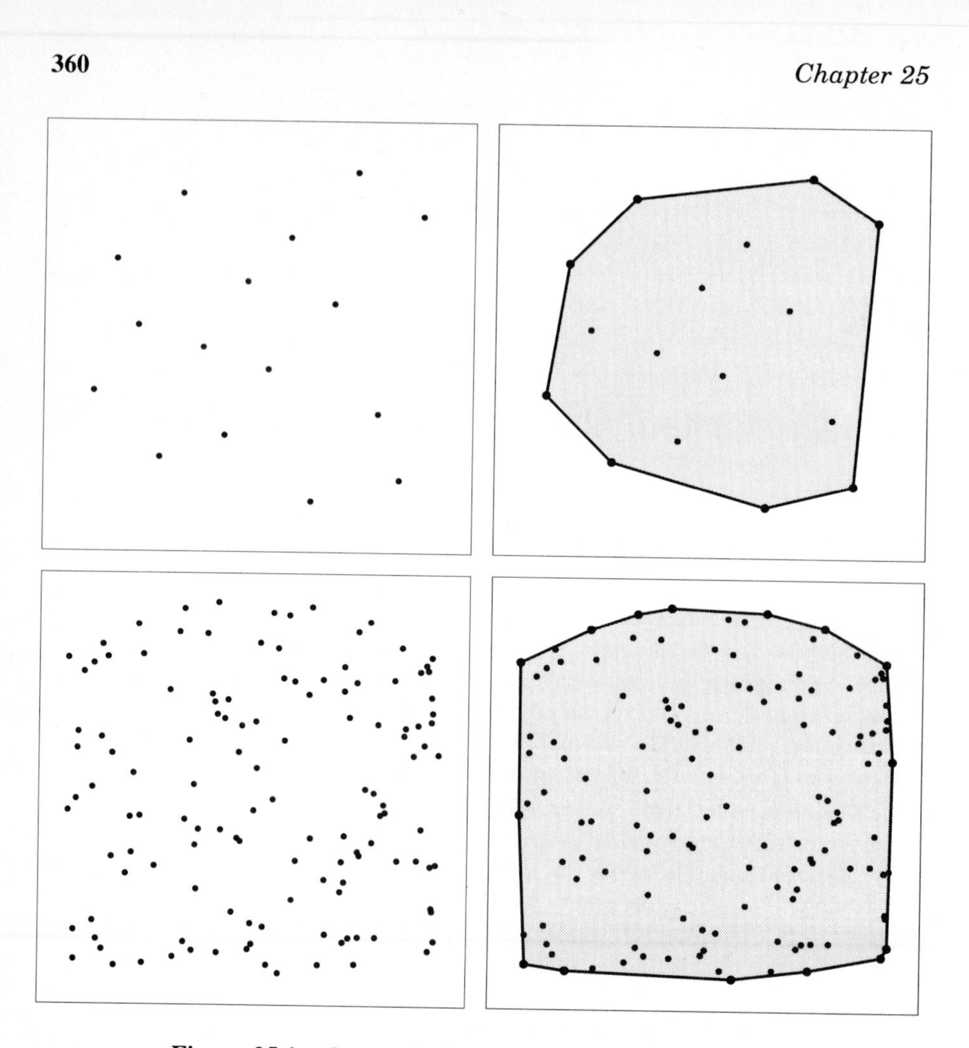

Figure 25.1 Convex hulls of the points of Figure 24.1.

falls somewhere in between these extremes, as we will see below. Some algorithms work well when there are many points on the convex hull; others work better when there are only a few.

A fundamental property of the convex hull is that any line outside the hull, when moved in any direction towards the hull, hits it at one of its vertex points. (This is an alternate way to define the hull: it is the subset of points from the point set that could be hit by a line moving in at some angle from infinity.) In particular, it's easy to find a few points guaranteed to be on the hull by applying this rule for horizontal and vertical lines: the points with the smallest and largest x and y coordinates are all on the convex hull. This fact is used as the starting point for the algorithms we consider.

Rules of the Game

The input to an algorithm for finding the convex hull is of course an array of points; we can use the *point* type defined in the previous chapter. The output is a polygon, also represented as an array of points with the property that tracing through the points in the order in which they appear in the array traces the outline of the polygon. On reflection, this might appear to require an extra ordering condition on the computation of the convex hull (why not just return the points on the hull in any order?), but output in the ordered form is obviously more useful, and it has been shown that the unordered computation is no easier to do. For all of the algorithms that we consider, it is convenient to do the computation *in place*: the array used for the original point set is also used to hold the output. The algorithms simply rearrange the points in the original array so that the convex hull appears in the first M positions, in order.

From the description above, it may be clear that computing the convex hull is closely related to sorting. In fact, a convex hull algorithm can be used to sort, in the following way. Given N numbers to sort, turn them into points (in polar coordinates) by treating the numbers as angles (suitably normalized) with a fixed radius for each point. The convex hull of this point set is an N-gon containing all of the points. Now, since the output must be ordered in the order in which the points appear on this polygon, it can be used to find the sorted order of the original values (remember that the input was unordered). This is not a formal proof that computing the convex hull is no easier than sorting, because, for example, the cost of the trigonometric functions required to convert the numbers into points on the polygon must be considered. Comparing convex hull algorithms (which involve trigonometric operations) to sorting algorithms (which involve comparisons between keys) is a bit like comparing apples to oranges, but even so it has been shown that any convex hull algorithm must require about $N \log N$ operations, the same as sorting (even though the operations allowed are likely to be quite different). It is helpful to view finding the convex hull of a set of points as a kind of "two-dimensional sort," since frequent parallels to sorting algorithms arise in the study of algorithms for finding the convex hull.

In fact, the algorithms we'll study show that finding the convex hull is no harder than sorting either: there are several algorithms that run in time proportional to $N \log N$ in the worst case. Many of the algorithms tend to use even less time on actual point sets, because their running time depends on how the points are distributed and on the number of points on the hull.

As with all geometric algorithms, we have to pay some attention to degenerate cases that are likely to occur in the input. For example, what is the convex hull of a set of points all of which fall on the same line segment? Depending upon the application, this could be all the points or just the two extreme points, or perhaps any set including the two extreme points would do. Though this seems an extreme example, it would not be unusual for more than two points to fall on one of the line

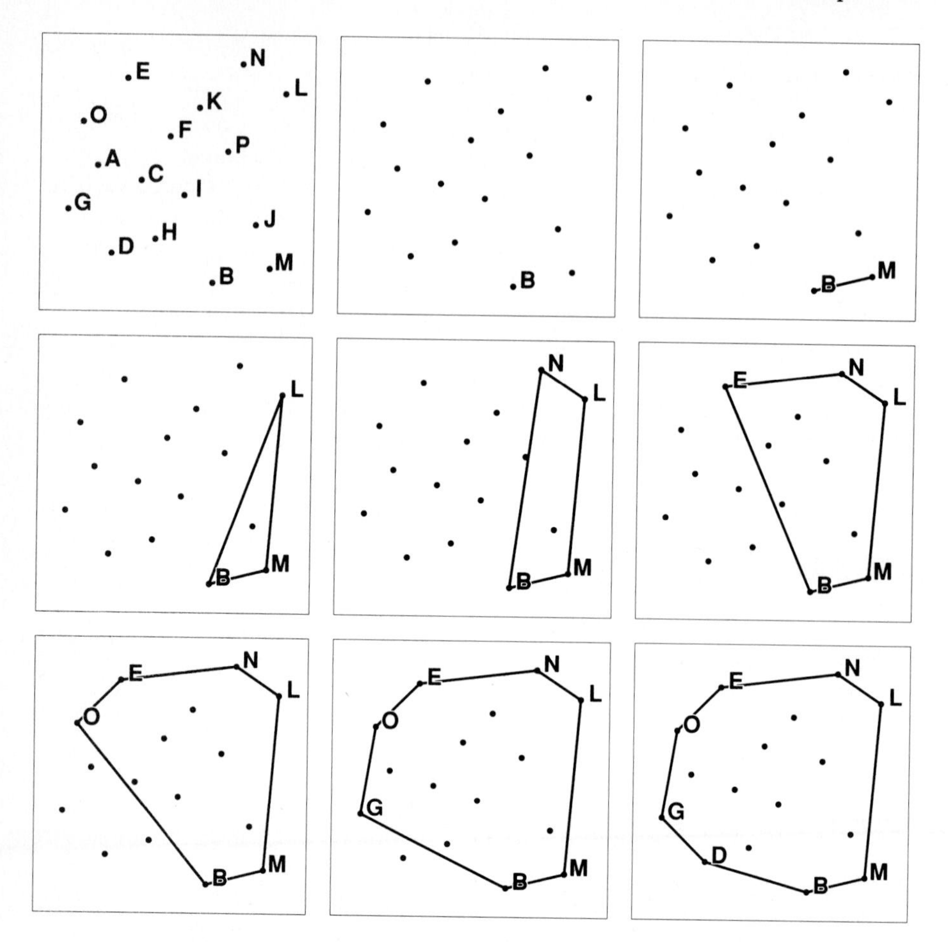

Figure 25.2 Package-wrapping.

segments defining the hull of a set of points. In the algorithms below, we won't insist that points falling on a hull edge be included, since this generally involves more work (though we will indicate how this could be done when appropriate). On the other hand, we won't insist that they be omitted either, since this condition could be tested afterwards if desired.

Package-Wrapping

The most natural convex hull algorithm, which parallels how a human would draw the convex hull of a set of points, is a systematic way to "wrap up" the set of points. Starting with some point guaranteed to be on the convex hull (say the one with the smallest *y* coordinate), take a horizontal ray in the positive direction and "sweep" it upward until hitting another point; this point must be on the hull. Then anchor at that point and continue "sweeping" until hitting another point, etc., until the "package" is fully "wrapped" (the beginning point is included again). Figure 25.2 shows how the hull is discovered in this way for our sample set of points. Point B has the minimum y coordinate and is the starting point. Then M is the first point hit by the sweeping ray, then L, etc.

Of course, we don't actually need to sweep through all possible angles; we just do a standard find-the-minimum computation to find the point that would be hit next. For each point to be included on the hull, we need to examine each point not yet included on the hull. Thus, the method is quite similar to selection sorting—we successively choose the "best" of the points not yet chosen, using a brute-force search for the minimum. The actual data movement involved is depicted in Figure 25.3: the Mth line of the table shows the situation after the Mth point is added to the hull.

The following program finds the convex hull of an array $p[1..N]$ of points, represented as described at the beginning of Chapter 24. The basis for this implementation is the function *theta*$(p1,p2$: *point*) developed in the previous chapter, which can be thought of as returning the angle between $p1$, $p2$ and the horizontal (though it actually returns a more easily computed number with the same ordering properties). Otherwise, the implementation follows directly from the discussion above. The array position $p[N+1]$ is used to hold a sentinel.

0	*1*	*2*	*3*	*4*	*5*	*6*	*7*	*8*	*9*	*10*	*11*	*12*	*13*	*14*	*15*	*16*
A	B	C	D	E	F	G	H	I	J	K	L	M	N	O	P	
B	A	C	D	E	F	G	H	I	J	K	L	M	N	O	P	
B	M	C	D	E	F	G	H	I	J	K	L	A	N	O	P	
B	M	L	D	E	F	G	H	I	J	K	C	A	N	O	P	
B	M	L	N	E	F	G	H	I	J	K	C	A	D	O	P	
B	M	L	N	E	F	G	H	I	J	K	C	A	D	O	P	
B	M	L	N	E	O	G	H	I	J	K	C	A	D	F	P	
B	M	L	N	E	O	G	H	I	J	K	C	A	D	F	P	
B	M	L	N	E	O	G	D	I	J	K	C	A	H	F	P	

Figure 25.3 Data movement in package-wrapping.

```
function wrap: integer;
  var i, min, M: integer;
      minangle, v: real;
      t: point;
  begin
  min:=1;
  for i:=2 to N do
    if p[i].y<p[min].y then min:=i;
  M:=0; p[N+1]:=p[min]; minangle:=0.0;
  repeat
    M:=M+1; t:=p[M]; p[M]:=p[min]; p[min]:=t;
    min:=N+1; v:=minangle; minangle:=360.0;
    for i:=M+1 to N+1 do
      if theta(p[M],p[i])>v then
        if theta(p[M],p[i])<minangle then
          begin min:=i; minangle:=theta(p[M],p[min]) end;
  until min=N+1;
  wrap:=M;
  end;
```

First, the point with the smallest y coordinate is found and copied into $p[N+1]$ in order to stop the loop, as described below. The variable M is maintained as the number of points so far included on the hull, and v is the current value of the "sweep" angle (the angle from the horizontal to the line between $p[M-1]$ and $p[M]$). The **repeat** loop puts the last point found into the hull by exchanging it with the Mth point, and uses the *theta* function from the previous chapter to compute the angle from the horizontal made by the line between that point and each of the points not yet included on the hull, searching for the one whose angle is smallest among those with angles greater than v. The loop stops when the first point (actually the copy of the first point put into $p[N+1]$) is encountered again.

This program may or may not return points which fall on a convex hull edge. This situation is encountered when more than one point has the same *theta* value with $p[M]$ during the execution of the algorithm; the implementation above returns the point first encountered among such points, even though there may be others closer to $p[M]$. When it is important to find points falling on convex hull edges, we can achieve this by changing *theta* to take into account the distance between the points given as its arguments and assign the closer point a smaller value when two points have the same angle.

The major disadvantage of package-wrapping is that in the worst case, when all the points fall on the convex hull, the running time is proportional to N^2 (like that of selection sort). On the other hand, the method has the attractive feature that it generalizes to three (or more) dimensions. The convex hull of a set of points in

k-dimensional space is the minimal convex polytope containing them all, where a convex polytope is defined by the property that any line connecting two points inside must itself lie inside. For example, the convex hull of a set of points in 3-space is a convex three-dimensional object with flat faces. It can be found by "sweeping" a plane until the hull is hit, then "folding" faces of the plane, anchoring on different lines on the boundary of the hull, until the "package" is "wrapped." (Like many geometric algorithms, it is rather easier to explain this generalization than to implement it!)

The Graham Scan

The next method we'll examine, invented by R. L. Graham in 1972, is interesting because most of the computation involved is for sorting: the algorithm includes a sort followed by a relatively inexpensive (though not immediately obvious) computation. The algorithm starts by constructing a simple closed polygon from the points using the method of the previous chapter: sort the points using as keys the *theta* function values corresponding to the angle from the horizontal made from the line connecting each point with an 'anchor' point $p[1]$ (the one with the lowest y coordinate), so that tracing $p[1]$, $p[2]$, ..., $p[N]$, $p[1]$ gives a closed polygon. For our example set of points, we get the simple closed polygon of the previous chapter. Note that $p[N]$, $p[1]$, and $p[2]$ are consecutive points on the hull; by sorting, we've essentially run the first iteration of the package-wrapping procedure (in both directions).

Computation of the convex hull is completed by proceeding around, trying to place each point on the hull and eliminating previously placed points that couldn't possibly be on the hull. For our example, we consider the points in the order B M J L N P K F I E C O A H G D; the first few steps are shown in Figure 25.4. At the beginning, we know because of the sort that B and M are on the hull. When J is encountered, the algorithm includes it on the trial hull for the first three points.

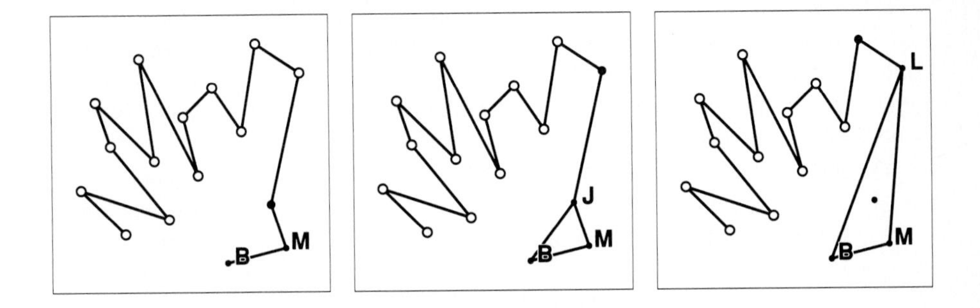

Figure 25.4 Start of Graham scan.

Then, when L is encountered, the algorithm finds out that J couldn't be on the hull (since, for example, it falls inside the triangle BML).

In general, testing which points to eliminate is not difficult. After each point has been added, we assume that we have eliminated enough points that what we have traced out so far could be part of the convex hull on the basis of the points so far seen. As we trace around, we expect to turn left at each hull vertex. If a new point causes us to turn *right*, then the point just added must be eliminated, since there exists a convex polygon containing it. Specifically, the test for eliminating a point uses the *ccw* procedure of the previous chapter, as follows. Suppose we have determined that $p[1..M]$ are on the partial hull determined on the basis of examining $p[1..i-1]$. When we come to examine a new point $p[i]$, we eliminate $p[M]$ from the hull if $ccw(p[M], p[M-1], p[i])$ is nonnegative. Otherwise, $p[M]$ could still be on the hull, so we don't eliminate it.

Figure 25.5 shows the completion of this process on our sample set of points. The situation as each new point is encountered is diagrammed: each new point is added to the partial hull so far constructed and is then used as a "witness" for the elimination of (zero or more) points previously considered. After L, N, and P are added to the hull, P is eliminated when K is considered (since NPK is a right turn), then F and I are added, leading to the consideration of E. At this point, I must be eliminated because FIE is a right turn, then F and K must be eliminated because KFE and NKE are right turns.. Thus more than one point can be eliminated during the "backup" procedure, perhaps several. Continuing in this way, the algorithm finally arrives back at B.

The initial sort guarantees that each point is considered in turn as a possible hull point, because all points considered earlier have a smaller *theta* value. Each line that survives the "eliminations" has the property that all points so far considered are on the same side of it, so that when we get back to $p[N]$, which also must be on the hull because of the sort, we have the complete convex hull of all the points.

As with the package-wrapping method, points on a hull edge may or may not be included, though there are two distinct situations that can arise with collinear points. First, if there are two points collinear with $p[1]$, then, as above, the sort using *theta* may or may not get them in order along their common line. Points out of order in this situation will be eliminated during the scan. Second, collinear points along the trial hull can arise (and not be eliminated).

Once the basic method is understood, the implementation is straightforward, though there are a number of details to attend to. First, the point with the maximum x value among all points with minimum y value is exchanged with $p[1]$. Next, *shellsort* is used to rearrange the points (any comparison-based sorting routine would do), modified as necessary to compare two points using their *theta* values with $p[1]$. After the sort, $p[N]$ is copied into $p[0]$ to serve as a sentinel in case $p[3]$ is not on the hull. Finally, the scan described above is performed. The following program finds the convex hull of the point set $p[1..N]$:

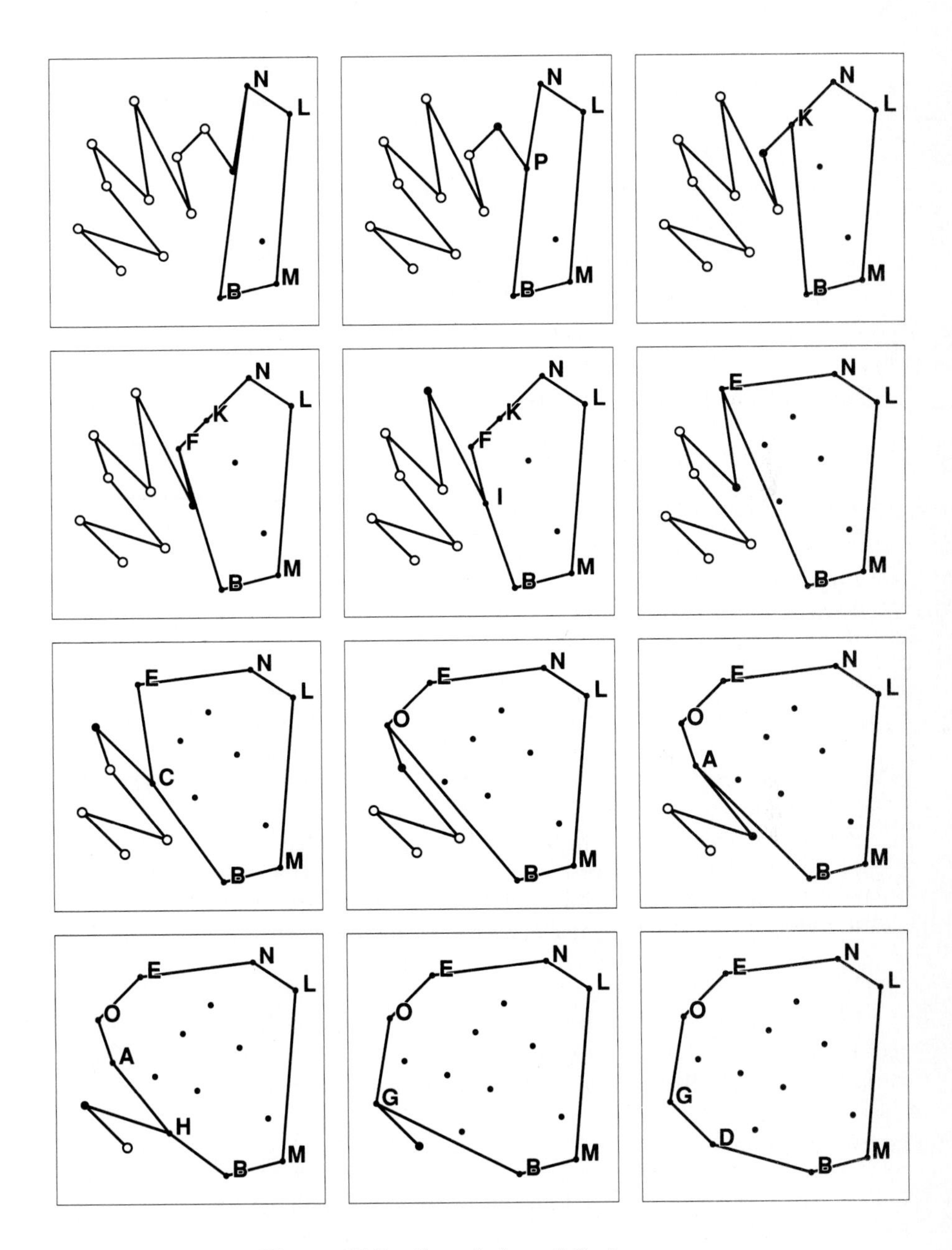

Figure 25.5 Completion of Graham scan.

```
function grahamscan: integer;
  var i,j,min,M: integer;
      l: line; t: point;
  begin
  min:=1;
  for i:=2 to N do
    if p[i].y<p[min].y then min:=i;
  for i:=1 to N do
    if (p[i].y=p[min].y) and (p[i].x>p[min].x) then min:=i;
  t:=p[1]; p[1]:=p[min]; p[min]:=t;
  shellsort;
  p[0]:=p[N];
  M:=3;
  for i:=4 to N do
    begin
    while ccw(p[M],p[M-1],p[i])>=0 do M:=M-1;
    M:=M+1;
    t:=p[M]; p[M]:=p[i]; p[i]:=t;
    end;
  grahamscan:=M;
  end;
```

The loop maintains a partial hull in *p*[*1..M*], as described above. For each new *i* value considered, *M* is decremented if necessary to eliminate points from the partial hull and then *p*[*i*] is exchanged with *p*[*M+1*] to (tentatively) add it to the partial hull. Figure 25.6 shows the contents of the *p* array each time a new point is considered for our example.

The reader may wish to check why it is necessary for the *min* computation to find the point with the lowest x coordinate among all points with the lowest y coordinate, the canonical form described in Chapter 24. As discussed above, another subtle point is to consider the effect of the fact that collinear points lead to equal *theta* values, and may not be sorted in the order in which they appear on the line, as one might have hoped.

One reason that this method is interesting to study is that it is a simple form of *backtracking*, the algorithm design technique of "try something, and if it doesn't work then try something else" that we'll revisit in Chapter 44.

Interior Elimination

Almost any convex hull method can be vastly improved by a simple techique which quickly disposes of most points. The general idea is simple: pick four points known to be on the hull, then throw out everything inside the quadrilateral

0	1	2	3	4	5	6	7	8	9	10	11	12	13	14	15	16
B	M	J	L	N	P	K	F	I	E	C	O	A	H	G	D	
B	M	L	J													
B	M	L	N	J												
B	M	L	N	P	J											
B	M	L	N	K	J	P										
B	M	L	N	K	F	P	J									
B	M	L	N	K	F	I	J	P								
B	M	L	N	E	F	I	J	P	K							
B	M	L	N	E	C	I	J	P	K	F						
B	M	L	N	E	O	I	J	P	K	F	C					
B	M	L	N	E	O	A	J	P	K	F	C	I				
B	M	L	N	E	O	A	H	P	K	F	C	I	J			
B	M	L	N	E	O	G	H	P	K	F	C	I	J	A		
B	M	L	N	E	O	G	D	P	K	F	C	I	J	A	H	

Figure 25.6 Data movement in the Graham scan.

formed by those four points. This leaves many fewer points to be considered by, say, the Graham scan or the package wrapping technique.

The four points known to be on the hull should be chosen with an eye towards any information available about the input points. Generally, it is best to adapt the choice of points to the distribution of the input. For example, if all x and y values within certain ranges are equally likely (a rectangular distribution), then choosing four points by scanning in from the corners (find the four points with the largest and smallest sum and difference of the two coordinates) turns out to eliminate nearly all the points. Figure 25.7 shows that this technique eliminates most points not on the hull in our two example point sets.

In an implementation of the interior elimination method, the "inner loop" for random point sets is the test of whether or not a given point falls within the test quadrilateral. This can be speeded up somewhat by using a rectangle with edges parallel to the x and y axes. The largest such rectangle which fits in the quadrilateral described above is easy to find from the coordinates of the four points defining the quadrilateral. Using this rectangle will eliminate fewer points from the interior, but the speed of the test more than offsets this loss.

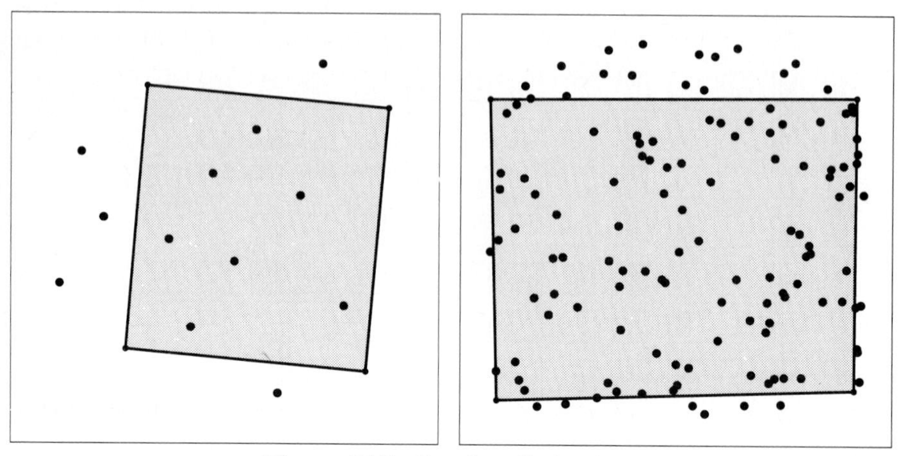

Figure 25.7 Interior elimination.

Performance Issues

As mentioned in the previous chapter, geometric algorithms are somewhat harder to analyze than algorithms in some of the other areas we've studied because the input (and the output) is more difficult to characterize. It often doesn't make sense to speak of "random" point sets: for example, as N gets large, the convex hull of points drawn from a rectangular distribution is extremely likely to be very close to the rectangle defining the distribution. The algorithms we've looked at depend on different properties of the point set distribution and are thus incomparable in practice, because to compare them analytically would require an understanding of very complicated interactions between little-understood properties of point sets. On the other hand, we can say some things about the performance of the algorithms that will help choosing one for a particular application.

Property 25.1 *After the sort, the Graham scan is a linear-time process.*

A moment's reflection is necessary to convince oneself that this is true, since there is a "loop-within-a-loop" in the program. However, it is easy to see that no point is "eliminated" more than once, so the code within that double loop is iterated fewer than N times. The total time required to find the convex hull using this method is $O(N \log N)$, but the "inner loop" of the method is the sort itself, which can be made efficient using techniques of Chapters 8-12. ■

Property 25.2 *If there are M vertices on the hull, then the "package-wrapping" technique requires about MN steps.*

First, we must compute $N - 1$ angles to find the minimum, then $N - 2$ to find the next, then $N - 3$, etc., so the total number of angle computations is $(N-1)+(N-2)+\cdots+(N-M+1)$, which is exactly equal to $MN - M(M-1)/2$. To compare this analytically with the Graham scan would require a formula for M in terms of N, a difficult problem in stochastic geometry. For a circular distribution (and some others) the answer is that M is $O(N^{1/3})$, and for values of N which are not large $N^{1/3}$ is comparable to $\log N$ (which is the expected value for a rectangular distribution), so this method will compete very favorably with the Graham scan. Of course, the N^2 worst case should always be taken into consideration. ■

Property 25.3 *The interior elimination method is linear, on the average.*

Full mathematical analysis of this method would require even more sophisticated stochastic geometry than above, but the general result is the same as that given by intuition: almost all the points fall inside the quadrilateral and are discarded—the number of points left over is $O(\sqrt{N})$. This is true even if the rectangle is used as described above. This makes the average running time of the whole convex hull algorithm proportional to N, since most points are examined only once (when they are thrown out). On the average, it doesn't matter much which method is afterwards, since so few points are likely to be left. However, to protect against the worst case (when all points are on the hull), it is prudent to use the Graham scan. This gives an algorithm which is almost sure to run in linear time in practice and is guaranteed to run in time proportional to $N \log N$. ■

The average-case result of Property 25.3 holds only for randomly distributed points in a rectangle, and in the worst case nothing is eliminated by the interior elimination method. However, for other distributions or for point sets with unknown properties, this method is still recommended because the cost is low (a linear scan through the points, with a few simple tests) and the possible savings is high (most of the points can be easily eliminated). The method also extends to higher dimensions.

It is possible to devise a recursive version of the interior elimination method: find extreme points and remove points on the interior of the defined quadrilateral as above, but then consider the remaining points as partitioned into subproblems which can be solved independently, using the same method. This recursive technique is similar to the Quicksort-like *select* procedure for selection discussed in Chapter 12. Like that procedure, it is vulnerable to an N^2 worst-case running time. For example, if all the original points are on the convex hull, then no points are thrown out in the recursive step. Like *select*, the running time is linear on the average (though it is not easy to prove this). But because so many points are eliminated in the first step, it is not likely to be worth the trouble to do further recursive decomposition in any practical application.

□

Exercises

1. Suppose you know in advance that the convex hull of a set of points is a triangle. Give an easy algorithm for finding the triangle. Answer the same question for a quadrilateral.
2. Give an efficient method for determining whether a point falls within a given convex polygon.
3. Implement a convex hull algorithm like insertion sort, using your method from the previous exercise.
4. Is it strictly necessary for the Graham scan to start with a point guaranteed to be on the hull? Explain why or why not.
5. Is it strictly necessary for the package-wrapping method to start with a point guaranteed to be on the hull? Explain why or why not.
6. Draw a set of points that makes the Graham scan for finding the convex hull particularly inefficient.
7. Does the Graham scan find the convex hull of the points that make up the vertices of *any* simple polygon? Explain why or give a counterexample showing why not.
8. What four points should be used for the interior elimination method if the input is assumed to be randomly distributed within a circle (using random polar coordinates)?
9. Empirically compare the Graham scan and the package-wrapping method for large point sets with both x and y equally likely to be between 0 and 1000.
10. Implement the interior elimination method and determine empirically how large N should be before one might expect fifty points to be left after the method is used on point sets with x and y equally likely to be between 0 and 1000.

26

Range Searching

☐ Given a set of points in the plane, it is natural to ask which of those points fall within some specified area. "List all cities within 50 miles of Princeton" is a question of this type which could reasonably be asked if a set of points corresponding to the cities of the U.S. were available. When the geometric shape is restricted to be a rectangle, the issue readily extends to non-geometric problems. For example, "list all those people between 21 and 25 with incomes between $60,000 and $100,000" asks which "points" from a file of data on people's names, ages, and incomes fall within a certain rectangle in the age-income plane.

Extension to more than two dimensions is immediate. If we want to list all stars within 50 light-years of the sun, we have a three-dimensional problem, and if we want the rich young people of the paragraph above to be tall and female as well, we have a four-dimensional problem. In fact, the dimension of such problems can get very high.

In general, we assume that we have a set of *records* with certain *attributes* that take on values from some ordered set. (This is sometimes called a *database*, though more specific and complete definitions have been developed for this important term.) Finding all records in a database that satisfy specified range restrictions on a specified set of attributes is called *range searching*,and is a difficult and important problem in practical applications. In this chapter, we'll concentrate on the two-dimensional geometric problem in which records are points and attributes are their coordinates, and then discuss appropriate generalizations.

The methods we'll look at are direct generalizations of methods we have seen for searching on single keys (in one dimension). We presume that many queries will be made on the same set of points, so the problem splits into two parts: we need a *preprocessing* algorithm, which builds the given points into a structure supporting efficient range searching, and a *range-searching* algorithm, which uses the structure to return points falling within any given (multidimensional) range. This separation makes different methods difficult to compare, since the total cost depends not only on the distribution of the points involved but also on the number and nature of the queries.

The range-searching problem in one dimension is to return all points falling within a specified interval. This can be done by sorting the points for preprocessing and then doing a binary search on the endpoints of the interval to return all the points that fall in between. Another solution is to build a binary search tree and then do a simple recursive traversal of the tree, returning points within the interval and ignoring parts of the tree outside the interval. The program required is a simple recursive tree traversal (see Chapter 4). If the left endpoint of the interval falls to the left of the point at the root, we (recursively) search the left subtree, and similarly for the right, checking each node we encounter to see whether its point falls within the interval:

```
type interval = record x1,x2: integer end;
procedure treerange(t: link; int: interval);
  var tx1,tx2: boolean;
  begin
  if t<>z then
    begin
    tx1:=t↑.key>=int.x1;
    tx2:=t↑.key<=int.x2;
    if tx1 then treerange(t↑.l,int);
    if tx1 and tx2
      then t↑.key is within the range
    if tx2 then treerange(t↑.r,int);
    end
  end;
```

This program could be made slightly more efficient by maintaining the interval *int* as a global variable rather than passing its unchanged values through the recursive calls. Figure 26.1 shows the points found when this program is run on a sample tree. Note that the points returned do not need to be connected in the tree.

Property 26.1 *One-dimensional range searching can be done with $O(N \log N)$ steps for preprocessing and $O(R+\log N)$ for range searching, where R is the number of points actually falling in the range.*

This follows directly from elementary properties of the search structures (see Chapters 14 and 15). A balanced tree could be used, if desired. ■

Our goal in this chapter will be to achieve these same running times for multidimensional range searching. The parameter R can be quite significant: given the facility to make range queries, a user could easily formulate queries that could require all or nearly all of the points. This type of query could reasonably be

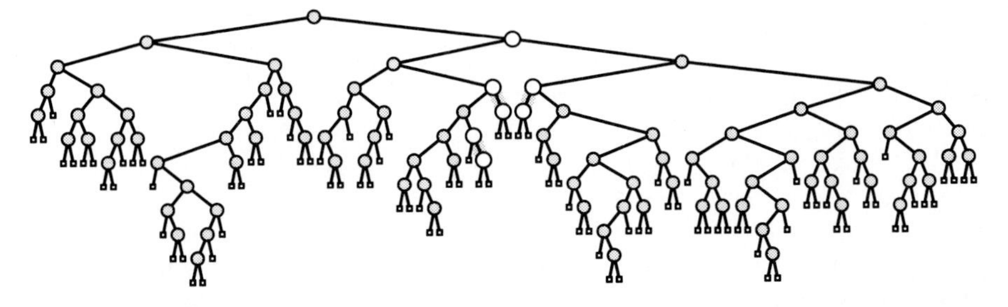

Figure 26.1 Range searching (one-dimensional) with a binary search tree.

expected to occur in many applications, but sophisticated algorithms are not necessary if *all* queries are of this type. The algorithms we consider are designed to be efficient for queries that are not expected to return a large number of points.

Elementary Methods

In two dimensions, our "range" is an area in the plane. For simplicity, we'll consider the problem of finding all points whose x coordinates fall within a given x-interval and whose y coordinates fall within a given y-interval: that is, we seek all points falling within a given rectangle. Thus, we'll assume a type *rectangle* which is a record of four integers, the horizontal and vertical interval endpoints. Our basic operation is to test whether a point falls within a given rectangle, so we'll assume a function *insiderect*(*p*: *point*; *rect*: *rectangle*) which checks this in the obvious way, returning *true* if p falls within *rect*. Our goal is to find all the points that fall within a given rectangle, using as few calls to *insiderect* as possible.

The simplest way to solve this problem is *sequential search*: scan through all the points, testing each to see if it falls within the specified range (by calling *insiderect* for each point). This method is in fact used in many database applications because it is easily improved by "batching" the range queries, testing for many different ones in the same scan through the points. In a very large database, where the data is on an external device and the time to read it is by far the dominating cost factor, this can be a very reasonable method: collect as many queries as will fit in internal memory and search for them all in one pass through the large external data file. If this type of batching is inconvenient or the database is somewhat smaller, however, there are much better methods available.

A simple first improvement to sequential search is direct application of a known one-dimensional method along one or more of the dimensions to be searched. Figure 26.2 shows an example of a search rectangle for our sample sets of points.

One way to proceed is to find the points whose x coordinates fall within the x range specified by the rectangle, then check the y coordinates of those points to determine whether or not they fall within the rectangle. Thus, points that cannot

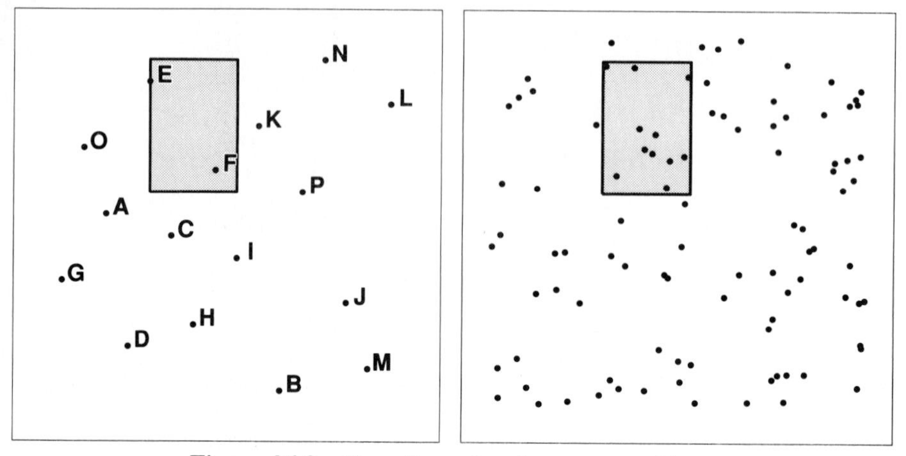

Figure 26.2 Two-dimensional range searching.

be within the rectangle because their x coordinates are out of range are never examined. This technique is called *projection*; obviously we could also project on y. For our example, we would check E C H F and I for an x projection, as described above, and we would check O E F K P N and L for a y projection.

If the points are uniformly distributed in a rectangular region, then it's trivial to calculate the average number of points checked. The fraction of points we would expect to find in a given rectangle is simply the ratio of the area of that rectangle to the area of the full region; the fraction of points we would expect to check for an x projection is the ratio of the width of the rectangle to the width of the region, and similarly for a y projection. For our example, using a 4-by-6 rectangle in a 16-by-16 region means that we would expect to find $3/32$ of the points in the rectangle, $1/4$ of them in an x projection, and $3/8$ of them in a y projection. Obviously, under such circumstances, it's best to project onto the axis corresponding to the narrower of the two rectangle dimensions. On the other hand, it's easy to construct situations in which the projection technique could fail miserably: for example, if the point set forms an "L" shape and the search is for a range that encloses only the point at the corner of the "L," then projection on either axis eliminates only half the points.

At first glance, it seems that the projection technique could be improved somehow to "intersect" the points that fall within the x range and the points that fall within the y range. Attempts to do this without examining in the worst case either all the points in the x range or all the points in the y range serve mainly to make one appreciate the more sophisticated methods we are about to study.

Grid Method

A simple but effective technique for maintaining proximity relationships among

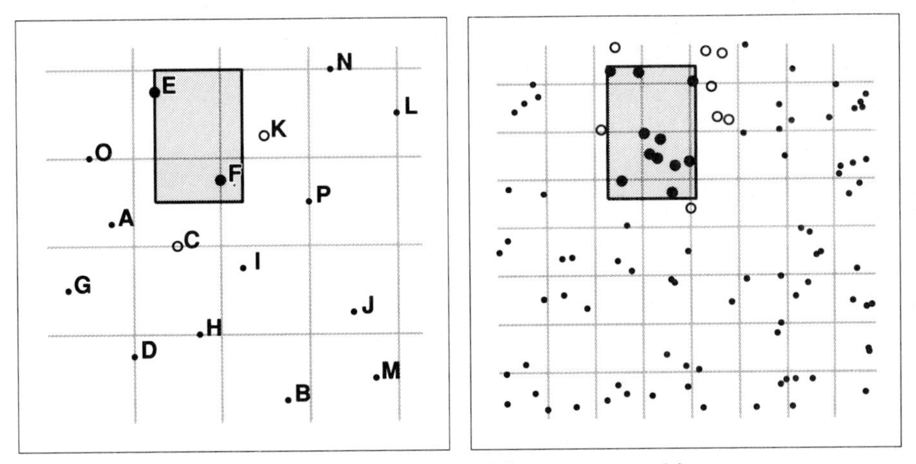

Figure 26.3 Grid method for range searching.

points in the plane is to construct an artificial grid that divides the area to be searched into small squares and keep short lists of points falling into each square. (This technique is used in archaeology, for example.) Then, when points lying within a given rectangle are sought, only the lists corresponding to squares that intersect the rectangle need to be searched. In our example, only E, C, F, and K are examined, as shown in Figure 26.3.

The main decision to be made in implementing this method is to determine the size of the grid: if it is too coarse, each grid square will contain too many points, and if it is too fine, there will be too many grid squares to search (most of which will be empty). One way to strike a balance between these two extremes is to choose the grid size so that the number of grid squares is a constant fraction of the total number of points. Then the number of points in each square is expected to be about equal to some small constant. For our small sample point set, using a 4 by 4 grid for a sixteen-point set means that each grid square is expected to contain one point.

Below is a straightforward implementation of a program to build the grid structure containing the points in an array $p[0..N]$ of points of the type described at the beginning of Chapter 24. The variable *size* is used to control how big the grid squares are and thus determine the resolution of the grid. Assume for simplicity that the coordinates of all the points fall between 0 and some maximum value *max*. Then *size* is taken to be the width of a grid square and there are $max/size$ by $max/size$ grid squares. To find which grid square a point belongs to, we divide its coordinates by *size*, as in the following implementation:

```
const maxG=20;
type link=↑node;
     node=record p: point; next: link end;
var grid: array[0..maxG,0..maxG] of link;
    size: integer;
    z: link;
procedure preprocess;
  procedure insert(p: point);
    var t: link;
    begin
    new(t); t↑.p:=p;
    t↑.next:=grid[p.x div size,p.y div size];
    grid[p.x div size,p.y div size]:=t;
    end;
  begin
  new(z);
  size:=1; while size*size<max*max/N do size:=size+size;
  for i:=0 to maxG do
    for j:=0 to maxG do grid[i,j]:=z;
  for i:=0 to N do insert(p[i]);
  end;
```

This program uses our standard linked-list representations, with dummy tail node z. Again, the variable *max* is assumed to be global, perhaps set to the maximum coordinate value encountered at the time the points are input.

As mentioned above, how to set the variable *size* depends on the number of points, the amount of memory available, and the range of coordinate values. Roughly, to get M points per grid square, *size* should be chosen to be the nearest integer to *max* divided by $\sqrt{N/M}$. This leads to about N/M grid squares. These estimates aren't accurate for small values of the parameters, but they are useful for most situations, and similar estimates can easily be formulated for specialized applications. The value need not be computed exactly—the implementation above makes *size* a power of two, which should make multiplication and division by *size* much more efficient in most programming environments.

The above implementation uses $M = 1$, a commonly used choice. If space is at a premium, a large value may be appropriate, but a smaller value is not likely to be useful except in specialized situations.

Now, most of the work for range searching is handled simply by indexing into the *grid* array, as follows:

```
procedure gridrange(rect: rectangle);
  var t: link;
      i,j: integer;
  begin
  for i:=(rect.x1 div size) to (rect.x2 div size) do
    for j:=(rect.y1 div size) to (rect.y2 div size) do
      begin
      t:=grid[i,j];
      while t<>z do
        begin
        if insiderect(t↑.p, rect)
          then point t↑.p is within the range
          t:=t↑.next
        end
      end
  end;
```

The running time of this program is proportional to the number of grid squares touched. Since we were careful to arrange things so that each grid square contains a constant number of points on the average, the number of grid squares touched is also proportional, on the average, to the number of points examined.

Property 26.2 *The grid method for range searching is linear in the number of points in the range, on the average, and linear in the total number of points in the worst case.*

If the number of points in the search rectangle is R, then the number of grid squares examined is proportional to R. The number of grid squares examined that do not fall completely inside the search rectangle is certainly less than a small constant times R, so the total running time (on the average) is linear in R. For large R, the number of points examined that don't fall in the search rectangle gets quite small: all such points fall in a grid square that intersects the edge of the search rectangle, and the number of such squares is proportional to $\sqrt{R}$ for large R. Note that this argument falls apart if the grid squares are too small (too many empty grid squares inside the search rectangle) or too large (too many points in grid squares on the perimeter of the search rectangle) or if the search rectangle is thinner than the grid squares (it could intersect many grid squares but have few points inside it). ■

The grid method works well if the points are well distributed over the assumed range, but badly if they are clustered together. (For example, all the points could fall in one grid box, which would mean that all the grid machinery gained nothing.) The method we examine next makes this worst case very unlikely by subdividing the space in a nonuniform way, adapting to the point set at hand.

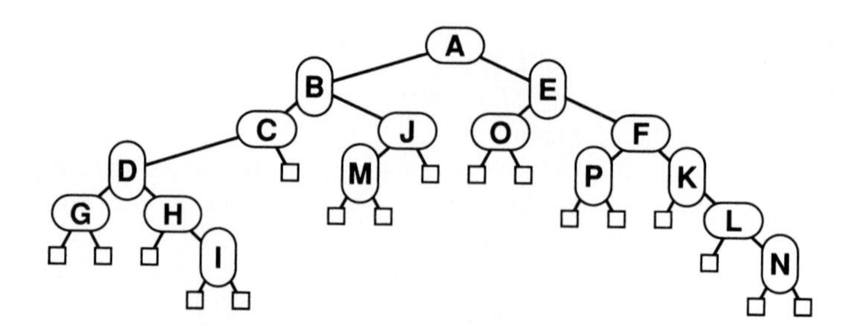

Figure 26.4 A two-dimensional (2D) tree.

Two-Dimensional Trees

Two-dimensional (2D) trees are dynamic, adaptable data structures that are very similar to binary trees but divide up a geometric space in a manner convenient for use in range searching and other problems. The idea is to build binary search trees with points in the nodes, using the y and x coordinates of the points as keys in a strictly alternating sequence.

The same algorithm is used to insert points into 2D trees as in normal binary search trees, but at the root we use the y coordinate (if the point to be inserted has a smaller y coordinate than the point at the root, go left; otherwise go right), then at the next level we use the x coordinate, then at the next level the y coordinate, etc., alternating until an external node is encountered. Figure 26.4 shows the 2D tree corresponding to our small sample set of points.

The significance of this technique is that it corresponds to dividing up the plane in a simple way: all the points below the point at the root go in the left subtree, all those above in the right subtree, then all the points above the point at the root and to the left of the point in the right subtree go in the left subtree of the right subtree of the root, etc.

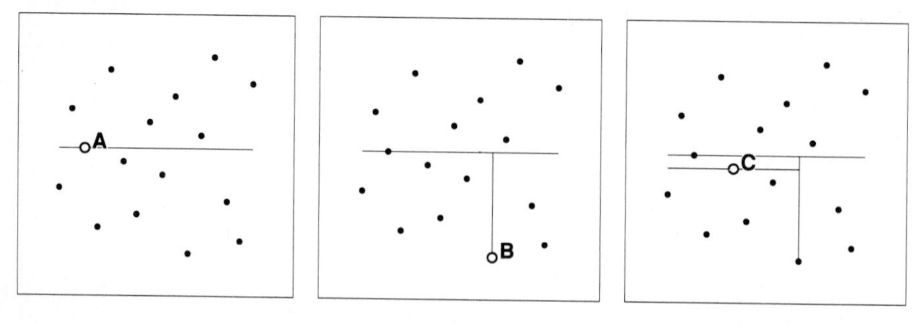

Figure 26.5 Subdividing the plane with a 2D tree: initial steps.

Figures 26.5 and 26.6 show how the plane is subdivided corresponding to the construction of the tree in Figure 26.4. First a horizontal line is drawn at the y-coordinate of A, the first node inserted. Then, since B is below A, it goes to the left of A in the tree, and the halfplane below A is divided with a vertical line at the x-coordinate of B (second diagram in Figure 26.5). Then, since C is below A, we go left at the root, and since it is to the left of B we go left at B, and divide the portion of the plane below A and to the left of B with a horizontal line at the y coordinate of C (third diagram in Figure 26.5). The insertion of D is similar, then E goes to the right of A since it is above it (first diagram of Figure 26.6), etc.

Every external node of the tree corresponds to some rectangle in the plane. Each region corresponds to an external node in the tree; each point lies on a horizontal or vertical line segment that defines the division made in the tree at that point.

The code to construct 2D trees is a straightforward modification of standard binary tree search to switch between x and y coordinates at each level.

```
type link=↑node;
        node=record p: point; l,r: link end;
var   t,head,z: link;
procedure treeinsert(p: point; t: link);
   var f: link;
          d,td: boolean;
   begin
   d:=true;
   repeat
      if d then td:=p.x<t↑.p.x
          else td:=p.y<t↑.p.y;
      f:=t;
      if td then t:=t↑.l else t:=t↑.r;
      d:= not d;
   until t=z;
   new(t); t↑.p:=p; t↑.l:=z; t↑.r:=z;
   if td then f↑.l:=t else f↑.r:=t;
   end;
```

As usual, we use a header node *head* with an artificial point (0, 0) which is "less than" all the other points so that the tree hangs off the right link of *head*, and an artificial node z is used to represent all the external nodes. The call *treeinsert(p,head)* inserts a new node containing p into the tree. A boolean variable d is toggled on the way down the tree to effect the alternating tests on x and y coordinates. Otherwise the procedure is identical to the standard procedure from Chapter 14.

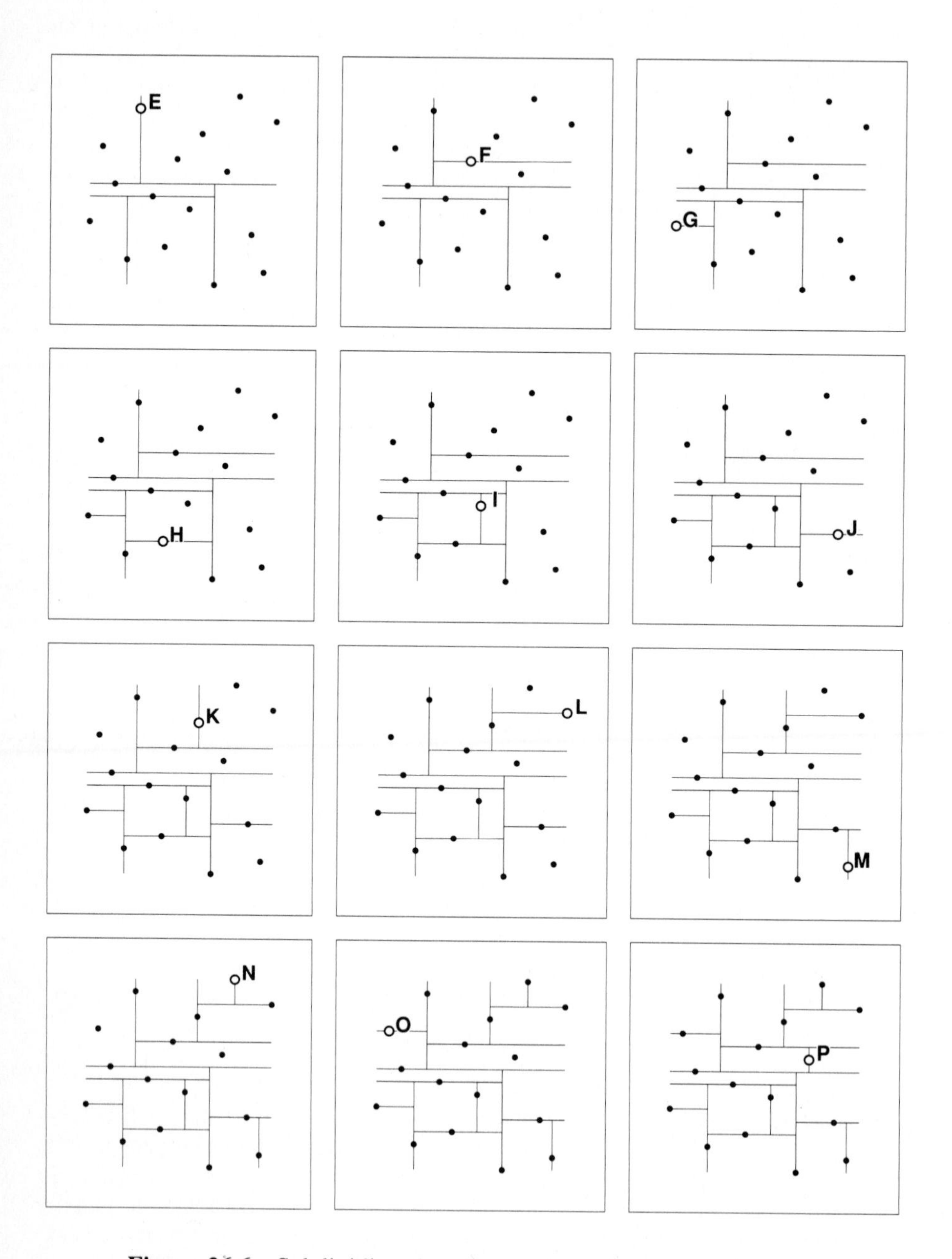

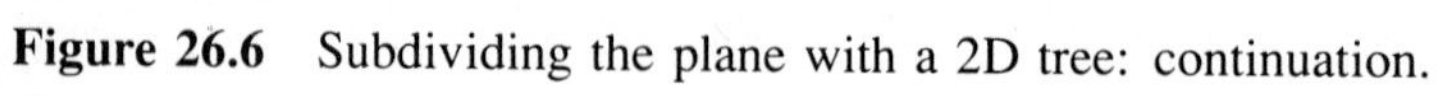
Figure 26.6 Subdividing the plane with a 2D tree: continuation.

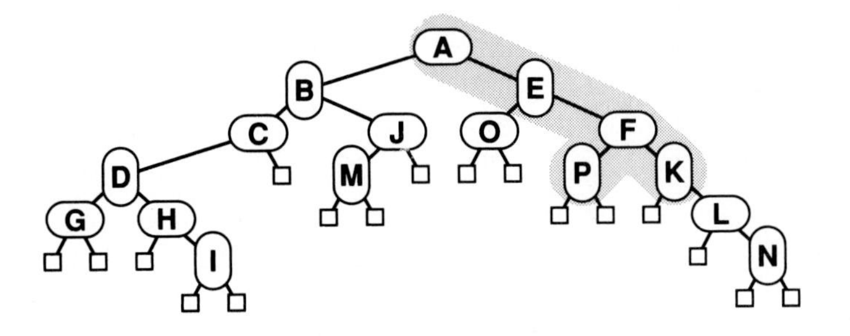

Figure 26.7 Range searching with a 2D tree.

Property 26.3 *Construction of a 2D tree from N random points requires 2N* ln *N comparisons, on the average.*

Indeed, for randomly distributed points, 2D trees have the same performance characteristics as binary search trees. Both coordinates act as random "keys". ∎

To do range searching using 2D trees, we first build the 2D tree from the points in the preprocessing phase:

```
procedure preprocess;
  function initialize: link;
    var t: link;
    begin
    p[0].x:=0; p[0].y:=0; p[0].info:=0;
    new(t); t↑.p:=p[0]; t↑.r:=z;
    initialize:=t
    end;
  begin
  new(z); head:=initialize;
  for i:=1 to N do treeinsert(p[i], head);
  end;
```

Then, for range searching, we test the point at each node against the range along the dimension used to divide the plane of that node. For our example, we begin by going right at the root and right at node E, since our search rectangle is entirely above A and to the right of E. Then, at node F, we must go down both subtrees, since F falls in the x range defined by the rectangle (note carefully that this is *not* the same as saying that F falls within the rectangle). Then the left subtrees of P and K are checked, corresponding to checking the areas of the plane that overlap the search rectangle. (See Figures 26.7 and 26.8.)

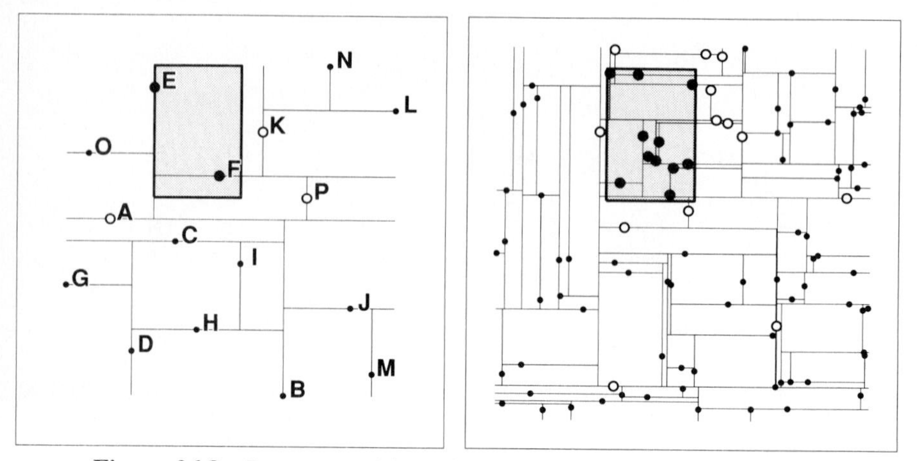

Figure 26.8 Range searching with a 2D tree (planar subdivision).

This process is easily implemented with a straightforward generalization of the 1D *range* procedure examined at the beginning of this chapter:

```
procedure range(t: link; rect: rectangle; d: boolean);
  var t1,t2,tx1,tx2,ty1,ty2: boolean;
  begin
  if t<>z then
    begin
    tx1:=rect.x1<t↑.p.x; tx2:=t↑.p.x<=rect.x2;
    ty1:=rect.y1<t↑.p.y; ty2:=t↑.p.y<=rect.y2;
    if d then begin t1:=tx1; t2:=tx2 end
         else  begin t1:=ty1; t2:=ty2 end;
    if t1 then range(t↑.l,rect, not d);
    if insiderect(t↑.p, rect)
       then point t↑.p is within the range
    if t2 then range(t↑.r,rect, not d);
    end
  end;
```

This procedure goes down both subtrees only when the dividing line cuts the rectangle, which should happen infrequently for relatively small rectangles. Figure 26.8 shows the planar subdivisions and the points examined for our two examples.

Property 26.4 *Range searching with a 2D tree seems to use about* $R + \log N$ *steps to find* R *points in reasonable ranges in a region containing* N *points.*

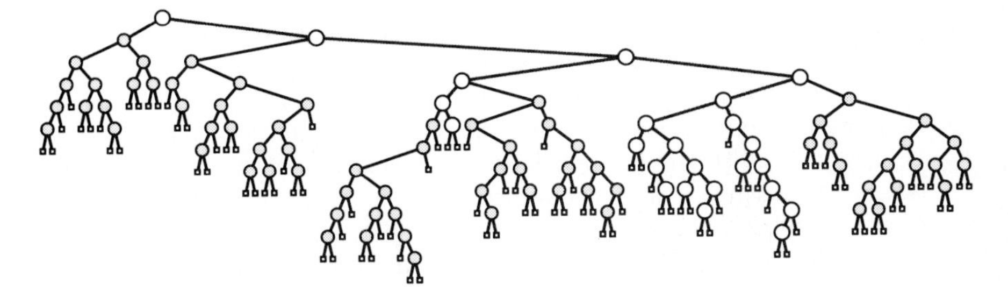

Figure 26.9 Range searching with a large 2D tree.

This method has yet to be analyzed, and the property stated is a conjecture based purely on empirical evidence. Of course, the performance (and the analysis) is as always very dependent on the type of range used. But the method is very competitive with the grid method, and somewhat less dependent on "randomness" in the point set. Figure 26.9 shows the 2D tree for our large example. ■

Multidimensional Range Searching

Both the grid method and 2D trees generalize directly to more than two dimensions: simple, straightforward extensions to the above algorithms immediately yield range-searching methods that work for more than two dimensions. However, the nature of multidimensional space calls for some caution and suggests that the performance characteristics of the algorithms might be difficult to predict for a particular application.

To implement the grid method for k-dimensional searching, we simply make *grid* a k-dimensional array and use one index per dimension. The main problem is to pick a reasonable value for *size*. This problem becomes quite obvious when large k is considered: what type of grid should we use for 10-dimensional search? The problem is that even if we use only three divisions per dimension, we need 3^{10} grid squares, most of which will be empty, for reasonable values of N.

The generalization from 2D to kD trees is also straightforward: simply cycle through the dimensions (as we did for two dimensions by alternating between x and y) while going down the tree. As before, in a random situation, the resulting trees have the same characteristics as binary search trees. Also as before, there is a natural correspondence between the trees and a simple geometric process. In three dimensions, branching at each node corresponds to cutting the three-dimensional region of interest with a plane; in general we cut the k-dimensional region of interest with a $(k-1)$-dimensional hyperplane.

If k is very large, there is likely to be a significant amount of imbalance in the kD trees, again because practical point sets can't be large enough to exhibit randomness over a large number of dimensions. Typically, all points in a subtree will have the same value across several dimensions, which leads to several one-way branches in the trees. One way to alleviate this problem is, rather than simply cycling through the dimensions, always to use the dimension that divides up the point set in the best way. This technique can also be applied to 2D trees. It requires that extra information (which dimension should be discriminated upon) be stored in each node, but it does relieve imbalance, especially in high-dimensional trees.

In summary, though it is easy to see how to generalize our programs for range searching to handle multidimensional problems, such a step should not be taken lightly for a large application. Large databases with many attributes per record can be very complicated objects indeed, and a good understanding of the characteristics of a database is often necessary in order to develop an efficient range-searching method for a particular application. This is a quite important problem which is still being actively studied.

□

Exercises

1. Write a nonrecursive version of the 1D *range* program given in the text.
2. Write a program to print out all points from a binary tree that do *not* fall in a specified interval.
3. Give the maximum and minimum number of grid squares that will be searched in the grid method as functions of the dimensions of the grid squares and the search rectangle.
4. Discuss the idea of avoiding the search of empty grid squares by using linked lists: each grid square could be linked to the next nonempty grid square in the same row and the next nonempty grid square in the same column. How would using such a scheme affect the grid square size to be used?
5. Draw the tree and the resulting subdivision of the plane if we build a 2D tree for our sample points starting with a vertical dividing line. (That is, call *range* with a third argument of *false* rather than *true*.)
6. Give a set of points leading to a worst-case 2D tree which has no nodes with two children; give the subdivision of the plane that results.
7. Describe how to modify each of the methods to return all points that fall within a given circle.
8. Of all search rectangles with the same area, what shape is likely to make each of the methods perform the worst?
9. Which method should be preferred for range searching when the points cluster together in large groups spaced far apart?
10. Draw the 3D tree that results when the points $(3,1,5)$, $(4,8,3)$, $(8,3,9)$, $(6,2,7)$, $(1,6,3)$, $(1,3,5)$, $(6,4,2)$ are inserted into an initially empty tree.

27

Geometric Intersection

A natural problem that arises frequently in applications involving geometric data is: "Given a set of N objects, do any two intersect?" The "objects" involved may be lines, rectangles, circles, polygons, or other types of geometric objects. For example, in a system for designing and processing integrated circuits or printed circuit boards, it is important to know that no two wires intersect to make a short circuit. In an industrial system for designing layouts to be executed by a numerically controlled cutting tool, it is important to know that no two parts of the layout intersect. In computer graphics, the problem of determining which of a set of objects is obscured from a particular viewpoint can be formulated as a geometric intersection problem on the projections of the objects onto the viewing plane. And in operations research, the mathematical formulation of many important problems leads naturally to geometric intersection problems.

The obvious solution to the intersection problem is to check each pair of objects to see if they intersect. Since there are about $N^2/2$ pairs of objects, the running time of this algorithm is proportional to N^2. For a few applications, this may not be a problem because other factors limit the number of objects to be processed. However, for many other applications, is not uncommon to deal with hundreds of thousands or even millions of objects. The brute-force N^2 algorithm is obviously inadequate for such applications. In this section, we'll study a general method for determining, in time proportional to $N \log N$, whether any two out of a set of N objects intersect; this method is based on algorithms presented by M. Shamos and D. Hoey in a seminal 1976 paper.

First, we'll consider an algorithm for returning all intersecting pairs among a set of lines that are constrained to be horizontal or vertical. This makes the problem easier in one sense (horizontal and vertical lines are relatively simple geometric objects), more difficult in another sense (returning all intersecting pairs is more difficult than simply determining whether one such pair exists). The implementation we'll develop applies binary search trees and the interval range-searching program of the previous chapter in a doubly recursive program.

Next, we'll examine the problem of determining whether any two of a set of N lines intersect, with no constraints on the lines. The same general strategy as used for the horizontal-vertical case can be applied. In fact, the same basic idea works for detecting intersections among many other types of geometric objects. However, for lines and other objects, the extension to return all intersecting pairs is somewhat more complicated than for the horizontal-vertical case.

Horizontal and Vertical Lines

To begin, we'll assume that all lines are either horizontal or vertical: the two points defining each line have either equal x coordinates or equal y coordinates, as in the sample sets of lines shown in Figure 27.1. (This is sometimes called *Manhattan geometry* because, Broadway to the contrary notwithstanding, the Manhattan street map consists mostly of horizontal and vertical lines.) Constraining lines to be horizontal or vertical is certainly a severe restriction, but this is far from a "toy" problem. Indeed, this restriction is often imposed in a particular application: for example, very large-scale integrated circuits are typically designed under this constraint. In the figure on the right, the lines are relatively short, as is typical in many applications, though one can usually count on encountering a few very long lines.

The general plan of the algorithm to find an intersection in such sets of lines is to imagine a horizontal scan line sweeping from bottom to top. Projected onto this scan line, vertical lines are points, and horizontal lines are intervals: as the scan line proceeds from bottom to top, points (representing vertical lines) appear and disappear, and horizontal lines are encountered periodically. An intersection is found when a horizontal line is encountered representing an interval on the scan

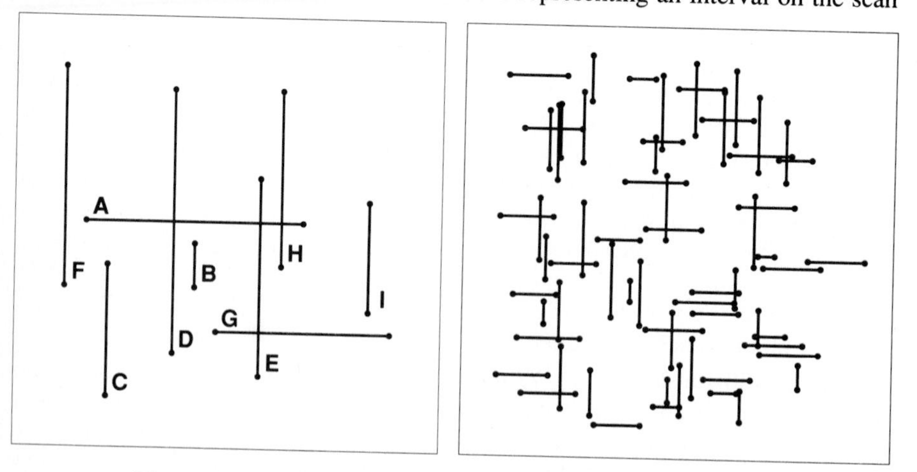

Figure 27.1 Two line intersection problems (Manhattan).

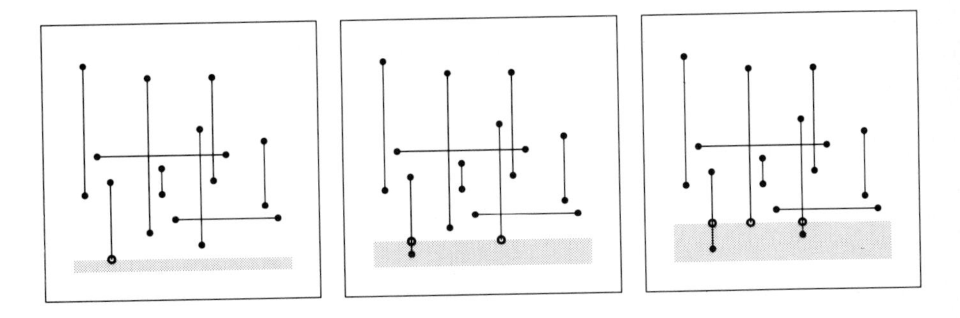

Figure 27.2 Scanning for intersections: initial steps.

line that contains a point representing a vertical line. Meeting the point means that the vertical line intersects the scan line, and the horizontal line lies on the scan line, so the horizontal and vertical lines must intersect. In this way, the two-dimensional problem of finding an intersecting pair of lines is reduced to the one-dimensional range-searching problem of the previous chapter.

Of course, it is not necessary actually to "sweep" a horizontal line all the way up through the set of lines; since we need to take action only when endpoints of the lines are encountered, we can begin by sorting the lines according to their y coordinate, then processing the lines in that order. If the bottom endpoint of a vertical line is encountered, we add the x coordinate of that line to the binary search tree (here called the x-tree); if the top endpoint of a vertical line is encountered, we delete that line from the tree; and if a horizontal line is encountered, we do an interval range search using its two x coordinates. As we'll see, some care is required to handle equal coordinates among line endpoints (by now the reader should be accustomed to encountering such difficulties in geometric algorithms).

Figure 27.2 shows the first few steps of scanning to find the intersections in the example on the left in Figure 27.1. The scan starts at the point with the lowest y coordinate, the lower endpoint of C. Then E is encountered, then D. The rest of the process is shown in Figure 27.3: the next line encountered is the horizontal line G, which is tested for intersection with C, D, and E (the vertical lines that intersect the scan line).

To implement the scan, we need only sort the line endpoints by their y coordinates. For our example, this gives the list

C E D G I B F C H B A I E D H F

Each vertical line appears twice in this list, each horizontal line appears once. For the purposes of the line intersection algorithm, this sorted list can be thought of as a sequence of *insert* (vertical lines when the bottom endpoint is encountered), *delete* (vertical lines when the top endpoint is encountered), and *range* (for the endpoints of horizontal lines) commands. All of these "commands" are simply calls on the standard binary tree routines from Chapters 14 and 26, using x coordinates as keys.

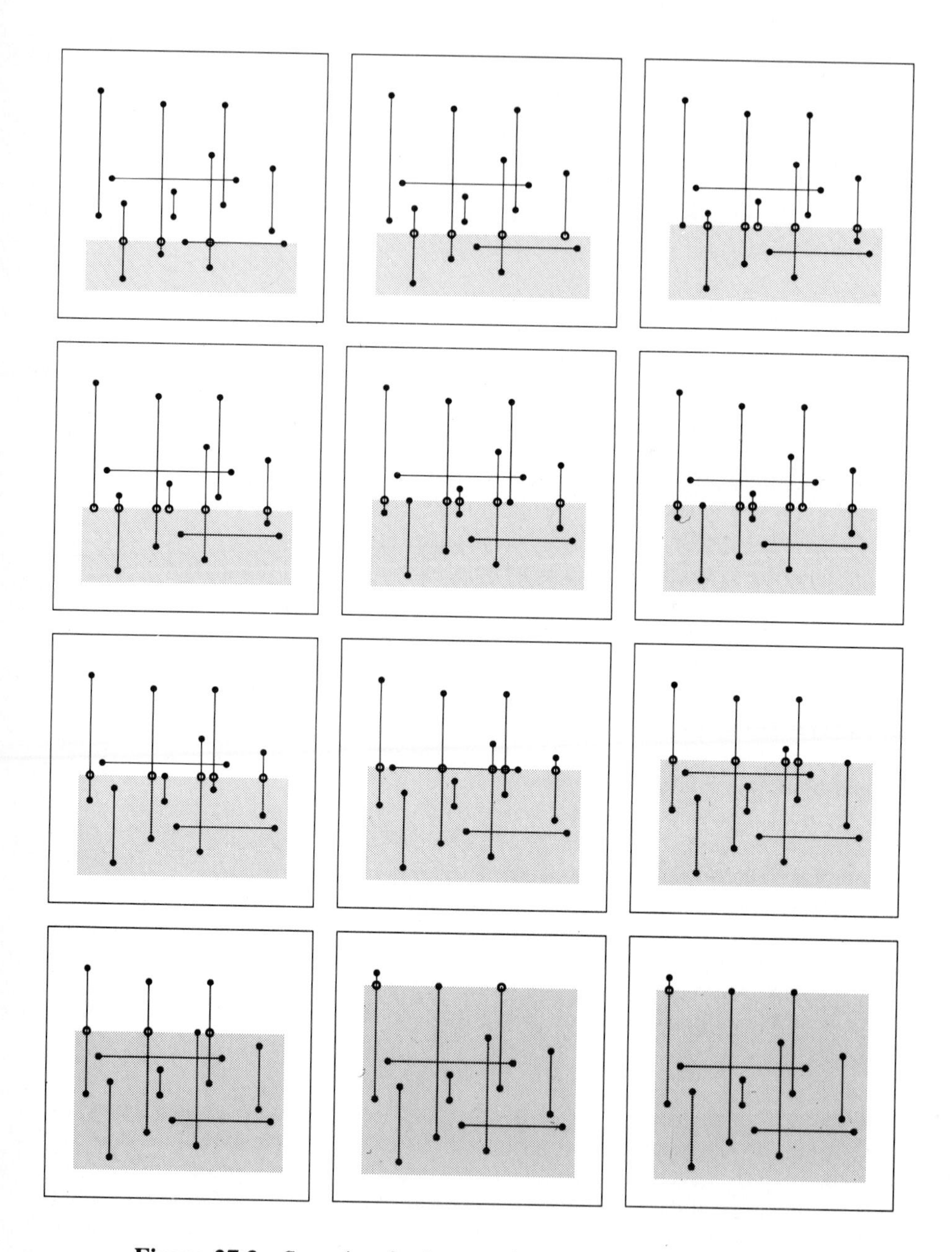

Figure 27.3 Scanning for intersections: completion of process.

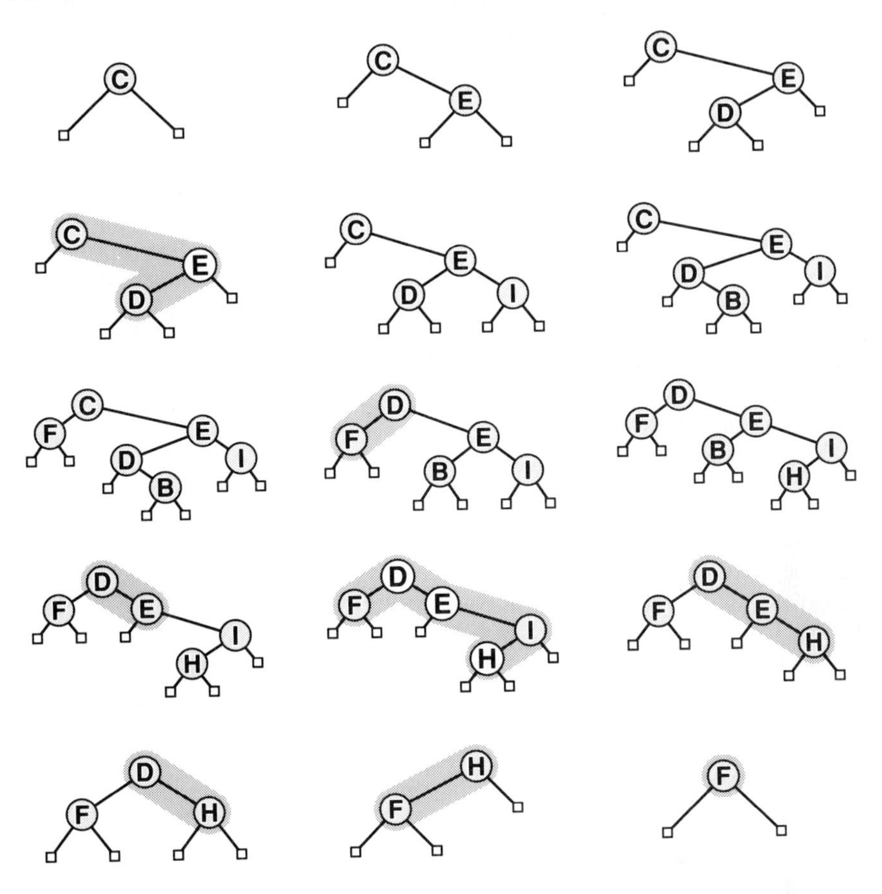

Figure 27.4 Data structure during scan: constructing the x-tree.

Figure 27.4 shows the x-tree construction process during the scan. Each node in the tree corresponds to a vertical line, but the key used during the construction of the tree is the x-coordinate. Since E is to the right of C, it is in C's right subtree, etc. The first line of Figure 27.4 corresponds to Figure 27.2 and the rest to Figure 27.3.

When a horizontal line is encountered, it is used to make a range search in the tree: all vertical lines in the range represented by the horizontal line correspond to intersections. In our example, the intersection between E and G is discovered, then I, B, and F are inserted. Then C is deleted, H inserted, and B deleted. At this point, A is encountered, and a range search for the interval defined by A is performed. This search discovers the intersections between A and D, E, and H. Next, the upper endpoints of I, E, D, and F are encountered and deleted, leading back to the empty tree.

Implementation

The first step in the implementation is to sort the line endpoints on their *y* coordinates. But since binary trees will be used to maintain the status of vertical lines with respect to the horizontal scan line, they may as well be used for the initial *y* sort! Specifically, we will use two "indirect" binary trees on the line set, one with header node *hy* and one with header node *hx*. The *y* tree will contain all the line endpoints, to be processed in order one at a time; the *x* tree will contain the lines that intersect the current horizontal scan line. We begin by initializing both *hx* and *hy* with 0 keys and pointers to a dummy external node *z*, as in *treeinitialize* in Chapter 14. Then the *hy* tree is constructed by inserting both *y* coordinates from vertical lines and the *y* coordinate of horizontal lines into the binary search tree with header node *hy*, as follows:

```
procedure buildytree;
  var x1,y1,x2,y2: integer;
  begin
  hy:=bstinitialize; N:=0;
  repeat
  N:=N+1;
    read(x1,y1,x2,y2); if eoln then readln;
    lines[N].p1.x:=x1; lines[N].p1.y:=y1;
    lines[N].p2.x:=x2; lines[N].p2.y:=y2;
    bstinsert(N,y1,hy);
    if y2<>y1 then bstinsert(N,y2,hy);
  until eof;
end;
```

This program reads in groups of four numbers that specify lines and puts them into the *lines* array and into the binary search tree on the *y* coordinate. The standard *bstinsert* routine from Chapter 14 is used, with the *y* coordinates as keys, and indices into the array of lines as the *info* field. For our example set of lines, the tree shown in Figure 27.5 is constructed.

Now, the sort on *y* is effected by a recursive inorder tree traversal routine (see Chapters 4 and 14). We visit the nodes in increasing *y* order by visiting all the nodes in the left subtree of the *hy* tree, then visiting the root, then visiting all the nodes in the right subtree of the *hy* tree. At the same time, we maintain a separate tree (rooted at *hx*) as described above, to simulate the operation of passing through a horizontal scan line:

```
procedure scan(next: link);
  var t,x1,x2,y1,y2: integer;
      int: interval;
  begin
  if next<>z then
    begin
    scan(next↑.l);
    x1:=lines[next↑.info].p1.x; y1:=lines[next↑.info].p1.y;
    x2:=lines[next↑.info].p2.x; y2:=lines[next↑.info].p2.y;
    if x2<x1 then begin t:=x2; x2:=x1; x1:=t end;
    if y2<y1 then begin t:=y2; y2:=y1; y1:=t end;
    if next↑.key=y1 then bstinsert(next↑.info,x1,hx);
    if next↑.key=y2 then
      begin
      bstdelete(next↑.info,x1,hx);
      int.x1:=x1; int.x2:=x2;
      bstrange(hx↑.r,int);
      end;
    scan(next↑.r)
    end
  end;
```

From the description above, it is rather straightforward to put together the code at the point where each node is "visited". First, the coordinates of the endpoint of the corresponding line are fetched from the *lines* array, indexed by the *info* field of the node. Then the *key* field in the node is compared against these coordinates to determine whether this node corresponds to the upper or the lower endpoint of the line: if it is the lower endpoint, it is inserted into the *hx* tree, and if it is the upper endpoint, it is deleted from the *hx* tree and a range search is performed. The

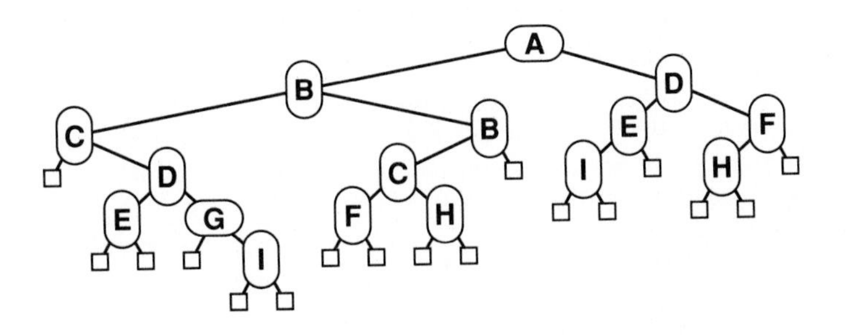

Figure 27.5 Sorting for scan using the *y*-tree.

implementation differs slightly from this description in that horizontal lines are actually inserted into the *hx* tree, then immediately deleted, and a range search for a one-point interval is performed for vertical lines. This makes the code properly handle the case of overlapping vertical lines, which are considered to "intersect."

This approach of intermixed application of recursive procedures operating on the x and y coordinates is quite important in geometric algorithms. Another example of this is the 2D tree algorithm of the previous chapter, and we'll see yet another example in the next chapter.

Property 27.1 *All intersections among N horizontal and vertical lines can be found in time proportional to* $N \log N + I$*, where I is the number of intersections.*

The tree manipulation operations take time proportional to $\log N$ on the average (if balanced trees were used, a $\log N$ worst case could be guaranteed), but the time spent in *bstrange* also depends on the total number of intersections. In general, the number of intersections can be quite large. For example, if we have $N/2$ horizontal lines and $N/2$ vertical lines arranged in a crosshatch pattern, then the number of intersections is proportional to N^2. ■

As with range searching, if it is known in advance that the number of intersections is very large, then some brute-force approach should be used. Typically, applications involve a "needle-in-haystack" kind of situation where a large set of lines is to be checked for a few possible intersections.

General Line Intersection

When lines of arbitrary slope are allowed, the situation can become more complicated, as illustrated in Figure 27.6. First, the various line orientations possible make it necessary to test explicitly whether certain pairs of lines intersect—we can't get by with a simple interval range test. Second, the ordering relationship between lines for the binary tree is more complicated than before, since it depends on the current y range of interest. Third, any intersections that do occur add new "interesting" y values that are likely to be different from the set of y values we get from the line endpoints.

It turns out that these problems can be handled in an algorithm with the same basic structure as given above. To simplify the discussion, we'll consider an algorithm for detecting whether or not there exists an intersecting pair in a set of N lines, and then we'll discuss how it can be extended to return all intersections.

As before, we first sort on y to divide the space into strips within which no line endpoints appear. Just as before, we proceed through the sorted list of points, adding each line to a binary search tree when its bottom point is encountered and deleting it when its top point is encountered. Just as before, the binary tree gives the order in which the lines appear in the horizontal "strip" between two consecutive y values. For example, in the strip between the bottom endpoint of D and the top endpoint of B in Figure 27.6, the lines should appear in the order F

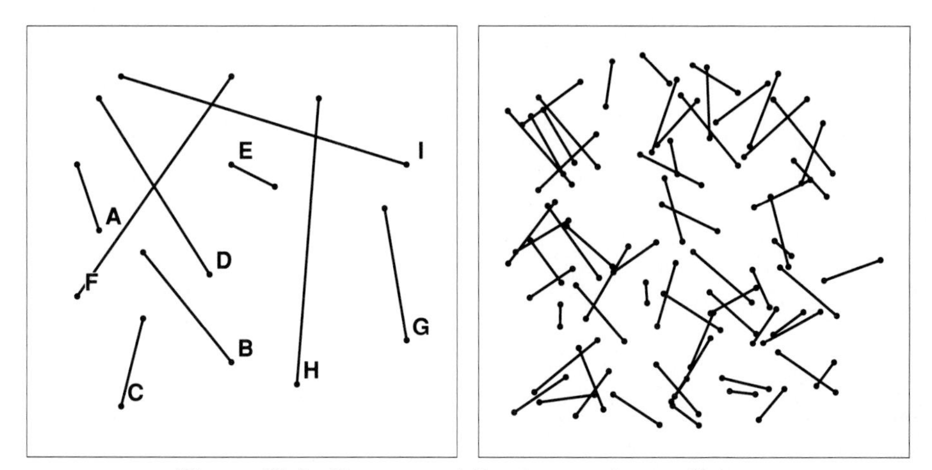

Figure 27.6 Two general line intersection problems.

B D H G. We assume that there are no intersections within the current horizontal strip of interest: our goal is to maintain this tree structure and use it to help find the first intersection.

To build the tree, we can't simply use x coordinates from line endpoints as keys (doing this would put B and D in the wrong order in the example above, for instance). Instead, we use a more general ordering relationship: a line x is defined to be to the right of a line y if both endpoints of x are on the same side of y as a point infinitely far to the right, *or* if y is to the left of x, with "left" defined analogously. Thus, in the diagram above, B is to the right of A and B is to the right of C (since C is to the left of B). If x is neither to the left nor to the right of y, then they must intersect. This generalized "line comparison" operation can be implemented using the *ccw* procedure of Chapter 24. Except for the use of this function whenever a comparison is needed, the standard binary search tree procedures (even balanced trees, if desired) can be used. Figure 27.7 shows the manipulation of the tree for our example between the time line C is encountered and the time line D is encountered. Each "comparison" performed during the tree-manipulation procedures is actually a line-intersection test: if the binary search tree procedure can't decide to go right or left, then the two lines in question must intersect, and we're finished.

But this is not the whole story, because this generalized comparison operation is not *transitive*. In the example above, F is to the left of B (because B is to the right of F) and B is to the left of D, but F is *not* to the left of D. It is essential to note this, because the binary tree deletion procedure assumes that the comparison operation is transitive: when B is deleted from the last tree in the above sequence, the tree shown in Figure 27.7 is formed without any explicit comparison of F and D. For our intersection-testing algorithm to work correctly, we must test explicitly

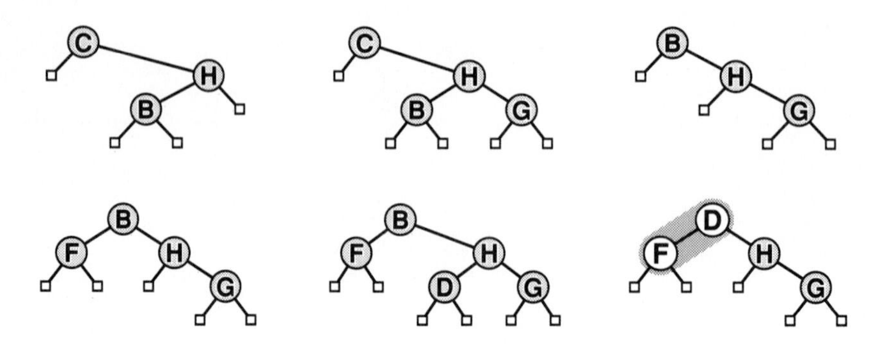

Figure 27.7 Data structure (x-tree) for general problem.

that comparisons are valid each time we change the tree structure. Specifically, every time we make the left link of node x point to node y, we explicitly test that the line corresponding to x is to the left of the line corresponding to y, according to the above definition, and similarly for the right. Of course, this comparison could result in the detection of an intersection, as it does in our example.

In summary, to test for an intersection among a set of N lines, we use the program above, but we remove the call to *range* and extend the binary tree routines to use the generalized comparison as described above. If there is no intersection, we'll start with a null tree and end with a null tree without finding any incomparable lines. If there is an intersection, then the two lines that intersect must be compared against each other at some point during the scanning process and the intersection will be discovered.

Once we've found an intersection, however, we can't simply press on and hope to find others, because the two lines that intersect should swap places in the ordering directly after the point of intersection. One way to handle this issue would be to use a priority queue instead of a binary tree for the y sort: initially put lines on the priority queue according to the y coordinates of their endpoints, then work the scan line up by successively taking the smallest y coordinate from the priority queue and doing a binary tree insert or delete as above. When an intersection is found, new entries are added to the priority queue for each line, using the intersection point as the lower endpoint for each.

Another way to find all intersections, which is appropriate if not too many are expected, is simply to remove one of the intersecting lines when an intersection is found. Then after the scan is completed, we know that all intersecting pairs must involve one of those lines, and we can use a brute-force method to enumerate all the intersections.

Property 27.2 *All intersections among N lines can be found in time proportional to $N \log N + I$, where I is the number of intersections.*

This follows directly from the discussion above. ■

An interesting feature of the above procedure is that it can be adapted just by changing the generalized comparison procedure to test for the existence of an intersecting pair among a set of more general geometric shapes. For example, if we implement a procedure that compares two rectangles whose edges are horizontal and vertical according to the trivial rule that rectangle x is to the left of rectangle y if the right edge of x is to the left of the left edge of y, then we can use the above method to test for intersection among a set of such rectangles. For circles, we can use the x coordinates of the centers for the ordering and explicitly test for intersection (for example, compare the distance between the centers to the sum of the radii). Again, if this comparison procedure is used in the above method, we have an algorithm for testing for intersection among a set of circles. The problem of returning all intersections in such cases is much more complicated, though the brute-force method mentioned in the previous paragraph will always work if few intersections are expected. Another approach that will suffice for many applications is simply to consider complicated objects as sets of lines and use the line-intersection procedure.

□

Exercises

1. How would you determine whether two triangles intersect? Squares? Regular n-gons for $n > 4$?

2. In the horizontal-vertical line-intersection algorithm, how many pairs of lines are tested for intersection in a set of lines with no intersections in the worst case? Give a diagram supporting your answer.

3. What happens when the horizontal-vertical line-intersection procedure is used on a set of lines with arbitrary slope?

4. Write a program to find the number of intersecting pairs among a set of N random horizontal and vertical lines, each line generated with two random integer coordinates between 0 and 1000 and a random bit to distinguish horizontal from vertical.

5. Give a method for testing whether or not a given polygon is simple (doesn't intersect itself).

6. Give a method for testing whether one polygon is totally contained within another.

7. Describe how you would solve the general line-intersection problem given the additional fact that the minimum separation between two lines is greater than the maximum length of the lines.

8. Give the binary tree structures that exist when the line intersection algorithm detects the intersections in the lines of Figure 27.6 rotated by 90 degrees.

9. Are the comparison procedures for circles and Manhattan rectangles described in the text transitive?

10. Write a program to find the number of intersecting pairs among a set of N random lines, each line generated with random integer coordinates between 0 and 1000.

28

Closest-Point Problems

☐ Geometric problems involving points on the plane usually involve implicit or explicit treatment of distances between the points. For example, a very natural problem which arises in many applications is the *nearest-neighbor* problem: find the point among a set of given points closest to a given new point. This seems to involve checking the distance from the given point to each point in the set, but much better solutions are possible. In this section we'll look at some other distance problems, a prototype algorithm, and a fundamental geometric structure called the *Voronoi diagram* that can be used effectively for a variety of such problems in the plane. Our approach will be to describe a general method for solving closest point problems through careful consideration of a prototype implementation for a simple problem.

Some of the problems that we consider in this chapter are similar to the range-searching problems of Chapter 26, and the grid and 2D tree methods developed there are suitable for solving the nearest-neighbor and other problems. The fundamental shortcoming of those methods, however, is that they rely on randomness in the point set: they have bad worst-case performance. Our aim in this chapter is to examine another general approach that has guaranteed good performance for many problems, no matter what the input. Some of the methods are too complicated for us to examine a full implementation, and they involve sufficient overhead that the simpler methods may do better when the point set is not large or when it is sufficiently well dispersed. However, the study of methods with good worst-case performance will uncover some fundamental properties of point sets that should be understood even if simpler methods are more suitable in specific situations.

The general approach we'll examine provides yet another example of the use of doubly recursive procedures to intertwine processing along the two coordinate directions. The two previous methods we've seen of this type (kD trees and line intersection) have been based on binary search trees; here the method is a "combine and conquer" method based on mergesort.

Closest-Pair Problem

The *closest-pair* problem is to find the two points that are closest together among a set of points. This problem is related to the nearest-neighbor problem; though it is not as widely applicable, it will serve us well as a prototype closest-point problem in that it can be solved with an algorithm whose general recursive structure is appropriate for other problems.

It would seem necessary to examine the distances between all pairs of points to find the smallest such distance: for N points this would mean a running time proportional to N^2. However, it turns out that we can use sorting to get by with examining only about $N \log N$ distances between points in the worst case (far fewer on the average) and get a worst-case running time proportional to $N \log N$ (far better on the average). In this section, we'll examine such an algorithm in detail.

The algorithm we'll use is based on a straightforward "divide-and-conquer" strategy. The idea is to sort the points on one coordinate, say the x coordinate, then use that ordering to divide the points in half. The closest pair in the whole set is either the closest pair in one of the halves or the closest pair with one member in each half. The interesting case, of course, is when the closest pair crosses the dividing line: the closest pair in each half can obviously be found by using recursive calls, but how can all the pairs on either side of the dividing line be checked efficiently?

Since the only information we seek is the closest pair of the point set, we need examine only points within distance *min* of the dividing line, where *min* is the smaller of the distances between the closest pairs found in the two halves. By itself, however, this observation isn't enough help in the worst case, since there could be many pairs of points very close to the dividing line; for example, all the points in each half could be lined up right next to the dividing line.

To handle such situations, it seems necessary to sort the points on y. Then we can limit the number of distance computations involving each point as follows: proceeding through the points in increasing y order, check if each point is inside the vertical strip containing all points in the plane within *min* of the dividing line. For each such point, compute the distance between it and any point also in the strip whose y coordinate is less than the y coordinate of the current point, but not more than *min* less. The fact that the distance between all pairs of points in each half is at least *min* means that only a few points are likely to be checked.

In the small set of points on the left in Figure 28.1, the imaginary vertical dividing line just to the right of F has eight points to the left, eight points to the right. The closest pair on the left half is AC (or AO), the closest pair on the right is JM. If the points are sorted on y, then the closest pair split by the line is found by checking the pairs HI, CI, FK (the closest pair in the whole point set), and finally EK. For larger point sets, the band that could contain a closest pair spanning the dividing line is narrower, as shown on the right in Figure 28.1.

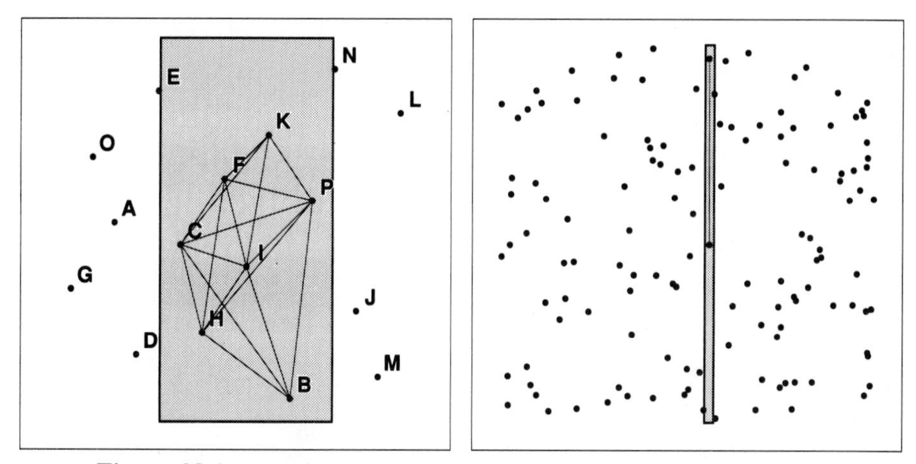

Figure 28.1 Divide-and-conquer approach to find the closest pair.

Though this algorithm is stated simply, some care is required to implement it efficiently: for example, it would be too expensive to sort the points on y within our recursive subroutine. We've seen several algorithms with running times described by the recurrence $C_N = 2C_{N/2} + N$, which implies that C_N is proportional to $N \log N$; if we were to do the full sort on y, then the recurrence would become $C_N = 2C_{N/2} + N \log N$, which implies that C_N is proportional to $N \log^2 N$ (see Chapter 6). To avoid this, we need to avoid the sort of y.

The solution to this problem is simple, but subtle. The *mergesort* method from Chapter 12 is based on dividing the elements to be sorted exactly as the points are divided above. We have two problems to solve and the same general method to solve them, so we may as well solve them simultaneously! Specifically, we'll write one recursive routine that both sorts on y *and* finds the closest pair. It will do so by splitting the point set in half, then calling itself recursively to sort the two halves on y and find the closest pair in each half, then merging to complete the sort on y and applying the procedure above to complete the closest-pair computation. In this way, we avoid the cost of doing an extra y sort by intermixing the data movement required for the sort with the data movement required for the closest-pair computation.

For the y sort, the split in half could be done in any way, but for the closest-pair computation, it's required that the points in one half all have smaller x coordinates than the points in the other half. This is easily accomplished by sorting on x before doing the division. In fact, we may as well use the same routine to sort on x! Once this general plan is accepted, the implementation is not difficult to understand.

As mentioned above, the implementation will use the recursive *sort* and *merge* procedures of Chapter 12. The first step is to modify the list structures to hold points instead of keys, and to modify *merge* to check a global variable *pass* to decide how to do its comparison. If *pass=1*, we should compare the x coordinates of the two points; if *pass=2* we compare the y coordinates of the two points. The implementation of this is straightforward:

```
function merge(a,b: link): link;
  var c: link; comp: boolean;
  begin
  c:=z;
  repeat
    if pass=1
      then comp:=a↑.p.x<b↑.p.x
      else comp:=a↑.p.y<b↑.p.y;
    if comp
      then begin c↑.next:=a; c:=a; a:=a↑.next end
      else begin c↑.next:=b; c:=b; b:=b↑.next end
  until c=z;
  merge:=z↑.next; z↑.next:=z;
  end;
```

The dummy node z which appears at the end of all lists is initialized to contain a "sentinel" point with artificially high x and y coordinates.

To compute distances, we use another simple procedure that checks that the distance between the two points given as arguments is less than the global variable *min*. If so, it resets *min* to that distance and saves the points in the global variables *cp1* and *cp2*:

```
procedure check(p1,p2: point);
  var dist: real;
  begin
  if (p1.y<>z↑.p.y) and (p2.y<>z↑.p.y) then
    begin
    dist:=sqrt((p1.x-p2.x)*(p1.x-p2.x)+(p1.y-p2.y)*(p1.y-p2.y));
    if dist<min then
      begin min:=dist; cp1:=p1; cp2:=p2 end;
    end;
  end;
```

Thus, the global *min* always contains the distance between *cp1* and *cp2*, the closest pair found so far.

The next step is to modify the recursive *sort* of Chapter 12 also to do the closest-point computation when *pass=2*, as follows:

```
function sort(c: link; N: integer): link;
  var a,b: link; i: integer;
      middle: real;
      p1,p2,p3,p4: point;
  begin
  if c↑.next=z then sort:=c else
    begin
    a:=c;
    for i:= 2 to N div 2 do c:=c↑.next;
    b:=c↑.next; c↑.next:=z;
    if pass=2 then middle:=b↑.p.x;
    c:=merge(sort(a,N div 2),sort(b,N-(N div 2)));
    sort:=c;
    if pass=2 then
      begin
      a:=c; p1:=z↑.p; p2:=z↑.p; p3:=z↑.p; p4:=z↑.p;
      repeat
        if abs(a↑.p.x-middle)<min then
          begin
          check(a↑.p,p1);
          check(a↑.p,p2);
          check(a↑.p,p3);
          check(a↑.p,p4);
          p1:=p2; p2:=p3; p3:=p4; p4:=a↑.p
          end;
        a:=a↑.next
      until a=z
      end
    end;
  end;
```

If *pass=1*, this is exactly the recursive mergesort routine of Chapter 12: it returns a linked list containing the points sorted on their x coordinates (because *merge* has been modified as described above to compare x coordinates on *pass1*). The magic of this implementation comes when *pass=2*. The program not only sorts on y (because *merge* has been modified as described above to compare y coordinates on *pass2*) but also completes the closest-point computation, as described in detail below.

First, we sort on x; then we sort on y and find the closest pair by invoking *sort* as follows:

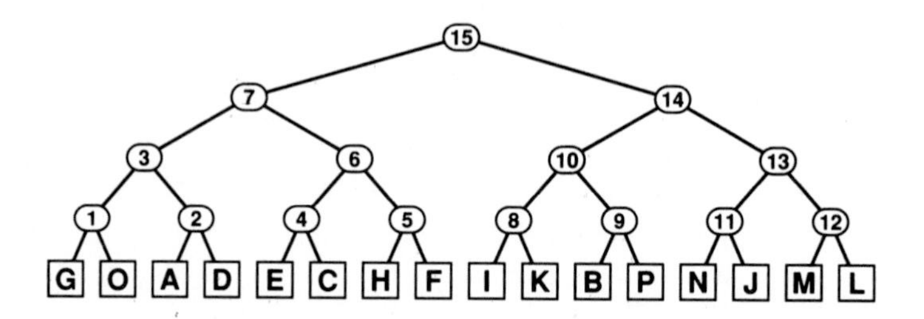

Figure 28.2 Recursive call tree for closest-pair computation.

```
new(z); z↑.next:=z;
z↑.p.x:=maxint; z↑.p.y:=maxint;
new(h); h↑.next:=readlist;
min:=maxint;
pass:=1; h↑.next:=sort(h↑.next,N);
pass:=2; h↑.next:=sort(h↑.next,N);
```

After these calls, the closest pair of points is found in the global variables *cp1* and *cp2*, which are managed by the *check* "find the minimum" procedure.

The crux of the implementation is the operation of *sort* when *pass=2*. *Before* the recursive calls the points are sorted on x: this ordering is used to divide the points in half and to find the x coordinate of the dividing line. *After* the recursive calls the points are sorted on y and the distance between every pair of points in each half is known to be greater than *min*. The ordering on y is used to scan the points near the dividing line; the value of *min* is used to limit the number of points to be tested. Each point within a distance of *min* of the dividing line is *check*ed against each of the previous four points found within a distance of *min* of the dividing line. This check is guaranteed to find any pair of points closer together than *min* with one member of the pair on either side of the dividing line. (This is an amusing geometric fact which the reader may wish to verify): We know that points that fall on the same side of the dividing line are spaced by at least *min*, so the number of points falling in any circle of radius *min* is limited.)

Figure 28.2 shows the recursive call tree describing the operation of this algorithm on our small set of points. An internal node in this tree represents a vertical line dividing the points in the left and right subtree. The nodes are numbered in the order in which the vertical lines are tried in the algorithm. This numbering corresponds to a postorder traversal of the tree because the computation involving the dividing line comes *after* the recursive calls in the program, and is simply another way of looking at the order in which merges are done during a recursive mergesort (see Chapter 12).

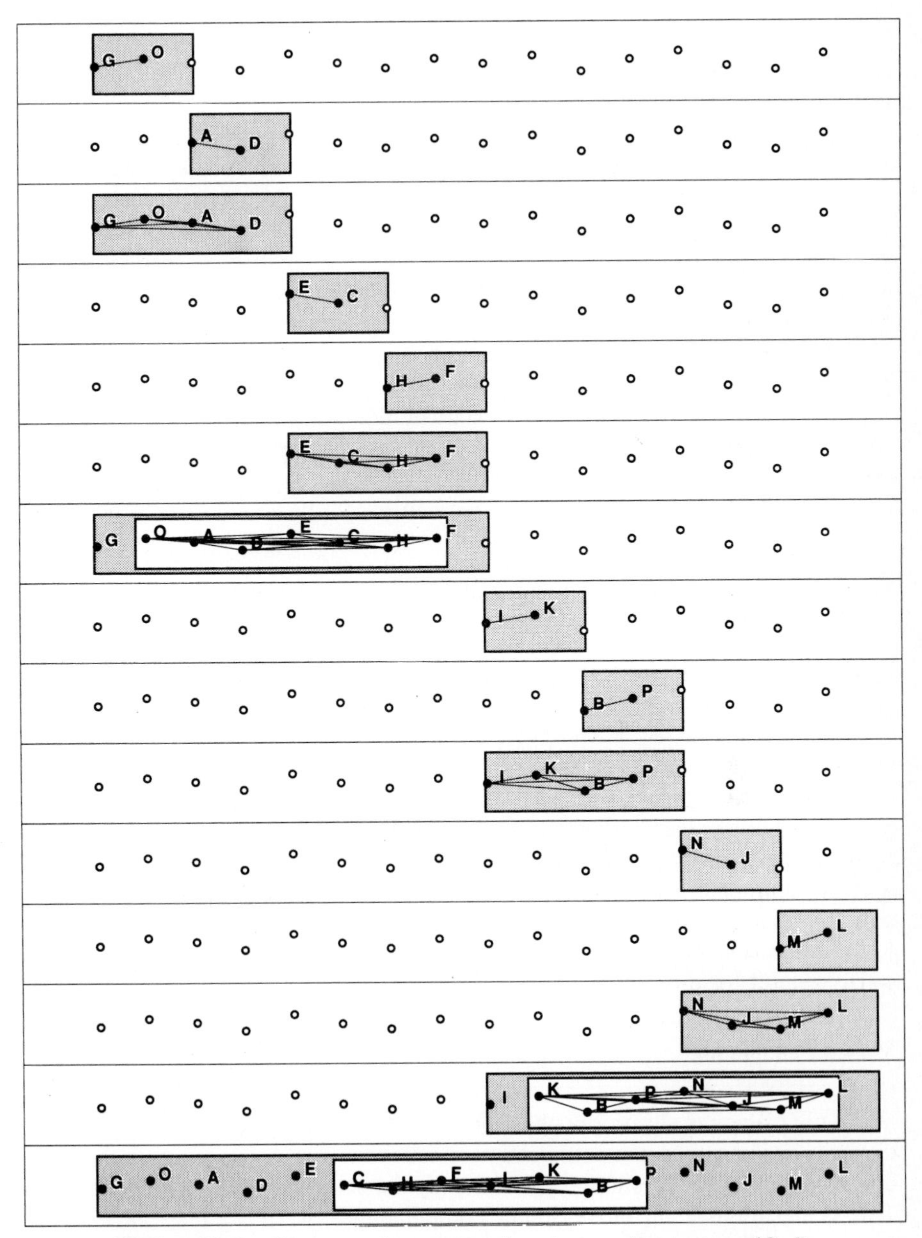

Figure 28.3 Closest-pair computation (x coordinate magnified).

Thus, first the line between G and O is tried and the pair GO is retained as the closest so far. Then the line between A and D is tried, but A and D are too far apart to change *min*. Then the line between O and A is tried and the pairs GD GA and OA all are successively closer pairs. It happens for this example that no closer pairs are found until FK, which is the last pair checked for the last dividing line tried.

The careful reader may have noticed that we have not implemented the pure divide-and-conquer algorithm described above—we don't actually compute the closest pair in the two halves, then take the better of the two. Instead, we get the closer of the two closest pairs simply by using a global variable for *min* during the recursive computation. Each time we find a closer pair, we can consider a narrower vertical strip around the current dividing line, no matter where we are in the recursive computation.

Figure 28.3 shows the process in detail. The x-coordinate in these diagrams is magnified to emphasize the x orientation of the process and to point out parallels with mergesort (see Chapter 12). We start by doing a y-sort on the four leftmost points G O A D, by sorting G O, then sorting A D, then merging. After the merge, the y-sort is complete, and we find the closest pair AO spanning the dividing line. Eventually, the points are sorted on their y-coordinate and the closest pair is computed.

Property 28.1 *The closest pair in a set of N points can be found in* $O(N \log N)$ *steps.*

Essentially, the computation is done in the time it takes to do two mergesorts (one on the x-coordinate, one on the y-coordinate) plus the cost of looking along the dividing line. This cost is also governed by the recurrence $T_N = T_{N/2} + N$ (see Chapter 6).

The general approach we've used here for the closest-pair problem can be used to solve other geometric problems. For example, another question of interest is the *all-nearest-neighbors* problem: for each point we want to find the point nearest to it. This problem can be solved using a program like the one above with extra processing along the dividing line to find, for each point, whether there is a point on the other side closer than its closest point on its own side. Again, the "free" y sort is helpful for this computation.

Voronoi Diagrams

The set of all points closer to a given point in a point set than to all other points in the set is an interesting geometric structure called the *Voronoi polygon* for the point. The union of all the Voronoi polygons for a point set is called its *Voronoi diagram*. This is the ultimate in closest-point computations: we'll see that most of the problems we face involving distances between points have natural and interesting

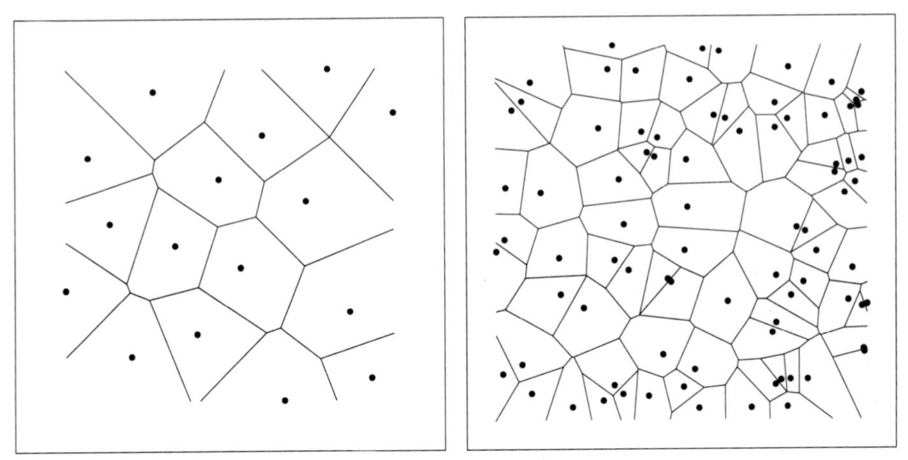

Figure 28.4 Voronoi Diagram.

solutions based on the Voronoi diagram. The diagrams for our sample point sets are shown in Figure 28.4.

The Voronoi polygon for a point is made up of the perpendicular bisectors of the segments linking the point to those points closest to it. Its actual definition is the other way around: the Voronoi polygon is defined to be perimeter of the set of all points in the plane closer to the given point than to any other point in the point set, and each edge on the Voronoi polygon separates a given point from one of the points "closest to" it.

The *dual* of the Voronoi diagram, shown in Figure 28.5, makes this correspon-

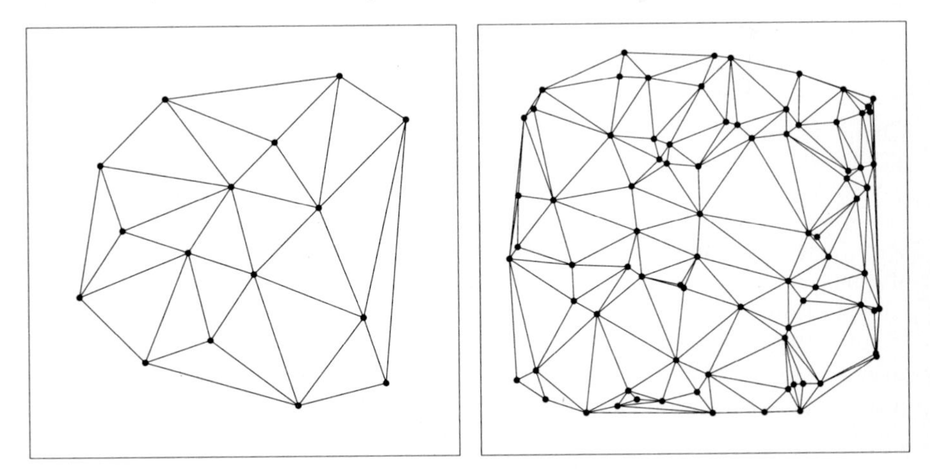

Figure 28.5 Delaunay Triangulation.

dence explicit: in the dual, a line is drawn between each point and all the points "closest to" it. This is also called the Delaunay triangulation. Put another way, x and y are connected in the Voronoi dual if their Voronoi polygons have an edge in common.

The Voronoi diagram and the Delaunay triangulation have many properties that lead to efficient algorithms for closest-point problems. The property that makes these algorithms efficient is that the number of lines in both the diagram and the dual is proportional to a small constant times N. For example, the line connecting the closest pair of points must be in the dual, so the problem of the previous section can be solved by computing the dual and then simply finding the minimum length line among the lines in the dual. Similarly, the line connecting each point to its nearest neighbor must be in the dual, so the all-nearest-neighbors problem reduces directly to finding the dual. The convex hull of the point set is part of the dual, so computing the Voronoi dual is yet another convex hull algorithm. We'll see yet another example in Chapter 31 of a problem which can be solved efficiently by first finding the Voronoi dual.

The defining property of the Voronoi diagram means that it can be used to solve the nearest-neighbor problem: to identify the nearest neighbor in a point set to a given point, we need only find out which Voronoi polygon the point falls in. It is possible to organize the Voronoi polygons in a structure like a 2D tree to allow this search to be done efficiently.

The Voronoi diagram can be computed using an algorithm with the same general structure as the closest-point algorithm above. The points are first sorted on their x coordinate. Then that ordering is used to split the points in half, leading to two recursive calls to find the Voronoi diagram of the point set for each half. At the same time, the points are sorted on y; finally, the two Voronoi diagrams for the two halves are merged together. As before, this merging (done with *pass=2*) can exploit the fact that the points are sorted on x before the recursive calls and that they are sorted on y and that the Voronoi diagrams for the two halves have been built after the recursive calls. However, even with these aids, the merge is quite a complicated task, and presentation of a full implementation would be beyond the scope of this book.

The Voronoi diagram is certainly the natural structure for closest-point problems, and understanding the characteristics of a problem in terms of the Voronoi diagram or its dual is certainly a worthwhile exercise. However, for many particular problems, a direct implementation based on the general schema given in this chapter may be suitable. This schema is powerful enough to compute the Voronoi diagram, so it is powerful enough for algorithms based on the Voronoi diagram, and it may admit to simpler, more efficient code, as we saw for the closest-pair problem.

□

Exercises

1. Write programs to solve the nearest-neighbor problem, first using the grid method, then using 2D trees.
2. Describe what happens when the closest-pair procedure is used on a set of points that fall on the same horizontal line, equally spaced.
3. Describe what happens when the closest-pair procedure is used on a set of points that fall on the same vertical line, equally spaced.
4. Give an algorithm that, given a set of $2N$ points, half with positive x coordinates, half with negative x coordinates, finds the closest pair with one member of the pair in each half.
5. Give the successive pairs of points assigned to *cp1* and *cp2* when the program in the text is run on the example points, but with A removed.
6. Test the effectiveness of making *min* global by comparing the performance of the implementation given to a purely recursive implementation for some large random point set.
7. Give an algorithm for finding the closest pair from a set of lines.
8. Draw the Voronoi diagram and its dual for the points A B C D E F from the sample point set.
9. Give a "brute-force" method (which might require time proportional to N^2) for computing the Voronoi diagram.
10. Write a program that uses the same recursive structure as the closest-pair implementation given in the text to find the convex hull of a set of points.

SOURCES for Geometric Algorithms

Much of the material described in this section has actually been developed quite recently. Many of the problems and solutions that we've discussed were presented by M. Shamos in 1975. Shamos' Ph.D. thesis treated a large number of geometric algorithms, stimulated much of the recent research, and eventually developed into the authoritative reference in the field, the book by Preparata and Shamos. The field is developing quickly: the book by Edelsbrunner describes many more recent research results.

For the most part, each of the geometric algorithms that we've discussed is described in its own original reference. The convex hull algorithms treated in Chapter 25 may be found in the papers by Jarvis, Graham, and Golin and Sedgewick. The range-searching methods of Chapter 26 come from Bentley and Friedman's survey article, which contains many references to original sources (of particular interest is Bentley's own original article on kD trees, written while he was an undergraduate). The treatment of the closest-point problems in Chapter 28 is based on Shamos and Hoey's 1976 paper, and the intersection algorithms of Chapter 27 are from their 1975 paper and the article by Bentley and Ottmann.

But the best route for someone interested in learning more about geometric algorithms is to implement some and run them to learn their properties and properties of the objects they manipulate.

J. L. Bentley, "Multidimensional binary search trees used for associative searching," *Communications of the ACM*, **18**, 9 (September, 1975).

J. L. Bentley and J.H. Friedman, "Data structures for range searching," *Computing Surveys*, **11**, 4 (December, 1979).

J. L. Bentley and T. Ottmann, "Algorithms for reporting and counting geometric intersections," *IEEE Transactions on Computing*, **C-28**, 9 (September, 1979).

H. Edelsbrunner, *Algorithms in Combinatorial Geometry*, Springer-Verlag, 1987.

M. Golin and R. Sedgewick, "Analysis of a simple yet efficient convex hull algorithm," in *4th Annual Symposium on Computational Geometry*, ACM, 1988.

R. L. Graham, "An efficient algorithm for determining the convex hull of a finite planar set," *Information Processing Letters*, **1** (1972).

R. A. Jarvis, "On the identification of the convex hull of a finite set of points in the plane," *Information Processing Letters*, **2** (1973).

F. P. Preparata and M. I. Shamos, *Computational Geometry: An Introduction*, Springer-Verlag, 1985.

M. I. Shamos and D. Hoey, "Closest-point problems," in *16th Annual Symposium on Foundations of Computer Science*, IEEE, 1975.

M. I. Shamos and D. Hoey, "Geometric intersection problems," in *17th Annual Symposium on Foundations of Computer Science*, IEEE, 1976.

Graph Algorithms

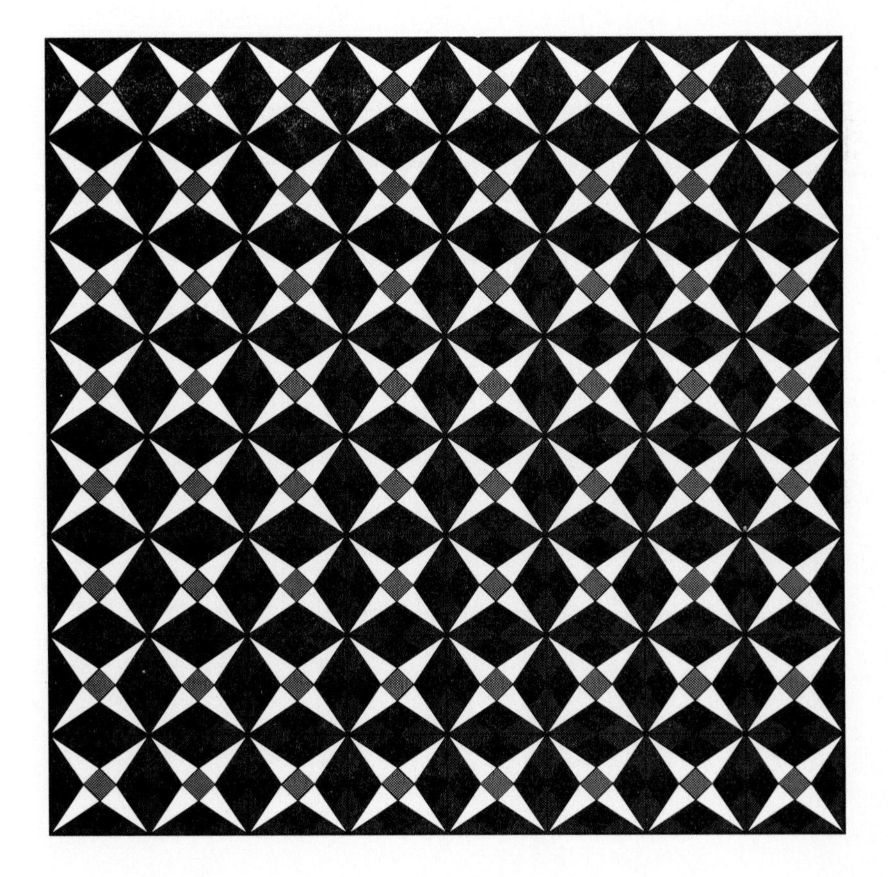

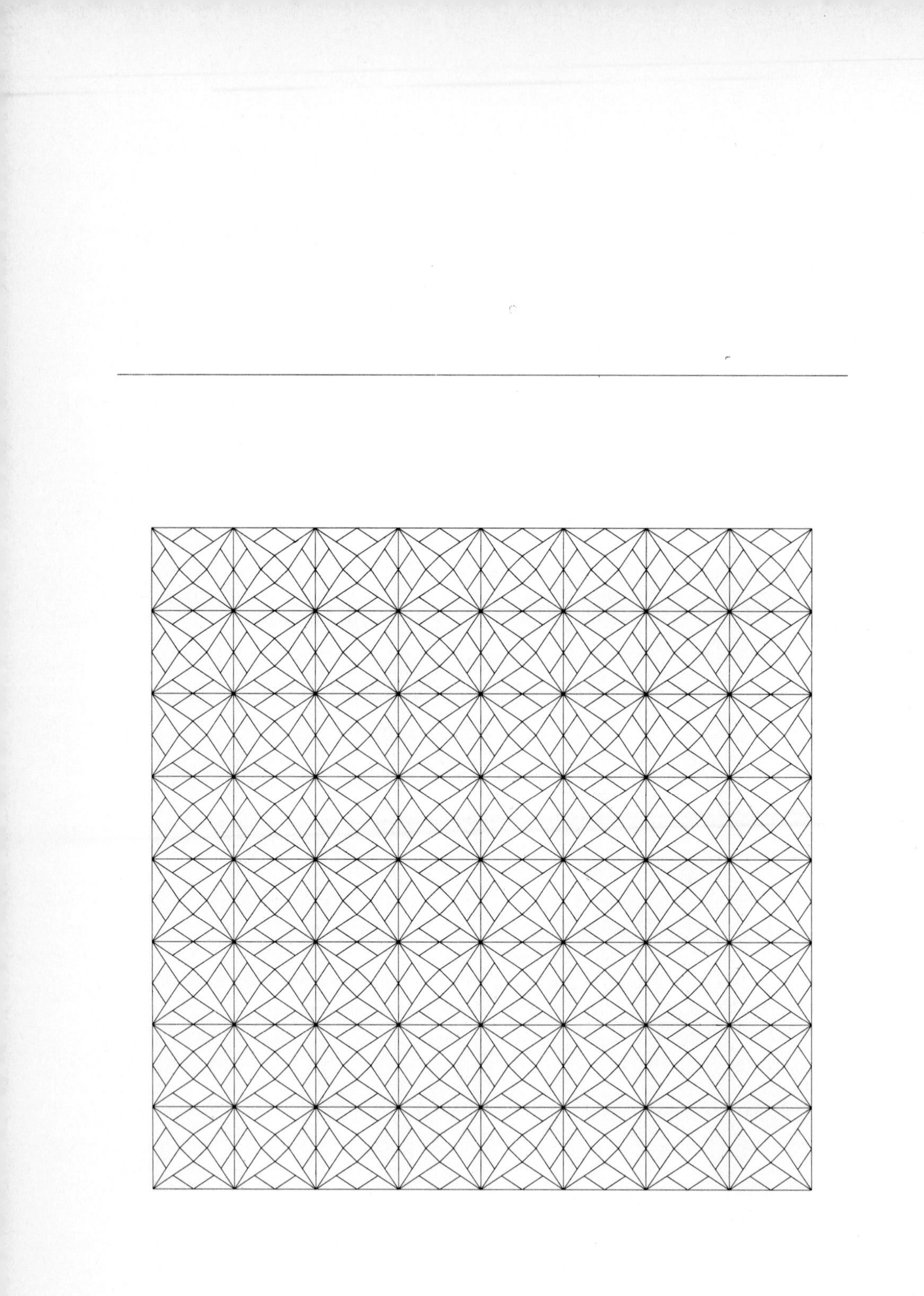

29

Elementary Graph Algorithms

☐ A great many problems are naturally formulated in terms of objects and connections between them. For example, given an airline route map of the eastern U.S., we might be interested in questions like: "What's the fastest way to get from Providence to Princeton?" Or we might be more interested in money than in time, and look for the cheapest way to get from Providence to Princeton. To answer such questions we need only information about interconnections (airline routes) between objects (towns).

Electric circuits are another obvious example where interconnections between objects play a central role. Circuit elements like transistors, resistors, and capacitors are intricately wired together. Such circuits can be represented and processed within a computer in order to answer simple questions like "Is everything connected together?" as well as complicated questions like "If this circuit is built, will it work?" Here, the answer to the first question depends only on the properties of the interconnections (wires), while the answer to the second requires detailed information about both the wires and the objects that they connect.

A third example is "job scheduling," where the objects are tasks to be performed, say in a manufacturing process, and interconnections indicate which jobs should be done before others. Here we might be interested in answering questions like "When should each task be performed?"

A *graph* is a mathematical object that accurately models such situations. In this chapter, we'll examine some basic properties of graphs, and in the next several chapters we'll study a variety of algorithms for answering questions of the type posed above.

Actually, we've already encountered graphs in previous chapters. Linked data structures are actually representations of graphs, and some of the algorithms we'll see for processing graphs are similar to algorithms we've already seen for processing trees and other structures. For example, the finite-state machines of Chapters 19 and 20 are represented with graph structures.

Graph theory is a major branch of combinatorial mathematics and has been studied intensively for hundreds of years. Many important and useful properties of graphs have been proved, but many difficult problems have yet to be resolved. Here we can only scratch the surface of what is known about graphs, covering enough to be able to understand the fundamental algorithms.

Like so many of the problem domains we've studied, graphs have only recently begun to be examined from an algorithmic point of view. Although some of the fundamental algorithms are quite old, many of the interesting ones have been discovered within the last ten years. Even trivial graph algorithms lead to interesting computer programs, and the nontrivial algorithms we'll examine are among the most elegant and interesting (though difficult to understand) algorithms known.

Glossary

A good deal of nomenclature is associated with graphs. Most of the terms have straightforward definitions, and it is convenient to put them in one place even though we won't be using some of them until later.

A *graph* is a collection of *vertices* and *edges*. Vertices are simple objects that can have names and other properties; an edge is a connection between two vertices. One can draw a graph by marking points for the vertices and drawing lines connecting them for the edges, but it must be borne in mind that the graph is defined independently of the representation. For example, the two drawings in Figure 29.1 represent the same graph. We define this graph by saying that it consists of the set of vertices A B C D E F G H I J K L M and the set of edges between these vertices AG AB AC LM JM JL JK ED FD HI FE AF GE .

For some applications, such as the airline route example above, it might not make sense to rearrange the vertices as in Figure 29.1 But for other applications, such as the electric circuit application above, it is best to concentrate only on the edges and vertices, independent of any particular geometric placement. And for

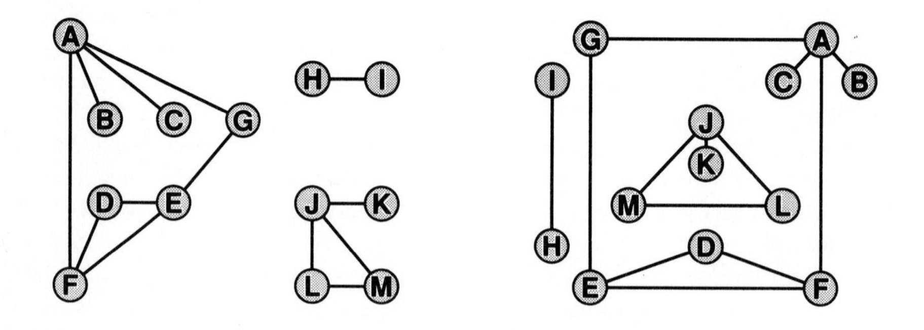

Figure 29.1 Two representations of the same graph.

still other applications, such as the finite-state machines in Chapters 19 and 20, no particular geometric placement of nodes is ever implied. The relationship between graph algorithms and geometric problems is discussed in further detail in Chapter 31. For now, we'll concentrate on "pure" graph algorithms that process simple collections of edges and nodes.

A *path* from vertex x to y in a graph is a list of vertices in which successive vertices are connected by edges in the graph. For example, BAFEG is a path from B to G in Figure 29.1. A graph is *connected* if there is a path from every node to every other node in the graph. Intuitively, if the vertices were physical objects and the edges were strings connecting them, a connected graph would stay in one piece if picked up by any vertex. A graph which is not connected is made up of *connected components*; for example, the graph in Figure 29.1 has three connected components. A *simple path* is a path in which no vertex is repeated. (For example, BAFEGAC is not a simple path.) A *cycle* is a path that is simple except that the first and last vertex are the same (a path from a point back to itself): the path AFEGA is a cycle.

A graph with no cycles is called a *tree* (see Chapter 4). A group of disconnected trees is called a *forest*. A *spanning tree* of a graph is a subgraph that contains all the vertices but only enough of the edges to form a tree. For example, the edges AB AD AF DE EG form a spanning tree for the large component of the graph in Figure 29.1, and Figure 29.2 shows a larger graph and one of its spanning trees.

Note that if we add any edge to a tree, it must form a cycle (because there is already a path between the two vertices it connects). Also, as we saw in Chapter 4, a tree on V vertices has exactly $V - 1$ edges. If a graph with V vertices has less than $V - 1$ edges, it can't be connected. If it has more that $V - 1$ edges, it

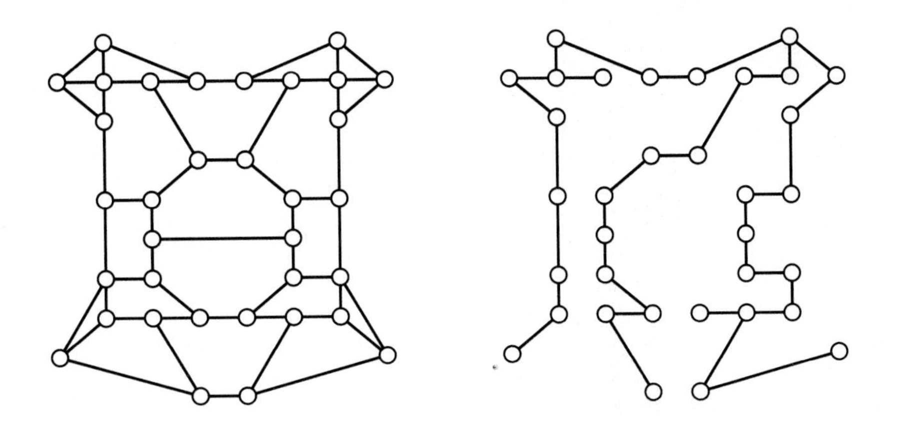

Figure 29.2 A large graph and a spanning tree for that graph.

must have a cycle. (But if it has exactly $V - 1$ edges, it need not be a tree.)

We'll denote the number of vertices in a given graph by V, the number of edges by E. Note that E can range anywhere from 0 to $\frac{1}{2}V(V - 1)$. Graphs with all edges present are called *complete* graphs; graphs with relatively few edges (say less than $V \log V$) are called *sparse*; graphs with relatively few of the possible edges missing are called *dense*.

The fundamental dependence of graph topology on two parameters makes the comparative study of graph algorithms somewhat more complicated than many algorithms we've studied, because more possibilities arise. For example, one algorithm may take about V^2 steps, while another algorithm for the same problem may take $(E + V)\log E$ steps. The second algorithm would be better for sparse graphs, but the first would be preferred for dense graphs.

Graphs as defined to this point are called *undirected graphs*, the simplest type of graph. We'll also be considering more complicated type of graphs in which more information is associated with the nodes and edges. In *weighted graphs* integers (*weights*) are assigned to each edge to represent, say, distances or costs. In *directed graphs*, edges are "one-way": an edge may go from x to y but not from y to x. Directed weighted graphs are sometimes called *networks*. As we'll discover, the extra information weighted and directed graphs contain makes them somewhat more difficult to manipulate than simple undirected graphs.

Representation

In order to process graphs with a computer program, we first need to decide how to represent them within the computer. We'll look at two commonly used representations; the choice between them depends primarily upon whether the graph is dense or sparse, although, as usual, the nature of the operations to be performed also plays an important role.

The first step in representing a graph is to map the vertex names to integers between 1 and V. The main reason for doing this is to make it possible to quickly access information corresponding to each vertex, using array indexing. Any standard searching scheme can be used for this purpose; for instance, we can translate vertex names to integers between 1 and V by maintaining a hash table or a binary tree that can be searched to find the integer corresponding to any given vertex name. Since we have already studied these techniques, we assume that a function *index* is available to convert from vertex names to integers between 1 and V and a function *name* to convert from integers to vertex names. To make our algorithms easy to follow, we use one-letter vertex names, with the ith letter of the alphabet corresponding to the integer i. Thus, though *name* and *index* are trivial to implement for our examples, their use makes it easy to extend the algorithms to handle graphs with real vertex names using techniques from Chapters 14–17.

The most straightforward representation for graphs is the so-called *adjacency matrix* representation. A V-by-V array of boolean values is maintained, with $a[x,y]$

	A	**B**	**C**	**D**	**E**	**F**	**G**	**H**	**I**	**J**	**K**	**L**	**M**
A	*1*	*1*	*1*	*0*	*0*	*1*	*1*	*0*	*0*	*0*	*0*	*0*	*0*
B	*1*	*1*	*0*	*0*	*0*	*0*	*0*	*0*	*0*	*0*	*0*	*0*	*0*
C	*1*	*0*	*1*	*0*	*0*	*0*	*0*	*0*	*0*	*0*	*0*	*0*	*0*
D	*0*	*0*	*0*	*1*	*1*	*1*	*0*	*0*	*0*	*0*	*0*	*0*	*0*
E	*0*	*0*	*0*	*1*	*1*	*1*	*1*	*0*	*0*	*0*	*0*	*0*	*0*
F	*1*	*0*	*0*	*1*	*1*	*1*	*0*	*0*	*0*	*0*	*0*	*0*	*0*
G	*1*	*0*	*0*	*0*	*1*	*0*	*1*	*0*	*0*	*0*	*0*	*0*	*0*
H	*0*	*0*	*0*	*0*	*0*	*0*	*0*	*1*	*1*	*0*	*0*	*0*	*0*
I	*0*	*0*	*0*	*0*	*0*	*0*	*0*	*1*	*1*	*0*	*0*	*0*	*0*
J	*0*	*0*	*0*	*0*	*0*	*0*	*0*	*0*	*0*	*1*	*1*	*1*	*1*
K	*0*	*0*	*0*	*0*	*0*	*0*	*0*	*0*	*0*	*1*	*1*	*0*	*0*
L	*0*	*0*	*0*	*0*	*0*	*0*	*0*	*0*	*0*	*1*	*0*	*1*	*1*
M	*0*	*0*	*0*	*0*	*0*	*0*	*0*	*0*	*0*	*1*	*0*	*1*	*1*

Figure 29.3 Adjacency matrix representation.

set to *true* if there is an edge from vertex x to vertex y and *false* otherwise. The adjacency matrix for the graph in Figure 29.1 is shown in Figure 29.3 (*1* means *true* and *0* means *false*.

Notice that each edge is really represented by two bits: an edge connecting x and y is represented by true values in both $a[x,y]$ and $a[y,x]$. While space can be saved by storing only half of this symmetric matrix, it is inconvenient to do this in Pascal and the algorithms are somewhat simpler with the full matrix. Also, it is usually convenient to assume that there's an "edge" from each vertex to itself, so $a[x,x]$ is set to 1 for x from 1 to V. (In some cases, it is more convenient to set the diagonal elements to 0; we're free to do so when appropriate.)

A graph is defined by a set of nodes and a set of edges connecting them. To take a graph as input, we need to settle on a format for reading in these sets. One possibility is to use the adjacency matrix itself as the input format, but, as we'll see, this is inappropriate for sparse graphs. Instead, we will use a more direct format: we first read in the vertex names, then pairs of vertex names (which define edges). As mentioned above, one easy way to proceed is to read the vertex names into a hash table or binary search tree and assign to each vertex name an integer for use in accessing vertex-indexed arrays like the adjacency matrix. The ith vertex read can be assigned the integer i. For simplicity in our programs, we first read in V and E, then the vertices and the edges. Alternatively, the input could be arranged with a delimiter separating the vertices from the edges, and the program could determine V and E from the input. (In our examples, we use the first V letters of the alphabet for vertex names, so the even simpler scheme of reading V and E, then E pairs of letters from the first V letters of the alphabet would work.) The order in which the edges appear is not important, since all orderings of the edges represent the same graph and result in the same adjacency matrix, as computed by the following program:

```
program adjmatrix(input, output);
const maxV=50;
var j, x, y, V, E: integer;
      a: array[1..maxV, 1..maxV] of boolean;
begin
readln(V, E);
for x:=1 to V do
   for y:=1 to V do a[x, y]:=false;
for x:=1 to V do a[x, x]:=true;
for j:=1 to E do
   begin
   readln(v1, v2);
   x:=index(v1); y:=index(v2);
   a[x, y]:=true; a[y, x]:=true
   end;
end.
```

The types of *v1* and *v2* are omitted from this program, as well as the code for *index*. These can be added in a straightforward manner, depending on the graph input representation desired. (For our examples, *v1* and *v2* could be of type *char* and *index* could be a simple function that uses the Pascal *ord* function.)

The adjacency-matrix representation is satisfactory only if the graphs to be processed are dense: the matrix requires V^2 bits of storage and V^2 steps just to initialize it. If the number of edges (the number of *1* bits in the matrix) is proportional to V^2, then this may be acceptable because about V^2 steps are required to read in the edges in any case. If the graph is sparse, however, just initializing the matrix could be the dominant factor in the running time of an algorithm. This might also be the best representation for some algorithms requiring more than V^2 steps for execution.

Let us now look at a representation that is more suitable for graphs that are not dense. In the *adjacency-structure* representation all the vertices connected to each vertex are listed on an *adjacency list* for that vertex. This can be easily done with linked lists, as shown in the program below which builds the adjacency structure for our sample graph. The linked lists are built as usual, with an artificial node z at the end (which points to itself). The artificial nodes for the beginning of the lists are kept in an array *adj* indexed by vertex. To add an edge connecting x to y to this representation of the graph, we add x to y's adjacency list and y to x's adjacency list:

```
program adjlist(input,output);
const maxV=1000;
type link=↑node;
       node=record v: integer; next: link end;
var j,x,y,V,E: integer;
     t,z: link;
     adj: array[1..maxV] of link;
begin
readln(V,E);
new(z); z↑.next:=z;
for j:=1 to V do adj[j]:=z;
for j:=1 to E do
  begin
  readln(v1,v2);
  x:=index(v1); y:=index(v2);
  new(t); t↑.v:=x; t↑.next:=adj[y]; adj[y]:=t;
  new(t); t↑.v:=y; t↑.next:=adj[x]; adj[x]:=t;
  end;
end.
```

If the edges appear in the order AG AB AC LM JM JL JK ED FD HI FE AF GE, the program above builds the adjacency list structure shown in Figure 29.4. Note again that each edge is represented twice: an edge connecting x and y is represented as a node containing x on y's adjacency list and as a node containing y on x's adjacency list. It is important to include both, since otherwise simple questions like "Which nodes are connected directly to node x?" could not be

Figure 29.4 An adjacency-structure representation.

answered efficiently.

For this representation, the order in which the edges appear in the input is quite important: it (along with the list insertion method used) determines the order in which the vertices appear on the adjacency lists. Thus, the same graph can be represented in many different ways in an adjacency-list structure. Indeed, it is difficult to predict what the adjacency lists will look like by examining just the sequence of edges, because each edge involves insertions into two adjacency lists.

The order in which edges appear on the adjacency list affects, in turn, the order in which edges are processed by algorithms. That is, the adjacency-list structure determines how various algorithms we'll be examining "see" the graph. While an algorithm should produce a correct answer no matter how the edges are ordered on the adjacency lists, it might get to that answer by quite different sequences of computations for different orders. And if there is more than one "correct answer," different input orders might lead to different output results.

Some simple operations are not supported by this representation. For example, one might want to delete a vertex, x, and all the edges connected to it. It's not sufficient to delete nodes from the adjacency list: each node on the adjacency list specifies another vertex whose adjacency list must be searched to delete a node corresponding to x. This problem can be corrected by linking together the two list nodes that correspond to a particular edge and making the adjacency lists doubly linked. Then if an edge is to be removed, both list nodes corresponding to that edge can be deleted quickly. Of course, these extra links are cumbersome to process, and they shouldn't be included unless operations like deletion are needed.

Such considerations also make it plain why we don't use a "direct" representation for graphs: a data structure that models the graph exactly, with vertices represented as allocated records and edge lists containing links to vertices instead of vertex names. How would one add an edge to a graph represented in this way?

Directed and weighted graphs are represented with similar structures. For directed graphs, everything is the same, except that each edge is represented just once: an edge from x to y is represented by a *true* value in $a[x,y]$ in the adjacency matrix or by the appearance of y on x's adjacency list in the adjacency structure. Thus an undirected graph can be thought of as a directed graph with directed edges going both ways between each pair of vertices connected by an edge. For weighted graphs, everything is again the same except that we fill the adjacency matrix with weights instead of boolean values (using some non-existent weight to represent *false*), or include in the adjacency structure a field for the edge weight in adjacency list records.

It is often necessary to associate other information with the vertices or nodes of a graph to allow it to model more complicated objects or to save bookkeeping information in complicated algorithms. Extra information associated with each vertex can be accommodated by using auxiliary arrays indexed by vertex number or by making *adj* an array of records in the adjacency structure representation.

Extra information associated with each edge can be put in the adjacency list nodes (or in an array *a* of records in the adjacency matrix representation), or in auxiliary arrays indexed by edge number (this requires numbering the edges).

Depth-First Search

At the beginning of this chapter, we saw several questions that arise immediately when processing a graph. Is the graph connected? If not, what are its connected components? Does the graph have a cycle? These and many other problems can be easily solved with a technique called *depth-first search*, which is a natural way to "visit" every node and check every edge in the graph systematically. We'll see in the chapters that follow that simple variations on a generalization of this method can be used to solve a variety of graph problems.

For now, we'll concentrate on the mechanics of examining every piece of the graph in an organized way. We start with a program for graphs represented with adjacency lists:

```
procedure listdfs;
  var id, k: integer;
      val: array[1..maxV] of integer;
  procedure visit(k: integer);
    var t: link;
    begin
    id:=id+1; val[k]:=id;
    t:=adj[k];
    while t<>z do
      begin
      if val[t↑.v]=0 then visit(t↑.v);
      t:=t↑.next
      end
    end;
  begin
  id:=0;
  for k:=1 to V do val[k]:=0;
  for k:=1 to V do
    if val[k]=0 then visit(k)
  end;
```

This program fills in an array *val*[*1..V*] as it visits every vertex of a graph. The array is initially set to all zeros, so *val*[*k*]=*0* indicates that vertex *k* has not yet been visited. The goal is to systematically visit all the vertices of the graph, setting the *val* entry for the *id*th vertex visited to *id*, for *id*= $1, 2, \ldots, V$. The program

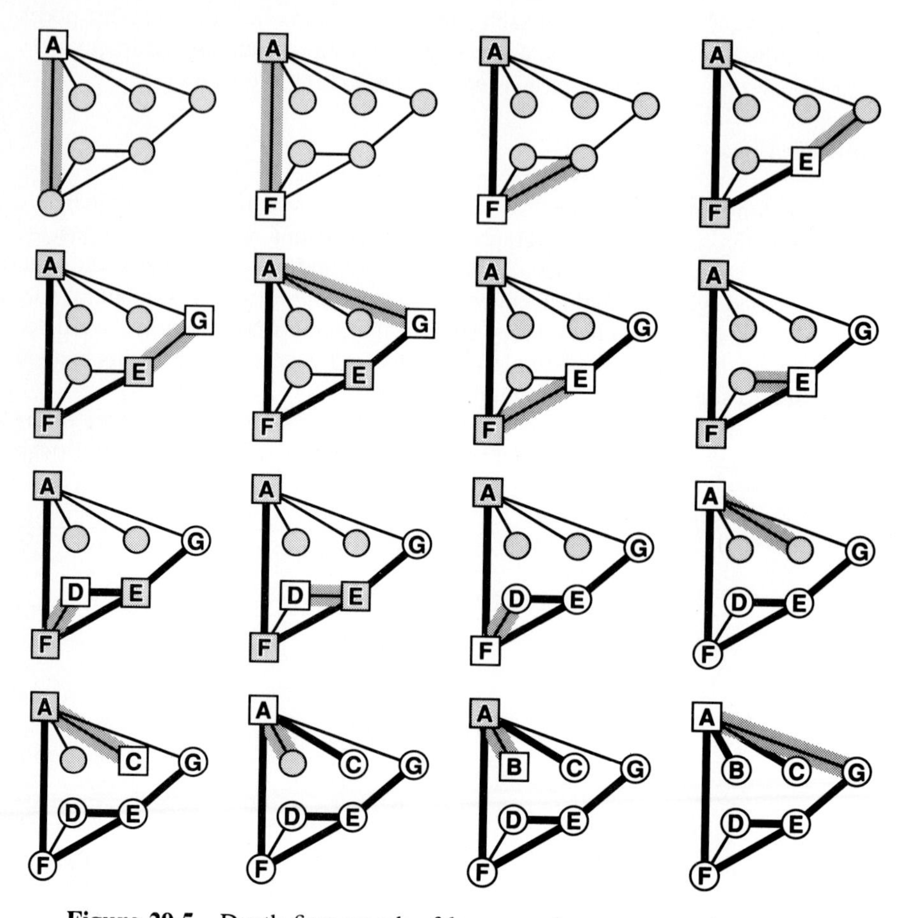

Figure 29.5 Depth-first search of large graph component (recursive).

uses a recursive procedure *visit* that visits all the vertices in the same connected component as the vertex given in the argument. To *visit* a vertex, we check all its edges to see if they lead to vertices that haven't yet been visited (as indicated by 0 *val* entries); if so, we *visit* them. First *visit* is called for the first vertex, which results in nonzero *val* values being set for all the vertices connected to that vertex. Then *dfs* scans through the *val* array to find a zero entry (corresponding to a vertex that hasn't been seen yet) and calls *visit* for that vertex, continuing in this way until all vertices have been visited.

Figure 29.5 traces through the operation of depth-first search on the large component in our sample graph, and shows how every edge in that component is touched as a result of the call *visit(1)* (after the adjacency lists shown in Figure

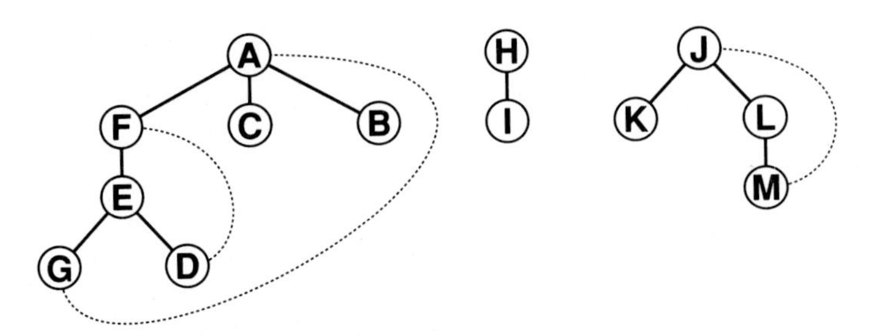

Figure 29.6 Depth-first search forest.

29.4 have been built). Actually, each edge is "touched" *twice*, since each edge is represented on both adjacency lists of the vertex it connects. There is one diagram in Figure 29.5 for each edge traversed (each time the link t is set to point to some node on some adjacency list). In each diagram, the "current" edge is shaded and the node whose adjacency list contains that edge is labeled with a square. In addition, each time a node is visited for the first time (corresponding to a new call to *visit*), the edge that led to that node is darkened. Nodes that have not been touched at all yet are shaded but not labeled and nodes for which the *visit* is complete are shaded and labeled.

The first edge traversed is AF, the first node on the first adjacency list. Then *visit* is called for node F and the edge FA is traversed, since A is the first node on F's adjacency list. But node A has a nonzero *val* entry at this point, so we take edge FE, the next entry on F's adjacency list. Then EG is traversed, then GE, since G and E are first on each other's list. Next, GA is traversed; this completes the *visit* of G, so the algorithm continues the *visit* of E and traverses EF, then ED. Then, the *visit* of D consists of traversing DE and DF, neither of which leads to a new node. Since D is the last node on E's adjacency list, the *visit* of E is thus complete, and the visit of F is then completed by traversing FD. Finally, we return to A, and traverse AC, CA, AB, BA, and AG.

Another way to follow the operation of depth-first search is to redraw the graph as indicated by the recursive calls during the *visit* procedure, as in Figure 29.6. Each connected component leads to a tree, called the *depth-first search tree* for the component. Traversing this tree in preorder gives the vertices of the graph in the order in which they are first encountered by the search; traversing the tree in postorder gives the vertices in the order in which their *visit* completes. It is important to note that this forest of depth-first search trees is simply another way of drawing the graph; all vertices and edges of the graph are examined by the algorithm.

Solid lines in Figure 29.6 indicate that the lower vertex was found by the algorithm to be on the edge list of the upper vertex and had not been visited at that time, so that a recursive call was made. Dotted lines correspond to edges to vertices that had already been visited, so the **if** test in *visit* failed, and the edge was not "followed" with a recursive call. These comments apply to the *first* time each edge is encountered; the **if** test in *visit* also guards against following the edge the *second* time it is encountered, as we saw in Figure 29.5.

A crucial property of these depth-first search trees for undirected graphs is that the dotted links always go from a node to some *ancestor* in the tree (another node in the same tree that is higher up on the path to the root). At any point during the execution of the algorithm, the vertices divide into three classes: those for which *visit* has finished, those for which *visit* has only partially finished, and those that haven't been seen at all. By definition of *visit*, we won't encounter an edge pointing to any vertex in the first class, and if we encounter an edge to a vertex in the third class, a recursive call will be made (so the edge will be solid in the depth-first search tree). The only vertices remaining are those in the second class. But these are precisely the vertices on the path from the current vertex to the root in the same tree, and any edge to any of them will correspond to a dotted link in the depth-first search tree.

Property 29.1 *Depth-first search of a graph represented with adjacency lists requires time proportional to $V + E$.*

We set each of the V *val* values (hence the V term), and we examine each edge twice (hence the E term). One could encounter an (extremely) sparse graph with $E < V$, but if isolated vertices are not allowed (for example, they could be removed in a preprocessing phase), we prefer to think of the running time of depth-first search as linear in the number of edges. ■

The same method can be applied to graphs represented with adjacency matrices by using the following *visit* procedure:

```
procedure visit(k: integer);
  var t: integer;
  begin
  id:=id+1; val[k]:=id;
  for t:=1 to V do
    if a[k,t] then
      if val[t]=0 then visit(t);
  end;
```

Traveling through an adjacency list translates to scanning through a row in the adjacency matrix, looking for *true* values (which correspond to edges). As before, any edge to a vertex that hasn't been seen before is "followed" via a recursive call. Now, the edges connected to each vertex are examined in a different order, so we get a different depth-first search forest, as shown in Figure 29.7. This underscores

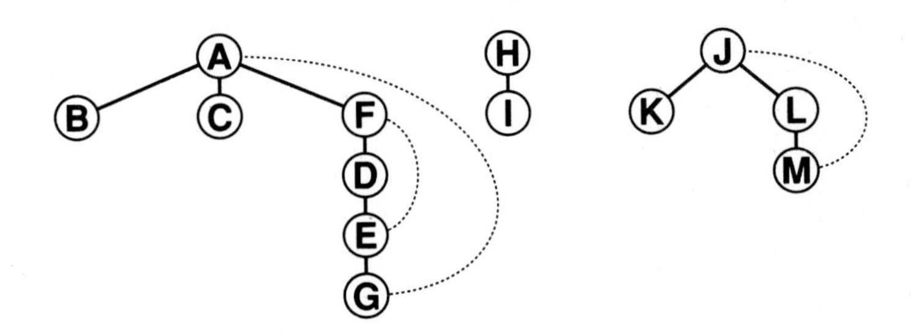

Figure 29.7 Depth-first search forest (matrix representation of the graph).

the point that the depth-first search forest is simply another representation of the graph, one whose particular structure depends both on the search algorithm and the internal representation used.

Property 29.2 *Depth-first search of a graph represented with an adjacency matrix requires time proportional to* V^2.

The proof of this fact is trivial: every bit in the adjacency matrix is checked. ■

Depth-first search immediately solves some basic graph-processing problems. For example, the procedure is based on finding the connected components in turn: the number of connected components is the number of times *visit* is called in the last line of the program. Testing if a graph has a cycle is also a trivial modification of the above program. A graph has a cycle if and only if a nonzero *val* entry is discovered in *visit*. That is, if we encounter an edge pointing to a vertex that we've already visited, then we have a cycle. Equivalently, all the dotted links in the depth-first search trees belong to cycles.

Nonrecursive Depth-First Search

Depth-first search in a graph is a generalization of tree traversal. When invoked on a tree, it is exactly equivalent to tree traversal; for graphs, it corresponds to traversing a tree that spans the graph and that is "discovered" as the search proceeds. As we have seen, the particular tree traversed depends on how the graph is represented.

The recursion in depth-first search can be removed by using a stack, in the same manner as we removed recursion from tree traversal in Chapter 5. For trees, we found that recursion removal led us to an alternate (rather simple) equivalent implementation, and we found a related nonrecursive traversal algorithm (level-order). For graphs, we'll see a similar evolution, ultimately leading (in Chapter 31) to a general-purpose graph-traversal algorithm.

Drawing upon our experience in Chapter 5, we move directly to a stack-based implementation:

```
procedure listdfs;
  var id,k: integer;
      val: array[1..maxV] of integer;
  procedure visit(k: integer);
    var t: link;
    begin
    push(k);
    repeat
      k:=pop;
      id:=id+1; val[k]:=id;
      t:=adj[k];
      while t<>z do
        begin
        if val[t↑.v]=0 then
          begin push(t↑.v); val[t↑.v]:=-1 end;
        t:=t↑.next
        end
    until stackempty
    end;
  begin
  id:=0; stackinit;
  for k:=1 to V do val[k]:=0;
  for k:=1 to V do
    if val[k]=0 then visit(k)
  end;
```

Vertices that have been touched but not yet visited are kept on a stack. To visit a vertex, we traverse its edges and push onto the stack any vertex that has not yet been visited and that is not already on the stack. In the recursive implementation, the bookkeeping for the "partially visited" vertices is hidden in the local variable t in the recursive procedure. We could implement this directly by maintaining pointers (corresponding to t) into the adjacency lists, and so on. Instead, we simply extend the meaning of the *val* entries to help mark vertices that are on the stack: vertices with zero *val* entries have not yet been encountered (as before), those with negative *val* entries are on the stack, and those with positive *val* entries have been visited (all the edges on their adjacency lists have been put on the stack).

Figure 29.8 traces the operation of this stack-based depth-first search procedure as the first four nodes of our sample graph are visited. Each diagram in this figure corresponds to a node visit: the node visited is drawn as a square, and all the edges on its adjacency list are shaded. As before, nodes not yet encountered are unlabeled and shaded, nodes for which the visit is complete are labeled and not shaded, and each node is connected by a darkened edge to the node that caused it

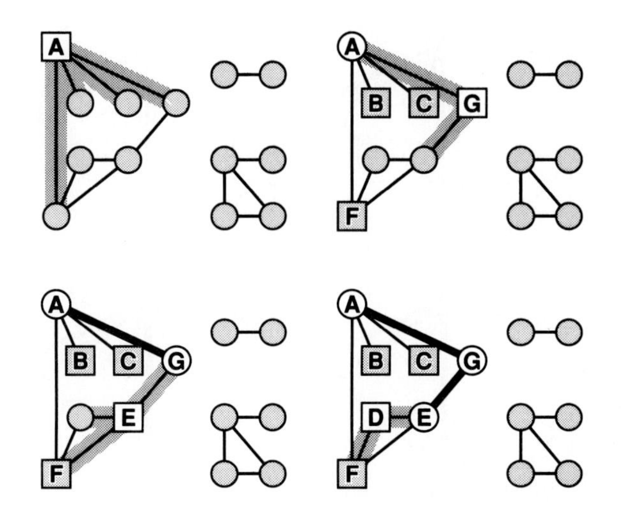

Figure 29.8 Start of depth-first search (graph traversal with a stack).

to be put on the stack. Nodes still on the stack are in squares.

The first node visited is A: the edges AF, AB, AC, and AG are traversed, and F, B, C, and G are put on the stack. Then G is popped (note that it is the *last* node on A's adjacency list), and edges GA and GE are traversed, resulting in E being pushed on the stack (A is not pushed again). Then EG, EF, and ED are traversed and D is pushed onto the stack, etc. Figure 29.9 shows the contents of the stack during the search, and Figure 29.10 shows the continuation of Figure 29.8.

The reader has doubtless noticed that the program above does *not* visit the edges and nodes in precisely the same order as the recursive implementation. That could be achieved, as in Chapter 5, by paying more careful attention to the order in which edges are added to the stack—in the above program we visit the edges on the adjacency list for each node in the reverse of the order in which they appear on the list, since we go through the list, pushing the nodes onto the stack. Thus, if we went through the list in *reverse* order, the nodes would be visited in the same

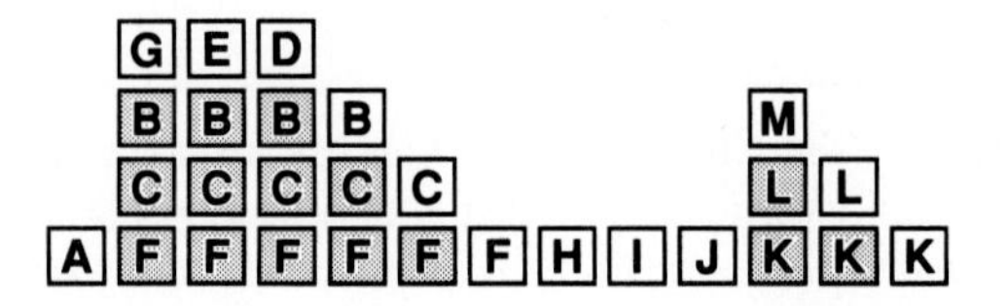

Figure 29.9 Contents of stack during depth-first search.

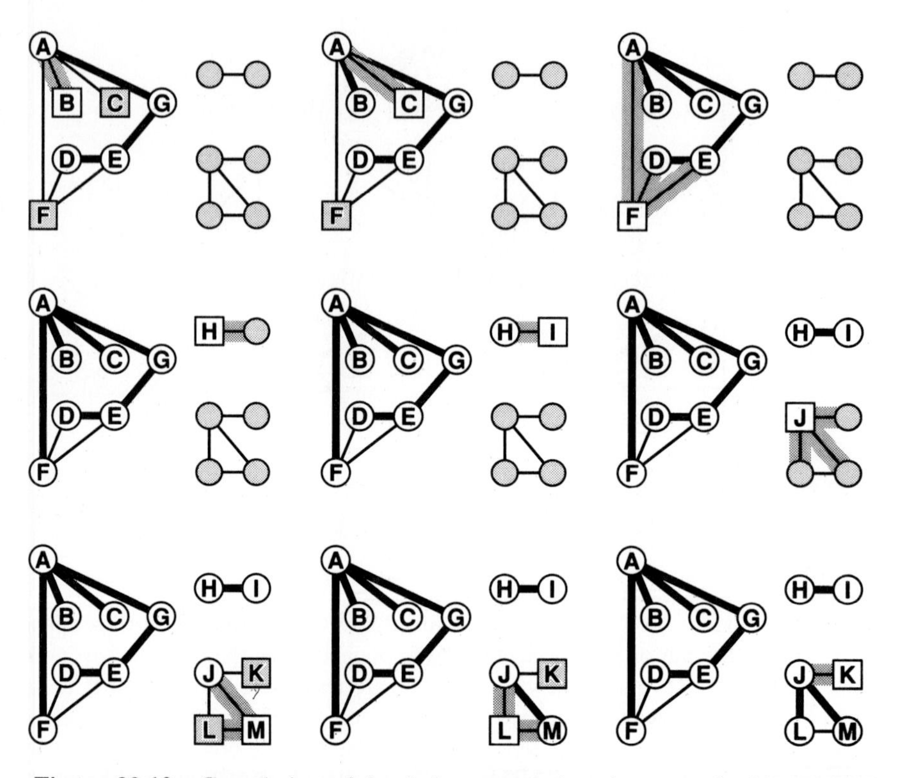

Figure 29.10 Completion of depth-first search (graph traversal with a stack).

order as in the recursive implementation. (This is the same effect as in Chapter 5, where the right subtree is pushed onto the stack before the left in the nonrecursive implementation.) However, for graphs, it is not necessary to adhere strictly to the order in which the edges are visited in the recursive implementation because, as we've seen, several other (arbitrary) factors affect the order in which edges are traversed. As we'll see in Chapter 31, there are a variety of options that allow us to use this flexibility to our advantage.

Breadth-First Search

Just as in tree traversal (see Chapter 4), we can use a *queue* rather than a stack as the data structure to hold vertices. This leads to a second classical graph-traversal algorithm called *breadth-first search*. To implement breadth-first search, we simply change stack operations to queue operations in the stack-based depth-first search program above, as follows:

```
procedure listbfs;
  var id,k: integer;
      val: array[1..maxV] of integer;
  procedure visit(k: integer);
    var t: link;
    begin
    put(k);
    repeat
      k:=get;
      id:=id+1; val[k]:=id;
      t:=adj[k];
      while t<>z do
        begin
        if val[t↑.v]=0 then
          begin put(t↑.v); val[t↑.v]:=-1 end;
        t:=t↑.next
        end
    until queueempty
    end;
  begin
  id:=0; queueinitialize;
  for k:=1 to V do val[k]:=0;
  for k:=1 to V do
    if val[k]=0 then visit(k)
  end;
```

Changing the data structure in this way affects the order in which the nodes are visited. For our small sample graph, the edges are visited in the order AF AC AB AG FA FE FD CA BA GE GA DF DE EG EF ED HI IH JK JL JM KJ LJ LM MJ ML. The contents of the queue during the traversal are shown in Figure 29.11.

As with depth-first search, we can define a forest from the edges that lead us for the first time to each node, as shown in Figure 29.12. Breadth-first search corresponds to traversing the trees in this forest in level order.

In both algorithms, we may think of the vertices as being divided into three classes: *tree* (or *visited*) vertices, those that have been taken off the data structure; *fringe* vertices, those adjacent to tree vertices but not yet visited; and *unseen* vertices, those that haven't been encountered at all yet. If each tree vertex is connected to the edge that caused it to be added to the data structure (the darkened edges in Figures 29.8 and 29.10), then these edges form a tree.

To search a connected component of a graph systematically (implement a *visit* procedure), we begin with one vertex on the fringe, all others unseen, and perform the following step until all vertices have been visited: "move one vertex (call it x)

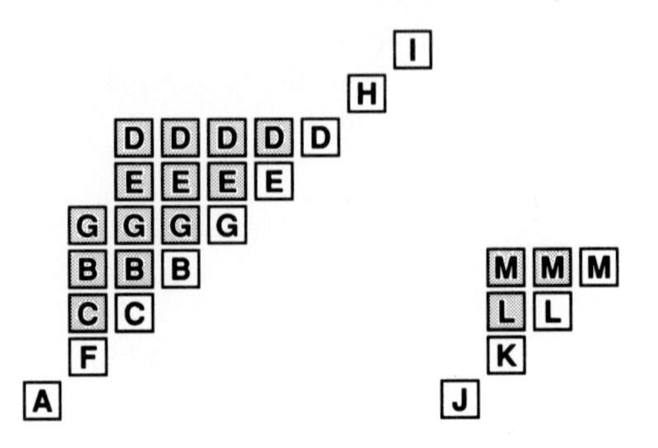

Figure 29.11 Contents of queue during breadth-first search.

from the fringe to the tree, and put any unseen vertices adjacent to x on the fringe." Graph traversal methods differ in how it is decided which vertex should be moved from the fringe to the tree. For depth-first search we want to choose the vertex from the fringe that was most recently encountered; this corresponds to using a stack to hold vertices on the fringe. For breadth-first search we want to choose the vertex from the fringe that was *least* recently encountered; this corresponds to using a queue to hold vertices on the fringe. In Chapter 31, we'll see the effect of using a *priority queue* for the fringe.

The contrast between depth-first and breadth-first search is quite evident when we consider a larger graph. Figure 29.13 shows the operation of depth-first search in a larger graph, one-third and two-thirds of the way through the process; Figure 29.14 is the corresponding picture for breadth-first search. In these diagrams, tree vertices and edges are blackened, unseen vertices are shaded, and fringe vertices are white.

In both cases, the search starts at the node at the bottom left. Depth-first search wends its way through the graph, storing on the stack the points where other paths

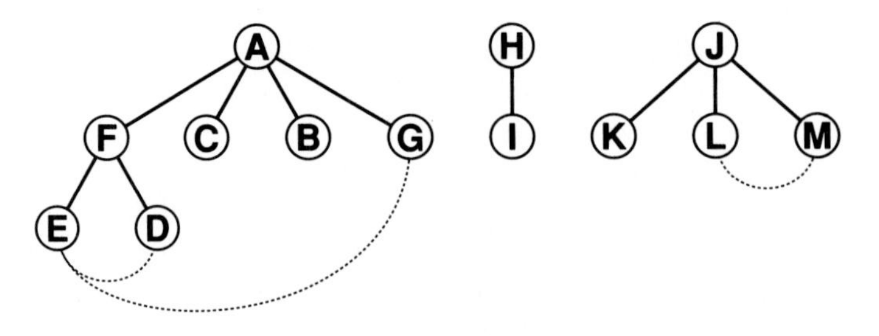

Figure 29.12 Breadth-first search forest.

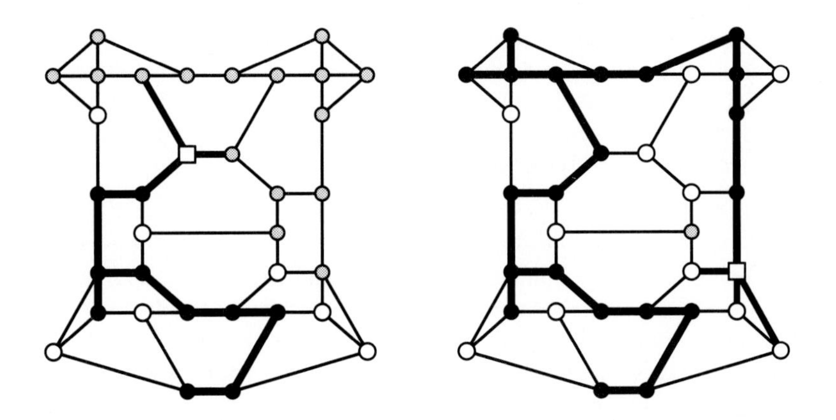

Figure 29.13 Depth-first search in a larger graph.

branch off; breadth-first search "sweeps through" the graph, using a queue to remember the frontier of visited places. Depth-first search "explores" the graph by looking for new vertices far away from the start point, taking closer vertices only when dead ends are encountered; breadth-first search completely covers the area close to the starting point, moving farther away only when everything close has been looked at. Again, the order in which the nodes are visited depends largely upon the order in which the edges appear in the input and upon the effects of this ordering on the order in which vertices appear on the adjacency lists.

Beyond these operational differences, it is interesting to reflect on the fundamental differences in the implementations of these methods. Depth-first search

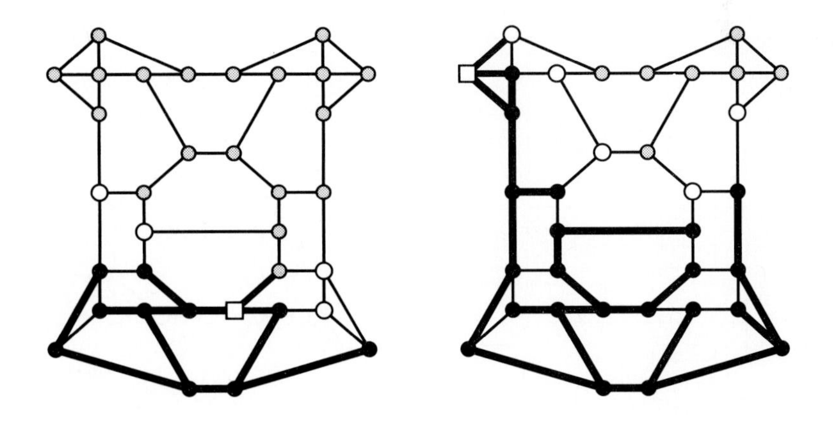

Figure 29.14 Breadth-first search in a larger graph.

is very simply expressed recursively (because its underlying data structure is a stack), and breadth-first search admits a very simple nonrecursive implementation (because its underlying data structure is a queue). In Chapter 31 we see that the actual underlying data structure for graph algorithms is a priority queue, and this implies a wealth of interesting properties and algorithms.

Mazes

Our systematic way of examining every vertex and edge of a graph has a distinguished history: depth-first search was first stated formally hundreds of years ago as a method for traversing mazes. For example, at left in Figure 29.15 is a popular maze, and at right is the graph constructed by putting a vertex at each point where there is more than one path to take, and then connecting the vertices according to the paths. This is significantly more complicated than early English garden mazes, which were constructed as paths through tall hedges. In these mazes, all walls were connected to the outer walls, so that gentlemen and ladies could stroll in and clever ones could find their way out by simply keeping their right hand on the wall (laboratory mice have reportedly learned this trick). When independent inside walls can occur, a more sophisticated strategy is necessary to get around in a maze, which leads us to depth-first search.

To use depth-first search to get from one place to another in a maze, we use *visit*, starting at the vertex on the graph corresponding to our starting point. Each time *visit* "follows" an edge via a recursive call, we walk along the corresponding path in the maze. The trick to getting around is that we must walk *back* along the path that we used to enter each vertex when *visit* finishes for that vertex. This puts us back at the vertex one step higher up in the depth-first search tree, ready to

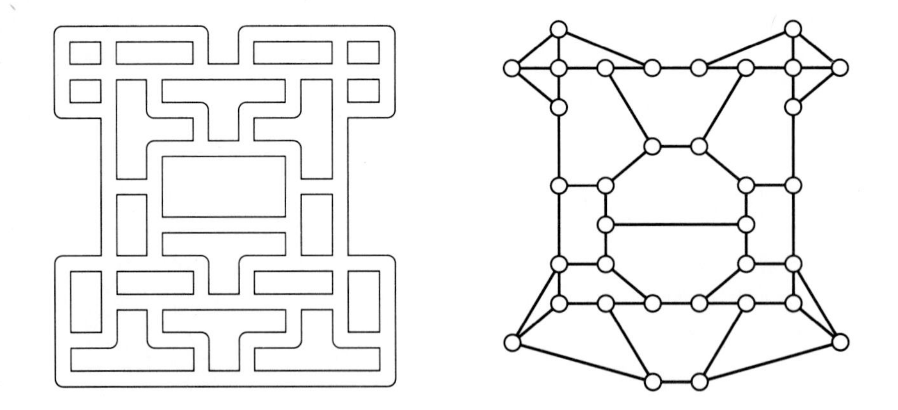

Figure 29.15 A maze and an associated graph.

follow its next edge. (This process exactly mimics traversal of the depth-first search tree for the graph.) Depth-first search is appropriate for one person looking for something in a maze because the "next place to look" is always close by; breadth-first search is more like a group of people looking for something by fanning out in all directions.

Perspective

In the chapters that follow we'll consider a variety of graph algorithms largely aimed at determining connectivity properties of both undirected and directed graphs. These algorithms are fundamental ones for processing graphs, but are only an introduction to the subject of graph algorithms. Many interesting and useful algorithms have been developed that are beyond the scope of this book, and many interesting problems have been studied for which good algorithms have not yet been found.

Some very efficient algorithms have been developed that are much too complicated to present here. For example, it is possible to determine efficiently whether or not a graph can be drawn on the plane without any intersecting lines. This problem is called the *planarity* problem, and no efficient algorithm for solving it was known until 1974, when R. E. Tarjan developed an ingenious (but quite intricate) algorithm for solving the problem in linear time, using depth-first search.

Some graph problems that arise naturally and are easy to state seem to be quite difficult, and no good algorithms are known to solve them. For example, no efficient algorithm is known for finding the minimum-cost tour that visits each vertex in a weighted graph. This problem, called the *traveling salesman problem*, belongs to a large class of difficult problems that we'll discuss in more detail in Chapter 45. Most experts believe that no efficient algorithms exist for these problems.

Other graph problems may well have efficient algorithms, though none have been found. An example of this is the *graph isomorphism* problem: determine whether two graphs could be made identical by renaming vertices. Efficient algorithms are known for this problem for many special types of graphs, but the general problem remains open.

In short, there is a wide spectrum of problems and algorithms for dealing with graphs. We certainly can't expect to solve every problem that comes along, and even some problems that appear to be simple are still baffling the experts. But many relatively easy problems do arise quite often, and the graph algorithms we will study serve well in a great variety of applications.

□

Exercises

1. Which undirected graph representation is most appropriate for determining quickly whether or not a vertex is isolated (is connected to no other vertices)?
2. Suppose depth-first search is used on a binary search tree and the right edge taken before the left out of each node. In what order are the nodes visited?
3. How many bits of storage are required to represent the adjacency matrix for an undirected graph with V nodes and E edges, and how many are required for the adjacency list representation?
4. Draw a graph that cannot be written down on a piece of paper without two edges crossing.
5. Write a program to delete an edge from a graph represented with adjacency lists.
6. Write a version of *adjlist* that keeps the adjacency lists in sorted order of vertex index. Discuss the merits of this approach.
7. Draw the depth-first search forests that result for the example in the text when *dfs* scans the vertices in reverse order (from V down to 1), for both representations.
8. Exactly how many times is *visit* called in the depth-first search of an undirected graph, in terms of the number of vertices V, the number of edges E, and the number of connected components C?
9. Give the adjacency lists produced if the edges for the sample graph are read in the reverse order of that used to make the structure in Figure 29.4.
10. Give the depth-first search forest for the sample graph in the text when the recursive *listdfs* routine is used on the adjacency list of the previous exercise.

30

Connectivity

☐ The fundamental depth-first search procedure in the previous chapter finds the connected components of a given graph; in this chapter we'll examine related algorithms and problems concerning other graph-connectivity properties.

After looking at a few direct applications of depth-first search to get connectivity information, we'll look at a generalization of connectivity called *biconnectivity*. Here we are interested in knowing if there is more than one way to get from one vertex in a graph to another. A graph is *biconnected* if and only if there are at least two different paths connecting each pair of vertices. Thus even if one vertex and all the edges touching it are removed, the graph is still connected. If it is important that a graph *be* connected for some application, it might also be important that it *stay* connected. The solution to this problem is a rather more complicated algorithm than the traversal algorithms of the previous chapter, but is still based upon depth-first search.

One particular version of the connectivity problem that arises frequently involves a dynamic situation where edges are added to the graph one by one, interspersed with queries as to whether or not two particular vertices belong to the same connected component. This is a well-studied problem, and we'll look in detail at two related "classical" algorithms for it. Not only are the methods simple and widely applicable, but they also illustrate how very difficult to analyze simple algorithms can be. The problem is sometimes called the "union-find" problem, a nomenclature that comes from the application of the algorithms to processing simple operations on sets of elements.

Connected Components

Any of the graph-traversal methods of the previous chapter can be used to find the connected components of a graph, since they're all based on the same general strategy of visiting all the nodes in a connected component before moving to the next. An easy way to print out the connected components is to modify one of

k	*1*	*2*	*3*	*4*	*5*	*6*	*7*	*8*	*9*	*10*	*11*	*12*	*13*
name[k]	*A*	*B*	*C*	*D*	*E*	*F*	*G*	*H*	*I*	*J*	*K*	*L*	*M*
val[k]	*1*	*7*	*6*	*5*	*3*	*2*	*4*	*8*	*9*	*10*	*11*	*12*	*13*
inval[k]	*-1*	*6*	*5*	*7*	*4*	*3*	*2*	*-8*	*9*	*-10*	*11*	*12*	*13*

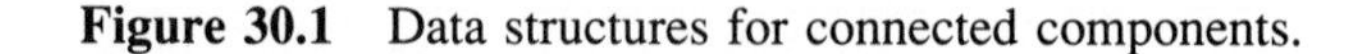

Figure 30.1 Data structures for connected components.

the recursive depth-first search programs to have *visit* print out the vertex being visited (say, by inserting *write(name(k))* just before exiting), then print out some indication that a new connected component is to start just before the (nonrecursive) call to *visit* in *dfs* (say, by inserting two *writeln* statements). This technique would produce the following output when *dfs* is used on the adjacency list representation of our sample graph (Figure 29.1):

G D E F C B A
I H
K M L J

Other variations, such as the adjacency-matrix version of *visit*, stack-based depth-first search, and breadth-first search, can compute the same connected components (of course), but the vertices will be printed out in a different order.

Extensions to do more complicated processing on the connected components are straightforward. For example, by simply inserting *inval*[*id*]=*k* after *val*[*k*]=*id* we get the "inverse" of the *val* array, whose *id*th entry is the index of the *id*th vertex visited. Vertices in the same connected components are contiguous in this array, the index of each new connected component given by the value of *id* each time *visit* is called in *dfs*. These values can be stored or used to mark delimiters in *inval* (for example, the first entry in each connected component could be made negative).

Figure 30.1 shows the values taken by these arrays for our example if the adjacency list version of *dfs* were modified in this way. Typically, it is worthwhile to use such techniques to divide a graph into its connected components for later processing by more sophisticated algorithms, in order to free them from the details of dealing with disconnected components.

Biconnectivity

It is sometimes useful to design more than one route between points on a graph, so as to handle possible failures at the connection points (vertices). Thus we can fly from Providence to Princeton even if New York is snowed in by going through Philadelphia instead. The main communications lines in an integrated circuit are

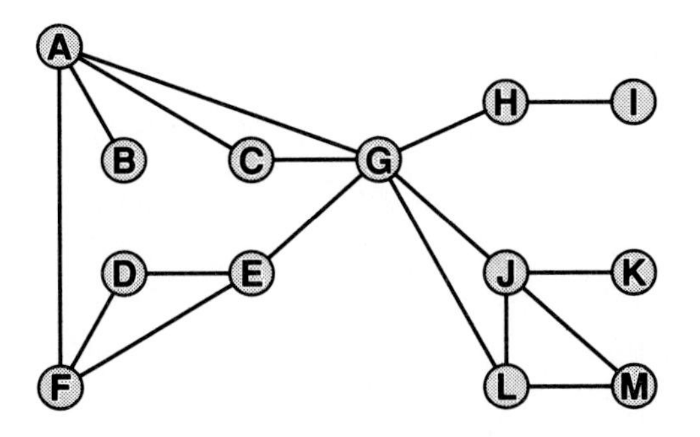

Figure 30.2 A graph that is not biconnected.

often biconnected, so that the rest of the circuit still can function if one component fails. Another application (not particularly realistic but a natural illustration of the concept) is to imagine a wartime situation where we can make it so that an enemy must bomb at least two stations in order to cut our rail lines.

An *articulation point* in a connected graph is a vertex that, if deleted, would break the graph into two or more pieces. A graph with no articulation points is said to be *biconnected.* In a biconnected graph, two distinct paths connect each pair of vertices. A graph that is not biconnected divides into *biconnected components,* sets of nodes mutually accessible via two distinct paths.

Figure 30.2 shows a graph that is connected but not biconnected. (This graph is obtained from the graph of the previous chapter by adding the edges GC, GH, JG, and LG. In our examples, we'll assume that these four edges are added at the end of the input in the order given, so that (for example) the adjacency lists are similar to those in Figure 29.4 with eight new entries in the lists to reflect the four new edges.) The articulation points of this graph are A (because it connects B to the rest of the graph), H (because it connects I to the rest of the graph), J (because it connects K to the rest of the graph), and G (because the graph would fall into three pieces if G were deleted). There are six biconnected components: {A C G D E F}, {G J L M}, and the individual nodes B, H, I, and K.

Determining the articulation points turns out to be a simple extension of depth-first search. To see this, consider the depth-first search tree for this graph, shown in Figure 30.3. Deleting node E does not disconnect the graph because G and D both have dotted links that point above E, giving alternate paths from them to F (E's parent in the tree). On the other hand, deleting G does disconnect the graph because there are no such alternate paths from L or H to E (G's parent).

A vertex x is not an articulation point if every child y has some node lower in the tree connected (via a dotted link) to a node higher in the tree than x, thus providing an alternate connection from x to y. This test doesn't quite work for the root of the depth-first search tree, since there are no nodes "higher in the tree."

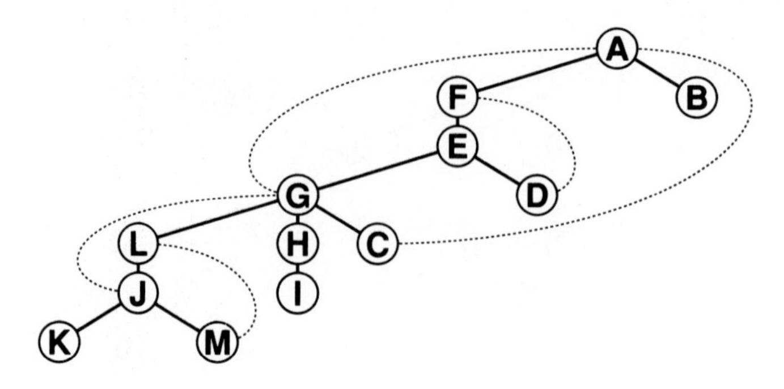

Figure 30.3 Depth-first search for biconnectivity.

The root is an articulation point if it has two or more children, since the only path connecting children of the root goes through the root. These tests are easily incorporated into depth-first search by changing the node-visit procedure into a function that returns the highest point in the tree (lowest *val* value) seen during the search, as follows:

```
function visit(k: integer): integer;
  var t: link;
        m,min: integer;
  begin
  id:=id+1; val[k]:=id; min:=id;
  t:=adj[k];
  while t<>z do
    begin
    if val[t↑.v]=0 then
      begin
      m:=visit(t↑.v);
      if m<min then min:=m;
      if m>=val[k] then write(name(k));
      end
    else  if val[t↑.v]<min then min:=val[t↑.v];
    t:=t↑.next
    end
  visit:=min;
  end;
```

This procedure recursively determines the highest point in the tree reachable (via a dotted link) from any descendant of vertex k, and uses this information to determine if k is an articulation point. Normally this calculation simply involves testing whether or not the minimum value reachable from a child is higher up in the tree. However, we need an extra test to determine whether k is the root of a depth-first search tree (or, equivalently, whether this is the first call to *visit* for the connected component containing k), since we're using the same recursive program for both cases. This test is properly performed outside the recursive *visit* and thus does not appear in the code above.

Property 30.1 *The biconnected components of a graph can be found in linear time.*

Although the program above simply prints out the articulation points, it is easily extended, as we did for connected components, to do additional processing on the articulation points and biconnected components. Since it is a depth-first search procedure, the running time is proportional to $V + E$. (A similar program based on an adjacency matrix would run in $O(V^2)$ steps. ■

In addition to the kinds of application mentioned above, where biconnectedness is used to improve reliability, it can be helpful in decomposing large graphs into manageable pieces. It is obvious that a very large graph may be processed one connected component at a time for many applications; it is somewhat less obvious but occasionally just as useful that a graph can sometimes be processed one biconnected component at a time.

Union-Find Algorithms

In some applications we wish to know simply whether or not a vertex x is connected to a vertex y in a graph; the actual path connecting them may not be relevant. This problem has been carefully studied in recent years; the efficient algorithms that have been developed are of independent interest because they can also be used for processing *sets* (collections of objects).

Graphs correspond to sets of objects in a natural way: vertices correspond to objects and edges mean "is in the same set as." Thus, the sample graph in the previous chapter corresponds to the sets {A B C D E F G}, {H I} and {J K L M}. Each connected component corresponds to a different set. For sets, we're interested in the fundamental question "is x in the same set as y?" This clearly corresponds to the fundamental graph question "is vertex x connected to vertex y?"

Given a set of edges, we can build an adjacency list representation of the corresponding graph and use depth-first search to assign to each vertex the index of its connected component, and so questions of the form "is x connected to y?" can be answered with just two array accesses and a comparison. The extra twist in the methods we consider here is that they are *dynamic*: they can accept new edges arbitrarily intermixed with questions and answer the questions correctly using the

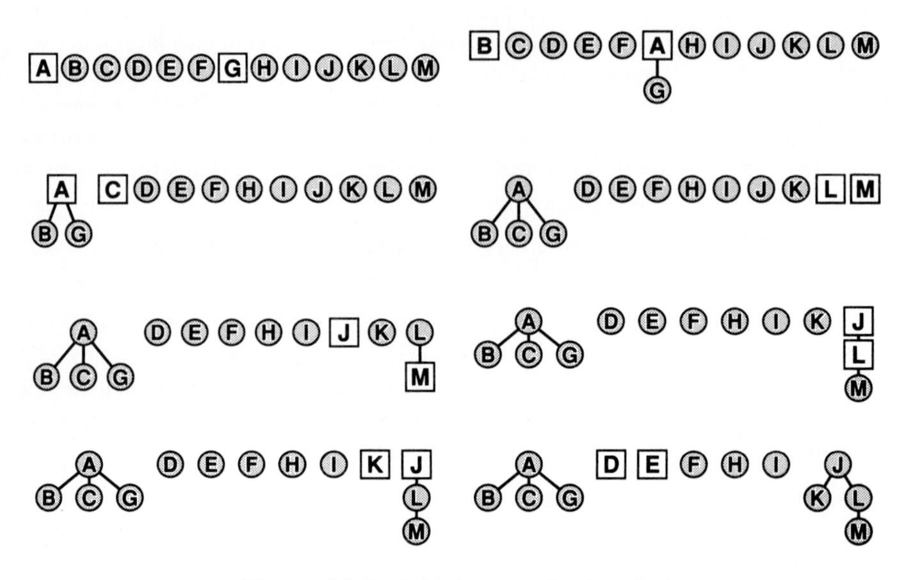

Figure 30.4 Initial steps of union-find.

information received. From the correspondence with the set problem, the addition of a new edge is called a *union* operation and the queries are called *find* operations.

Our objective is to write a function that can check if two vertices x and y are in the same set (or, in the graph representation, the same connected component) and, if not, can put them in the same set (put an edge between them in the graph). Instead of building a direct adjacency-list or other representation of the graph, we'll gain efficiency by using an internal structure specifically oriented towards supporting the *union* and *find* operations. This internal structure will be a *forest of trees*, one for each connected component. We need to be able to find out if two vertices belong to the same tree and to be able to combine two trees into one. It turns out that both of these operations can be implemented efficiently.

To illustrate how this algorithm works, we'll look at the forest constructed when the edges from the sample graph of Figure 30.1 are processed in the order AG AB AC LM JM JL JK ED FD HI FE AF GE GC GH JG LG. The first seven steps are shown in Figure 30.4. Initially, all nodes are in separate trees. Then the edge AG causes a two-node tree to be formed with A at the root. (This choice is arbitrary —we could equally well have put G at the root.) The edges AB and AC add B and C to this tree in the same way. Then the edges LM, JM, JL, and JK build a tree containing J, K, L, and M that has a slightly different structure (note that JL contributes nothing, since LM and JM put L and J in the same component).

Figure 30.5 shows the completion of the process. The edges ED, FD, and HI build two more trees, leaving a forest with four trees. This forest indicates that the

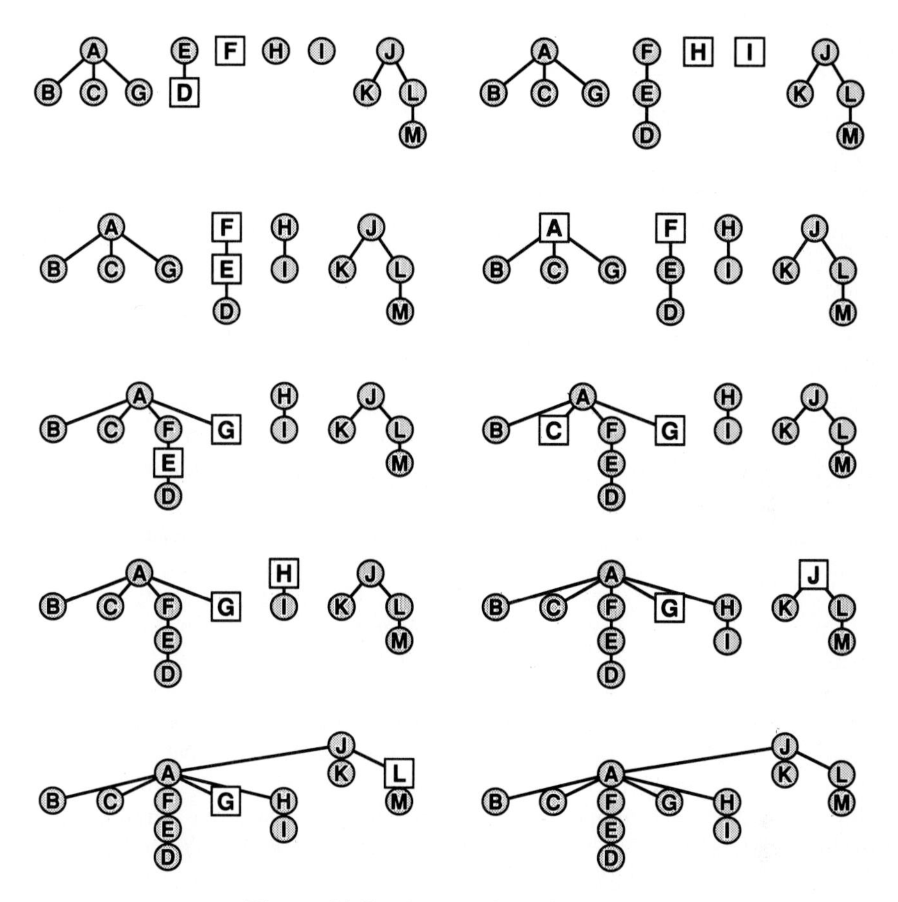

Figure 30.5 Completion of union-find.

edges processed to this point describe a graph with four connected components, or, equivalently, that the set union operations processed to this point have led to four sets {A B C G}, {J K L M}, {D E F} and {H I}. Now the edge FE doesn't contribute anything to the structure, since F and E are in the same component, but the edge AF combines the first two trees; then GE and GC don't contribute anything, but GH and JG result in everything being combined into one tree.

It must be emphasized that, unlike depth-first search trees, the only relationship between these union-find trees and the underlying graph with the given edges is that they divide the vertices into sets in the same way. For example, there is no correspondence between the paths that connect nodes in the trees and the paths that connect nodes in the graph.

The *union* and *find* operations are very easily implemented by using the "parent link" representation for the trees (see Chapter 4):

```
function find(x,y: integer; union: boolean): boolean;
  var i,j: integer;
  begin
  i:=x; while dad[i]>0 do i:=dad[i];
  j:=y; while dad[j]>0 do j:=dad[j];
  if union and (i<>j) then dad[j]:=i;
  find:=(i<>j)
  end;
```

The array *dad*[*1..V*] contains, for each vertex, the index of its parent (with a 0 entry for nodes that are at the root of a tree). To find the parent of a vertex j, we simply set j:=*dad*[j], and to find the root of the tree to which j belongs, we repeat this operation until we reach 0.

The function *find* returns *true* if the two given vertices are in the same component. If they are not in the same component and the *union* flag is set, they are put into the same component. The method is simple. Use the *dad* array to get to the root of the tree containing each vertex, then check to see if the roots are the same. To merge the tree rooted at j with the tree rooted at i, we simply set *dad*[j]=i.

Figure 30.6 shows the contents of the data structure during this process. As usual, we assume that functions *index* and *name* are available to translate between vertex names and integers between 1 and V: each table entry is the name of the corresponding *dad* array entry. For example, one would test whether a vertex named x is in the same component as a vertex named y (without introducing an edge between them) with the function call *find*(*index*(x),*index*(y),*false*).

The algorithm described above has bad worst-case performance because the trees formed can be degenerate. For example, taking in order the edges AB BC CD DE EF FG GH HI IJ ... YZ produces a long chain with Z pointing to Y, Y pointing to X, etc. This kind of structure takes time proportional to V^2 to build, and has time proportional to V for an average equivalence test.

Several methods have been suggested to deal with this problem. One natural method, which may have already occurred to the reader, is to try to do the "right" thing when merging two trees, rather than arbitrarily setting *dad*[j]=i. When a tree rooted at i is to be merged with a tree rooted at j, one of the nodes must remain a root and the other (and all its descendants) must go one level down in the tree. To minimize the distance to the root for the most nodes, it makes sense to take as the root the node with more descendants. This idea, called *weight balancing*, is easily implemented by maintaining the size of each tree (number of descendants of the root) in the *dad* array entry for each root node, encoded as a nonpositive number so that the root node can be detected when traveling up the tree in *find*.

	A	B	C	D	E	F	G	H	I	J	K	L	M
AG							A						
AB		A					A						
AC		A	A				A						
LM		A	A				A						L
JM		A	A				A					J	L
JL		A	A				A					J	L
JK		A	A				A				J	J	L
ED		A	A	E			A				J	J	L
FD		A	A	E	F		A				J	J	L
HI		A	A	E	F		A		H		J	J	L
FE		A	A	E	F		A		H		J	J	L
AF		A	A	E	F	A	A		H		J	J	L
GE		A	A	E	F	A	A		H		J	J	L
GC		A	A	E	F	A	A		H		J	J	L
GH		A	A	E	F	A	A	A	H		J	J	L
JG	J	A	A	E	F	A	A	A	H		J	J	L
LG	J	A	A	E	F	A	A	A	H		J	J	L

Figure 30.6 Union-find data structure.

Ideally, we would like every node to point directly to the root of its tree. No matter what strategy we use, however, achieving this ideal would require examining at least all the nodes in one of the two trees to be merged, and this could be quite a lot compared to the relatively few nodes on the path to the root that *find* usually examines. But we can approach the ideal by making all the nodes we do examine point to the root! This seems at first blush a drastic step, but it is easy to do, and there is nothing sacrosanct about the structure of these trees: if they can be modified to make the algorithm more efficient, we should do so. This method, called *path compression*, is easily implemented by making another pass through each tree after the root has been found and setting the *dad* entry of each vertex encountered along the way to point to the root.

The combination of weight balancing and path compression ensures that the algorithms will run very quickly. The following implementation shows that the extra code involved is a small price to pay to guard against degenerate cases.

```
function find(x,y: integer; union: boolean): boolean;
  var i,j,t: integer;
  begin
  i:=x; while dad[i]>0 do i:=dad[i];
  j:=y; while dad[j]>0 do j:=dad[j];
  while dad[x]>0 do
    begin t:=x; x:=dad[x]; dad[t]:=i end;
  while dad[y]>0 do
    begin t:=y; y:=dad[y]; dad[t]:=j end;
  if union and (i<>j) then
    if dad[j]<dad[i]
      then begin dad[j]:=dad[j]+dad[i]-1; dad[i]:=j end
      else begin dad[i]:=dad[i]+dad[j]-1; dad[j]:=i end;
  find:=(i<>j)
  end;
```

The *dad* array is assumed to be initialized to 0. (We'll assume in later chapters that this is done in a separate procedure *findinit*.) Figure 30.7 shows the first eight steps when this method is applied to our example data, and Figure 30.8 shows the completion of the process. The average path length of the resulting tree is $31/13 \approx 2.38$, as compared to $38/13 \approx 2.92$ for Figure 30.5. For the

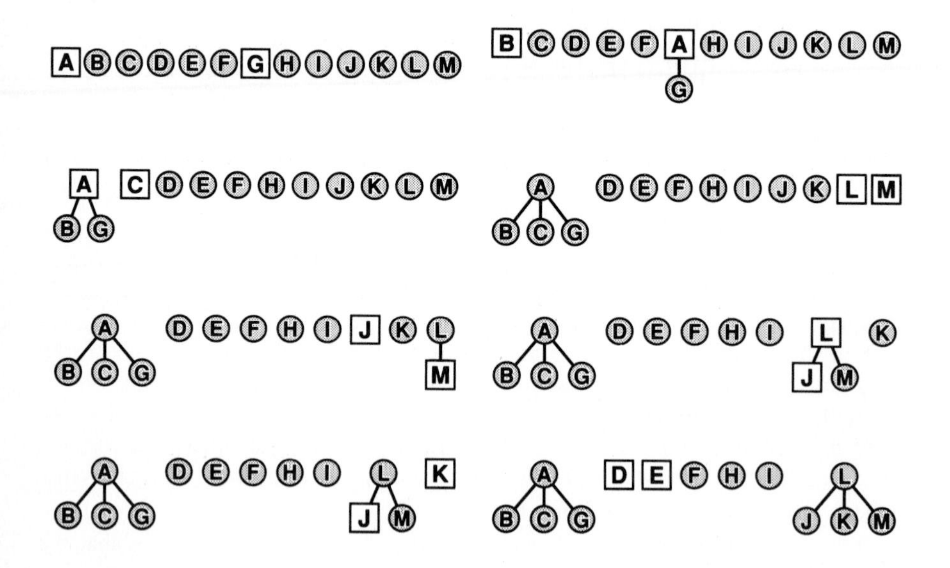

Figure 30.7 Initial steps of union-find (weighted, with path compression).

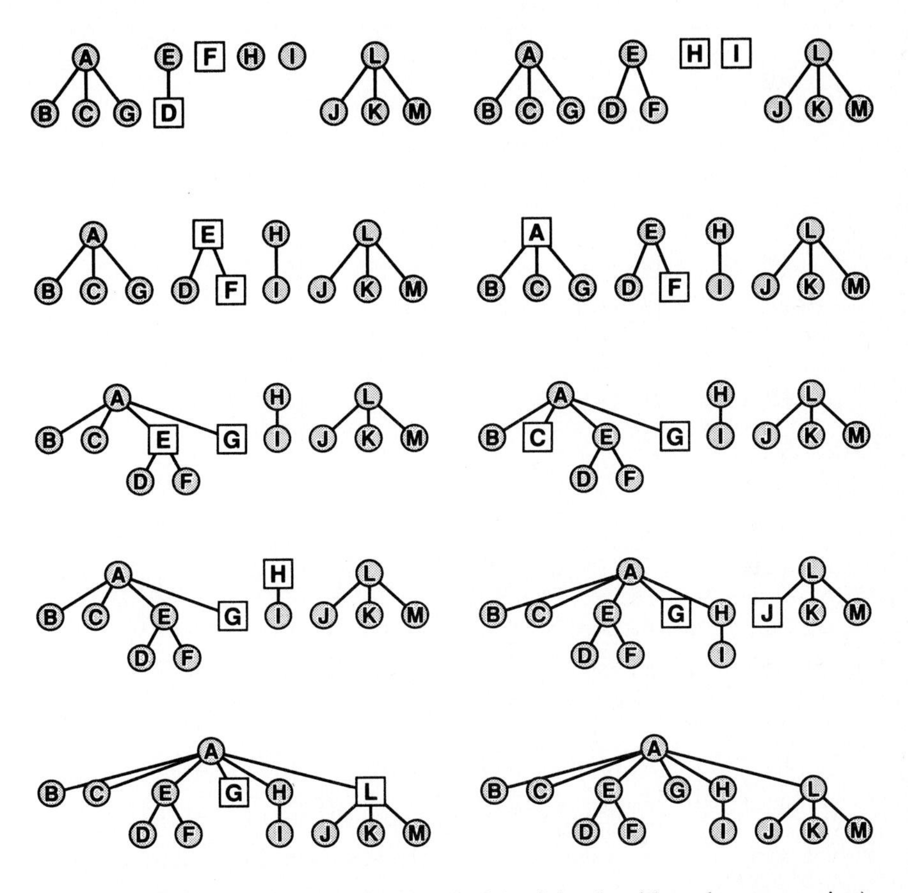

Figure 30.8 Completion of union-find (weighted, with path compression).

first five edges, the resulting forest is the same as in Figure 30.4; the last three edges, however, give a "flat" tree containing J, K, L, and M because of the weight balancing rule. The forests are actually so flat in this example that all vertices involved in *union* operations are at the root or just below—path compression is not used. Path compression would make the trees flatter still. For example, if the last *union* were FJ rather than GJ, then F would also be a child of A at the end.

Figure 30.9 gives the contents of the *dad* array as this forest is constructed. For clarity in this table, each positive entry i is replaced by the ith letter of the alphabet (the name of the parent), and each negative entry is complemented to give a positive integer (the weight of the tree).

Several other techniques have been developed to avoid degenerate structures. For example, path compression has the disadvantage of requiring another pass up

	A	B	C	D	E	F	G	H	I	J	K	L	M
AG	1						A						
AB	2	A					A						
AC	3	A	A				A						
LM	3	A	A				A					1	L
JM	3	A	A				A			L		2	L
JL	3	A	A				A			L		2	L
JK	3	A	A				A			L	L	3	L
ED	3	A	A	E	1		A			L	L	3	L
FD	3	A	A	E	2	E	A			L	L	3	L
HI	3	A	A	E	2	E	A	1	H	L	L	3	L
FE	3	A	A	E	2	E	A	1	H	L	L	3	L
AF	6	A	A	E	A	E	A	1	H	L	L	3	L
GE	6	A	A	E	A	E	A	1	H	L	L	3	L
GC	6	A	A	E	A	E	A	1	H	L	L	3	L
GH	8	A	A	E	A	E	A	A	H	L	L	3	L
JG	12	A	A	E	A	E	A	A	H	L	L	A	L
LG	12	A	A	E	A	E	A	A	H	L	L	A	L

Figure 30.9 Union-find data structure (weighted, with path compression).

through the tree. Another technique, called *halving*, makes each node point to its grandparent on the way up the tree. Still another technique, *splitting*, is like halving, but is applied only to every other node on the search path. Either of these can be used in combination with weight balancing or with *height balancing*, which is similar but uses tree height instead of tree size to decide which way to merge trees.

How is one to choose from among all these methods? And exactly how "flat" are the trees produced? Analysis for this problem is quite difficult because the performance depends not only on the V and E parameters, but also on the number of *find* operations and, what's worse, on the order in which the *union* and *find* operations appear. Unlike sorting, where the actual files arising in practice are quite often close to "random," it's hard to see how to model graphs and request patterns that might appear in practice. For this reason, algorithms that do well in the worst case are normally preferred for union-find (and other graph algorithms), though this may be an over-conservative approach.

Even if only the worst case is being considered, analyzing union-find algorithms is extremely complex and intricate. This can be seen even from the nature of the results, which nonetheless do give us clear indications of how the algorithms will perform in a practical situation.

Property 30.2 *If either weight balancing or height balancing is used in combination with compression, halving, or splitting, then the total number of operations required to build up a structure using E edges is almost (but not quite) linear.*

Precisely, the number of operations required is proportional to $E\alpha(E)$, where $\alpha(E)$ is a function that grows so slowly that $\alpha(E) < 4$ unless E is so large that taking $\lg E$, then taking lg of the result, then taking lg of that result, and repeating up to 16 times still gives a number bigger than 1. This is a stunningly large number; for all practical purposes, it is safe to assume that the average amount of time to execute each *union* and *find* operation is constant. This result is due to R. E. Tarjan, who further showed that *no* algorithm for this problem (from a certain general class) can do better that $E\alpha(E)$, so that this function is intrinsic to the problem. ■

An important practical application of union-find algorithms is to determine whether a graph with V vertices and E edges is connected in space proportional to V (and almost linear time). This is an advantage over depth-first search in some situations: here we don't need to ever store the edges. Thus connectivity of a graph with thousands of vertices and millions of edges can be determined with one quick pass through the edges.

□

Exercises

1. Give the articulation points and the biconnected components of the graph formed by deleting GJ and adding IK to our sample graph.
2. Draw the depth-first search tree for the graph described in Exercise 1.
3. What is the minimum number of edges required to make a biconnected graph with V vertices?
4. Write a program to print out the biconnected components of a graph.
5. Draw the union-find forest constructed for the example in the text, but assume that *find* is changed to set $a[i]=j$ rather than $a[j]=i$.
6. Solve the previous exercise, assuming further that path compression is used.
7. Draw the union-find forests constructed for the edges AB BC CD DE EF ... YZ, assuming first that weight balancing without path compression is used, then that path compression without weight balancing is used.
8. Solve the previous exercise, assuming that both path compression and weight balancing are used.
9. Implement the union-find variants described in the text, and empirically determine their comparative performance for 1000 *union* operations with both arguments random integers between 1 and 100.
10. Write a program to generate a random connected graph on V vertices by generating random pairs of integers between 1 and V. Estimate how many edges are needed to produce a connected graph as a function of V.

31

Weighted Graphs

☐ We often want to model practical problems using graphs in which *weights* or *costs* are associated with each edge. In an airline map where edges represent flight routes, these weights might represent distances or fares. In an electric circuit where edges represent wires, the length or cost of the wire are natural weights to use. In a job-scheduling chart, weights can represent time or cost of performing tasks or of waiting for tasks to be performed.

Questions entailing minimizing costs naturally arise for such situations. In this chapter, we'll examine algorithms for two such problems in detail: "find the lowest-cost way to connect all of the points," and "find the lowest-cost path between two given points." The first, which is obviously useful for graphs representing something like an electric circuit, is called the *minimum spanning tree* problem; the second, which is obviously useful for graphs representing something like an airline route map, is called the *shortest-path* problem. These problems are representative of a variety of problems that arise on weighted graphs.

Our algorithms involve searching through the graph, and sometimes our intuition is supported by thinking of the weights as distances: we speak of "the closest vertex to x," etc. In fact, this bias is built into the nomenclature for the shortest-path problem. Despite this, it is important to remember that the weights need not be proportional to a distance at all; they might represent time or cost or something else entirely different. When the weights actually *do* represent distances, other algorithms may be appropriate. This issue is discussed in further detail at the end of the chapter.

Figure 31.1 shows a sample weighted undirected graph. It is obvious how to represent weighted graphs: in the adjacency-matrix representation, the matrix can contain edge weights rather than boolean values, and in the adjacency-structure representation, a field can be added to each list element (which represents an edge) for the weights. We assume that all of the weights are positive. Some algorithms can be adapted to handle negative weights, but they become significantly more

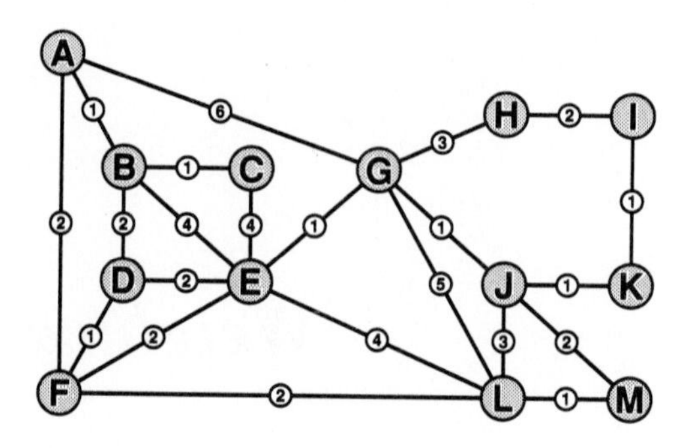

Figure 31.1 A weighted undirected graph.

complicated. In other cases, negative weights change the nature of the problem in an essential way and require far more sophisticated algorithms than those considered here. For an example of the type of difficulty that can arise, consider the situation where the sum of the weights of the edges around a cycle is negative: an infinitely short path could be generated by simply spinning around the cycle.

Several "classical" algorithms have been developed for the minimum-spanning-tree and shortest-path problems. These methods are among the most well-known and most heavily used algorithms in this book. As we have seen before when studying old algorithms, the classical methods provide a general approach, but modern data structures help provide compact and efficient implementations. In this chapter, we'll see how to use priority queues in a generalization of the graph-traversal methods of Chapter 29 to solve both problems efficiently for sparse graphs; we'll see the relationship of this to the classical methods for dense graphs; and we'll look at a method for the minimum-spanning-tree problem that uses an entirely different approach.

Minimum Spanning Tree

A *minimum spanning tree* of a weighted graph is a collection of edges connecting all the vertices such that the sum of the weights of the edges is at least as small as the sum of the weights of any other collection of edges connecting all the vertices. The minimum spanning tree need not be unique: Figure 31.2 shows three minimum spanning trees for our sample graph. It's easy to prove that the "collection of edges" in the definition above must form a spanning tree: if there's any cycle, some edge in the cycle can be deleted to give a collection of edges that still connects the vertices but has a smaller weight.

We saw in Chapter 29 that many graph-traversal procedures compute a spanning tree for the graph. How can we arrange things for a weighted graph so that the tree computed is the one with the lowest total weight? There are several ways to do so, all based on the following general property of minimum spanning trees.

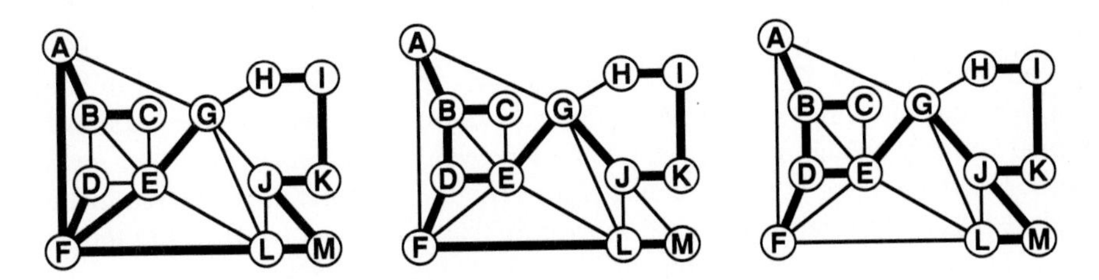

Figure 31.2 Minimum spanning trees.

Property 31.1 *Given any division of the vertices of a graph into two sets, the minimum spanning tree contains the shortest of the edges connecting a vertex in one of the sets to a vertex in the other set.*

For example, dividing the vertices in our sample graph into the sets {A B C D} and {E F G H I J K L M} implies that DF must be in any minimum spanning tree. This property is easy to prove by contradiction. Call the shortest edge connecting the two sets s, and assume that s is not in the minimum spanning tree. Then consider the graph formed by adding s to the purported minimum spanning tree. This graph has a cycle; in that cycle some other edge besides s must connect the two sets. Deleting this edge and adding s gives a shorter spanning tree, and this contradicts the assumption that s is not in the minimum spanning tree. ■

Thus we can build the minimum spanning tree by starting with any vertex and always taking next the vertex "closest" to the vertices already taken. In other words, we find the edge of lowest weight among those edges that connect vertices already on the tree to vertices not yet on the tree, then add to the tree that edge and the vertex it leads to. (In case of a tie, any of the edges involved in the tie will do.) Property 31.1 guarantees that each edge added is part of the minimum spanning tree.

Figure 31.3 illustrates the first four steps when this strategy is used for our example graph, starting with node A. The vertex "closest" to A (connected with an edge of lowest weight) is B, so AB is in the minimum spanning tree. Of all the edges touching AB, the edge BC is of lowest weight, so it is added to the tree and vertex C is visited next. Then, the closest vertex to A, B, or C is now D, so BD is added to the tree. The completion of the process is shown below, after we discuss of the implementation, in Figure 31.5.

How do we actually implement this strategy? By now the reader has surely recognized the basic structure of tree, fringe, and unvisited vertices that characterized the depth-first and breadth-first search strategies in Chapter 29. It turns out that the same method works, using a *priority queue* (instead of a stack or a queue) to hold the vertices in the fringe.

Priority-First Search

Recall from Chapter 29 that graph searching can be described in terms of dividing the vertices into three sets: *tree* vertices, whose edges have all been examined; *fringe* vertices, which are on a data structure waiting for processing; and *unseen* vertices, which haven't been touched at all. The fundamental graph-search method we use is based on the step "move one vertex (call it x) from the fringe to the tree, then put on the fringe any unseen vertices adjacent to x." We use the term *priority-first search* to refer to the general strategy of using a priority queue to decide which vertex to take from the fringe. This permits a great deal of flexibility. As we'll see, several classical algorithms (including both depth-first search and breadth-first search) differ only in the choice of priority.

For computing the minimal spanning tree, the priority of each vertex on the fringe should be the length of the shortest edge connecting it to the tree. Figure 31.4 shows the contents of the priority queue during the construction process depicted in Figures 31.3 and 31.5. For clarity, the items in the queue are shown in sorted order. This "sorted list" implementation of priority queues might be appropriate for small graphs, but heaps should be used for large graphs to ensure that all operations can be completed in $O(\log N)$ steps (see Chapter 11).

First, we consider sparse graphs with an adjacency-list representation. As mentioned above, we add a weight field w to the *edge* record (and modify the input code to read in weights as well). Then, using a priority queue for the fringe, we have the following implementation:

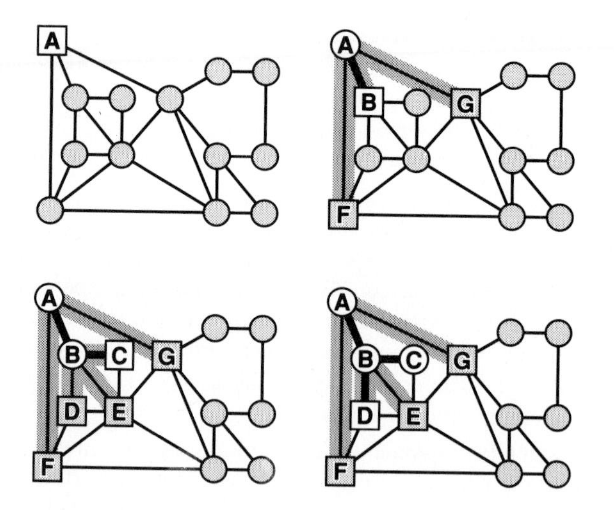

Figure 31.3 Initial steps of constructing a minimum spanning tree.

```
procedure listpfs;
  var id,k: integer;
      val: array[1..maxV] of integer;
  procedure visit(k: integer);
    var t: link;
    begin
    if pqupdate(k,unseen) then dad[k]:=0;
    repeat
      id:=id+1;
      k:=pqremove; val[k]:=-val[k];
      if val[k]=unseen then val[k]:=0;
      t:=adj[k];
      while t<>z do
        begin
        if val[t↑.v]<0 then
          if pqupdate(t↑.v,priority) then
            begin val[t↑.v]:=-(priority); dad[t↑.v]:=k end;
        t:=t↑.next
        end
    until pqempty
    end;
  begin
  id:=0; pqinitialize;
  for k:=1 to V do val[k]:=-unseen;
  for k:=1 to V do
    if val[k]=-unseen then visit(k)
  end;
```

To compute the minimum spanning tree, replace both occurrences of *priority* by *t↑.w*. The procedures *pqinitialize* and *pqremove* are priority queue utility routines as described in Chapter 11. The function *pqupdate* is an easily implemented addition to that set of routines whose purpose is to ensure that the given vertex appears on the queue with at least the given priority: if the vertex is not on the queue, a *pqinsert* is done, and if the vertex is there but has a larger priority value, then *pqchange* is used to change the priority. If any change is made (either insertion or priority change), then *pqupdate* returns *true*. This allows the above program to keep the *val* and *dad* arrays current. The *val* array itself could actually hold the priority queue in "indirect" fashion; in the above program we have separated out the priority queue operations for clarity.

Beyond changing the data structure to a priority queue, this is virtually the same program we used for depth-first and breadth-first search, with two exceptions. First, extra action is necessary when an edge is encountered that is already on the fringe: for depth-first and breadth-first search, such edges are ignored, but in the above program we need to check whether the new edge lowers the priority. This ensures that, as desired, we always visit next the vertex in the fringe that is closest to the tree. Second, this program explicitly keeps track of the tree by maintaining the *dad* array that stores the parent of each node in the priority-first search tree (the name of the node that caused it to be moved from the fringe to the tree). Also, for each node k on the tree, $val[k]$ is the weight of the edge between k and $dad[k]$. Vertices on the fringe are marked with negative *val* values as before; unseen vertices are marked with the sentinel $-unseen$ rather than 0. The reason for this change in convention will become apparent below.

Figure 31.5 completes the construction of the minimum spanning tree for our example. As usual, vertices on the fringe are labeled squares, tree vertices are labeled circles, and unvisited vertices are unlabeled circles. Tree edges are indicated with thick black lines, and the shortest edge connecting each vertex on the fringe to the tree is shaded. The reader is encouraged to follow through the construction of the tree using these diagrams and Figure 31.4. In particular, note how vertex G is added to tree after being on the fringe during several steps. Initially, the distance from G to the tree is 6 (because of edge GA). After L is added to the tree, GL brings the distance down to 5, then, after E is added, the distance finally goes down to 1, and G is added to the tree *before* J. Figure 31.6 shows the construction of a minimum spanning tree for our large "maze" graph, with edge lengths used as weights.

Property 31.2 *Priority-first search on sparse graphs computes the minimum spanning tree in $O((E+V)\log V)$ steps.*

Property 31.1 above applies: the two sets of nodes in question are the visited nodes and the unvisited ones. At each step, we pick the shortest edge from a visited node to a fringe node (there are no edges from visited nodes to unseen nodes). Thus, by Property 31.1, every edge that is picked is on the minimum spanning tree. The priority queue contains only vertices; if it is implemented as a heap (see Chapter 11) then each operation requires $O(\log V)$ steps. Each vertex leads to an insertion, and each edge leads to a *pqchange* operation. ∎

		C 1										
		D 2	D 2			M 1						
	B 1	F 2	F 2	F 1	L 2	E 2	E 2					
	F 2	E 4	E 4	E 2	E 2	J 3	J 2	G 1	J 1	K 1	I 1	
A *	G 6	G 6	G 6	G 6	G 6	G 5	G 5	J 2	H 3	H 3	H 3	H 2

Figure 31.4 Contents of priority queue during minimum spanning tree construction.

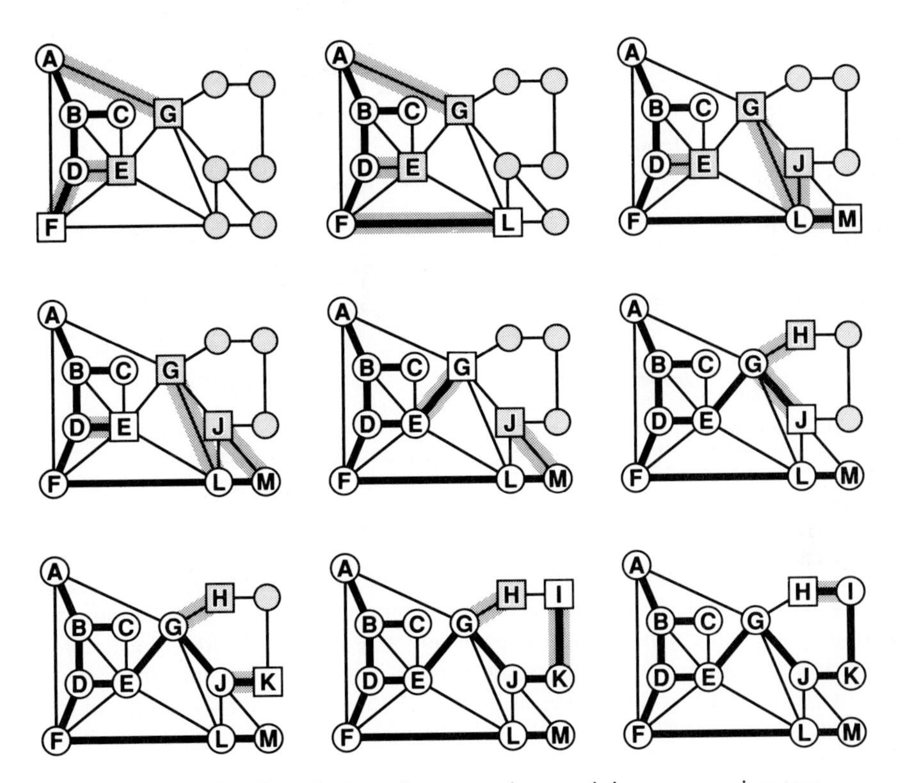

Figure 31.5 Completion of constructing a minimum spanning tree.

We'll see below that this method also solves the shortest path problem with appropriate choice of priority. Also, we'll see how a different implementation of the priority queue can give a V^2 algorithm, which is appropriate for dense graphs. This is equivalent to an old algorithm that dates at least to 1957: for minimum spanning trees it is generally attributed to R. Prim, and for shortest paths it is generally attributed to E. Dijkstra. For consistency, we refer to those solutions (for dense graphs) as "Prim's algorithm" and "Dijkstra's algorithm" respectively; we refer to the method above (for sparse graphs) as the "priority-first search solution."

Priority-first search is a proper generalization of breadth-first and depth-first search, because these methods can be derived through appropriate priority settings. Recall that *id* increases from 1 to V during the execution of the algorithm and thus can be used to assign unique priorities to the vertices. If we change the two occurrences of *priority* in *listpfs* to $V-id$, we get depth-first search, because newly encountered nodes have the highest priority. If we use *id* for *priority* we get breadth-first search, because old nodes have the highest priority. These priority assignments make the priority queues operate like stacks and queues.

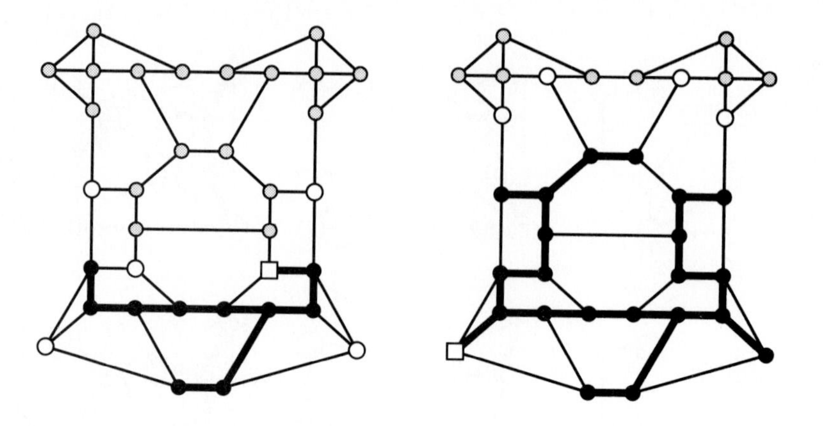

Figure 31.6 Constructing a large minimum spanning tree.

Kruskal's Method

A completely different approach to finding the minimum spanning tree is simply to add edges one at a time, at each step using the shortest edge that does not form a cycle. Put another way, the algorithm starts with an N-tree forest: for N steps, it combines two trees (using the shortest edge possible) until there is just one tree left. This algorithm dates at least to 1956 and is generally attributed to J. Kruskal.

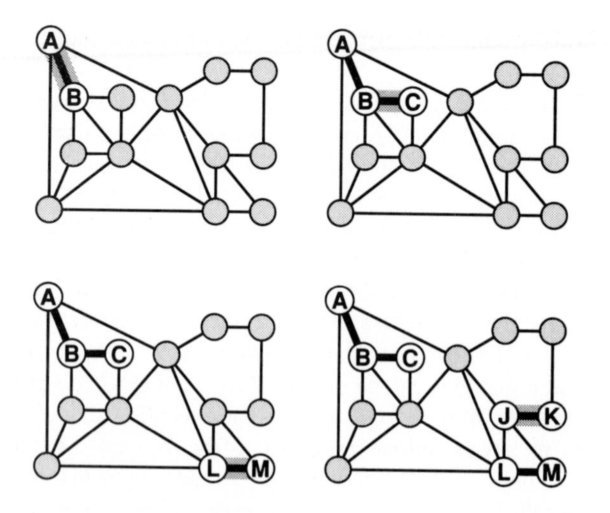

Figure 31.7 Initial steps of Kruskal's algorithm.

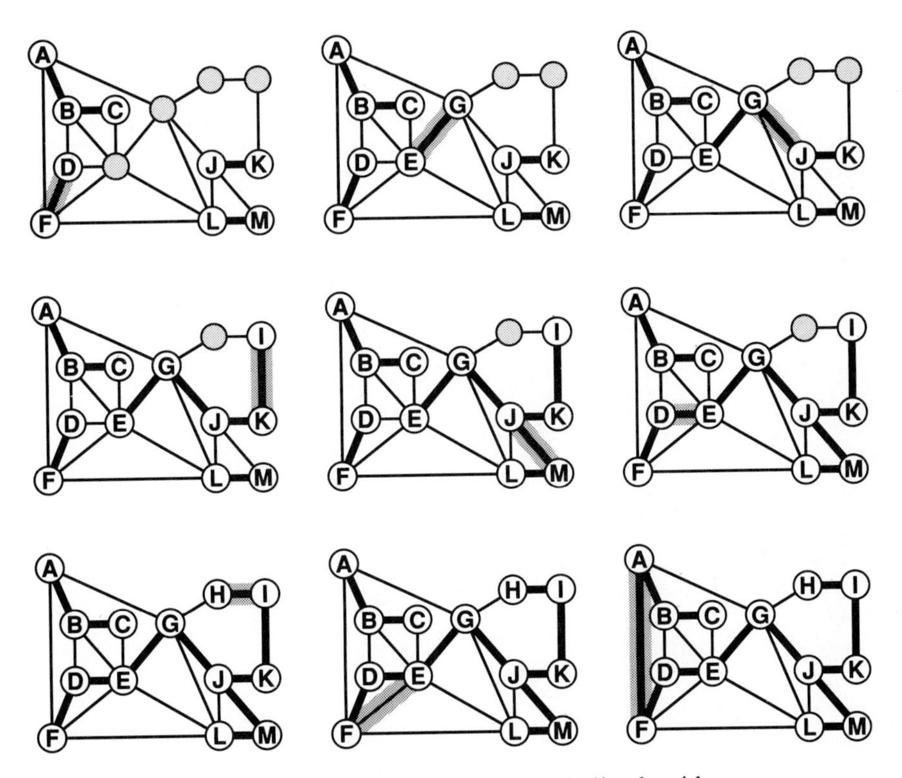

Figure 31.8 Completion of Kruskal's algorithm.

Figures 31.7 and 31.8 show the operation of this algorithm on our sample graph. The first edges chosen are those of shortest length (1) in the graph. Then edges of length 2 are tried; in particular note that FE is considered but not included, because it forms a cycle with edges already known to be in the tree. The disconnected components gradually evolve into a tree, in contrast to priority-first search, where the tree "grows" one edge at a time.

The implementation of Kruskal's algorithm can be pieced together from programs we've already studied. First, we need to consider the edges, one at a time, in increasing order of their weight. One possibility is simply to sort them, but it turns out to be better to use a priority queue, mainly because we may not need to examine all the edges. This is discussed in more detail below. Second, we need to be able to test whether a given edge, if added to the edges taken so far, causes a cycle. The union-find structures discussed in the previous chapter are designed for precisely this task.

Now, the appropriate data structure for the graph is simply an array *edges* with one entry for each edge. This could easily be built from either the adjacency-list or adjacency-matrix representation of the graph with depth-first search or some simpler procedure. In the program below, however, we just fill this array directly from the input. The indirect priority queue procedures *pqconstruct* and *pqremove* from Chapter 11 are used to maintain the priority queue, using the weight (w) fields in the *edge* array for priorities, and the *findinit* and *find* procedures from Chapter 30 are used to test for cycles. The program below simply calls the procedure *edgefound* for each edge in the spanning tree; with slightly more work a *dad* array or other representation could be computed.

```
program kruskal(input,output);
const maxV=50; maxE=2500;
type edge=record v1,v2,w: integer end;
var i,j,m,x,y,V,E: integer;
      edges: array[0..maxE] of edge;
begin
readln(V,E);
for j:=1 to E do
   begin
   readln(c,d,edges[j].w);
   edges[j].v1:=index(c);
   edges[j].v2:=index(d);
   end;
findinit; pqconstruct; i:=0;
repeat
   m:=pqremove; x:=edges[m].v1; y:=edges[m].v2;
   if find(x,y,true) then
      begin
      edgefound(x,y);
      i:=i+1
      end
until pqempty or (i=V-1);
end.
```

Note that there are two ways in which the process can terminate. If we find $V - 1$ edges, then we have a tree and can stop. If the priority queue runs out first, then we have examined all the edges without finding a spanning tree; this will happen if the graph is not connected. The running time of this program is dominated by the time spent processing edges in the priority queue.

Property 31.3 *Kruskal's algorithm computes the minimum spanning tree of a graph in $O(E \log E)$ steps.*

The correctness of this algorithm also follows from Property 31.1. The two sets

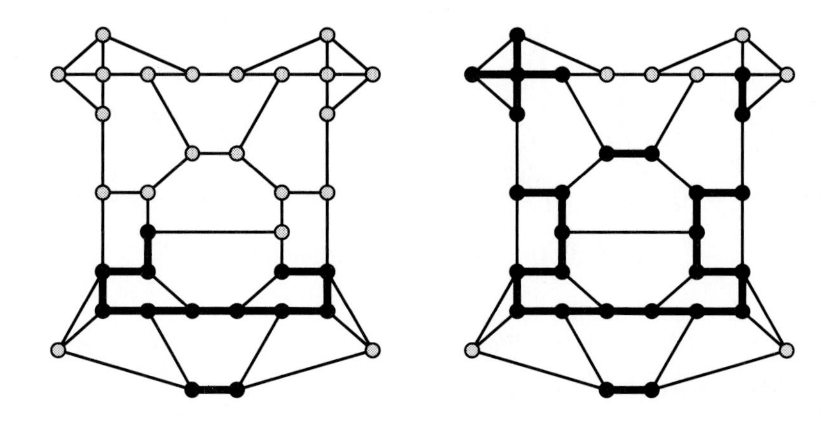

Figure 31.9 Finding a large minimum spanning tree with Kruskal's algorithm.

of vertices in question are those attached to edges chosen for the tree and those not yet touched. Each edge added is the shortest between vertices in these two sets. The worst case is a graph that is not connected, so that all the edges must be examined. Even for a connected graph, the worst case is the same, because the graph might consist of two clusters of vertices all connected together by very short edges, and only one edge that is very long connecting the two clusters. Then the longest edge in the graph is in the minimum spanning tree, but it will be the last edge out of the priority queue. For typical graphs, we might expect the spanning tree to be complete (it has only $V - 1$ edges) well before getting to the longest edge in the graph, but it does always take time proportional to E to build the priority queue initially (see Property 11.2). ▪

Figure 31.9 shows the construction of a larger minimum spanning tree with Kruskal's algorithm. This diagram clearly shows how the method picks all the short edges first: it will add the longer (diagonal) edges last.

Rather than using priority queues, one could simply sort the edges by weight initially, then process them in order. Also, the cycle testing can be done in time proportional to $E \log E$ with a much simpler strategy than union-find, to give a minimum spanning tree algorithm that always takes $E \log E$ steps. This is the method proposed by Kruskal, but we refer to the modernized version above, which uses priority queues and union-find structures, as "Kruskal's algorithm."

Shortest Path

The *shortest-path* problem is to find the path in a weighted graph connecting two given vertices x and y with the property that the sum of the weights of all the

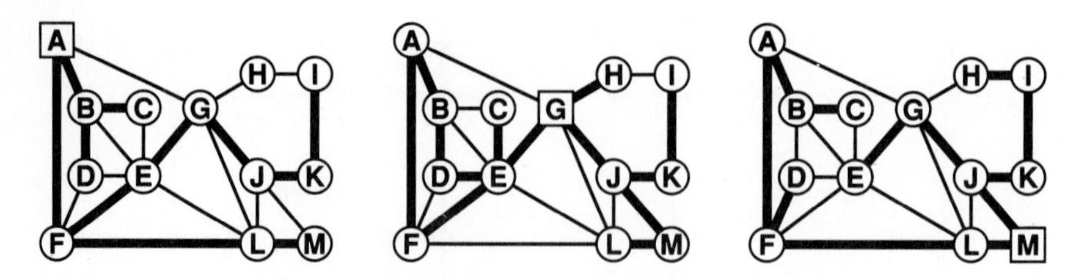

Figure 31.10 Shortest-path spanning trees.

edges is minimized over all such paths.

If the weights are all 1, then the problem is still interesting: it is to find the path containing the minimum number of edges that connects x and y. Moreover, we've already considered an algorithm that solves the problem: breadth-first search. It is easy to prove by induction that breadth-first search starting at x will first visit all vertices that can be reached from x with one edge, then all vertices that can be reached from x with two edges, etc., visiting all vertices that can be reached with k edges before encountering any that require $k + 1$ edges. Thus, when y is first encountered, the shortest path from x has been found because no shorter paths reached y (cf. Figure 29.14).

In general, the path from x to y could touch all the vertices, so we usually consider the problem of finding the shortest paths connecting a given vertex x with *all* of the other vertices in the graph. Again, it turns out that the problem is simple

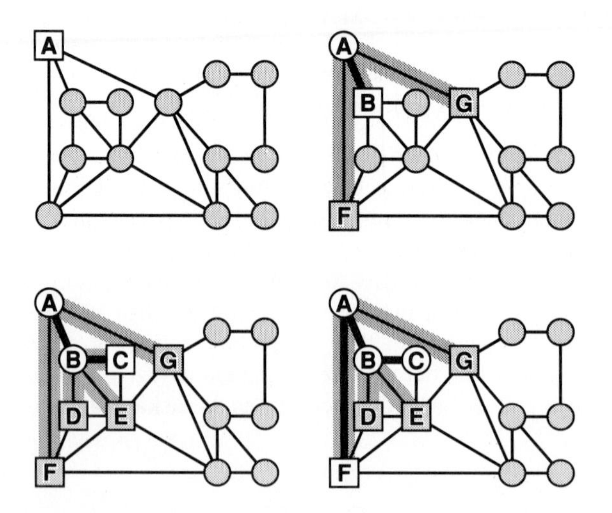

Figure 31.11 Initial steps of constructing a shortest path tree.

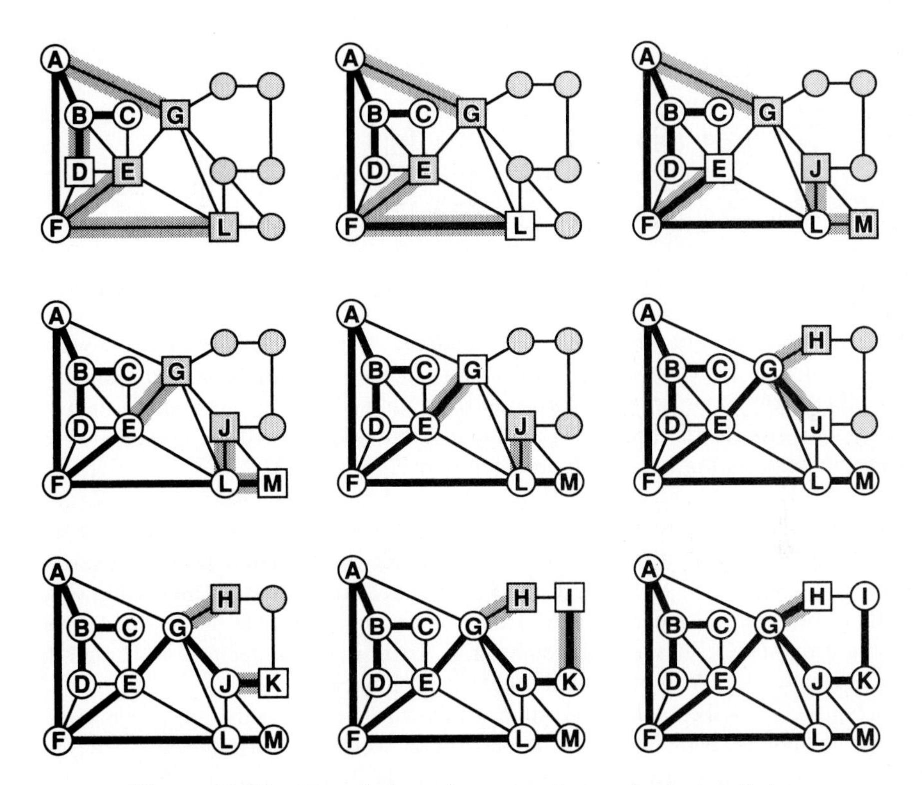

Figure 31.12 Completion of constructing a shortest path tree.

to solve with the priority-first graph-traversal algorithm given above.

If we draw the shortest path from x to every other vertex in the graph, then we clearly get no cycles, and we have a spanning tree. Each vertex leads to a different spanning tree; for example, Figure 31.10 shows the shortest-path spanning trees for vertices A, G, and M in the example graph we've been using.

The priority-first-search solution to this problem is virtually identical to the solution for the minimum spanning tree: we build the tree for vertex x by adding, at each step, the vertex on the fringe that is closest to x (before, we added the one closest to the *tree*). To find which fringe vertex is closest to x, we use the *val* array: for each tree vertex k, $val[k]$ is the distance from that vertex to x, using the shortest path (which must be comprised of tree nodes). When k is added to the tree, we update the fringe by going through k's adjacency list. For each node t on the list, the shortest distance to x through k from $t\uparrow.v$ is $val[k]+t\uparrow.w$. Thus, the algorithm is trivially implemented by using this quantity for *priority* in the priority-first graph-traversal program.

Figure 31.11 shows the first four steps in the construction of the shortest-

		C 2										
		F 2	F 2	D 3		E 4						
	B 1	D 3	D 3	L 4	L 4	M 5	M 5					
	F 2	E 5	E 5	E 4	E 4	G 6	G 5	G 5	J 6	K 7	I 8	
A *	G 6	G 6	G 6	G 6	G 6	J 7	J 7	J 7	H 8	H 8	H 8	H 8

Figure 31.13 Contents of priority queue during shortest-path tree construction.

path spanning tree for vertex A in our example. First we visit the closest vertex to A, which is B. Then both C and F are distance 2 from A, so we visit them next (in whatever order the priority queue returns them, in this case C then F). The completion of the process is shown in Figure 31.12, and the contents of the priority queue during the search are shown in Figure 31.13.

Next, D can be attached at F or at B to get a path of distance 3 to A. (The algorithm attaches D to B because B was put on the tree before F; thus D was already on the fringe when F was put on the tree and F didn't provide a shorter path to A.) Then L E M G J K I and H are added to the tree, in increasing order of their minimum distance from A. Thus, for example, H is the farthest node from A: the path AFEGH has a total weight of 8. There is no shorter path to H, and the shortest path from A to every other node is no longer.

Figure 31.14 shows the final values of the *dad* and *val* arrays for our example. Thus the shortest path from A to H has a total weight of 8 (found in *val*[*8*], the entry for H) and goes from A to F to E to G to H (found by tracing backwards in the *dad* array, starting at H). Note that this program depends on the *val* entry for the root being zero, the convention that we adopted for *listpfs*.

Property 31.4 *Priority-first search on sparse graphs computes the shortest path tree in* $O((E+V)\log V)$ *steps.*

Correctness of the algorithm can be proven in a manner similar to Property 31.1. With the priority queue implemented using heaps as in Chapter 12, priority-first search can *always* be guaranteed to achieve the stated time bound, no matter what priority rule is used. ■

k	*1*	*2*	*3*	*4*	*5*	*6*	*7*	*8*	*9*	*10*	*11*	*12*	*13*
name(dad[k])		*A*	*B*	*B*	*F*	*A*	*E*	*G*	*K*	*G*	*I*	*F*	*L*
val[k]	*0*	*1*	*2*	*3*	*4*	*2*	*5*	*8*	*8*	*6*	*7*	*4*	*5*

Figure 31.14 Shortest-path spanning tree representation.

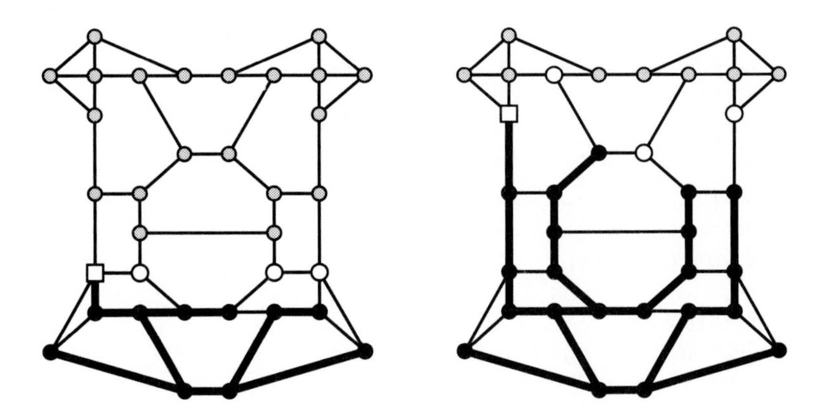

Figure 31.15 Constructing a large shortest-path tree.

Below, we'll see how a different implementation of the priority queue can give a V^2 algorithm that is appropriate for dense graphs. For the shortest-path problem, this reduces to a method that dates at least to 1959 and is generally attributed to E. Dijkstra.

Figure 31.15 shows a larger shortest-path tree. As before, the lengths of the edges are used as weights in this graph, so the solution amounts to finding the path of minimum length from the bottom left node to every other node. Later, we discuss an improvement that may be appropriate for such graphs. But even in this graph, it might be appropriate to use other values for the weights: for example, if this graph represents a maze (see Chapter 29), the weight of a edge could represent the distance in the maze itself, not the shortcuts drawn in the graph.

Minimum Spanning Tree and Shortest Paths in Dense Graphs

For a graph represented with an adjacency matrix, it is best to use an unordered array representation for the priority queue in order to achieve a V^2 running time for any priority-first graph-traversal algorithm. This is done by combining the loop to update the priorities and the loop to find the minimum: each time we remove a vertex from the fringe, we pass through all the vertices, updating their priority if necessary and keeping track of the minimum value found. This provides a linear algorithm for priority-first search (and thus the minimum-spanning-tree and shortest-path problems) for dense graphs.

Specifically, we maintain the priority queue in the *val* array (this could also be done in *listpfs*, as discussed above) but we implement the priority-queue operations directly rather than using heaps. As above, the sign of a *val* entry tells whether the corresponding vertex is on the tree or the priority queue. All vertices start on the priority queue and have the sentinel priority *unseen*. To change the priority of a vertex, we simply assign the new priority to the *val* entry for that vertex. To remove the highest-priority vertex, we scan through the *val* array to find the vertex with the largest negative (closest to 0) *val* value, then complement its *val* entry. After making these mechanical changes to the *listpfs* program we have been using, we are left with the following compact program:

```
procedure matrixpfs;
  var k,min,t: integer;
  begin
  for k:=1 to V do
    begin val[k]:=-unseen; dad[k]:=0 end;
  val[0]:=-(unseen+1);
  min:=1;
  repeat
    k:=min; val[k]:=-val[k]; min:=0;
    if val[k]=unseen then val[k]:=0;
    for t:=1 to V do
      if val[t]<0 then
        begin
        if (a[k,t]<>0) and (val[t]<-(priority)) then
          begin val[t]:=-(priority); dad[t]:=k end;
        if val[t]>val[min] then min:=t;
        end
    until min=0;
  end;
```

Note that *unseen* must be slightly less than *maxint*, since a value one higher is used as a sentinel to find the minimum, and the negative of this value must be representable.

If we store weights in the adjacency matrix and use *a[k,t]* for *priority* in this program, we get Prim's algorithm for finding the minimum spanning tree; if we use *val[k]+a[k,t]* for *priority* we get Dijkstra's algorithm for the shortest-path problem. As above, if we include the code to maintain *id* as the number of vertices so far searched and use $V-id$ for *priority*, we get depth-first search; if we use *id* we get breadth-first search. This program differs from the priority-first-search program we've been working with for sparse graphs only in the graph representation used (adjacency matrix instead of adjacency list) and the priority-queue implementation (unordered array instead of indirect heap).

Property 31.5 *The minimum-spanning-tree and shortest-path problems can be solved in linear time on dense graphs.*

It is immediate from inspection of the program that the worst-case running time is proportional to V^2. Each time a vertex is visited, a pass through the V entries in its row in the adjacency matrix accomplishes the dual purpose of testing all adjacent edges and updating and finding the next minimum value in the priority queue. Thus, the running time is linear when E is proportional to V^2. ■

We have discussed three programs for the minimal spanning tree problem with quite different performance characteristics: the priority-first search method (page 455), Kruskal's algorithm (page 460), and Prim's algorithm (page 466). Prim's algorithm is likely to be the fastest of the three for some graphs, Kruskal's for some others, the priority-first search method for still others. As described above, the worst case for the priority-first search method is $(E+V)\log V$, while the worst case for Prim's is V^2 and the worst case for Kruskal's is $E \log E$. But it is unwise to choose between the algorithms on the basis of these formulas because "worst-case" graphs are unlikely to occur in practice. In fact, the priority-first-search method and Kruskal's method are both likely to run in time proportional to E for graphs that arise in practice: the first because most edges do not require a priority queue adjustment that takes $\log V$ steps, and the second because the longest edge in the minimum spanning tree is probably short enough that not many edges are taken off the priority queue. For sparse graphs, priority-first search is likely to be fastest because it is likely to be working with a small priority queue. Of course, Prim's method also runs in time proportional to about E for dense graphs (but it shouldn't be used for sparse graphs).

Geometric Problems

Suppose that we are given N points in the plane and we want to find the shortest set of lines connecting all the points. This is a geometric problem called the *Euclidean minimum spanning tree* problem. It can be solved using the graph algorithm given above, but it seems clear that the geometry provides enough extra structure to allow much more efficient algorithms to be developed.

The way to solve the Euclidean problem using the algorithm given above is to build a complete graph with N vertices and $N(N-1)/2$ edges, one edge connecting each pair of vertices weighted with the distance between the corresponding points. Then the minimum spanning tree can be found with *matrixpfs* dense graphs in time proportional to N^2.

It has been proven that it is possible to do better. The point is that the geometric structure makes most of the edges in the complete graph irrelevant to the problem, and we can eliminate most of them before even starting to construct the minimum spanning tree. In fact, it has been proven that the minimum spanning tree is a subset of the graph derived by taking only the edges from the dual of the Voronoi diagram (see Chapter 28). We know that this graph has a number of edges proportional to N, and both Kruskal's algorithm and the priority-first search method work efficiently on such sparse graphs. In principle, then, we could compute the Voronoi dual (which takes time proportional to $N \log N$), then run either Kruskal's algorithm or the priority-first-search method to get a Euclidean minimum-spanning-tree algorithm that runs in time proportional to $N \log N$. But writing a program to compute the Voronoi dual is quite a challenge even for an experienced programmer, and so this approach is likely to prove impracticable.

Another approach, which can be used for random point sets, is to take advantage of the distribution of the points to limit the number of edges included in the graph, as in the grid method used in Chapter 26 for range searching. If we divide up the plane into squares such that each square is likely to contain about $\lg N/2$ points, and then include in the graph only the edges connecting each point to the points in the neighboring squares, then we are very likely (though not guaranteed) to get all the edges in the minimum spanning tree, which would mean that Kruskal's algorithm or the priority-first search method would finish the job efficiently.

It is interesting to reflect on the relationship between graph and geometric algorithms brought out by the problem posed in the previous paragraphs. It is certainly true that many problems can be formulated either as geometric problems or as graph problems. If the actual physical placement of objects is a dominating characteristic, then the geometric algorithms of Chapters 24–28 may be appropriate, but if interconnections between objects are of fundamental importance, then the graph algorithms of this section may be better. The Euclidean minimum spanning tree seems to fall at the interface between these two approaches (the input involves geometry and the output involves interconnections) and the development of simple, straightforward methods for this and related problems remains an important but elusive goal.

Another place where geometric and graph algorithms interact is the problem of finding the shortest path from x to y in a graph whose vertices are points in the plane and whose edges are lines connecting the points. The maze graph we have been using can be viewed as such a graph. The solution to this problem is simple: use priority-first search, setting the priority of each fringe vertex encountered to the distance in the tree from x to the fringe vertex (as in the algorithm given) *plus* the Euclidean distance from the fringe vertex to y. Then we stop when y is added to the tree. This method will very quickly find the shortest path from x to y by always going towards y, while the standard graph algorithm has to "search" for y. Going from one corner to another of a large maze graph might require examining a number of nodes proportional to $\sqrt{V}$, while the standard algorithm must examine virtually all the nodes.

□

Exercises

1. Give another minimum spanning tree for the example graph at the beginning of the chapter.

2. Give an algorithm to find the *minimum spanning forest* of a connected graph (each vertex must be touched by some edge, but the resulting graph need not be connected).

3. Is there a graph with V vertices and E edges for which the priority-first solution to the minimum-spanning-tree problem could require time proportional to $(E + V) \log V$? Give an example or explain your answer.

4. Suppose we maintain the priority queue as a sorted list in the general graph traversal implementations. What would be the worst-case running time, to within a constant factor? When would this method be appropriate, if at all?

5. Give counterexamples that show why the following "greedy" strategy doesn't work for either the shortest-path or the minimum-spanning-tree problems: "at each step visit the unvisited vertex closest to the one just visited."

6. Give the shortest path trees for the other nodes in the example graph.

7. Describe how you would find the minimum spanning tree of an extremely large graph (too large to fit in main memory).

8. Write a program to generate random connected graphs with V vertices, then find the minimum spanning tree and shortest path tree for some vertex. Use random weights between 1 and V. How do the weights of the trees compare for different values of V?

9. Write a program to generate random complete weighted graphs with V vertices by simply filling in an adjacency matrix with random numbers between 1 and V. Run empirical tests to determine which method finds the minimum spanning tree faster for $V = 10, 25, 100$: Prim's or Kruskal's.

10. Give a counterexample to show why the following method for finding the Euclidean minimum spanning tree doesn't work: "Sort the points on their x coordinates, then find the minimum spanning trees of the first half and the second half, then find the shortest edge that connects them."

32

Directed Graphs

☐ Directed graphs are graphs in which the edges connecting the nodes are one-way; this added structure makes it more difficult to determine various properties. Processing such graphs is akin to traveling around in a city with many one-way streets or to traveling around in a country where airlines rarely run round-trip routes: getting from one point to another in such situations can be a challenge indeed.

Often the edge direction reflects some type of precedence relationship in the application being modeled. For example, a directed graph might be used to model a manufacturing line: nodes correspond to jobs to be done and an edge exists from node x to node y if the job corresponding to node x must be done before that corresponding to node y. How do we decide when to perform each of the jobs so that none of these precedence relationships are violated?

In this chapter, we'll look at depth-first search for directed graphs, as well as algorithms for computing the *transitive closure* (which summarizes connectivity information), for *topological sorting* and for computing *strongly connected components* (which have to do with precedence relationships).

As mentioned in Chapter 29, representations for directed graphs are simple extensions (restrictions, actually) of representations for undirected graphs. In the adjacency-list representation, each edge appears only once: the edge from x to y is represented as a list node containing y in the linked list corresponding to x. In the adjacency-matrix representation, we need to maintain a full V-by-V matrix, with a 1 bit in row x and column y (but not necessarily in row y and column x) if there is an edge from x to y.

A directed graph similar to the undirected graph we've been considering is drawn in Figure 32.1. This graph consists of the edges AG AB CA LM JM JL JK ED DF HI FE AF GE GC HG GJ LG IH ML. The order in which the vertices appear when specifying edges is now significant: the notation AG describes an edge that points from A to G, but *not* from G to A. But it is possible to have two

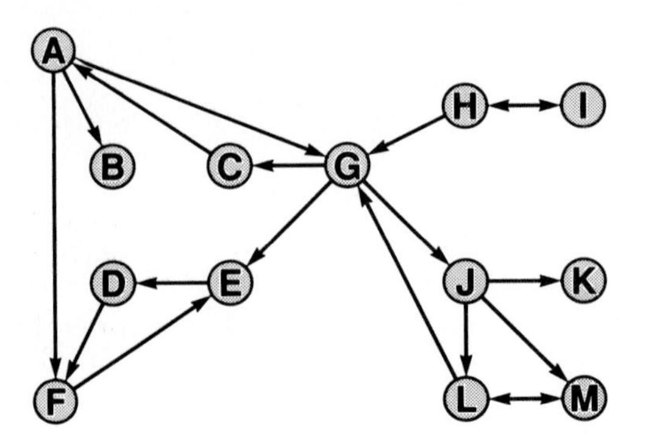

Figure 32.1 A directed graph.

edges between two nodes, one in either direction (Figure 32.1 has both HI and IH and both LM and ML).

Note that no difference could be perceived in these representations between an undirected graph and a directed graph with two opposite directed edges for each edge in the undirected graph. Thus, some of algorithms in this chapter can be considered generalizations of algorithms in previous chapters.

Depth-First Search

The depth-first search algorithm of Chapter 29 works properly for directed graphs exactly as given. In fact, its operation is a little more straightforward than for undirected graphs because we needn't be concerned with double edges between nodes unless they're explicitly included in the graph. However, the search trees have a somewhat more complicated structure. For example, Figure 32.2 shows the depth-first search structure that describes the operation of the recursive algorithm of Chapter 29 on our sample graph. As before, this is a redrawn version of the graph: solid edges correspond to those edges that were actually used to visit vertices via recursive calls and dotted edges correspond to those edges pointing to vertices that had already been visited at the time the edge was considered. The nodes are visited in the order A F E D B G J K L M C H I.

Note that the directions on the edges make this depth-first search forest quite different from the depth-first search forests for undirected graphs. For example, even though the original graph was connected, the depth-first search structure defined by the solid edges is not connected: it is a forest, not a tree.

For undirected graphs, we had only one kind of dotted edge, one that connected a vertex with some ancestor in the tree. For directed graphs, there are three kinds of dotted edges: *up* edges, which point from a vertex to some ancestor in the tree, *down* edges, which point from a vertex to some descendant in the tree, and *cross*

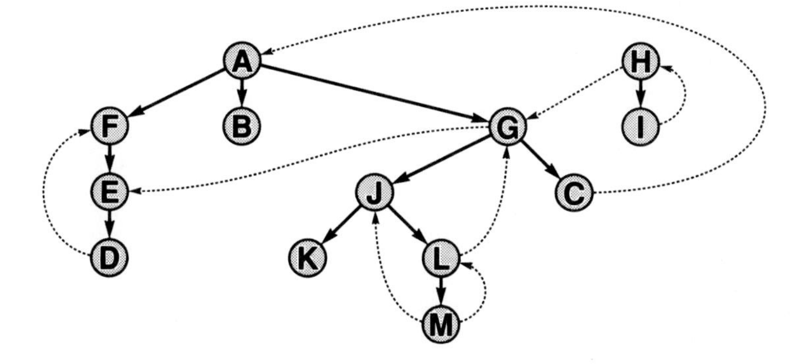

Figure 32.2 Depth-first search forest for a directed graph.

edges, which point from a vertex to another vertex that is neither a descendant nor an ancestor in the tree.

As with undirected graphs, we're interested in connectivity properties of directed graphs. We would like to be able to answer questions like "Is there a *directed path* from vertex x to vertex y (a path that follows edges only in the indicated direction)?" and "Which vertices can we get to from vertex x with a directed path?" and "Is there a directed path from vertex x to vertex y *and* a directed path from y to x?" Just as with undirected graphs, we'll be able to answer such questions by appropriately modifying the basic depth-first search algorithm, though the various different types of dotted edges make the modifications somewhat more complicated.

Transitive Closure

In undirected graphs, simple connectivity gives the vertices that can be reached from a given vertex by traversing edges from the graph: they are all those in the same connected component. Similarly, for directed graphs, we're often interested in the set of vertices that can be reached from a given vertex by traversing edges from the graph in the indicated direction. The problem for directed graphs is rather more complicated than simple connectivity.

It is easy to prove that the recursive *visit* procedure from the depth-first search method in Chapter 29 visits all the nodes that can be reached from the start node. Thus, if we modify that procedure to print out the nodes it is visiting (say, by inserting *write*(*name*(k)) just upon entering), we are printing out all the nodes that can be reached from the start node. But note carefully that it is *not* necessarily true that each tree in the depth-first search forest contains all the nodes that can be reached from the root of that tree: in our example, all the nodes in the graph can be reached from H, not just I. To get all the nodes that can be visited from each node, we simply call *visit* V times, once for each node:

```
for k:=1 to V do
  begin
  id:=0;
  for j:=1 to V do val[j]:=0;
  visit(k);
  writeln
  end;
```

This program produces the following output for the directed graph in Figure 32.1. As before, the ordering of the vertex names on each line is an artifact of the particular graph representation and search procedure used. The *set* of nodes on each line is a structural property of the graph itself: each line has those nodes reachable via a directed path from the first node on the line.

```
A F E D B G J K L M C
B
C A F E D B G J K L M
D F E
E D F
F E D
G J K L M C A F E D B
H G J K L M C A F E D B I
I H G J K L M C A F E D B
J K L G C A F E D B M
K
L G J K M C A F E D B
M L G J K C A F E D B
```

For undirected graphs, this computation would produce a table with the property that each line corresponding to the nodes in a connected component lists all the nodes in that component. The table above has a similar property: certain of the lines list identical sets of nodes. Below we shall examine the generalization of connectedness that explains this property.

As usual, we could add code to do extra processing rather than merely writing out the table. One operation we might want to perform is to add an edge directly from x to y if there is some way to get from x to y. The graph that results from adding all edges of this nature to a directed graph is called the *transitive closure* of the graph. Normally, a large number of edges will be added and the transitive closure is likely to be dense, so an adjacency-matrix representation is called for. This is an analogue to connected components in an undirected graph; once we've performed this computation once, then we can quickly answer questions like "Is there a way to get from x to y?"

Property 32.1 *Depth-first search can be used to compute the transitive closure of a directed graph in $O(V(E+V))$ steps for a sparse graph, and in $O(V^3)$ steps for a dense graph.*

This follows immediately from basic properties in Chapter 29: we perform the depth-first search procedure there for each of the V vertices in the graph. The same result holds for breadth-first search: as mentioned above, the order in which we visit the nodes is not particularly relevant for this problem. ■

There is a remarkably simple nonrecursive program for computing the transitive closure of a graph represented with an adjacency matrix:

```
for y:=1 to V do
  for x:=1 to V do
    if a[x,y] then
      for j:=1 to V do
        if a[y,j] then a[x,j]:=true;
```

S. Warshall invented this method in 1962, using the simple observation that "if there's a way to get from node x to node y and a way to get from node y to node j, then there's a way to get from node x to node j." The trick is to make this observation a little stronger, so that the computation can be done in only one pass through the matrix, to wit: "if there's a way to get from node x to node y *using only nodes with indices less than* y and a way to get from node y to node j then there's a way to get from node x to node j *using only nodes with indices less than* $y+1$." The above program is a direct implementation of this.

	A	B	C	D	E	F	G	H	I	J	K	L	M
A	1	1	0	0	0	1	1	0	0	0	0	0	0
B	0	1	0	0	0	0	0	0	0	0	0	0	0
C	1	0	1	0	0	0	0	0	0	0	0	0	0
D	0	0	0	1	0	1	0	0	0	0	0	0	0
E	0	0	0	1	1	0	0	0	0	0	0	0	0
F	0	0	0	0	1	1	0	0	0	0	0	0	0
G	0	0	1	0	1	0	1	0	0	1	0	0	0
H	0	0	0	0	0	0	1	1	1	0	0	0	0
I	0	0	0	0	0	0	0	1	1	0	0	0	0
J	0	0	0	0	0	0	0	0	0	1	1	1	1
K	0	0	0	0	0	0	0	0	0	0	1	0	0
L	0	0	0	0	0	0	1	0	0	0	0	1	1
M	0	0	0	0	0	0	0	0	0	0	0	1	1

	A	B	C	D	E	F	G	H	I	J	K	L	M
A	1	1	0	0	0	1	1	0	0	0	0	0	0
B	0	1	0	0	0	0	0	0	0	0	0	0	0
C	1	1	1	0	0	1	1	0	0	0	0	0	0
D	0	0	0	1	0	1	0	0	0	0	0	0	0
E	0	0	0	1	1	0	0	0	0	0	0	0	0
F	0	0	0	0	1	1	0	0	0	0	0	0	0
G	0	0	1	0	1	0	1	0	0	1	0	0	0
H	0	0	0	0	0	0	1	1	1	0	0	0	0
I	0	0	0	0	0	0	0	1	1	0	0	0	0
J	0	0	0	0	0	0	0	0	0	1	1	1	1
K	0	0	0	0	0	0	0	0	0	0	1	0	0
L	0	0	0	0	0	0	1	0	0	0	0	1	1
M	0	0	0	0	0	0	0	0	0	0	0	1	1

Figure 32.3 Initial stages of Warshall's algorithm.

	A	B	C	D	E	F	G	H	I	J	K	L	M
A	1	1	1	1	1	1	1	0	0	1	1	1	1
B	0	1	0	0	0	0	0	0	0	0	0	0	0
C	1	1	1	1	1	1	1	0	0	1	1	1	1
D	0	0	0	1	1	1	0	0	0	0	0	0	0
E	0	0	0	1	1	1	0	0	0	0	0	0	0
F	0	0	0	1	1	1	0	0	0	0	0	0	0
G	1	1	1	1	1	1	1	0	0	1	1	1	1
H	1	1	1	1	1	1	1	1	1	1	1	1	1
I	1	1	1	1	1	1	1	1	1	1	1	1	1
J	0	0	0	0	0	0	0	0	0	1	1	1	1
K	0	0	0	0	0	0	0	0	0	0	1	0	0
L	1	1	1	1	1	1	1	0	0	1	1	1	1
M	0	0	0	0	0	0	0	0	0	0	0	1	1

	A	B	C	D	E	F	G	H	I	J	K	L	M
A	1	1	1	1	1	1	1	0	0	1	1	1	1
B	0	1	0	0	0	0	0	0	0	0	0	0	0
C	1	1	1	1	1	1	1	0	0	1	1	1	1
D	0	0	0	1	1	1	0	0	0	0	0	0	0
E	0	0	0	1	1	1	0	0	0	0	0	0	0
F	0	0	0	1	1	1	0	0	0	0	0	0	0
G	1	1	1	1	1	1	1	0	0	1	1	1	1
H	1	1	1	1	1	1	1	1	1	1	1	1	1
I	1	1	1	1	1	1	1	1	1	1	1	1	1
J	1	1	1	1	1	1	1	0	0	1	1	1	1
K	0	0	0	0	0	0	0	0	0	0	1	0	0
L	1	1	1	1	1	1	1	0	0	1	1	1	1
M	1	1	1	1	1	1	1	0	0	1	1	1	1

Figure 32.4 Final stages of Warshall's algorithm.

Warshall's method converts the adjacency matrix for a graph into the adjacency matrix for its transitive closure. One way to follow the algorithm is to think of it as setting values in the matrix a whole row at a time The processing for column y consists of replacing each row with a 1 bit in column y by the "or" of itself and *row* y. Figure 32.3 shows the initial matrix for our sample graph and the state of the matrix after the first two columns and half the third have been processed: only row C is affected to this point. Figure 32.4 shows the matrix before the last few columns are to be processed and the final result (the transitive closure).

Property 32.1 *Warshall's algorithm finds the transitive closure in* $O(V^3)$ *steps.*

If the matrix is initially filled with 1 bits, this is obvious by inspection because of the three nested loops. But even a sparse graph can quickly reduce to this case—for example if the first node is connected to every other node, then the matrix becomes filled with 1 bits before y is set to 2. ■

For very large graphs, this computation can be organized so that the operations on bits can be done a computer word at a time, which will lead to significant savings in many environments.

All Shortest Paths

The transitive closure of an unweighted graph (directed or not) answers the question "Is there a path from x to y?" for all pairs x, y of vertices. For weighted graphs (directed or not) one might want to build a table allowing one to find the shortest path from x to y for all pairs of vertices. This is the *all-pairs shortest-path problem.* For example, in the weighted graph shown in Figure 32.5, we want to know just by accessing this table that the shortest path from M to K is of length 6 and the shortest path from J to F is of length 10, etc.

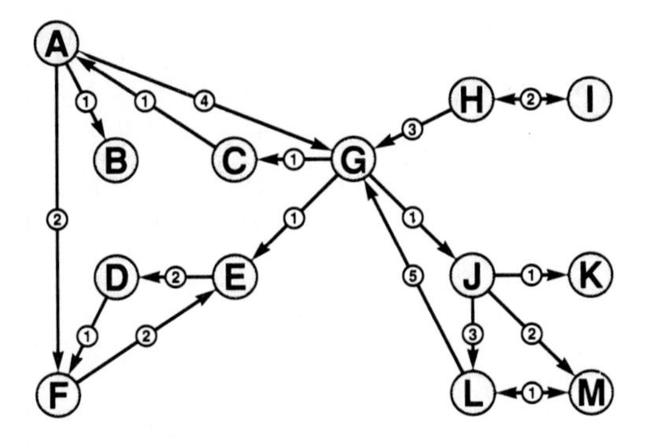

Figure 32.5 A weighted directed graph.

As above, the shortest-path algorithm from the previous chapter finds the shortest path from the start vertex to each of the other vertices, so we need only run that procedure V times, saturating at each vertex. This gives an algorithm that runs in $O(EV + V^2 \log V)$ steps. But it is also possible to use a method just like Warshall's method, which is generally attributed to R. W. Floyd:

```
for y:=1 to V do
  for x:=1 to V do
    if a[x,y]>0 then
      for j:=1 to V do
        if a[y,j]>0 then
          if (a[x,j]=0) or (a[x,y]+a[y,j]<a[x,j])
            then a[x,j]:=a[x,y]+a[y,j];
```

The structure of the algorithm is precisely the same as in Warshall's method. Instead of using "or" to keep track of the paths, we do a little computation for each edge to determine if it is part of a new short path: "the shortest way from node x to node j *using only nodes with indices less than* $y + 1$ is either the shortest way from node x to node j *using only nodes with indices less than* y, or, if it is shorter, the shortest way from x to y plus the distance from y to j." As usual, a matrix entry of 0 corresponds to the absence of the indicated edge; the program could be simplified somewhat (all the comparisons against zero removed) by using a sentinel value corresponding to an edge of infinitely large weight.

Property 32.2 *Floyd's algorithm solves the all-pairs shortest-path problem in* $O(V^3)$ *steps.*

This follows by the same reasoning as for Property 32.1. ■

	A	B	C	D	E	F	G	H	I	J	K	L	M
A	0	1	0	0	0	2	4	0	0	0	0	0	0
B	0	0	0	0	0	0	0	0	0	0	0	0	0
C	1	0	0	0	0	0	0	0	0	0	0	0	0
D	0	0	0	0	0	1	0	0	0	0	0	0	0
E	0	0	0	2	0	0	0	0	0	0	0	0	0
F	0	0	0	0	2	0	0	0	0	0	0	0	0
G	0	0	1	0	1	0	0	0	0	1	0	0	0
H	0	0	0	0	0	0	3	0	1	0	0	0	0
I	0	0	0	0	0	0	0	1	0	0	0	0	0
J	0	0	0	0	0	0	0	0	0	0	1	3	2
K	0	0	0	0	0	0	0	0	0	0	0	0	0
L	0	0	0	0	0	0	5	0	0	0	0	0	1
M	0	0	0	0	0	0	0	0	0	0	0	1	0

	A	B	C	D	E	F	G	H	I	J	K	L	M
A	0	1	0	0	0	2	4	0	0	0	0	0	0
B	0	0	0	0	0	0	0	0	0	0	0	0	0
C	1	2	0	0	0	3	5	0	0	0	0	0	0
D	0	0	0	0	0	1	0	0	0	0	0	0	0
E	0	0	0	2	0	0	0	0	0	0	0	0	0
F	0	0	0	0	2	0	0	0	0	0	0	0	0
G	0	0	1	0	1	0	0	0	0	1	0	0	0
H	0	0	0	0	0	0	3	0	1	0	0	0	0
I	0	0	0	0	0	0	0	1	0	0	0	0	0
J	0	0	0	0	0	0	0	0	0	0	1	3	2
K	0	0	0	0	0	0	0	0	0	0	0	0	0
L	0	0	0	0	0	0	5	0	0	0	0	0	1
M	0	0	0	0	0	0	0	0	0	0	0	1	0

Figure 32.6 Initial stages of Floyd's algorithm.

Figures 32.6 and 32.7 show in more detail the performance of Floyd's algorithm on our example, arranged exactly as Figures 32.3 and 32.4 for comparison. The zero entries in the various matrices, corresponding to the absence of a path between the two index vertices, are identical for the two algorithms. The non-zero entries in the matrices for Warshall's algorithm denote the existence of a path between the two vertices; for Floyd's algorithm they give the length of the shortest such path yet discovered. The actual shortest path can also be computed by using a matrix version of our *dad* array of previous chapters: set the entry in row x and column j to the name of the previous vertex in the shortest path from x to j (vertex y in the inner loop of the code above).

	A	B	C	D	E	F	G	H	I	J	K	L	M
A	6	1	5	6	4	2	4	0	0	5	6	8	7
B	0	0	0	0	0	0	0	0	0	0	0	0	0
C	1	2	6	7	5	3	5	0	0	6	7	9	8
D	0	0	0	5	3	1	0	0	0	0	0	0	0
E	0	0	0	2	5	3	0	0	0	0	0	0	0
F	0	0	0	4	2	5	0	0	0	0	0	0	0
G	2	3	1	3	1	4	6	0	0	1	2	4	3
H	5	6	4	6	4	7	3	2	1	4	5	7	6
I	6	7	5	7	5	8	4	1	2	5	6	8	7
J	0	0	0	0	0	0	0	0	0	0	1	3	2
K	0	0	0	0	0	0	0	0	0	0	0	0	0
L	7	8	6	8	6	9	5	0	0	6	7	9	1
M	0	0	0	0	0	0	0	0	0	0	0	1	0

	A	B	C	D	E	F	G	H	I	J	K	L	M
A	6	1	5	6	4	2	4	0	0	5	6	8	7
B	0	0	0	0	0	0	0	0	0	0	0	0	0
C	1	2	6	7	5	3	5	0	0	6	7	9	8
D	0	0	0	5	3	1	0	0	0	0	0	0	0
E	0	0	0	2	5	3	0	0	0	0	0	0	0
F	0	0	0	4	2	5	0	0	0	0	0	0	0
G	2	3	1	3	1	4	6	0	0	1	2	4	3
H	5	6	4	6	4	7	3	2	1	4	5	7	6
I	6	7	5	7	5	8	4	1	2	5	6	8	7
J	10	11	9	11	9	12	8	0	0	9	1	3	2
K	0	0	0	0	0	0	0	0	0	0	0	0	0
L	7	8	6	8	6	9	5	0	0	6	7	2	1
M	8	9	7	9	7	10	6	0	0	7	8	1	2

Figure 32.7 Final stages of Floyd's algorithm.

Topological Sorting

Cyclic graphs arise in many applications involving directed graphs. If, however, the graph in Figure 32.1 modeled a manufacturing line, then it would imply, say, that job A must be done before job G, which must be done before job C, which must be done before job A. But such a situation is inconsistent: for this and many other applications, directed graphs with no *directed cycles* (cycles with all edges pointing the same way) are called for. Such graphs are called *directed acyclic graphs*, or *dags* for short. Dags may have many cycles if the directions on the edges are not taken into account; their defining property is simply that one should never get in a cycle by following edges in the indicated direction. Figure 32.7 shows a dag similar to the directed graph in Figure 32.1, with a few edges removed or directions switched in order to remove cycles. The edge list for this graph is the same as for the connected graph of Chapter 30, but here, again, the order in which the vertices are given when the edge is specified makes a difference.

Dags really are quite different from general directed graphs: in a sense, they are part tree, part graph. We can certainly take advantage of their special structure when processing them. Viewed from any vertex, a dag looks like a tree; put another way, the depth-first search forest for a dag has no up edges. Figure 32.9 gives the depth-first search forest that describes the operation of *dfs* on the dag in Figure 32.8.

A fundamental operation on dags is processing the vertices of the graph in such an order that no vertex is processed before any vertex that points to it. For example, the nodes in the above graph could be processed in the following order:

J K L M A G H I F E D B C

If edges were drawn with the vertices in these positions, they would all go from left to right. As mentioned above, this has obvious application, for example, in

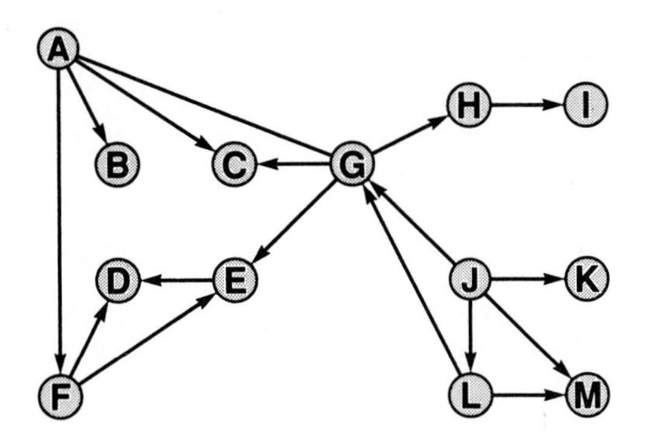

Figure 32.8 A directed acyclic graph.

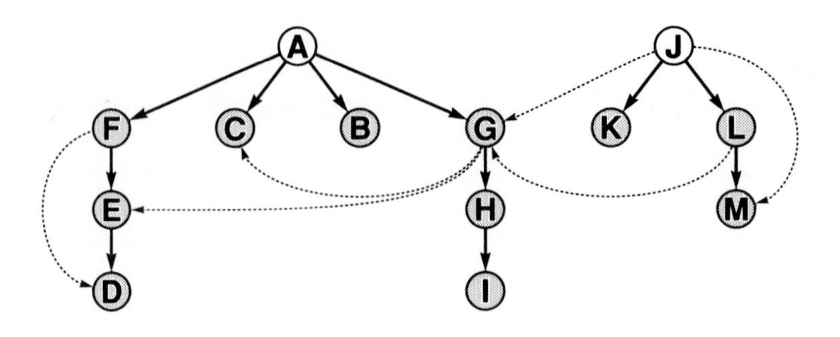

Figure 32.9 Depth-first search in a dag.

graphs that represent manufacturing processes, for it gives a specific way to proceed within the constraints represented by the graph. This operation is called *topological sorting*, because it involves ordering the vertices of the graph.

In general, the vertex order produced by a topological sort is not unique. For example, the order

A J G F K L E M B H C I D

is a legal topological ordering for our example (and there are many others). In the manufacturing application mentioned, this situation arises when one job has no direct or indirect dependence on another and thus they can be performed in either order.

It is occasionally useful to interpret the edges in a graph the other way around: to say that an edge directed from x to y means that vertex x "depends" on vertex y. For example, the vertices might represent terms to be defined in a programming language manual (or a book on algorithms!) with an edge from x to y if the definition of x uses y. In this case, it would be useful to find an ordering with the property that every term is defined before it is used in another definition. This corresponds to positioning the vertices in a line so that edges all go from right to left. A *reverse topological order* for our sample graph is:

D E F C B I H G A K M L J

The distinction here is not crucial: performing a reverse topological sort on a graph is equivalent to performing a topological sort on the graph obtained by reversing all the edges.

But we've already seen an algorithm for reverse topological sorting, the standard recursive depth-first search procedure of Chapter 29! When the input graph is a dag, simply changing *visit* to print out the vertex visited just before *exiting* (for example by inserting *write*(*name*[*k*]) right at the end) causes *dfs* to print out the vertices in reverse topological order. A simple induction argument proves that this works: we print out the name of each vertex *after* we've printed out the names of all the vertices that it points to. When *visit* is changed in this way and run on our example, it prints out the vertices in the reverse topological order given above. Printing out the vertex name on exit from this recursive procedure is exactly equivalent to putting the vertex name on a stack on entry, then popping it and printing it on exit. There is no reason to use an explicit stack in this case, since the mechanism for recursion provides it automatically; we will, however, need a stack for the more difficult problem to be considered next.

Strongly Connected Components

If a graph contains a *directed cycle* (if we can get from a node back to itself by following edges in the indicated direction), then it is not a dag and it can't be topologically sorted: whichever vertex on the cycle is printed out first will have another vertex that points to it that hasn't yet been printed out. The nodes on the cycle are mutually accessible in the sense that there is a way to get from every node on the cycle to another node on the cycle and back. On the other hand, even though a graph may be connected, it is unlikely that any node can be reached from any other via a directed path. In fact, the nodes divide themselves into sets called *strongly connected components* with the property that all nodes within a component are mutually accessible, but there is no way to get from a node in one component to a node in another component and back. The strongly connected components of the directed graph in Figure 32.1 are two single nodes B and K, one pair of nodes H I, one triple of nodes D E F, and one large component with six nodes A C G J L M. For example, vertex A is in a different component from vertex F because though there is a path from A to F, there is no way to get from F to A.

The strongly connected components of a directed graph can be found using a variant of depth-first search, as the reader may have come to expect. The method we'll examine was discovered by R. E. Tarjan in 1972. Since it is based on depth-first search, it runs in time proportional to $V + E$, but it is actually quite an ingenious method. It requires only a few simple modifications to our basic *visit* procedure, but before Tarjan presented it, no linear-time algorithm was known for this problem, even though many people had worked on it.

The modified version of depth-first search we use to find the strongly connected components of a graph is quite similar to the program we studied in Chapter 30 for finding biconnected components. The recursive *visit* function given below uses the same *min* computation to find the highest vertex reachable (via an up link) from any descendant of vertex k, but uses the value of *min* in a slightly different way to write out the strongly connected components:

```
function visit(k: integer): integer;
  var t: link;
      m, min: integer;
  begin
  id:=id+1; val[k]:=id; min:=id;
  stack[p]:=k; p:=p+1;
  t:=adj[k];
  while t<>z do
    begin
    if val[t↑.v]=0
      then m:=visit(t↑.v)
      else m:=val[t↑.v];
    if m<min then min:=m;
    t:=t↑.next
    end;
  if min=val[k] then
    begin
    repeat
      p:=p-1; write(name(stack[p]));
      val[stack[p]]:=V+1
    until stack[p]=k;
    writeln
    end;
  visit:=min;
  end;
```

This program pushes the vertex names onto a stack on entry to *visit*, then pops them and prints them on exit from visiting the last member of each strongly connected component. The point of the computation is the test whether *min=val*[*k*] at the end: if so, all vertices encountered since entry (except those already printed out) belong to the same strongly connected component as *k*. As usual, this program can easily be modified to do more sophisticated processing than simply writing out the components.

Property 32.3 *The strongly connected components of a graph can be found in linear time.*

A fully rigorous proof that the algorithm above computes the strongly connected components is beyond the scope of this book, but we can sketch the main ideas. The method is based on two observations that we've already made in other contexts. First, once we reach the end of a call to *visit* for a vertex, then we won't encounter any more vertices in the same strongly connected component (because all the vertices that can be reached from that vertex have been processed, as we noted above for topological sorting). Second, the up links in the tree provide a second path from one vertex to another and bind together the strong components. As with the algorithm in Chapter 30 for finding articulation points, we keep track of the highest ancestor reachable via one up link from all descendants of each node. Now, if a vertex x has no descendants or up links in the depth-first search tree, or if it has a descendant in the depth-first search tree with an up link that points to x and no descendants with up links that point higher up in the tree, then it and all its descendants (except those vertices satisfying the same property and their descendants) comprise a strongly connected component. Thus, in the depth-first search tree of Figure 32.2, nodes B and K satisfy the first condition (so they represent strongly connected components themselves) and nodes F (representing F E D), H (representing H I), and A (representing A G J L M C) satisfy the second condition. The members of the component represented by A are found by deleting B K F and their descendants (they appear in previously discovered components). Every descendant y of x that does not satisfy this same property has some descendant that has an up link pointing higher than y in the tree. There is a path from x to y down through the tree; and a path from y to x can be found by going down from y to the vertex with the up link that reaches past y, then continuing the same process until x is reached. A crucial extra twist is that once we're done with a vertex, we give it a high *val*, so that cross links to that vertex are ignored. ■

This program provides a deceptively simple solution to a relatively difficult problem. It is certainly testimony to the subtleties involved in searching directed graphs, subtleties that can be handled (in this case) by a carefully crafted recursive program.

□

Exercises

1. Give the adjacency matrix for the transitive closure of the dag in Figure 32.8.
2. What would be the result of running the transitive closure algorithms on an undirected graph that is represented with an adjacency matrix?
3. Write a program to determine the number of edges in the transitive closure of a given directed graph, using the adjacency list representation.
4. Discuss how Warshall's algorithm compares with the transitive closure algorithm derived by using the depth-first search technique described in the text, but using the adjacency-matrix form of *visit* and removing the recursion.
5. Give the topological ordering produced for the dag given in Figure 32.8 when the suggested method is used with an adjacency matrix representation, but *dfs* scans the vertices in reverse order (from V down to 1) when looking for unvisited vertices.
6. Does the shortest-path algorithm from Chapter 31 work for directed graphs? Explain why or give an example for which it fails.
7. Write a program to determine whether or not a given directed graph is a dag.
8. How many strongly connected components are there in a dag? In a graph with a directed cycle of size V?
9. Use your programs from Chapters 29 and 30 to produce large random directed graphs with V vertices. How many strongly connected components do such graphs tend to have?
10. Write a program that is functionally analogous to *find* from Chapter 30, but maintains *strongly* connected components of the *directed* graph described by the input edges. (This is not an easy problem; you certainly won't be able to get a program as efficient as *find*.)

33

Network Flow

☐ Weighted directed graphs are useful models for several types of applications involving commodities flowing through an interconnected network. Consider, for example, a network of oil pipes of varying sizes, interconnected in complex ways, with switches controlling the direction of flow at junctions. Suppose further that the network has a single source (say, an oil field) and a single destination (say, a large refinery) to which all of the pipes ultimately connect. What switch settings will maximize the amount of oil flowing from source to destination? Complex interactions involving material flow at junctions make this *network flow problem* a nontrivial one to solve.

This same general setup can be used to describe traffic flowing along highways, materials flowing through factories, etc. Many different versions of the problem have been studied, corresponding to many different practical situations where it has been applied. There is clearly strong motivation to find an efficient algorithm for these problems.

This type of problem lies at the interface between computer science and the field of *operations research*. Operations researchers are generally concerned with mathematical modeling of complex systems for the purpose of (preferably optimal) decision-making. Network flow is a typical example of an operations research problem; we'll briefly touch upon some others in Chapters 42–45.

In Chapter 43, we'll study *linear programming*; a general approach to solving the complex mathematical equations that typically result from operations research models. For specific problems, such as the network flow problem, better algorithms are possible. In fact, we'll see that the classical solution to the network flow problem is closely related to the graph algorithms that we have been examining, and that it is rather easy to develop a program to solve the problem using the algorithmic tools we have developed. But this problem is one that is still actively being studied: unlike many of the problems we've looked at, the "best" solution has not yet been found and new algorithms are still being discovered.

The Network Flow Problem

Consider the idealized drawing of a small network of oil pipes shown in Figure 33.1. The pipes are of fixed capacity proportional to their size and oil can flow only downhill (from top to bottom). Furthermore, switches at each junction control how much oil goes in each direction. No matter how the switches are set, the system reaches a state of equilibrium when the amount of oil flowing into the system on the top is equal to the amount flowing out at the bottom (this is the quantity that we want to maximize) and when the amount of oil flowing in at each junction is equal to the amount of oil flowing out. We measure both flow and pipe capacity in terms of integral units (say, gallons per second).

It is not immediately obvious that the switch settings can really affect the total maximum flow: Figure 33.1 illustrates that they can. First, suppose that the switch controlling pipe AB is opened, filling that pipe, the pipe BD, and nearly filling DF, as shown in the left diagram in the figure. Next suppose that pipe AC is opened and switch C is set to close pipe CD and open pipe CE (perhaps the operator of switch D has informed the operator of switch C that he can't handle much more because of the load from B). The resulting flow is shown in the middle diagram of the figure: pipes BD and CE are full. Now, the flow could be increased some by sending enough through the path ACDF to fill pipe DF, but there is a better solution, as shown in the third diagram. By changing switch B to redirect enough flow to fill BE, we open up enough capacity in pipe DF to allow switch C to fully open pipe CD. The total flow into and out of the network is increased by finding the proper switch settings.

Our challenge is to develop an algorithm that can find the "proper" switch settings for any network. Furthermore, we want to be assured that no other switch setting will give a higher flow.

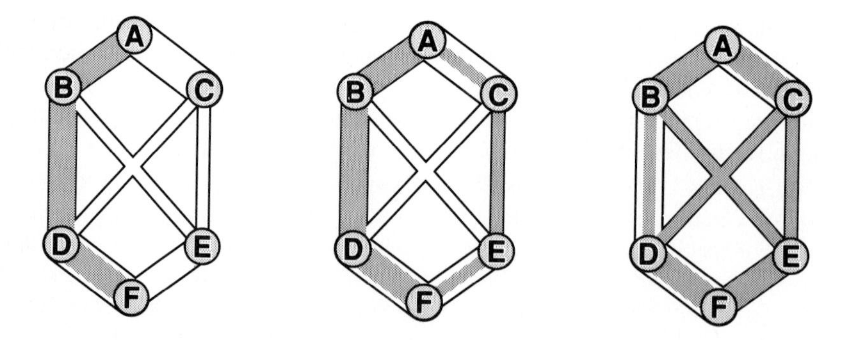

Figure 33.1 Maximum flow in a simple network.

This situation can obviously be modeled by a directed graph, and it turns out that the programs we have studied can apply. Define a *network* as a weighted directed graph with two distinguished vertices: one with no edges pointing in (the *source*), one with no edges pointing out (the *sink*). The weights on the edges, which we assume to be non-negative, are called the *edge capacities*. Now, a *flow* is defined as another set of weights on the edges such that the flow on each edge is equal to or less than the capacity, and the flow into each vertex is equal to the flow out of that vertex. The *value* of the flow is the flow into the source (or out of the sink). The *network flow problem* is to find a flow with maximum value for a given network.

Networks can obviously be represented with either the adjacency-matrix or adjacency-list representations we have used for graphs in previous chapters. Instead of a single weight, two weights are associated with each edge, the *size* and the *flow*. These can be represented as two fields in an adjacency-list node, as two matrices in the adjacency-matrix representation, or two fields within a single record in either representation. Even though networks are directed graphs, the algorithms we'll be examining need to traverse edges in the "wrong" direction, so we use an undirected graph representation: if there is an edge from x to y with size s and flow f, we also keep an edge from y to x with size $-s$ and flow $-f$. In an adjacency list representation, it is necessary to maintain links connecting the two list nodes that represent each edge, so that when we change the flow in one we can update it in the other.

Ford-Fulkerson Method

The classical approach to the network flow problem was developed by L. R. Ford and D. R. Fulkerson in 1962. They gave a method to improve any legal flow (except, of course, the maximum). Starting with a zero flow, we apply the method repeatedly. As long as the method can be applied, it produces an increased flow; if it can't be applied, the maximum flow has been found. In fact, the flow in Figure 33.1 was developed using this method; we now reexamine it in terms of the graph representation shown in Figure 33.2.

For simplicity, we omit the arrows, since they all point down. The methods we consider are not restricted to graphs that can be drawn with all edges pointing in one direction. We use such graphs because they provide good intuition for understanding network flow in terms of liquids flowing in pipes.

Consider any directed (down) path through the network (from source to sink). Clearly, the flow can be increased by at least the smallest amount of unused capacity on any edge on the path, by increasing the flow in all edges on the path by that amount. In the left diagram in Figure 33.2, this rule is applied along the path ABDF; then in the center diagram, it is applied along the path ACEF.

As mentioned above, we could then apply the rule along the path ACDF, creating a situation where all directed paths through the network have at least one

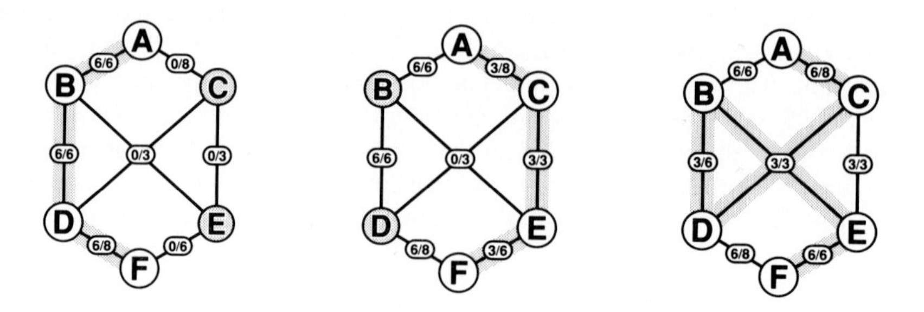

Figure 33.2 Finding the maximum flow in a network.

edge filled to capacity. But there is another way to increase the flow: we can consider arbitrary paths through the network that can contain edges that point the "wrong way" (from sink to source along the path). The flow can be increased along such a path by *increasing* the flow on edges from source to sink and *decreasing* the flow on edges from sink to source by the same amount. In our example, the flow through the network can be increased by 3 along the path ACDBEF, as shown in the third diagram in Figure 33.2. As described above, this corresponds to adding 3 units of flow through AC and CD, then diverting 3 units at switch B from BD to BE and EF. We don't lose any flow in DF because 3 of the units that used to come from BD now come from CD.

To simplify terminology, we'll call edges that flow from source to sink along a particular path *forward* edges and edges that flow from sink to source *backward* edges. Notice that the amount by which the flow can be increased is limited by the minimum of the unused capacities in the forward edges and the minimum of the flows in the backward edges. Put another way, in the new flow, at least one of the forward edges along the path becomes full or at least one of the backward edges along the path becomes empty. Furthermore, the flow can't be increased on any path containing a full forward edge or an empty backward edge.

The paragraph above gives a method for increasing the flow on any network, *provided* that a path with no full forward edges or empty backward edges can be found. The crux of the Ford-Fulkerson method is the observation that if no such path can be found then the flow is maximal.

Property 33.1 *If every path from the source to the sink in a network has a full forward edge or an empty backward edge, then the flow is maximal.*

To prove this fact, first go through the graph and identify the first full forward or empty backward edge on every path. This set of edges *cuts* the graph in two parts. (In our example, the edges AB, CD, and CE comprise such a cut.) For any cut of the network into two parts, we can measure the flow "across" the cut: the total of the flow on the edges that go from the source to the sink. In general, edges may go both ways across the cut: to get the flow across the cut, the total of the flow on the edges going the other way must be subtracted. Our example cut has a value of 12, which is equal to the total flow for the network. It turns out that whenever the cut flow equals the total flow, we know not only that the flow is maximal, but also that the cut is minimal (that is, every other cut has at least as high a "crossflow"). This is called the *maxflow-mincut theorem*: the flow couldn't be any larger (otherwise the cut would have to be larger also), and no smaller cuts exist (otherwise the flow would have to be smaller also). We omit details of this proof. ■

Network Searching

The Ford-Fulkerson method described above may be summarized as follows: "start with zero flow everywhere and increase the flow along any path from source to sink with no full forward edges or empty backward edges, continuing until there are no such paths in the network." But this is not an *algorithm* in the usual sense, since the method for finding paths is not specified, and any path at all could be used. For example, one might base the method on the intuition that the longer the path, the more the network is filled up, and thus that long paths should be preferred. But the (classical) example shown in Figure 33.3 demonstrates that some care must be exercised.

In this network, if the first path chosen is ABCD, then the flow is increased by only one. Then the second path chosen might be ACBD, again increasing the flow by one, and leaving a situation identical to the initial situation, except that the flows on the outside edges are increased by one. Any algorithm that chose those two paths (for example, one that looks for long paths) would continue with this strategy, thus requiring 1000 pairs of iterations before the maximum flow is found. If the numbers on the sides were a billion, then two billion iterations would be used. Obviously, this is an undesirable situation, since the paths ABC and ADC give the maximum flow in just two steps. For the algorithm to be useful, we must

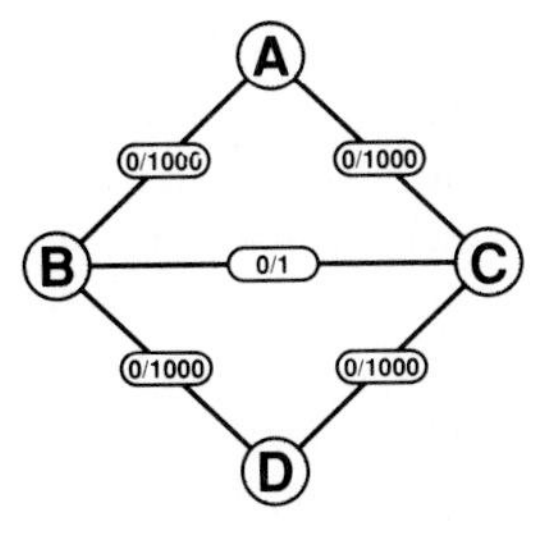

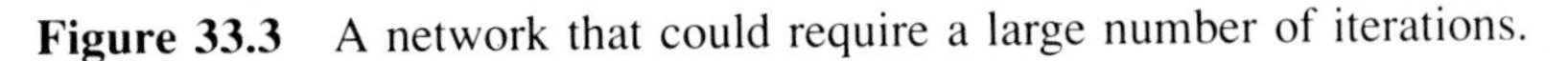
Figure 33.3 A network that could require a large number of iterations.

avoid having the running time so dependent on the magnitude of the capacities. Fortunately, this problem is easily eliminated:

Property 33.2 *If the shortest available path from source to sink is used in the Ford-Fulkerson method, then the number of paths used before the maximum flow is found in a network of V vertices and E edges must be less than VE.*

This fact was proven by Edmonds and Karp in 1972. Details of the proof are beyond the scope of this book. ▪

In other words, a good plan is simply to use an appropriately modified version of breadth-first search to find the path. The bound given in Property 33.2 is a worst-case bound: a typical network is likely to require many fewer steps.

With the priority graph traversal method of Chapter 31, we can implement another method suggested by Edmonds and Karp: *find the path through the network that increases the flow by the largest amount.* This can be achieved simply by using a variable for *priority* (whose value is set appropriately) in either the adjacency list or the adjacency matrix "priority-first search" methods of Chapter 31. For the matrix representation, the following statements compute the priority, and the code for the list representation is similar:

```
if size[k,t]>0
   then priority:=size[k,t]-flow[k,t]
   else priority:=-flow[k,t];
if priority>val[k] then priority:=val[k];
```

Then, since we want to take the node with the *highest* priority value, we must either reorient the priority-queue mechanisms in those programs to return the maximum instead of the minimum or use them as is with *priority* set to *maxint−1−priority* (and the process reversed when the value is removed). Also, we modify the priority-first search procedure to take the source and sink as arguments, then to start each search at the source and stop when a path to the sink has been found. If such a path is not found, the partial priority search tree defines a mincut for the network; otherwise the flow can be improved. Finally, the *val* for the source should be set to *maxint* before the search is started, to indicate that any amount of flow can be achieved at the source (though this is immediately restricted by the total capacity of all the pipes leading directly out of the source).

With *matrixpfs* implemented as described in the previous paragraph, finding the maximum flow is actually quite simple, as shown by the following program:

```
repeat
  matrixpfs(1,V);
  y:=V; x:=dad[V];
  while x<>0 do
    begin
    flow[x,y]:=flow[x,y]+val[V];
    flow[y,x]:=-flow[x,y];
    y:=x; x:=dad[y]
    end
until val[V]=1-maxint
```

This program assumes an adjacency-matrix representation is used for the network. As long as *matrixpfs* can find a path that increases the flow (by the maximum amount), we trace back through the path (using the *dad* array constructed by *matrixpfs*) and increase the flow as indicated. If V remains unseen after some call to *matrixpfs*, then a mincut has been found and the algorithm terminates.

As we have seen, the algorithm first increases the flow along the path ABDF, then along ACEF, then along ACDBEF. We see (finally) why the method does not use the choice ACDF for the third path: that would increase the flow by only one unit, not the three units available with the longer path. (Note that the breadth-first "shortest-path-first" method of Property 33.2 would make this choice. Then one more iteration would find the maximum flow.)

Though this algorithm is easily implemented and is likely to work well for networks arising in practice, its analysis is quite complicated. First, as usual, *matrixpfs* requires V^2 steps in the worst case; alternatively we could use *listpfs* to run in time proportional to $(E + V)\log V$ per iteration, though the algorithm is likely to run somewhat faster than this, since it stops when it reaches the sink. But how many iterations are required?

Property 33.3 *If the path from source to sink that increases the flow by the largest amount is used in the Ford-Fulkerson method, then the number of paths used before the maximum flow is found in a network is less than* $1 + \log_{M/M-1} f^*$ *where* f^* *is the cost of the flow and* M *is the maximum number of edges in a cut of the network.*

Again, proof of this fact, first given by Edmonds and Karp, is quite beyond the scope of this book. This quantity is certainly complicated to compute, but is unlikely to be large for real networks. ■

We mention this property to indicate not how long the algorithm might take on an actual network, but rather the complexity of the analysis. Actually, this problem has been quite widely studied, and complicated algorithms with much better worst-case bounds have been developed. However, the Edmonds-Karp algorithm as implemented above is likely to be difficult to beat for networks arising in practical applications. Figure 33.4 shows the algorithm operating on a larger network.

Figure 33.4 Finding the maximum flow in a larger network.

The network flow problem can be extended in several ways, and many variations have been studied in some detail because they are important in actual applications. For example, the *multicommodity flow problem* involves introducing into the network multiple sources, sinks, and types of material. This makes the problem much more difficult and requires more advanced algorithms than those considered here: for example, no analogue to the max-flow min-cut theorem is known to hold for the general case. Other extensions to the network flow problem include placing capacity constraints on vertices (easily handled by introducing artificial edges to handle these capacities), allowing undirected edges (also easily handled by replacing undirected edges by pairs of directed edges), and introducing lower bounds on edge flows (not so easily handled). If we make the realistic assumption that pipes have associated costs as well as capacities, then we have the *min-cost* flow problem, a quite difficult problem from operations research.

□

Exercises

1. Give an algorithm to solve the network flow problem for the case that the network forms a tree if the sink is removed.

2. What paths are traced by the algorithm referred to in Property 33.3 when finding the maximum flow in the network obtained by adding edges from B to C and E to D, both with weight 3?

3. Draw the priority search trees computed on each call to *matrixpfs* for the example discussed in the text.

4. Give the contents of the *flow* matrix after each call to *matrixpfs* for the example discussed in the text.

5. True or false: no algorithm can find the maximum flow without examining every edge in the network.

6. What happens to the Ford-Fulkerson method when the network has a directed cycle?

7. Give a simplified version of the Edmonds-Karp bound for the case that all the capacities are $O(1)$.

8. Give a counterexample that shows why depth-first search is not appropriate for the network flow problem.

9. Implement the breadth-first search solution to the network flow problem, using *sparsepfs*.

10. Write a program to find maximum flows in random networks with V nodes and about $10V$ edges. How many calls to *sparsepfs* are made for $V = 25, 50, 100$?

34

Matching

A problem that often arises is to "pair up" objects according to preference relationships that are likely to conflict. For example, a quite complicated system has been set up in the U.S. to place graduating medical students into hospital residency positions. Each student lists several hospitals in order of preference, and each hospital lists several students in order of preference. The problem is to assign students to positions in a fair way, respecting all the stated preferences. A sophisticated algorithm is required because the best students are likely to be preferred by several hospitals and the best hospital positions are likely to be preferred by several students. It's not even clear that each hospital position can be filled by a student whom the hospital has listed or that student can be assigned to a position that the student has listed, let alone that the order in the preference lists can be respected. This frequently occurs, in fact after the algorithm has done the best that it can, there is a last minute scramble among unmatched hospitals and students to complete the process.

This example is a special case of a difficult fundamental problem on graphs that has been widely studied. Given a graph, a *matching* is a subset of the edges in which no vertex appears more than once. That is, each vertex touched by one of the edges in the matching is paired with the other vertex of that edge, but some vertices may be left unmatched. Even if we insist that a matching covers as many vertices as possible, in the sense that none of the edges not in the matching should connect unmatched vertices, different ways of choosing the edges could lead to different numbers of leftover (unmatched) vertices.

Of particular interest is a *maximum matching*, which contains as many edges as possible or, equivalently, minimizes the number of unmatched vertices. The best we can hope to do is to have a set of edges in which each vertex appears exactly once (such a matching in a graph with $2V$ vertices would have V edges), but it is not always possible to achieve this.

Figure 34.1 shows a maximum matching (the shaded edges) for our sample

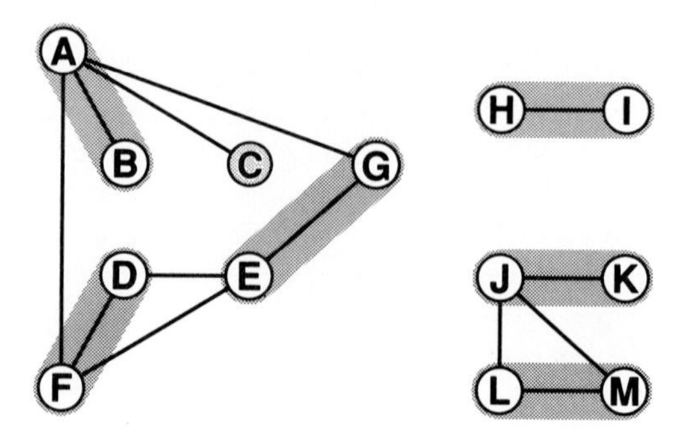

Figure 34.1 A maximum matching (shaded edges).

graph. With 13 vertices, we can't do better than a matching with six edges. But simple algorithms for finding matchings will have difficulty even on this example. For example, one method one might try would be to take eligible edges for the matching as they appear in depth-first search (cf. Figure 29.7). For the example in Figure 34.1, this would give the five edges AF EG HI JK LM, not a maximum matching. Also, as just mentioned, it is not even easy to tell how many edges there are in a maximum matching for a given graph. For example, note that there is no three-edge matching for the subgraph consisting of just the six vertices A through F and the edges connecting them. While it is often very easy to get a large matching on a big graph (for example, it is not difficult to find a maximum matching for the "maze" graph of Chapter 29), developing an algorithm to find the maximum matching for any graph is a difficult task indeed, as indicated by counterexamples such as these.

For the medical-student matching problem described above, the students and hospitals correspond to nodes in the graph; their preferences to edges. If they assign values to their preferences (perhaps using the time-honored "1–10" scale), then we have the *weighted matching* problem: given a weighted graph, find a set of edges in which no vertex appears more than once such that the sum of the weights on the edges in the set chosen is maximized. Below we'll see another alternative, where we respect the order in the preferences but do not require (perhaps arbitrary) values to be assigned to them.

The matching problem has attracted a great deal of attention among mathematicians because of its intuitive nature and its wide applicability. Its solution in the general case involves intricate and beautiful combinatorial mathematics quite beyond the scope of this book. Our intent here is to provide the reader with an appreciation for the problem by considering some interesting special cases while at the same time developing some useful algorithms.

Bipartite Graphs

The example mentioned above, matching medical students to residencies, is certainly representative of many other matching applications. For example, we might be matching men and women for a dating service, job applicants to available positions, courses to available hours, or members of Congress to committee assignments. The graphs arising in such cases are called *bipartite graphs*, defined as graphs in which all edges go between two sets of nodes. That is, the nodes divide into two sets and no edges connect two nodes in the same set. (Obviously, we wouldn't want to "match" one job applicant to another or one committee assignment to another.) An example of a bipartite graph is shown in Figure 34.2. The reader might be amused to search for a maximum matching in this graph.

In an adjacency-matrix representation for bipartite graphs, one can achieve obvious savings by including only rows for one set and only columns for the other set. In an adjacency-list representation, no particular saving suggests itself except to name the vertices intelligently so that it is easy to tell which set a vertex belongs to.

In our examples, we use letters for nodes in one set, numbers for nodes in the other. The maximum matching problem for bipartite graphs can be simply expressed in this representation: "Find the largest subset of a set of letter-number pairs with the property that no two pairs have the same letter or number." Finding the maximum matching for the bipartite graph in Figure 34.2 corresponds to solving this puzzle on the pairs E5 A2 A1 C1 B4 C3 D3 B2 A4 D5 E3 B1.

It is an interesting exercise to attempt to find a direct solution to the matching problem for bipartite graphs. The problem seems easy at first glance, but subtleties quickly become apparent. Certainly there are far too many pairings to try all possibilities: a solution to the problem must be clever enough to try only a few of the possible ways to match the vertices.

The solution we'll examine is an indirect one: to solve a particular instance of the matching problem, we'll construct an instance of the network flow problem,

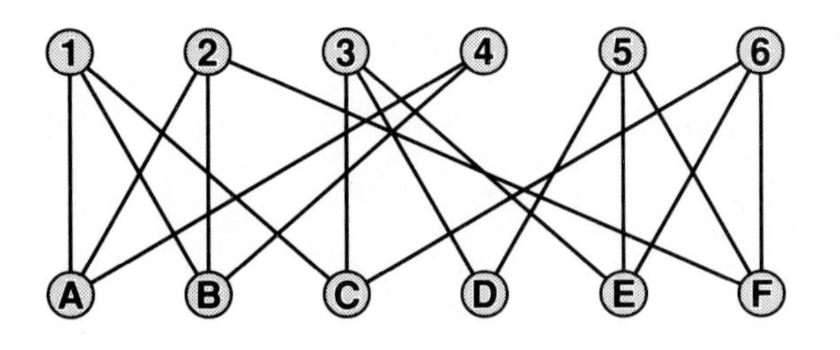

Figure 34.2 A bipartite graph.

use the algorithm from the previous chapter, then use the solution to the network flow problem to solve the matching problem. That is, we *reduce* the matching problem to the network flow problem. Reduction is a method of algorithm design somewhat akin to the use of a library subroutine by a systems programmer. It is of fundamental importance in the theory of advanced combinatorial algorithms (see Chapter 40). For the moment, reduction will provide us with an efficient solution to the bipartite matching problem.

The construction is straightforward: given an instance of bipartite matching, construct an instance of network flow by creating a source vertex with edges pointing to all the members of one set in the bipartite graph, then make all the edges in the bipartite graph point from that set to the other, then add a sink vertex pointed to by all the members of the other set. All of the edges in the resulting graph are given a capacity of one.

Figure 34.3 shows what happens when we construct a network flow problem from the bipartite graph of Figure 34.2, then use the network flow algorithm of the previous chapter. Note that the bipartite property of the graph, the direction of the flow, and the fact that all capacities are one force each path through the network to correspond to an edge in a matching: in the example, the paths found in the first four steps correspond to the partial matching A1 B2 C3 D5. Each time the network flow algorithm calls *pfs* it either finds a path that increases the flow by

Figure 34.3 Using network flow to find a maximum matching in a bipartite graph.

one or terminates.

In the fifth step, all forward paths through the network are full, and the algorithm must use backward edges. The path found in this step is the path 4B2F. This path clearly increases the flow in the network, as described in the previous chapter. In the present context, we can think of the path as a set of instructions to create a new partial matching (with one more edge) from the current one. This construction follows in a natural way from tracing through the path in order: "4A" means add A4 to the matching, "B2" means remove B2, and "2F" means add F2 to the matching. Thus, after this path is processed, we have the matching A1 B4 D3 D5 E6 F2; equivalently, the flow in the network is given by full pipes in the edges connecting those nodes. The algorithm finishes by making the matching F6; all the pipes leaving the source and entering the sink full, so that we have a maximum matching.

The proof that the matching is exactly those edges filled to capacity by the maxflow algorithm is straightforward. First, the network flow always gives a legal matching: since each vertex has an edge of capacity one either coming in (from the sink) or going out (to the source), at most one unit of flow can go through each vertex, which implies that each vertex will be included at most once in the matching. Second, no matching can have more edges, since any such matching would lead directly to a better flow than that produced by the maxflow algorithm.

Thus, to compute the maximum matching for a bipartite graph we simply format the graph so as to be suitable for input to the network flow algorithm of the previous chapter. Of course, the graphs presented to the network flow algorithm in this case are much simpler than the general graphs the algorithm is designed to handle, and it turns out that the algorithm is somewhat more efficient for this case.

Property 34.1 *A maximum matching in a bipartite graph can be found in $O(V^3)$ steps if the graph is dense or in $O(V(E+V)\log V)$ steps if the graph is sparse.*

The construction ensures that each call to *pfs* adds one edge to the matching, so we know that there are at most $V/2$ calls to *pfs* during the execution of the algorithm. Thus, the time taken is proportional to a factor of V greater than the time for a single search as discussed in Chapter 31. ■

Stable Marriage Problem

The example given at the beginning of this chapter, involving medical students and hospitals, is obviously taken quite seriously by the participants. But the method we'll examine for doing the matching is perhaps better understood in terms of a somewhat whimsical model of the situation. We assume that we have N men and N women who have expressed mutual preferences (each man must say exactly how he feels about each of the N women and vice versa). The problem is to find a set of N marriages that respects everyone's preferences.

A	B	C	D	E
2	1	2	1	5
5	2	3	3	3
1	3	5	2	2
3	4	4	4	1
4	5	1	5	4

1	2	3	4	5
E	D	A	C	D
A	E	D	B	B
D	B	B	D	C
B	A	C	A	E
C	C	E	E	A

Figure 34.4 Preference lists for the stable marriage problem.

How should the preferences be expressed? One method would be to use the "1–10" scale, each side assigning an absolute score to certain members of the other sex. This makes the marriage problem the same as the weighted matching problem, a relatively difficult problem to solve. Furthermore, use of absolute scales in itself can lead to inaccuracies, since people's scales will be inconsistent (one woman's 10 might be another woman's 7). A more natural way to express the preferences is to have each person list in order of preference all the people of the opposite sex. Figure 34.4 shows a set of preference lists that might exist among a set of five women and five men. As usual (and to protect the innocent!), we assume that hashing or some other method has been used to translate actual names to single digits for women and single letters for men.

Clearly, these preferences often conflict: for example, both A and C list 2 as their first choice, and nobody seems to want 4 very much (but someone must get her). The problem is to engage all the women to all the men in such a way as to respect all their preferences as much as possible, then perform N marriages in one grand ceremony. In developing a solution, we must assume that anyone assigned to someone less than their first choice will be disappointed and will always prefer anyone higher up on the list. A set of marriages is called *unstable* if two people who are not married both prefer each other to their spouses. For example, the assignment A1 B3 C2 D4 E5 is unstable because A prefers 2 to 1 and 2 prefers A to C. Thus, acting according to their preferences, A would leave 1 for 2 and 2 would leave C for A (leaving 1 and C with little choice but to get together).

Finding a stable configuration seems on the face of it a difficult problem, since there are so many possible assignments. Even determining whether a configuration is stable is not simple, as the reader may discover by looking (before reading the next paragraph) for the unstable couple in the example above after the new matches A2 and C1 have been made. In general, there are many different stable assignments for a given set of preference lists, and we need to find only one. (Finding *all* stable assignments is a much more difficult problem.)

One possible algorithm for finding a stable configuration might be to remove unstable couples one at a time. However, not only is this process slow because of the time required to determine stability, but it also does not even necessarily terminate! For example, after A2 and C1 have been matched in the example above, B and 2 make an unstable couple, which leads to the configuration A3 B2 C1 D4 E5. In this arrangement, B and 1 make an unstable couple, which leads to the configuration A3 B1 C2 D4 E5. Finally, A and 1 make an unstable configuration that leads back to the original configuration. An algorithm that attempts to solve the stable marriage problem by removing stable pairs one by one is bound to get caught in this type of loop.

We'll look instead at an algorithm that tries to build stable pairings systematically using a method based on what might happen in the somewhat idealized "real-life" version of the problem. The idea is to have each man, in turn, become a "suitor" and seek a bride. Obviously, the first step in his quest is to propose to the first woman on his list. If she is already engaged to a man whom she prefers, then our suitor must try the next woman on his list, continuing until he finds a woman who is not engaged or who prefers him to her current fiancée. If this women is not engaged, then she becomes engaged to the suitor and the next man becomes the suitor. If she is engaged, then she breaks the engagement and becomes engaged to the suitor (whom she prefers). This leaves her old fiancée with nothing to do but become the suitor once again, starting where he left off on his list. Eventually he finds a new fiancée, but another engagement may need to be broken. We continue in this way, breaking engagements as necessary, until some suitor finds a woman who has not yet been engaged.

This method may model what happens in some 19th-century novels, but some careful examination is required to show that it produces a stable set of assignments. Figure 34.5 shows the sequence of events for the initial stages of the process for our example. First, A proposes to 2 (his first choice) and is accepted; then B proposes to 1 (his first choice) and is accepted; then C proposes to 2, is turned down, proposes to 3 and is accepted, as depicted in the third diagram.

Each diagram shows the sequence of events when a new man sets out as the suitor to seek a fiancée. Each line gives the "used" preference list for the corresponding man, and each link is labeled with an integer telling when that link was used by that man to propose to that woman. This extra information is useful in tracking the sequence of proposals when D and E become the suitors: When D proposes to 1, we have our first broken engagement, since 1 prefers D to B. Then B becomes the suitor and proposes to 2; this gives our second broken engagement, since 2 prefers B to A. Then A becomes the suitor and proposes to 5, which leaves a stable situation. But this stability is only temporary! The reader might wish to trace through the sequence of proposals made when E becomes the suitor: things don't settle down until after eight proposals are made. Note that E takes on the

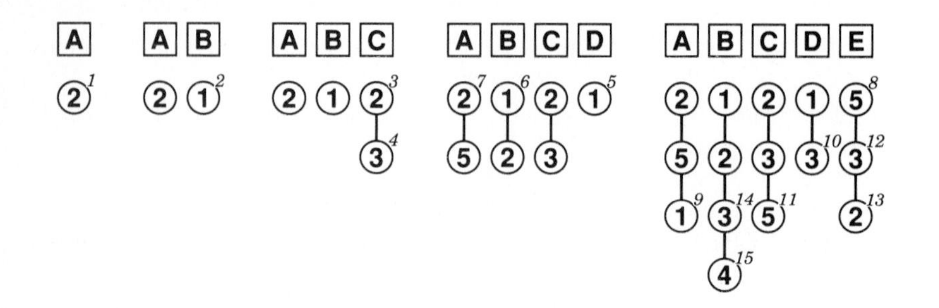

Figure 34.5 Solving the stable marriage problem.

suitor role twice in the process.

The first step in the implementation is to design the data structures to be used for the preference lists. These are both simple linear lists as abstract data structures, but as we've learned from the examples in Chapter 3 and elsewhere, proper choice of representation can directly impact performance. Also in this case, different structures are appropriate for the men and the women, since they use the preference lists in different ways.

The men simply go through their preference lists in order, so any linear-list implementation could be used. Since the preference lists are all of the same length, it is simplest to use a straightforward implementation as a two-dimensional array. For example, *prefer*[*m*,*w*] will be the wth woman in the preference list of the mth man. In addition, we need to keep track of how far each man has progressed on his list. This can be handled with a one-dimensional array *next*, initialized to zero, with *next*[*m*]+*1* the index of the next woman on man m's preference list; her identifier is found in *prefer*[*m*,*next*[*m*]+*1*].

For each woman, we need to keep track of her fiancée (*fiancée*[*w*] will be the man engaged to woman w) and we need to be able to answer the question "Is man s preferable to *fiancée*[*w*]?" This could be done by searching the preference list sequentially until either s or *fiancée*[*w*] is found, but this method would be rather inefficient if they're both near the end. What is called for is the "inverse" of the preference list: *rank*[*w*,*s*] is the index of man s on woman w's preference list. For the example above, we have that *rank*[*1*,*1*] is 2, since A is second on 1's preference list, *rank*[*5*,*4*] is 1, since D is fourth on 5's preference list, etc.

The suitability of suitor s can be very quickly determined by testing whether *rank*[*w*,*s*] is less than *rank*[*w*,*fiancée*[*w*]]. These arrays are easily constructed directly from the preference lists. To get things started, we use a "sentinel" man 0 as the initial suitor and put him at the end of all the women's preference lists.

With the data structures initialized in this way, the implementation as described above is straightforward:

```
for m:=1 to N do
  begin
  s:=m;
  repeat
    next[s]:=next[s]+1; w:=prefer[s,next[s]];
    if rank[w,s]<rank[w,fiancee[w]] then
      begin t:=fiancee[w]; fiancee[w]:=s; s:=t end;
  until s=0;
  end;
```

Each iteration starts with an unengaged man and ends with an unengaged woman. The **repeat** loop must terminate because every man's list contains every woman and each iteration of the loop involves incrementing some man's list, and thus an unengaged woman must be encountered before any man's list is exhausted. The set of engagements produced by the algorithm is stable because every woman whom any man prefers to his fiancée is engaged to someone that she prefers to him.

Property 34.2 *The stable marriage problem can be solved in linear time.*

As just mentioned, each iteration of the loop increments some man's preference list. In the worst case, all the list entries are examined (but no entries are examined twice). Actually, the algorithm might take much less time than it would take to build the lists, because a stable configuration could be found well before all the lists are exhausted. This type of consideration leads to a number of interesting analytic problems. ■

There are several obvious built-in biases in this algorithm. First, the men go through the women on their lists in order, while the women must wait for the "right man" to come along. This bias may be corrected (in a somewhat easier manner than in real life) by interchanging the order in which the preference lists are input. This produces the stable configuration 1E 2D 3A 4C 5B, where every women gets her first choice except 5, who gets her second. In general, there may be many stable configurations: it can be shown that this one is "optimal" for the women, in the sense that no other stable configuration will give any woman a better choice from her list. (Of course, the first stable configuration for our example is optimal for the men.)

Another feature of the algorithm that seems to be biased is the order in which the men become the suitor: is it better to be the first man to propose (and therefore be engaged at least for a little while to your first choice) or the last (and therefore have a reduced chance to suffer the indignities of a broken engagement)? The answer is that this is not a bias at all: the order in which the men become the suitor doesn't matter. As long as each man makes proposals and each woman accepts according to their lists, the same stable configuration results.

Advanced Algorithms

The two special cases we've examined give some indication of the intricacies of the matching problem. Though these specific algorithms are useful in several actual applications, as indicated, many other applications may require solution of more general problems.

Among the more general problems that have been studied in some detail are: the maximum matching problem for general (not necessarily bipartite) graphs; weighted matching for bipartite graphs, where edges have weights and a matching with maximum total weight is sought; and weighted matching for general graphs.

Weighted matching for bipartite graphs and similar generalizations can be handled to the extent that algorithms for generalizations of the network flow problem are known. General graphs, however, are quite another story. (The stable marriage problem might be characterized as a way of avoiding the weighted matching problem for general graphs by redefining the problem.) Treating the many techniques that have been tried for matching on general graphs would fill an entire volume: it is one of the most extensively studied problems in graph theory.

□

Exercises

1. Find all the matchings with five edges for the bipartite graph in Figure 34.2.
2. Use the algorithm given in the text to find maximum matchings for random bipartite graphs with 50 vertices and 100 edges. About how many edges are in the matchings?
3. Construct a bipartite graph with six nodes and eight edges that has a three-edge matching, or prove that none exists.
4. Suppose that vertices in a bipartite graph represent jobs and people and that each person is to be assigned to *two* jobs. Will reduction to network flow give an algorithm for this problem? Prove your answer.
5. Modify the network flow program of Chapter 33 to take advantage of the special structure of the 0-1 networks that arise for bipartite matching.
6. Write an efficient program to determine whether an assignment for the marriage problem is stable.
7. Is it possible for two men to get their last choice in the stable marriage algorithm? Prove your answer.
8. Construct a set of preference lists for $N = 4$ for the stable marriage problem where everyone gets his second choice, or prove that no such set exists.
9. Give a stable configuration for the stable marriage problem for the case where the preference lists for men and women are all the same: in ascending order.
10. Run the stable marriage program for $N = 50$, using random permutations for preference lists. About how many proposals are made during the execution of the algorithm?

SOURCES for Graph Algorithms

There are several textbooks on graph algorithms, but the reader should be forewarned that there is a great deal to be learned about graphs, that they still are not fully understood, and that they are traditionally studied from a mathematical (as opposed to an algorithmic) standpoint. Thus, many references give more rigorous and deeper coverage of much more difficult topics than we do here.

Many of the topics we've treated here are covered in the books by Mehlhorn and Tarjan. Both of these are basic references that give careful treatments of basic and advanced graph algorithms, with extensive references to the recent literature. Another source for further material is the book by Papadimitriou and Steiglitz. Though most of that book is about much more advanced topics (for example, there is a full treatment of matching in general graphs), it covers of many of the algorithms that we've discussed, including pointers to further reference material.

The application of depth-first search to solve graph connectivity and other problems is the work of R. E. Tarjan, whose original paper merits further study. The many variants on algorithms for the union-find problem of Chapter 30 are ably categorized and compared by van Leeuwen and Tarjan. The algorithms for shortest paths and minimum spanning trees in dense graphs in Chapter 31 are quite old, but the original papers by Dijkstra, Prim, and Kruskal still make interesting reading. Our treatment of the stable marriage problem in Chapter 34 is based on the entertaining account given by Knuth.

E. W. Dijkstra, "A note on two problems in connexion with graphs," *Numerische Mathematik*, **1** (1959).

D. E. Knuth, *Marriages stables*, Les Presses de l'Université de Montréal, Montréal, 1976.

J. R. Kruskal Jr., "On the shortest spanning subtree of a graph and the traveling salesman problem," *Proceedings AMS*, **7**, 1 (1956).

K. Mehlhorn, *Data Structures and Algorithms 2: NP-Completeness and Graph Algorithms*, Springer-Verlag, Berlin, 1984.

C. H. Papadimitriou and K. Steiglitz, *Combinatorial Optimization: Algorithms and Complexity*, Prentice-Hall, Englewood Cliffs, NJ, 1982.

R. C. Prim, "Shortest connection networks and some generalizations," *Bell System Technical Journal*, **36** (1957).

R. E. Tarjan, "Depth-first search and linear graph algorithms," *SIAM Journal on Computing*, **1**, 2 (1972).

R. E. Tarjan, *Data Structures and Network Algorithms*, Society for Industrial and Applied Mathematics, Philadelphia, PA, 1983.

J. van Leeuwen and R. E. Tarjan, "Worst-case analysis of set-union algorithms," *Journal of the ACM*, 1986.

Mathematical Algorithms

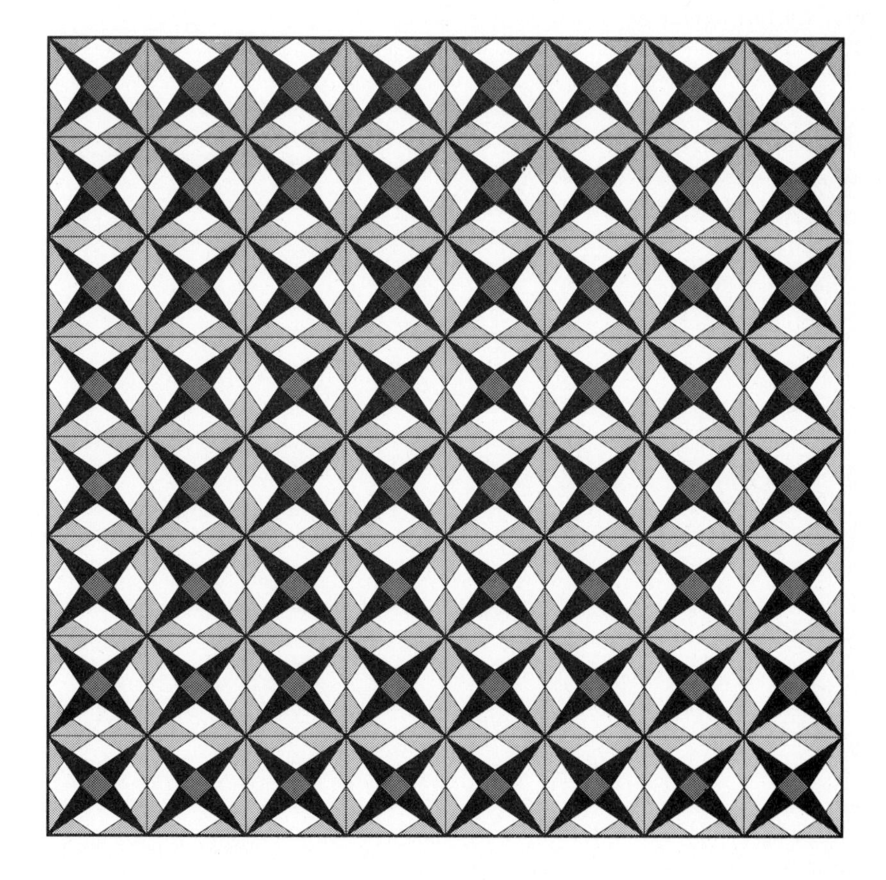

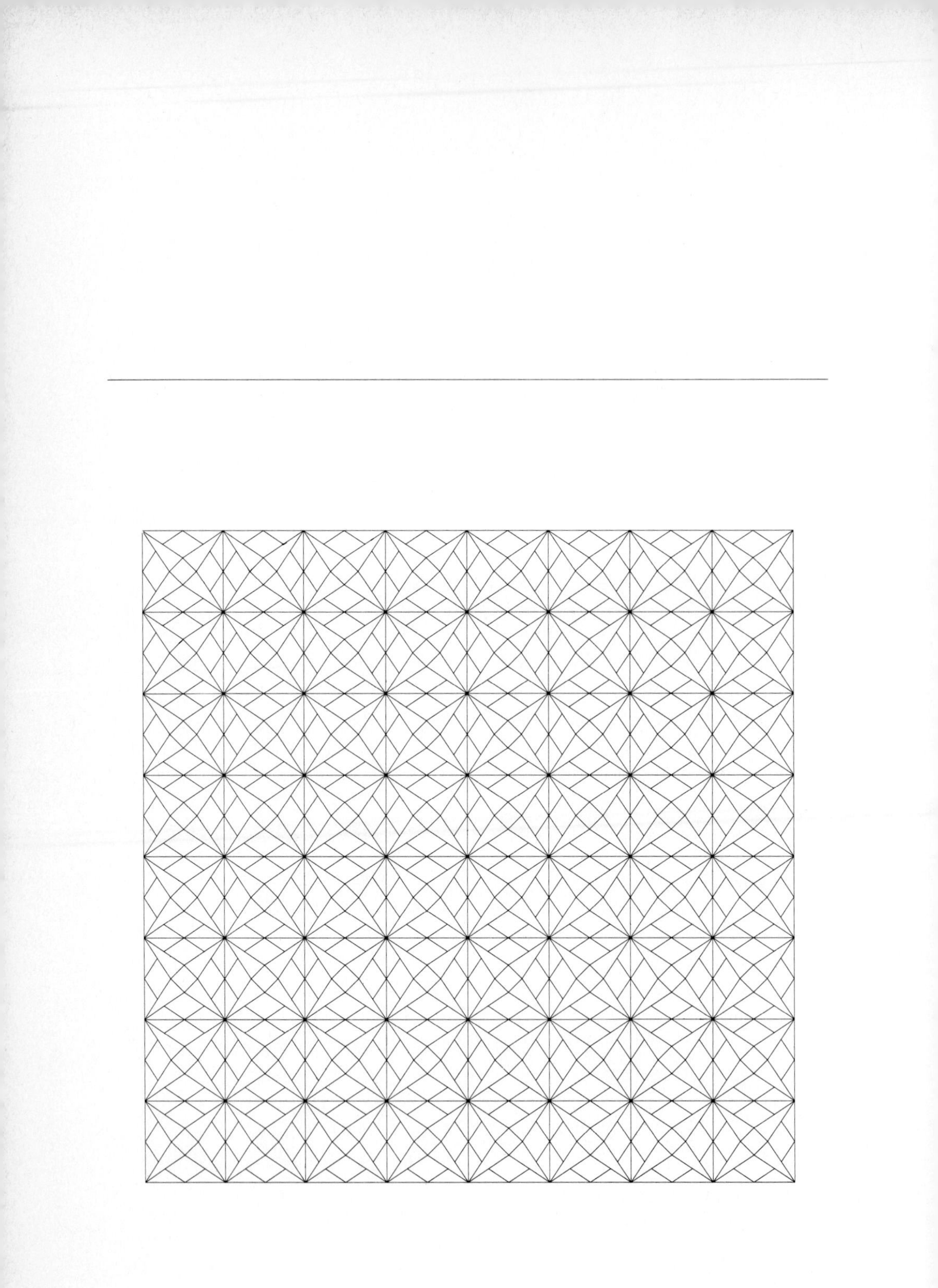

35

Random Numbers

☐ Our next set of algorithms will be methods for using a computer to generate random numbers. Though we have encountered random numbers in various contexts throughout the book, let's begin by trying to get a better idea of exactly what they are.

Often, in conversation, people use the term *random* when they really mean *arbitrary*. When one asks for an *arbitrary* number, one is saying that one doesn't really care what number one gets: almost any number will do. By contrast, a *random* number is a precisely defined mathematical concept: every number should be equally likely to occur. A random number will satisfy someone who needs an arbitrary number, but not the other way around.

For the phrase "every number should be equally likely to occur" to make sense, we must restrict the numbers to be used to some finite domain. You can't have a random integer, only a random integer in some range; you can't have a random real number, only a random fraction in some range to some fixed precision.

It is almost always the case that not just one random number, but a *sequence* of random numbers is needed (otherwise an arbitrary number might do). Here's where the mathematics comes in: it's possible to prove many facts about properties of sequences of random numbers. For example, in a very long sequence of random numbers from a small domain, we can expect to see each value about the same number of times. Random sequences model many natural situations, and a great deal is known about their properties. To be consistent with current usage, we'll refer to numbers from random sequences as random numbers.

There's no way to produce true random numbers on a computer (or any deterministic device). Once the program is written, the numbers it will produce can be deduced, so how could they be random? The best we can hope to do is to write programs which produce sequences of numbers having many of the same properties as random numbers. Such numbers are commonly called *pseudo-random* numbers: they're not really random, but they can be useful as approximations to random numbers, in much the same way as floating-point numbers are useful as approximations to real numbers. (Sometimes it's convenient to make a further distinction: in some situations, a few properties of random numbers are crucial while others are irrelevant. In such situations, one can generate *quasi-random* numbers that are sure to have the properties of interest but are unlikely to have other properties of random numbers. For some applications, quasi-random numbers are provably preferable to pseudo-random numbers.)

It's easy to see that approximating the property "each number is equally likely to occur" in a long sequence is not enough. For example, each number in the range [1,100] appears once in the sequence (1,2,. . .,100), but that sequence is unlikely to be useful as an approximation to a random sequence. In fact, in a random sequence of length 100 of numbers in the range [1,100], it is likely that a few numbers will appear more than once and a few will not appear at all. If this doesn't happen in a sequence of pseudo-random numbers, then there is something wrong with the random number generator. Many sophisticated tests based on specific observations like this have been devised for random number generators, in order to test whether a long sequence of pseudo-random numbers has some property that random numbers would. The random-number generators that we will study do very well in such tests. In this chapter, we'll look in detail at one of the most important such tests, the χ^2 (*chi-square*) test.

We have been (and will be) talking exclusively about *uniform* random numbers, with each value equally likely. It is also common to deal with random numbers which obey some other distribution in which some values are more likely than others. Pseudo-random numbers with non-uniform distributions are usually obtained by performing some operations on uniformly distributed ones. Most of the applications in this book use uniform random numbers. As we'll see, it is difficult enough to be convinced that the numbers we generate have "all" the properties of random numbers; this problem is compounded significantly when other types of distributions are involved.

Applications

We have encountered many applications in this book in which random numbers are useful. A few of them are outlined here. One obvious application is in *cryptography*, where the major goal is to encode a message so that it can't be read by anyone but the intended recipient. As we saw in Chapter 23, one way to do this is to make the message look random by using a pseudo-random sequence to encode the message in such a way that the recipient can use the same pseudo-random sequence to decode it.

Another area in which random numbers have been widely used is *simulation*. A typical simulation involves a large program which models some aspect of the real world: random numbers are natural for the input to such programs. Even if true random numbers are not needed, simulations typically need many arbitrary numbers for input, and these are conveniently provided by a random number generator.

When a very large amount of data is to be analyzed, it is sometimes sufficient to process only a very small subset of that data, chosen according to random *sampling*. Such applications are widespread, the most prominent being national political opinion polls.

Often it is necessary to make a choice when all factors under consideration seem to be equal. The national draft lottery of the '70s or the mechanisms used on college campuses to decide which students get the choice dormitory rooms are examples of using random numbers for *decision making*. In this way, the responsibility for the decision is given to "fate" (or the computer).

Readers of this book are likely to find themselves using random numbers extensively for simulation: to provide random or arbitrary inputs to programs. Another use we have seen is in algorithms which gain efficiency by using random numbers to do sampling or to aid in decision making. Prime examples of this are Quicksort (see Chapter 9) and the Rabin-Karp string searching method (see Chapter 19).

Linear Congruential Method

The best-known method for generating random numbers, which has been used almost exclusively since it was introduced by D. Lehmer in 1951, is the so-called *linear congruential* method. If *seed* contains some arbitrary number, then the following statement fills up an array with N random numbers using this method:

```
a[0]:=seed;
for i:=1 to N do
  a[i]:=(a[i-1]*b+1) mod m;
```

That is, to get a new random number, take the previous one, multiply it by a constant b, add 1 and take the remainder when divided by a second constant m. The result is always an integer between 0 and $m-1$. This is attractive for use on computers because the **mod** function is usually trivial to implement: if we ignore overflow on the arithmetic operations, then most computer hardware throws away the bits that overflowed and thus effectively performs a **mod** operation with m equal to one more than the largest integer that can be represented in the computer word. Again, the numbers aren't really random; the program just produces numbers that we hope will *appear* random to some other process.

Simple as it may seem, the linear congruential random-number generator has been the subject of volumes of detailed and difficult mathematical analysis. This work gives us some guidance in choosing the constants *seed*, b and m. Some "common-sense" principles apply, but in this case common sense isn't enough to ensure good random numbers. First, m should be large: it can be the computer word size, as mentioned above, but it needn't be quite that large if that's inconvenient (see the implementation below). It will normally be convenient to make m a power of 10 or 2. Second, b shouldn't be too large or too small: a safe choice is to use a number with one digit less than m. Third, b should be an arbitrary constant with no particular pattern in its digits, *except* that it should end with $\cdots x21$, with x even: this last requirement is admittedly peculiar, but it prevents the occurrence of some possibly troublesome cases that have been uncovered by the mathematical analysis.

The rules described above were developed by D.E. Knuth, whose textbook covers the subject in some detail. Knuth shows that these choices make the linear congruential method produce good random numbers which pass several sophisticated statistical tests. The most serious potential problem, which can quickly become apparent, is that the generator can get caught in a cycle and produce numbers it has already produced much sooner than it should. For example, the choice *b=19*, *m=381*, with *seed=0*, produces the sequence 0, 1, 20, 0, 1, 20, ..., a not-very-random sequence of integers between 0 and 380. Unfortunately, not all such difficulties are as easy to spot, so one is well-advised to follow the guidelines suggested by Knuth, thus avoiding many of the subtle traps he discovered.

Any initial value can be used to get the random number generator started with no particular effect (except, of course, that different initial values will give rise to different random sequences). Often, it is unnecessary to store the whole sequence as in the program above. Rather, we simply maintain a global variable a, initialized with some value, then updated by the computation *a:=(a*b+1)* **mod** *m*.

In Pascal (and many other programming languages) we're still one step away from a working implementation because we're not allowed to ignore overflow: it's defined to be an error condition that can lead to unpredictable results. Suppose that our computer has a 32-bit word, and we choose *m=100000000*, *b=31415821*, and, initially, *a=1234567*. All of these values are comfortably less than the largest integer that can be represented, but the first *a*b+1* operation causes overflow. The part of the product that causes the overflow is not relevant to our computation—we're only interested in the last eight digits. The trick is to avoid overflow by breaking the multiplication up into pieces. To multiply p by q, we write $p = 10^4 p_1 + p_0$ and $q = 10^4 q_1 + q_0$, so the product is

$$\begin{aligned} pq &= (10^4 p_1 + p_0)(10^4 q_1 + q_0) \\ &= 10^8 p_1 q_1 + 10^4 (p_1 q_0 + p_0 q_1) + p_0 q_0. \end{aligned}$$

Now, we only want eight digits for the result, so we can ignore the first term and the first four digits of the second term. This leads to the following program:

```
program random(input,output);
const m=100000000; m1=10000; b=31415821;
var i,a,N: integer;
function mult(p,q: integer): integer;
  var p1,p0,q1,q0: integer;
  begin
  p1:=p div m1; p0:=p mod m1;
  q1:=q div m1; q0:=q mod m1;
  mult:=(((p0*q1+p1*q0) mod m1)*m1+p0*q0) mod m;
  end;
function random: integer;
  begin
  a:=(mult(a,b)+1) mod m;
  random:=a;
  end;
begin
read(N,a);
for i:=1 to N do writeln(random)
end.
```

The function *mult* in this program computes $p*q$ **mod** m, with no overflow as long as m is less than half the largest integer that can be represented. The technique obviously can be applied with $m=m1*m1$ for other values of $m1$.

When run with the input $N = 10$ and $a = 1234567$, this program writes the following ten numbers: 35884508, 80001069, 63512650, 43635651, 1034472, 87181513, 6917174, 209855, 67115956, 59939877. There is some obvious non-randomness in these numbers: for example, the last digits cycle through the digits 0–9. It is easy to prove from the formula that this will happen. Generally speaking, the digits on the right are not particularly random, a fact that is the source of a common and serious mistake in the use of linear congruential random number generators. The following is a bad program for producing random numbers in the range $[0, r-1]$:

```
function randombad(r: integer): integer;
  begin
  a:=(mult(b,a)+1) mod m;
  randombad:=a mod r;
  end;
```

The non-random digits on the right are the only digits that are used, so the resulting sequence has few of the desired properties. This problem is easily fixed by using the digits on the *left* instead. We want to compute a number between 0 and $r-1$ by computing $a*r$ **div** m, but, again, overflow must be circumvented, as in the following implementation:

```
function randomint(r: integer): integer;
  begin
  a:=(mult(a,b)+1) mod m;
  randomint:=((a div m1)*r) div m1
  end;
```

Another common technique is to generate random real numbers between 0 and 1 by treating the above numbers as fractions with the decimal point to the left. This can be implemented by simply returning the real value a/m rather than the integer a. Then a user can get an integer in the range $[0,r)$ by simply multiplying this value by r and truncating to the nearest integer. Or, a random real number between 0 and 1 might be exactly what is needed.

Additive Congruential Method

Another method for generating random numbers is based on the *linear feedback shift registers* used for early encryption machines. The idea is to start with a register filled with some arbitrary pattern (not all zeros), then shift it right (say) a step at a time, filling in vacated positions from the left with a bit determined by the contents of the register.

Figure 35.1 shows a simple four-bit linear feedback shift register, with the new bit taken as the "exclusive or" of the two rightmost bits. For example, if the register is initially filled with the pattern 1111, then it will contain 0111 after one step: the bits 111 shift right one position and the leftmost bit becomes 0 because the rightmost two were the same. Continuing, the register contents in successive steps are 0011, 0001, 1000, 0100, 0010, 1001, 1100, 0110, 1011, 0101, 1010, 1101, 1110, 1111. Once we get back to the initial pattern, the process obviously cycles.

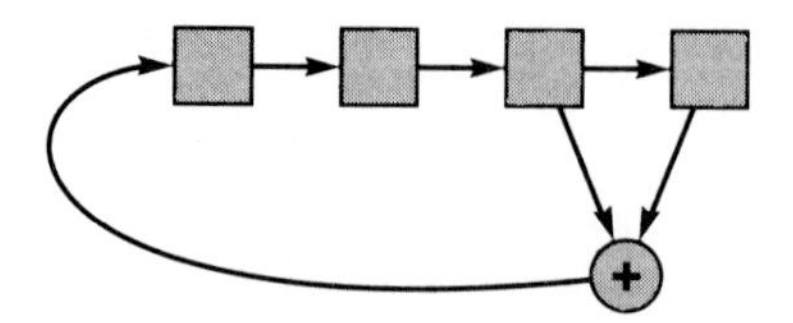

Figure 35.1 A four-bit linear feedback shift register.

Notice that all possible nonzero bit patterns occur: the starting value repeats after 15 steps. However, if we change the "tap" positions (the bits used for feedback) to 0 and 2 (numbering from the right) rather than 0 and 1, then we get the sequence 1111, 0111, 0011, 1001, 1100, 1110, 1111, not the full cycle. We would like to guarantee that we always get a full cycle.

In general, for n-bit linear feedback shift registers, it is possible to arrange things so that the cycle length is $2^n - 1$. Thus, for large n, such registers make good random number generators. Typically, one might use $n = 31$ or $n = 63$. As with the linear congruential method, the mathematical properties of these registers have been studied extensively. For example, much is known about the choices of "tap" positions which lead to the generation of all bit patterns for registers of various sizes. For example, for $n = 31$, tapping positions 0 and either 4, 7, 8, 14, 19, 25, 26, or 29 would work.

The successive contents of the register aren't useful as a random sequence, because all but one of the bits overlap in each successive pair. Rather, one should think of this device as producing a sequence of random bits (the leftmost bit of the register), in our example 100010011010111. As mentioned in Chapter 23, such devices are useful in cryptography because they can generate long sequences of bits from small keys.

Another interesting fact is that the calculation can be done a word at a time, rather than a bit at a time, according to the same recursion formula. In our example, if we take the bitwise "exclusive or" of two successive words, we get the word that appears three places later in the list. This leads us to a random-number generator easily implemented on a general-purpose computer. Using a feedback register with bits b and c tapped corresponds to using the recursion: $a[k]= (a[k-b]+a[k-c])$ **mod** m. To keep the correspondence with the shift register model, the "+" in this recursion should be a bitwise "exclusive or." However, it has been shown that good random numbers are likely to be produced even if normal integer addition is used. This is called the *additive congruential* method.

The rightmost bit of the numbers in an additive congruential generator behaves just like the bits in the corresponding linear feedback shift register, so the number of steps before the method begins to repeat is at least as great as the cycle length. Beyond this fact, few specific results have been developed about the numbers produced by such generators: the evidence in their support is primarily empirical (they pass the statistical tests).

To implement an additive congruential generator, we need to keep a table of size c which always has the c most recently generated numbers. The computation proceeds by replacing one of the numbers in the table by the sum of two of the other numbers in the table. Initially, the table should be filled with numbers that are not too small and not too large. (One easy way to get these numbers is to use a simple linear congruential generator!) Knuth recommends the choices $b=31$, $c=55$. Thus, we need to keep track of the 55 most recently generated numbers.

The appropriate data structure for this is a queue (see Chapter 3), but since it is of fixed size, we just use an array of that size, indexed by a "circular" pointer, as in the following implementation:

```
procedure randinit(s: integer);
  begin
  a[0]:=s; j:=0;
  repeat j:=j+1; a[j]:=(mult(b,a[j-1])+1) mod m until j=54;
  end;
function randomint(r: integer): integer;
  begin
  j:=(j+1) mod 55;
  a[j]:=(a[(j+23) mod 55]+a[(j+54) mod 55]) mod m;
  randomint:=((a[j] div m1)*r) div m1
  end;
```

The global variable a has been replaced by a full table plus a pointer (j) into it. This large amount of "global state" is a disadvantage of this generator in some applications, but it is also an advantage because it leads to an extremely long cycle (at least $2^{55} - 1$ even if the modulus m is small).

The function *randomint* returns a random integer between *0* and $r-1$. Of course, it can easily be changed, just as above, to a function which returns a random real number between 0 and 1 ($a[j]/m$).

Testing Randomness

Often one can detect that a sequence is not random, but certifying that a sequence *is* random is a difficult task indeed. As mentioned above, no sequence produced by a computer can in fact be random, but we can get a sequence that exhibits many of the properties of random numbers. Unfortunately, it is often impossible to articulate exactly which properties of random numbers are important for a particular application. Furthermore, it is always a good idea to perform some kind of test on a random number generator to be sure that no degenerate situations have turned up. Random number generators can be very, very good, but when they are bad they are horrid.

Many tests have been developed for determining whether a sequence shares various properties with a truly random sequence. Most of these tests have a substantial basis in mathematics, and it would be well beyond the scope of this book to examine them in detail. However, one statistical test, the χ^2 (chi-square) test, is fundamental in nature, quite easy to implement, and useful in several applications, so we'll examine it more carefully.

The idea of the χ^2 test is to check whether or not the numbers produced are spread out reasonably. If we generate N positive numbers less than r, then we'd expect to get about N/r numbers of each value. But—and this is the essence of the matter—the frequencies of occurrence of all the values should not be exactly the same: that wouldn't be random! It turns out that calculating whether or not a sequence of numbers is distributed as well as a random sequence is very simple, as in the following program:

```
function chisquare(N,r,s: integer): real;
  var i,t: integer;
      f: array[0..rmax] of integer;
  begin
  randinit(s);
  for i:=0 to rmax do f[i]:=0;
  for i:=1 to N do
    begin
    t:=randomint(r);
    f[t]:=f[t]+1;
    end;
  t:=0; for i:=0 to r-1 do t:=t+f[i]*f[i];
  chisquare:=((r*t/N) - N);
  end;
```

We simply calculate the sum of the squares of the frequencies of occurrence of each value, scaled by the expected frequency, and then subtract off the size of the sequence. This number, the "χ^2 statistic," may be expressed mathematically as

$$\chi^2 = \frac{\sum_{0 \le i < r} (f_i - N/r)^2}{N/r}.$$

If the χ^2 statistic is close to r, then the numbers are random; if it is too far away, then they are not. The notions of "close" and "far away" can be more precisely defined: tables exist that tell exactly how to relate the statistic to properties of random sequences. For the simple test that we're performing, the statistic should be within $2\sqrt{r}$ of r. This is valid if N is greater than about $10r$; to be sure, the test should be tried a few times, since it could be wrong about one out of ten times.

This test is so simple to implement that it should probably be included with every random-number generator, just to ensure that nothing unexpected can cause serious problems. All the "good" generators that we have discussed pass this test; the "bad" ones do not. Using the above generators to generate a thousand numbers less than 100, we get a χ^2 statistic of 100.8 for the linear congruential method and 105.4 for the additive congruential method, both certainly well within

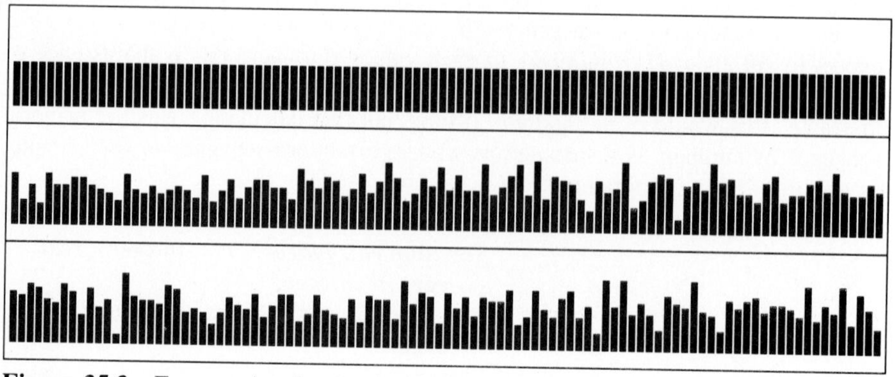

Figure 35.2 Frequencies for three generators: right bits, left bits, bad multiplier.

20 of 100. But for the "bad" generator which uses the right-hand bits from the linear congruential generator, the statistic is 0 (why?) and for a linear congruential method with a bad multiplier (101011) the statistic is 77.8, which is significantly out of range. Figure 35.2 shows the frequencies of occurrence of each of the values less than 100 for the three versions of the linear congruential method just mentioned. While using the left bits (top chart) is obviously bad, it's not easy to notice the difference between the middle and bottom frequency charts: the χ^2 statistic provides a way to identify the bad multiplier.

Implementation Notes

A number of facilities are commonly added to make a random-number generator useful for a variety of applications. It is usually desirable to set up the generator as a function that is initialized and then called repeatedly, returning a different random number each time. Another possibility is to call the random number generator once and have it fill up an array with all the random numbers needed for a particular computation. In either case, it is desirable that the generator produce the same sequence on successive calls (for initial debugging or comparison of programs on the same inputs) and produce an arbitrary sequence (for later debugging). These facilities all involve manipulating the "state" retained between calls by the random number generator. This can be very inconvenient in some programming environments. The additive generator has the disadvantage of having a relatively large state (the array of recently produced words), but has the advantage of having such a long cycle that it is probably not necessary for each user to initialize it.

A conservative way to protect against eccentricities in a random-number generator is to combine two generators. (The use of a linear congruential generator to initialize the table for an additive congruential generator is an elementary example of this.) An easy way to implement a combination generator is to have the first generator fill a table and the second choose random table positions to fetch numbers to output (and store new numbers from the first generator).

When debugging a program that uses a random-number generator, it is usually a good idea to use a trivial or degenerate generator at first, such as one which always returns 0 or one which returns numbers in order.

As a rule, random number generators are fragile and need to be treated with respect. It's difficult to be sure that a particular generator is good without investing an enormous amount of effort in the various statistical tests. The moral is: do your best to use a good generator, based on the mathematical analysis and the experience of others; just to be sure, examine the numbers to make sure that they "look" random; if anything goes wrong, blame the random-number generator!

□

Exercises

1. Write a program to generate random four-letter words (collections of letters). Estimate how many words your program will generate before it repeats a word.
2. How would you simulate generating random numbers by throwing two dice and taking their sum, with the added complication that the dice are nonstandard (say, painted with the numbers 1,2,3,5,8, and 13)?
3. Show the sequence of patterns produced by a linear feedback shift register like that in Figure 35.1, but with the tap positions at the first and last bits. Assume the initial pattern is 1111.
4. Why wouldn't the "or" or "and" function (instead of the "exclusive or" function) work for linear feedback shift registers?
5. Write a program to produce a random two-dimensional image. (Example: generate random bits, write a "*" when 1 is generated, " " when 0 is generated. Another example: use random numbers as coordinates in a two-dimensional cartesian system, write a "*" at addressed points.)
6. Use an additive congruential random-number generator to generate 1000 positive integers less than 1000. Design and apply a test to determine whether or not they're random.
7. Use a linear congruential generator with parameters of your own choosing to generate 1000 positive integers less than 1000. Design and apply a test to determine whether or not they're random.
8. Why would it be unwise to use, for example, $b=3$ and $c=6$ in the additive congruential generator?
9. What is the value of the χ^2 statistic for a degenerate generator which always returns the same number?
10. Describe how you would generate random numbers with m greater than the computer word size.

36

Arithmetic

☐ Though the situation is beginning to change, the *raison d'etre* of many computer systems is their capability for doing fast, accurate numerical calculations. Computers have built-in capabilities to perform arithmetic on integers and floating-point representations of real numbers; for example, Pascal allows numbers to be of type *integer* or *real*, with all of the normal arithmetic operations defined on both types. The actual method used for arithmetic operations on numbers is part of the architecture of the computer, and not our concern here (though a fast new algorithm for, say, multiplying two 32-bit integers could be of enormous importance indeed). Instead, we will discuss some of the algorithms that come into play when the operations must be performed on more complicated mathematical objects.

In this chapter, we'll look at Pascal implementations of algorithms for addition and multiplication of polynomials, long integers and matrices. Elementary algorithms for these problems are familiar and straightforward, but it is worthwhile to think through how basic data structures apply for particular situations. Our emphasis is less on applications than usual; we consider instead the computational complexity of fundamental arithmetic problems because arithmetic provides an excellent example of how proper application of algorithmic thinking can produce sophisticated methods which are substantially more efficient (asymptotically) than elementary methods. Such research is interesting because of its historical context: algorithms for doing elementary arithmetic operations such as addition, multiplication, and division have a very long history, dating back to the origins of algorithm studies in the work of the Arabic mathematician al-Khowarizmi, with roots going even further back to the Greeks and the Babylonians.

We'll see that, polynomial multiplication and especially matrix multiplication provide classic examples of the power of the divide-and-conquer paradigm. Unfortunately, the resulting algorithms (with the significant exception of the method discussed in Chapter 41) are hardly practical; elementary methods which use basic data structures judiciously are best except for outrageously large problems.

Polynomial Arithmetic

Suppose we wish to write a program that adds two polynomials: we would like it to perform calculations like

$$(1 + 2x - 3x^3) + (2 - x) = 3 + x - 3x^3.$$

In general, suppose we wish our program to be able to compute $r(x) = p(x)+q(x)$, where p and q are polynomials with N coefficients. Doing this is trivial with an array representation. We represent the polynomial $p(x) = p_0+p_1x+\cdots+p_{N-1}x^{N-1}$ by the **array** *p[0..N−1]* with $p[j] \equiv p_j$, etc. Then addition is nothing more than the one-line program

```
for i:=0 to N-1 do r[i]:=p[i]+q[i];
```

As usual in Pascal (cf. Chapter 3), we must decide ahead of time how large N might get, since the sizes of the p, q, and r arrays have to be set to the maximum size anticipated.

The program above shows that addition is quite trivial once the array representation for polynomials has been chosen; other operations are also easily coded. For example, the following code fragment implements polynomial multiplication:

```
for i:=0 to 2*(N-1) do r[i]:=0;
for i:=0 to N-1 do
  for j:=0 to N-1 do
    r[i+j]:=r[i+j]+p[i]*q[j];
```

The declaration of r has to accommodate twice as many coefficients for the product. Each of the N coefficients of p is multiplied by each of the N coefficients of q, so the running time of this algorithm is clearly quadratic in the number of coefficients.

As we saw in Chapter 3, an advantage of representing a polynomial by an array containing its coefficients is that it's easy to reference any coefficient directly; a disadvantage is that space may have to be saved for more numbers than necessary. For example, the program above couldn't reasonably be used to multiply

$$(1 + x^{10000})(1 + 2x^{10000}) = 1 + 3x^{10000} + 2x^{20000},$$

even though the input involves only four coefficients and the output only three.

Alternatively, we could represent polynomials using linked lists and add them as follows:

```
function add(p,q: link): link;
  var t: link;
  begin
  t:=z;
  repeat
    new(t↑.next); t:=t↑.next;
    t↑.c:=p↑.c+q↑.c;
    p:=p↑.next; q:=q↑.next
  until (p=z) and (q=z);
  t↑.next:=z; add:=z↑.next
  end;
```

The input polynomials are represented by linked lists with one list element per coefficient; the output polynomial is built by the *add* procedure. The manipulations with links are quite similar to programs we've seen in Chapters 3, 8, 14, 29, and elsewhere in this book.

As it stands, the program above is no real improvement over the array representation, except that it finesses the lack of dynamic arrays in Pascal (at the cost of space for a link per coefficient). However, as suggested by the example above, we can take advantage of the possibility that many of the coefficients may be zero. We can make list nodes represent only the nonzero terms of the polynomial by also including the degree of the term represented within the list node, so that each list node contains values of c and j to represent cx^j. It is then convenient to separate out the function of creating a node and adding it to a list, as follows:

```
type link = ↑node;
       node = record c: real; j: integer; next: link end;
function listadd(t: link; c: real; j: integer): link;
  begin
  new(t↑.next); t:=t↑.next;
  t↑.c:=c; t↑.j:=j;
  listadd:=t;
  end;
```

The *listadd* function creates a new node, gives it the specified fields, and links it into a list after node t. To make it possible to process the polynomials in an organized way, the list nodes can be kept in increasing order of degree of the term represented.

Now the *add* function becomes more interesting, since it has to perform an addition only for terms whose degrees match and then make sure that no term with a zero coefficient is output:

```
function add(p,q: link): link;
  begin
  t:=z; z↑.j:=N+1;
  repeat
    if (p↑.j=q↑.j) and (p↑.c+q↑.c<>0.0) then
      begin
      t:=listadd(t,p↑.c+q↑.c,p↑.j);
      p:=p↑.next; q:=q↑.next
      end
    else if p↑.j<q↑.j then
      begin t:=listadd(t,p↑.c,p↑.j); p:=p↑.next end
    else if q↑.j<p↑.j then
      begin t:=listadd(t,q↑.c,q↑.j); q:=q↑.next end;
  until (p=z) and (q=z);
  t↑.next:=z; add:=z↑.next
  end;
```

These refinements are worthwhile for processing "sparse" polynomials with many zero coefficients, because they mean that the space and the time required to process the polynomials will be proportional to the number of coefficients, not the degree of the polynomial. Similar savings are available for other operations on polynomials, for example multiplication, but one should exercise caution because the polynomials may become significantly less sparse after a number of such operations are performed. The array representation is better if there are only a few terms with zero coefficients, or if the degree is not high. We assume this representation for simplicity in describing more algorithms on polynomials given below.

A polynomial can involve not just one but several variables. For example, one might need to process polynomials such as

$$1 + wx^2 + y^6z + w^{25}x^{50}y^{99}z^{38} + x^{1000}z^{1000}.$$

The linked-list representation is definitely called for in such cases; the alternative (multidimensional arrays) would require too much space. It is not difficult to extend the *listadd* program above (for example) to handle such polynomials.

Polynomial Evaluation and Interpolation

Let us consider how to compute the value of a given polynomial at a given point. For example, to evaluate

$$p(x) = x^4 + 3x^3 - 6x^2 + 2x + 1$$

for any given x, one could compute x^4, then compute and add $3x^3$, etc. This method requires recomputation of the powers of x; alternatively, we could save the powers of x as they are computed, but this requires extra storage.

A simple method which avoids recomputation and uses no extra space is known as *Horner's rule*: by alternating the multiplication and addition operations appropriately, a degree-N polynomial can be evaluated using only $N - 1$ multiplications and N additions. The parenthesization

$$p(x) = x(x(x(x + 3) - 6) + 2) + 1$$

makes the order of computation obvious:

```
y:=p[N];
for i:=N-1 downto 0 do y:=x*y+p[i];
```

We have already used a version of this method in a very important practical application, computing hash functions of long keys (see Chapter 16).

A more complicated problem is to evaluate a given polynomial at many different points. Different algorithms are appropriate depending on how many evaluations are to be done and whether or not they are to be done simultaneously. If a very large number of evaluations is to be done, it may be worthwhile to do some "precomputing" which can slightly reduce the cost of later evaluations. Note that Horner's method requires about N^2 multiplications to evaluate a degree-N polynomial at N different points. Much more sophisticated methods have been designed which can solve the problem in $N(\log N)^2$ steps, and in Chapter 41 we'll see a method that uses only $N \log N$ multiplications for a specific set of N points of interest.

If the given polynomial has only one term, then the polynomial evaluation problem reduces to the *exponentiation* problem: compute x^N. Horner's rule in this case degenerates to a trivial algorithm which requires $N - 1$ multiplications. To see how we can do much better, consider the following sequence for computing x^{32}:

$$x, x^2, x^4, x^8, x^{16}, x^{32}.$$

Each term is obtained by squaring the previous term, so only five multiplications are required, not 31.

The "successive-squaring" method can easily be extended to general N if computed values are saved. For example, x^{55} can be computed from the above values with four more multiplications:

$$x^{55} = x^{32}x^{16}x^4x^2x^1.$$

In general, the binary representation of N can be used to choose which computed values to use. (In the example, since $55 = (110111)_2$, all but x^8 are used.) The successive squares can be computed and the bits of N tested within the same loop. Two methods are available that implement this using only one "accumulator," like Horner's method. One algorithm involves scanning the binary representation of N from left to right, starting with 1 in the accumulator. At each step, square the accumulator and also multiply by x when there is a 1 in the binary representation of N. The following sequence of values is computed by this method for $N = 55$:

$$1, 1, x, x^2, x^3, x^6, x^{12}, x^{13}, x^{26}, x^{27}, x^{54}, x^{55}.$$

Another well-known algorithm works similarly, but scans N from right to left. This problem is a standard introductory programming exercise. Though it hardly seems of practical interest to be able to compute such large numbers, we'll see below in our discussion of large integers that this method plays a role in implementing the public-key cryptosystems of Chapter 23.

The "inverse" problem to the problem of evaluating a polynomial of degree N at N points simultaneously is the problem of *polynomial interpolation*: given a set of N points $x_1, x_2, \ldots, x_N$ and associated values $y_1, y_2, \ldots, y_N$, find the unique polynomial of degree $N-1$ which has

$$p(x_1) = y_1, \; p(x_2) = y_2, \; \ldots, p(x_N) = y_N.$$

The interpolation problem is to find the polynomial, given a set of points and values. The evaluation problem is to find the values, given the polynomial and the points. (The problem of finding the points, given the polynomial and the values, is *root-finding*.)

The classic solution of the interpolation problem is given by Lagrange's interpolation formula, which is often used as a proof that a polynomial of degree $N-1$ is completely determined by N points:

$$p(x) = \sum_{1 \le j \le N} y_j \prod_{\substack{1 \le i \le N \\ i \ne j}} \frac{x - x_i}{x_j - x_i}.$$

This formula seems formidable at first but is actually quite simple. For example, the polynomial of degree 2 which has $p(1) = 3$, $p(2) = 7$, and $p(3) = 13$ is given by

$$p(x) = 3\frac{x-2}{1-2}\frac{x-3}{1-3} + 7\frac{x-1}{2-1}\frac{x-3}{2-3} + 13\frac{x-1}{3-1}\frac{x-2}{3-2}$$

which simplifies to

$$x^2 + x + 1.$$

For x from $x_1, x_2, \ldots, x_N$, the formula is constructed so that $p(x_k) = y_k$ for $1 \le k \le N$, since the product evaluates to 0 unless $j = k$, when it evaluates to 1. In the example, the last two terms are zero when $x = 1$, the first and last terms are zero when $x = 2$, and the first two terms are zero when $x = 3$.

Converting a polynomial from the form described by Lagrange's formula to our standard coefficient representation is not at all straightforward. At least N^2 operations seem to be required, since there are N terms in the sum, each consisting of a product with N factors. Actually, it takes some cleverness to achieve a quadratic algorithm, since the factors are not just numbers but polynomials of degree N. On the other hand, each term is very similar to the previous one. The reader might be interested to discover how to take advantage of this to achieve a quadratic algorithm. This exercise leaves one with an appreciation for the non-trivial nature of writing an efficient program to perform the calculation implied by a mathematical formula.

As with polynomial evaluation, there are more sophisticated methods which can solve the problem in $N(\log N)^2$ steps, and in Chapter 41 we'll see a method that uses only $N \log N$ multiplications for a specific set of N points of interest.

Polynomial Multiplication

Our first sophisticated arithmetic algorithm is for the problem of *polynomial multiplication*: given two polynomials $p(x)$ and $q(x)$, compute their product $p(x)q(x)$. As noted at the beginning of this chapter, polynomials of degree $N-1$ can have N terms (including the constant) and their product has degree $2N-2$ and as many as $2N-1$ terms. For example,

$$(1+x+3x^2-4x^3)(1+2x-5x^2-3x^3)=(1+3x-6x^3-26x^4+11x^5+12x^6).$$

The naive algorithm for this problem given at the beginning of this chapter requires N^2 multiplications for polynomials of degree $N-1$: each of the N terms of $p(x)$ are multiplied by each of the N terms of $q(x)$.

To improve upon the naive algorithm, we "divide and conquer." One way to split a polynomial in two is to divide the coefficients in half: given a polynomial of degree $N-1$ (with N coefficients), we can split it into two polynomials with $N/2$ coefficients (assume that N is even): use the $N/2$ low-order coefficients for one polynomial and the $N/2$ high-order coefficients for the other. For $p(x) = p_0 + p_1x + \cdots + p_{N-1}x^{N-1}$, define

$$p_l(x) = p_0 + p_1x + \cdots + p_{N/2-1}x^{N/2-1};$$
$$p_h(x) = p_{N/2} + p_{N/2+1}x + \cdots + p_{N-1}x^{N/2-1}.$$

Then, splitting $q(x)$ in the same way, we have:

$$p(x) = p_l(x) + x^{N/2}p_h(x),$$
$$q(x) = q_l(x) + x^{N/2}q_h(x).$$

Now, in terms of the smaller polynomials, the product is given by:

$$p(x)q(x) = p_l(x)q_l(x) + (p_l(x)q_h(x) + q_l(x)p_h(x))x^{N/2} + p_h(x)q_h(x)x^N.$$

(We used this same split in Chapter 35 to avoid overflow.)

Now, the point of these manipulations is that only *three* multiplications are necessary to compute these products (not four, as it would seem from the above formula) because if we compute $r_l(x) = p_l(x)q_l(x)$, $r_h(x) = p_h(x)q_h(x)$, and $r_m(x) = (p_l(x) + p_h(x))(q_l(x) + q_h(x))$, we can get the product $p(x)q(x)$ by computing

$$p(x)q(x) = r_l(x) + (r_m(x) - r_l(x) - r_h(x))x^{N/2} + r_h(x)x^N.$$

The savings to be achieved are perhaps not evident from this small example. The method is based on the fact that polynomial addition requires a linear algorithm, and brute-force polynomial multiplication is quadratic, so it's worthwhile to do a few (easy) additions to save one (difficult) multiplication. We'll look more closely below at the savings this method can yield.

For the example given above, with $p(x) = 1 + x + 3x^2 - 4x^3$ and $q(x) = 1 + 2x - 5x^2 - 3x^3$, we have

$$\begin{aligned} r_l(x) &= (1 + x)(1 + 2x) = 1 + 3x + 2x^2, \\ r_h(x) &= (3 - 4x)(-5 - 3x) = -15 + 11x + 12x^2, \\ r_m(x) &= (4 - 3x)(-4 - x) = -16 + 8x + 3x^2. \end{aligned}$$

Thus, $r_m(x) - r_l(x) - r_h(x) = -2 - 6x - 11x^2$, and the product is computed as the sum of three terms according to the above formula:

$$\begin{aligned} p(x)q(x) = (1 + 3x + 2x^2) & \\ + (-2 - 6x - 11x^2)x^2 & \\ + (-15 + 11x + 12x^2)x^4 & \\ = 1 + 3x - 6x^3 - 26x^4 + 11x^5 + 12x^6. & \end{aligned}$$

This divide-and-conquer approach solves a polynomial multiplication problem of size N by solving three subproblems of size $N/2$, using some polynomial addition to set up the subproblems and to combine their solutions. (If $N = 1$, then the product is just the scalar product of the two constant coefficients.) Thus, this procedure is easily described as a recursive program:

```
function mult(p,q: array[0..N-1] of real;
                         N: integer) : array [0..2*N-2] of real;
var pl,ql,ph,qh,t1,t2: array [0..(N  div 2)-1] of real;
    rl,rm,rh: array[0..N-1] of real;
    i,N2: integer;
begin
if N=1 then mult[0]:=p[0]*q[0]
else
  begin
  N2:=N div 2;
  for i:=0 to N2-1 do
    begin pl[i]:=p[i]; ql[i]:=q[i] end;
  for i:=N2 to N-1 do
    begin ph[i-N2]:=p[i]; qh[i-N2]:=q[i] end;
  for i:=0 to N2-1 do t1[i]:=pl[i]+ph[i];
  for i:=0 to N2-1 do t2[i]:=ql[i]+qh[i];
  rm:=mult(t1,t2,N2);
  rl:=mult(pl,ql,N2);
  rh:=mult(ph,qh,N2);
  for i:=0 to N-2 do mult[i]:=rl[i]
  mult[N-1]:=0;
  for i:=0 to N-2 do mult[N+i]:=rh[i]
  for i:=0 to N-2 do
    mult[N2+i]:=mult[N2+i]+rm[i]-(rl[i]+rh[i]);
  end;
end.
```

Although the above code is a succinct description of this method, it is unfortunately not a legal Pascal program because functions can't dynamically declare arrays. This problem can be handled in Pascal by representing the polynomials as linked lists; we leave this as an exercise for the reader. The program above assumes that N is a power of two, though the details for general N can be worked out easily. The main complications are to make sure that the recursion terminates properly and that the polynomials are divided properly when N is odd.

Why is this divide-and-conquer method an improvement? To find the answer, we need to solve a basic recurrence formula just slightly more complicated than those in Chapter 6.

Property 36.1 *Two polynomials of degree N can be multiplied using about $N^{1.58}$ multiplications.*

From the recursive program, it is clear that the number of integer multiplications required to multiply two polynomials of size N is the same as the number of multiplications to multiply three pairs of polynomials of size $N/2$. (Note that, for example, no multiplications are required to compute $r_h(x)x^N$, just data movement.) If M_N is the number of multiplications required to multiply two polynomials of size N, we have

$$M_N = 3M_{N/2} + 1, \qquad \text{for } N \geq 2 \text{ with } M_1 = 1.$$

Thus $M(2) = 3$, $M(4) = 9$, $M(8) = 27$, etc. As in Chapter 6, if we take $N = 2^n$, then we can repeatedly apply the recurrence to itself to find the solution:

$$M(2^n) = 3M(2^{n-1}) = 3^2M(2^{n-2}) = 3^3M(2^{n-3}) = \cdots = 3^nM(1) = 3^n.$$

$$\begin{aligned} M_{2^n} &= 3M_{2^{n-1}} \\ &= 3^2M_{2^{n-2}} \\ &= 3^3M_{2^{n-3}} \\ &\vdots \\ &= 3^nM_{2^0} + n \\ &= 3^n. \end{aligned}$$

If $N = 2^n$, then $3^n = 2^{(\lg 3)n} = 2^{n \lg 3} = N^{\lg 3}$. Although this solution is exact only for $N = 2^n$, it works out in general that

$$M_N \approx N^{\lg 3} \approx N^{1.58},$$

which is a substantial savings over the N^2 naive method. ■

Note that if we had used all four multiplications in the simple divide-and-conquer method, then it would have the same performance as the elementary method because the recurrence would be $M(N) = 4M(N/2)$ with the solution $M(2^n) = 4^n = N^2$.

This method nicely illustrates the divide-and-conquer technique, but it is seldom used in practice because a much better divide-and-conquer method is known, which we'll study in Chapter 41. This method gets by with dividing the original into only two subproblems, with a little extra processing. This leads to our standard $M_N = 2M_{N/2}+N$ divide-and-conquer recurrence for the number of multiplications required, and yields the solution that M_N is about $N \lg N$.

Arithmetic Operations with Large Integers

A large integer may be treated as a polynomial, with restrictions on the coefficients. For example, the 28-digit integer

0120200103110001200004012314

might correspond to the polynomial

$$x^{26} + 2x^{25} + 2x^{23} + x^{20} + 3x^{18} + x^{17} + x^{16} + x^{12} + 2x^{11} + 4x^{6} + x^{4} + 2x^{3} + 3x^{2} + x + 4.$$

That is, the number is the value of the polynomial at $x = 10$. Conversely, any polynomial of degree $N - 1$ with positive coefficients less than 10 corresponds precisely to an N-digit integer.

Thus, we can use polynomial operations to manipulate large integers. That is, we simply represent integers with arrays, then use the polynomial manipulation routines just developed as if the arrays represented polynomials. For example, to multiply two 100-digit numbers, we could use the algorithm above to compute a 200-digit result. The flaw in this strategy is that the coefficients in the result are not likely to be less than 10. This can be adjusted in a single pass: starting at *i=0*, add *p*[*i*] **div** *10* to *p*[*i+1*], replace *p*[*i*] by *p*[*i*] **mod** *10* and increment *i*, continuing until no nonzero coefficients are left.

A larger radix than 10 could be used; for example, the above 28-digit number could also correspond to the polynomial

$$120x^{6} + 2001x^{5} + 311x^{4} + x^{3} + 2000x^{2} + 401x + 2314.$$

Evaluating this polynomial at $x = 10000$ would yield the integer. This allows the integer to be represented with less memory (a quarter as much in this example), but it opens the possibility of overflow in the coefficients during some intermediate operation. The mathematics of just how large a radix can be used has been carefully worked out, but in practice, there's little harm in taking the conservative approach of using a small radix.

For the RSA cryptosystem (Chapter 23), we need not only to multiply large integers, but also exponentiate and divide. Specifically, we need to compute $M^p \bmod N$ when M, p, and N are all large integers. This is not a simple computation to perform, but we can sketch a method. First, exponentiation can be done with successive multiplications, as described above, so it suffices to consider how to compute $M_1 M_2 \bmod N$ when M_1, M_2, and N are all large integers. The key to performing the modulus computation is to compute $10^i \bmod N$ for all 10^i smaller than the largest integer to be encountered. Then any particular modulus will be a linear combination of these values. A larger radix will lower the amount of computation necessary. In mathematical terms, this method corresponds to computing, say,

$$0120200103110001200004012314 \bmod N$$

by computing

$$120(x^6 \bmod N) + 2001(x^5 \bmod N) + 311(x^4 \bmod N) + (x^3 \bmod N) + 2000(x^2 \bmod N) + 401(x \bmod N) + 2314$$

for $x = 10000$. The $10000^i \bmod N$ values can be computed ahead of time and stored in a table, or computed incrementally as needed, as in the Rabin-Karp string-searching algorithm in Chapter 19.

Matrix Arithmetic

Implementation of basic operations on matrices involves similar considerations to those discussed above for polynomials. For example, adding two matrices is trivial because matrix addition is term-by-term addition, just as for polynomials.

Matrix multiplication is also straightforward. If r is the product of p and q, then element $r[i,j]$ is the *dot product* of the ith row of p with the jth column of q. The dot product is simply the sum of the N term-by-term multiplications $p[i,1]*q[1,j]+p[i,2]*q[2,j]+\cdots p[i,N-1]*q[N-1,j]$, as in the following program:

```
for i:=0 to N-1 do
  for j:=0 to N-1 do
    begin
    t:=0.0;
    for k:=0 to N-1 do t:=t+p[i,k]*q[k,j];
    r[i,j]:=t
    end;
```

The reader may wish to use the following example when checking the code above:

$$\begin{pmatrix} 1 & 3 & -4 \\ 1 & 1 & -2 \\ -1 & -2 & 5 \end{pmatrix} \begin{pmatrix} 8 & 3 & 0 \\ 3 & 10 & 2 \\ 0 & 2 & 6 \end{pmatrix} = \begin{pmatrix} 17 & 25 & -18 \\ 11 & 9 & -10 \\ -14 & -13 & 26 \end{pmatrix}.$$

Each of the N^2 elements in the result matrix is computed with N multiplications, so about N^3 operations are required to multiply together two N-by-N matrices.

As with polynomials, *sparse* matrices (those with many zero elements) can be processed in a much more efficient manner using a linked-list representation. To keep the two-dimensional structure intact, each nonzero matrix element is represented by a list node containing a value and two links, one pointing to the next nonzero element in the same row and the other pointing to the next nonzero element in the same column. Implementing addition for sparse matrices represented in this way is similar to our implementation for sparse polynomials, but is complicated by the fact that each node appears on two lists.

The most famous application of the divide-and-conquer technique to an arithmetic problem is Strassen's method for matrix multiplication. We won't go into the details here, but we can sketch the method, since it is very similar in concept to the polynomial multiplication method just studied.

The straightforward method for multiplying two N-by-N matrices requires N^3 scalar multiplications, since each of the N^2 elements in the product matrix is obtained by N multiplications. Strassen's method is to divide the size of the problem in half; this corresponds to dividing each of the matrices into quarters, each $N/2$ by $N/2$. The remaining problem is equivalent to multiplying 2-by-2 matrices. Just as we were able to reduce the number of multiplications required from four to three by combining terms in the polynomial multiplication problem, Strassen was able to find a way to combine terms to reduce the number of multiplications required for the 2-by-2 matrix multiplication problem from 8 to 7. The rearrangement and the terms required are quite complicated.

Property 36.2 *Two N-by-N matrices can be multiplied using about $N^{2.81}$ multiplications.*

From the discussion above, the number of multiplications required for matrix multiplication using Strassen's method is defined by the divide-and-conquer recurrence

$$M(N) = 7M(N/2)$$

which has the solution

$$M(N) \approx N^{\lg 7} \approx N^{2.81},$$

as above. ■

This result was quite surprising when it first appeared in 1968, since it had previously been thought that N^3 multiplications were absolutely necessary for matrix multiplication. The problem has been studied very intensively in recent years, and methods slightly better than Strassen's have been found. The "best" algorithm for matrix multiplication has still not been found, and this is one of the most famous outstanding problems of computer science.

It is important to note that we have been counting multiplications only. Before choosing an algorithm for a practical application, the costs of the extra additions and subtractions for combining terms and the costs of the recursive calls must also be considered. These costs may depend heavily on the particular implementation or computer used. But this overhead clearly makes Strassen's method less efficient than the standard method for small matrices. Even for large matrices, in terms of the number of data items input, Strassen's method really represents an improvement only from $N^{1.5}$ to $N^{1.41}$. This improvement is hard to notice except for very large N. For example, N would have to be more than a million for Strassen's method to use one-fourth as many multiplications as the standard method, even though the overhead per multiplication is likely to be four times as large. Thus the algorithm is a theoretical, not practical, contribution.

□

Exercises

1. Polynomials also can be represented in the form $r_0(x - r_1)(x - r_2)\ldots(x - r_N)$. How would you multiply two polynomials in this representation?

2. Write a Pascal program that multiplies sparse polynomials, using a linked list representation with no nodes for terms with zero coefficients.

3. Write a Pascal procedure that sets the value of the element in the ith row and jth column of a sparse matrix to v, assuming that the matrix is represented in a linked-list representation with no nodes for zero entries.

4. Give a method for evaluating a polynomial with known roots $r_1, r_2, \ldots, r_N$, and compare your method with Horner's method.

5. Write a program to evaluate polynomials using Horner's method, where the polynomials are represented with linked lists. Be sure that your program works efficiently for sparse polynomials.

6. Write an N^2 program to do Lagrangian interpolation.

7. Can x^{55} be computed with fewer than nine multiplications? If so, say which ones; if not, say why not.

8. List all the polynomial multiplications performed when the divide-and-conquer polynomial multiplication method *mult* described in the text is used to square $1 + x + x^2 + x^3 + x^4 + x^5 + x^6 + x^7 + x^8$.

9. The *mult* method could be made more efficient for sparse polynomials by returning zero if all coefficients of either input are zero. About how many multiplications (to within a constant factor) would such a program use to square $1 + x^N$?

10. Implement a working version of *mult* which uses a linked-list representation, and empirically determine a value of N for which it is faster than the brute-force method (using the same representation).

37

Gaussian Elimination

□ One of the most fundamental scientific computations is the solution of systems of simultaneous equations. The basic algorithm for solving systems of equations, *Gaussian elimination*, is relatively simple and has changed little in the 150 years since it was invented. This algorithm has come to be well understood, especially in the past twenty years, so that it can be used with some confidence that it will efficiently produce accurate results.

Gaussian elimination is an example of an algorithm that will surely be available in most computer installations; indeed, it is a primitive in several computer languages, notably APL and Basic. However, the basic algorithm is easy to understand and implement, and special situations arise in which it is desirable to implement a modified version of the algorithm rather than work with a standard subroutine. Also, the method deserves to be studied as one of the most important numeric methods in use today.

As with the other mathematical material we have studied so far, our treatment will highlight only the basic principles and will be self-contained. Familiarity with linear algebra is not required to understand the basic method. We'll develop a simple Pascal implementation that might be easier to use than a library subroutine for simple applications. However, we'll also see examples of difficulties which could arise. Certainly for a large or important application, an expertly tuned implementation is called for, as well as some familiarity with the underlying mathematics.

A Simple Example

Suppose that we have three variables x, y and z and the following three equations:

$$x + 3y - 4z = 8,$$
$$x + y - 2z = 2,$$
$$-x - 2y + 5z = -1.$$

Our goal is to compute the values of the variables which simultaneously satisfy the equations. Depending on the particular equations, there may be no solution to this problem (for example, if two of the equations are contradictory, such as $x+y=1$, $x+y=2$) or there may be many solutions (for example, if two equations are the same or if there are more variables than equations). We'll assume that the number of equations and variables is the same, and we'll look at an algorithm that will find a unique solution if one exists.

To make it easier to extend the formulas to cover more than just three points, we'll begin by renaming the variables, using subscripts:

$$\begin{aligned} x_1 + 3x_2 - 4x_3 &= 8, \\ x_1 + x_2 - 2x_3 &= 2, \\ -x_1 - 2x_2 + 5x_3 &= -1. \end{aligned}$$

To avoid writing down variables repeatedly, it is convenient to use matrix notation to express the simultaneous equations. The above equations are exactly equivalent to the matrix equation

$$\begin{pmatrix} 1 & 3 & -4 \\ 1 & 1 & -2 \\ -1 & -2 & 5 \end{pmatrix} \begin{pmatrix} x_1 \\ x_2 \\ x_3 \end{pmatrix} = \begin{pmatrix} 8 \\ 2 \\ -1 \end{pmatrix}.$$

There are several operations that can be performed on such equations that do not alter the solution:

Interchange equations: Clearly, the order in which the equations are written down doesn't affect the solution. In the matrix representation, this operation corresponds to interchanging rows in the matrix (and in the vector on the right-hand side).

Rename variables: This corresponds to interchanging columns in the matrix representation. (If columns i and j are switched, then variables x_i and x_j must also be switched.)

Multiply equations by a constant: Again, in the matrix representation, this corresponds to multiplying a row in the matrix (and the corresponding element in the vector on the right-hand side) by a constant.

Add two equations and replace one of them by the sum.

It takes a little thought to convince oneself that these operations, especially the last, will not affect the solution. For example, we get a system of equations equivalent to the one above by replacing the second equation by the difference between the first two:

$$\begin{pmatrix} 1 & 3 & -4 \\ 0 & 2 & -2 \\ -1 & -2 & 5 \end{pmatrix} \begin{pmatrix} x_1 \\ x_2 \\ x_3 \end{pmatrix} = \begin{pmatrix} 8 \\ 6 \\ -1 \end{pmatrix}.$$

Notice that this eliminates x_1 from the second equation. Similarly, we can eliminate x_1 from the third equation by replacing that equation by the sum of the first and third:

$$\begin{pmatrix} 1 & 3 & -4 \\ 0 & 2 & -2 \\ 0 & 1 & 1 \end{pmatrix} \begin{pmatrix} x_1 \\ x_2 \\ x_3 \end{pmatrix} = \begin{pmatrix} 8 \\ 6 \\ 7 \end{pmatrix}.$$

Now the variable x_1 is eliminated from all but the first equation. By systematically proceeding in this way, we can transform the original system of equations into a system with the same solution that is much easier to solve. For our example, this requires only one more step that combines two of the operations above: replacing the third equation by the difference between the second and twice the third. This makes all of the elements below the main diagonal zero, and systems of equations of this form are particularly easy to solve. The simultaneous equations that result in our example are:

$$\begin{aligned} x_1 + 3x_2 - 4x_3 &= 8, \\ 2x_2 - 2x_3 &= 6, \\ -4x_3 &= -8. \end{aligned}$$

Now the third equation can be solved immediately: $x_3 = 2$. If we substitute this value into the second equation, we can compute the value of x_2:

$$\begin{aligned} 2x_2 - 4 &= 6, \\ x_2 &= 5. \end{aligned}$$

Similarly, substituting these two values in the first equation allows the value of x_1 to be computed:

$$\begin{aligned} x_1 + 15 - 8 &= 8, \\ x_1 &= 1, \end{aligned}$$

which completes the solution of the equations.

This example illustrates the two basic phases of Gaussian elimination. The first is the *forward-elimination* phase, where the original system is transformed, by systematically eliminating variables from equations, into a system with all zeros below the diagonal. This process is sometimes called *triangulation.* The second phase is the *backward-substitution phase*, where the values of the variables are computed using the triangulated matrix produced by the first phase.

Outline of the Method

In general, we want to solve a system of N equations in N unknowns:

$$\begin{aligned} a_{11}x_1 + a_{12}x_2 + \cdots + a_{1N}x_N &= b_1, \\ a_{21}x_1 + a_{22}x_2 + \cdots + a_{2N}x_N &= b_2, \\ &\vdots \\ a_{N1}x_1 + a_{N2}x_2 + \cdots + a_{NN}x_N &= b_N. \end{aligned}$$

These equations are written in matrix form as a single matrix equation:

$$\begin{pmatrix} a_{11} & a_{12} & \dots & a_{1N} \\ a_{21} & a_{22} & \dots & a_{2N} \\ \vdots & & & \\ a_{N1} & a_{N2} & \dots & a_{NN} \end{pmatrix} \begin{pmatrix} x_1 \\ x_2 \\ \vdots \\ x_N \end{pmatrix} = \begin{pmatrix} b_1 \\ b_2 \\ \vdots \\ b_N \end{pmatrix}$$

or simply $Ax = b$, where A represents the matrix, x represents the variables, and b represents the right-hand sides of the equations. Since the rows of A are manipulated along with the elements of b, it is convenient to regard b as the $(N+1)$st column of A and use an N-by-$(N+1)$ array to hold both.

Now the forward-elimination phase can be summarized as follows: first eliminate the first variable in all but the first equation by adding the appropriate multiple of the first equation to each of the other equations, then eliminate the second variable in all but the first two equations by adding the appropriate multiple of the second equation to each of the third through Nth equations, then eliminate the third variable in all but the first three equations, etc. To eliminate the ith variable in the jth equation (for j between $i+1$ and N) we multiply the ith equation by a_{ji}/a_{ii} and subtract it from the jth equation. This process is more succinctly described by the following code fragment:

```
for i:=1 to N do
  for j:=i+1 to N do
    for k:=N+1 downto i do
      a[j,k]:=a[j,k]-a[i,k]*a[j,i]/a[i,i];
```

The code consists of three nested loops, and the total running time is essentially proportional to N^3. The third loop goes backwards so as to avoid destroying *a*[*j,i*] before it is used to adjust the values of other elements in the same row.

The code fragment in the above paragraph is too simple to be quite right: *a*[*i,i*] might be zero, so division by zero could occur. This is easily fixed, however, since we can exchange any row (from *i+1* to *N*) with the *i*th row to make *a*[*i,i*] non-zero in the outer loop. If no such row can be found, then the matrix is *singular*: there is no unique solution. (Our program could report this explicitly, or we could let the error eventually surface as a division by zero.) We need to add code to the fragment above to find a lower row with a non-zero element in the *i*th column, then exchange that row with the *i*th row. The element *a*[*i,i*] that is eventually used to eliminate the non-zero elements below the diagonal in the *i*th column is called the *pivot*.

In fact, it is advisable to do slightly more than just find a row with a non-zero entry in the *i*th column. It's best to use the row (from *i+1* to *N*) whose entry in the *i*th column is largest in absolute value. The reason is that severe computational errors can arise if the pivot value used to scale a row is very small. If $a[i,i]$ is very small, then the scaling factor $a[j,i]/a[i,i]$ used to eliminate the ith variable from the jth equation (for j from $i+1$ to N) will be very large. In fact, it could get so large as to dwarf the actual coefficients $a[j,k]$, to the point where the $a[j,k]$ value becomes distorted by "round-off error."

Put simply, numbers that differ greatly in magnitude can't be accurately added or subtracted in the floating-point number system commonly used to represent real numbers, and using a small pivot greatly increases the likelihood that such operations will have to be performed. Using the largest value in the ith column from rows $i+1$ to N will ensure that the scaling factor is always less than 1, and will prevent this type of error. One might contemplate looking beyond the ith column to find a large element, but it has been shown that accurate answers can be obtained without resorting to this extra complication.

The following code for the forward-elimination phase of Gaussian elimination is a straightforward implementation of this process. For each i from 1 to N, we scan down the ith column to find the largest element (in rows past the ith). The row containing this element is exchanged with the ith, and then the ith variable is eliminated in the equations $i+1$ to N exactly as before:

```
procedure eliminate;
  var i,j,k,max: integer;
      t: real;
  begin
  for i:=1 to N do
    begin
    max:=i;
    for j:=i+1 to N do
      if abs(a[j,i])>abs(a[max,i]) then max:=j;
    for k:=i to N+1 do
      begin t:=a[i,k]; a[i,k]:=a[max,k]; a[max,k]:=t end;
    for j:=i+1 to N do
      for k:=N+1 downto i do
        a[j,k]:=a[j,k]-a[i,k]*a[j,i]/a[i,i];
    end
  end;
```

In some algorithms it is required that the pivot $a[i,i]$ be used to eliminate the ith variable from every equation but the ith (not just the $(i+1)$st through the Nth). This process is called *full pivoting*; for forward elimination we only do part of this work, and hence the process is called *partial pivoting*. We'll examine a full-pivoting algorithm in Chapter 43.

After the forward-elimination phase has been completed, the array a has all zeros below the diagonal and the backward substitution phase can be executed. The code for this is even more straightforward:

```
procedure substitute;
  var j,k: integer;
      t: real;
  begin
  for j:=N downto 1 do
    begin
    t:=0.0;
    for k:=j+1 to N do t:=t+a[j,k]*x[k];
    x[j]:=(a[j,N+1]-t)/a[j,j]
    end
  end;
```

A call to *eliminate* followed by a call to *substitute* computes the solution in the N-element array x. Division by zero could still occur for singular matrices—a "library" routine would check for this explicitly. Actually, most library routines do much more extensive checking, as discussed further below.

Property 37.1 *A system of N simultaneous equations in N unknowns can be solved using about $N^3/3$ multiplications and additions.*

The running time of *substitute* is $O(N^2)$, so most of the work is done in *eliminate*. Inspection of that routine shows that for each value of i, the k loop is iterated $N - i + 2$ times and the j loop $N - i$ times; this means that the inner loop is executed $\sum_{1 \le i \le N}(N - i + 2)(N - i) = N^3/3 + O(N^2)$ times. The value of $-a[j,i]/a[i,i]$ can be computed outside the k loop, so the inner loop consists of one multiplication and one addition. ■

An alternate way to proceed after forward elimination has created all zeros below the diagonal is to use precisely the same method to produce all zeros above the diagonal: first make the last column zero except for $a[N,N]$ by adding the appropriate multiple of $a[N,N]$, then do the same for the next-to-last column, etc. That is, we do "partial pivoting" again, but on the other "part" of each column, working backwards through the columns. After this process (called *Gauss-Jordan reduction*) is complete, only diagonal elements are non-zero, which yields a trivial solution. However, the number of arithmetic operations used in this process is much larger than for back substitution.

Computational errors are a prime source of concern in Gaussian elimination. As mentioned above, we should be wary of situations in which the magnitudes of the coefficients differ greatly. Using the largest available element in the column for partial pivoting ensures that large coefficients won't be created arbitrarily in the pivoting process, but it is not always possible to avoid severe errors. For example, very small coefficients turn up when two different equations have coefficients that are quite close to one another. It is actually possible to determine in advance, however, whether such problems will cause inaccurate answers in the solution. Each matrix has an associated numerical quantity called the *condition number* which can be used to estimate the accuracy of the computed answer. A good library subroutine for Gaussian elimination will compute the condition number of the matrix as well as the solution, so that the accuracy of the solution can be known. Full treatment of the issues involved here is beyond the scope of this book.

Gaussian elimination with partial pivoting using the largest available pivot is "guaranteed" to produce results with very small computational errors. There are quite carefully worked out mathematical results which show that the calculated answer is quite accurate, except for ill-conditioned matrices (which might be more indicative of problems in the system of equations than in the method of solution). The algorithm has been the subject of fairly detailed theoretical studies, and can be recommended as a computational procedure of very wide applicability.

Variations and Extensions

The method just described is most appropriate for N-by-N matrices with most of the N^2 elements non-zero. As we've seen for other problems, special techniques are appropriate for *sparse* matrices in which most of the elements are zero. This situation corresponds to systems of equations in which each equation has only a few terms.

If the non-zero elements have no particular structure, then the linked-list representation discussed in Chapter 36 is appropriate, with one node for each non-zero matrix element, linked together by both row and column. The standard method can be implemented for this representation, with the usual extra complications due to the need to create and destroy non-zero elements. This technique is unlikely to be worthwhile if one can afford the memory to hold the whole matrix, since it is much more complicated than the standard method. Also, sparse matrices become substantially less sparse during the Gaussian elimination process.

Some matrices not only have just a few non-zero elements but also have a simple structure, so that linked lists are not necessary. The most common example of this is a "band" matrix, in which the non-zero elements all fall very close to the diagonal. In such cases, the inner loops of the Gaussian elimination algorithms need be iterated only a few times, so that the total running time (and storage requirement) is proportional to N, not N^3.

An interesting special case of a band matrix is a "tridiagonal" matrix, in which only elements directly on, directly above, or directly below the diagonal are nonzero. For example, the general form of a tridiagonal matrix for $N = 5$ is:

$$\begin{pmatrix} a_{11} & a_{12} & 0 & 0 & 0 \\ a_{21} & a_{22} & a_{23} & 0 & 0 \\ 0 & a_{32} & a_{33} & a_{34} & 0 \\ 0 & 0 & a_{43} & a_{44} & a_{45} \\ 0 & 0 & 0 & a_{54} & a_{55} \end{pmatrix}$$

For such matrices, forward elimination and backward substitution each reduce to a single **for** loop:

```
for i:=1 to N-1 do
   begin
   a[i+1,N+1]:=a[i+1,N+1]-a[i,N+1]*a[i+1,i]/a[i,i];
   a[i+1,i+1]:=a[i+1,i+1]-a[i,i+1]*a[i+1,i]/a[i,i]
   end;
for j:=N downto 1 do
   x[j]:=(a[j,N+1]-a[j,j+1]*x[j+1])/a[j,j];
```

For forward elimination, only the case *j=i+1* and *k=i+1* needs to be included, since *a[i,k]=0* for *k>i+1*. (The case $k = i$ can be skipped since it sets to 0 an array element which is never examined again—this same change could be made to straight Gaussian elimination.)

Property 37.2 *A tridiagonal system of simultaneous equations can be solved in linear time.*

Of course, a two-dimensional array of size N^2 wouldn't be used for a tridiagonal matrix. The storage required for the above program can be made linear in N by maintaining four arrays instead of the a matrix: one for each of the three nonzero diagonals and one for the $(N + 1)$st column. Note that this program doesn't necessarily pivot on the largest available element, so there is no insurance against division by zero or the accumulation of computational errors. For some types of tridiagonal matrices that arise frequently, however, it can be proven that this is not a reason for concern. ■

Gauss-Jordan reduction can be implemented with full pivoting to replace a matrix by its *inverse* in one sweep through it. The inverse of a matrix A, written A^{-1}, has the property that a system of equations $Ax = b$ can be solved just by performing the matrix multiplication $x = A^{-1}b$. Still, N^3 operations are required to compute x given b. However, there is a way to preprocess a matrix and "decompose" it into component parts that makes it possible to solve the corresponding system of equations with any given right-hand side in time proportional to N^2, a savings of a factor of N over using Gaussian elimination each time. Roughly, this involves remembering the operations performed on the $(N + 1)$st column during the forward elimination phase, so that the result of forward elimination on a new $(N+1)$st column can be computed efficiently and back-substitution then performed as usual.

Solving systems of linear equations has been shown to be computationally equivalent to multiplying matrices, so there exist algorithms (for example, Strassen's matrix multiplication algorithm) which can solve systems of N equations in N variables in time proportional to $N^{2.81\ldots}$. As with matrix multiplication, using such a method is not worthwhile unless very large systems of equations are to be processed routinely (if then). As before, the actual running time of Gaussian elimination in terms of the number of inputs is $N^{3/2}$, which is difficult to improve upon in practice.

□

Exercises

1. Give the matrix produced by the forward-elimination phase of Gaussian elimination (*eliminate*) when used to solve the equations $x+y+z = 6$, $2x+y+3z = 12$, and $3x + y + 3z = 14$.

2. Give a system of three equations in three unknowns for which the naive triply nested **for** loop implementation of forward elimination fails, even though there is a solution.

3. What is the storage requirement for Gaussian elimination on an N-by-N matrix with only $3N$ nonzero elements?

4. Describe what happens when *eliminate* is used on a matrix with a row of all zeros.

5. Describe what happens when *eliminate* then *substitute* are used on a matrix with a column of all zeros.

6. About how many arithmetic operations are used in Gauss-Jordan reduction?

7. If we interchange columns in a matrix, what is the effect on the corresponding simultaneous equations?

8. How would you test for contradictory equations when using *eliminate*? How about identical equations?

9. Of what use would Gaussian elimination be on a system of M equations in N unknowns, with $M < N$? What if $M > N$?

10. Give an example showing the need for pivoting on the largest available element, using a mythical primitive computer in which numbers can be represented with only two significant digits (all numbers must be of the form $x.y \times 10^z$ for single-digit integers x, y, and z).

38

Curve Fitting

The term *curve fitting* (or *data fitting*) is used to describe the general problem of finding a function which matches a set of observed values at a set of given points. Specifically, given the points

$$x_1, x_2, \ldots, x_N$$

and the corresponding values

$$y_1, y_2, \ldots, y_N,$$

the goal is to find a function (perhaps of a specified type) such that

$$f(x_1) = y_1, f(x_2) = y_2, \ldots, f(x_N) = y_N$$

and such that $f(x)$ assumes "reasonable" values at other data points. It could be that the x's and y's are related by some unknown function and our goal is to find that function, but, in general, the definition of what is "reasonable" depends upon the application. We'll see that it is often easy to identify "unreasonable" functions.

Curve fitting has obvious application in the analysis of experimental data and has many other uses as well. For example, it can be used in computer graphics to produce curves that "look nice" without the overhead of storing a large number of points to be plotted. A related application is the use of curve fitting to provide a fast algorithm for computing the value of a known function at an arbitrary point: keep a short table of exact values, curve-fit to find other values.

Two principal methods are used to approach this problem. The first is *interpolation*: a smooth function is to be found which exactly matches the given values at the given points. The second method, *least-squares data fitting*, is used when the values given may not be exact and a function is sought which matches them as well as possible.

Polynomial Interpolation

We've already seen one method for solving the data-fitting problem: if f is known to be a polynomial of degree $N - 1$, then we have the polynomial interpolation problem of Chapter 36. Even if we have no particular knowledge about f, we could solve the data-fitting problem by letting $f(x)$ be the interpolating polynomial of degree $N - 1$ for the given points and values. This could be computed using methods outlined in Chapter 36, but there are many reasons not to use polynomial interpolation for data fitting. For one thing, a fair amount of computation is involved (advanced $N(\log N)^2$ methods are available, but elementary techniques are quadratic). Computing a polynomial of degree 100 (for example) seems overkill for interpolating a curve through 100 points.

The main drawback of polynomial interpolation is that high-degree polynomials are relatively complicated functions that may have unexpected properties not well suited to the function being fitted. A result from classical mathematics (the Weierstrass approximation theorem) tells us that it is possible to approximate any reasonable function with a polynomial (of sufficiently high degree). Unfortunately, polynomials of very high degree tend to fluctuate wildly. It turns out that, even though most functions are closely approximated almost everywhere on a closed interval by an interpolation polynomial, there are always some places where the approximation is terrible. Furthermore, this theory assumes that the data values are exact values from some unknown function, but it is often the case that the given data values are only approximate. If the y's were approximate values from some unknown low-degree polynomial, we would hope that the coefficients for the high-degree terms in the interpolating polynomial would be zero. It doesn't usually work out this way; instead, the interpolating polynomial tries to use the high-degree terms to help achieve an exact fit. These effects make interpolating polynomials inappropriate for many curve-fitting applications.

Spline Interpolation

Still, low-degree polynomials are simple curves that are easy to work with analytically, and they are widely used for curve fitting. The trick is to abandon the idea of trying to make *one* polynomial go through all the points and instead use different polynomials to connect adjacent points, piecing them together smoothly. An elegant special case of this, which also involves relatively straightforward computation, is called *spline interpolation.*

A spline is a mechanical device used by draftsmen to draw aesthetically pleasing curves: the draftsman fixes a set of points (*knots*) on his drawing, then bends a flexible strip of plastic or wood (the *spline*) around them and traces it to produce the curve. Spline interpolation is the mathematical equivalent of this process and results in the same curve. Figure 38.1 shows a spline through ten knots.

Figure 38.1 A spline through ten knots.

It can be shown from elementary mechanics that the shape assumed by the spline between two adjacent knots is a third-degree (cubic) polynomial. Translated to our data-fitting problem, this means that we should consider the curve to be $N-1$ different cubic polynomials

$$s_i(x) = a_i x^3 + b_i x^2 + c_i x + d_i, \qquad i = 1, 2, \ldots, N-1,$$

with $s_i(x)$ defined as the cubic polynomial to be used in the interval between x_i and x_{i+1}.

The spline can be represented in the obvious way as four one-dimensional arrays (or a 4-by-$(N-1)$ two-dimensional array). Creating a spline consists of computing the necessary a, b, c, d coefficients from the given x points and y values. The physical constraints on the spline correspond to simultaneous equations which can be solved to yield the coefficients.

For example, we obviously must have $s_i(x_i) = y_i$ and $s_i(x_{i+1}) = y_{i+1}$ for $i = 1, 2, \ldots, N-1$ because the spline must touch the knots. Not only does the spline touch the knots, but also it curves smoothly around them with no sharp bends or kinks. Mathematically, this means that the first derivatives of the spline polynomials must be equal at the knots ($s'_{i-1}(x_i) = s'_i(x_i)$ for $i = 2, 3, \ldots, N-1$). In fact, it turns out that the second derivatives of the polynomials must also be equal at the knots. These conditions give a total of $4N-6$ equations in the $4(N-1)$ unknown coefficients. Two more conditions need to be specified to describe the situation at the endpoints of the spline. Several options are available; we'll use the so-called "natural" spline which derives from $s''_1(x_1) = 0$ and $s''_{N-1}(x_N) = 0$. These conditions give a full system of $4N-4$ equations in $4N-4$ unknowns, which could be solved using Gaussian elimination to calculate all the coefficients that describe the spline.

This same spline can be computed somewhat more efficiently, however, because there are actually only $N-2$ "unknowns": most of the spline conditions are redundant. For example, suppose that p_i is the value of the second derivative of the spline at x_i, so that $s''_{i-1}(x_i) = s''_i(x_i) = p_i$ for $i = 2, \ldots, N-1$, with $p_1 = p_N = 0$. If the values of $p_1, \ldots, p_N$ are known, then all of the a, b, c, d coefficients can be computed for the spline segments, since we have four equations in four unknowns for each spline segment: for $i = 1,2,\ldots,N-1$, we must have

$$s_i(x_i) = y_i$$
$$s_i(x_{i+1}) = y_{i+1}$$
$$s_i''(x_i) = p_i$$
$$s_i''(x_{i+1}) = p_{i+1}.$$

The x and y values are given; to fully determine the spline, we need only compute the values of $p_2,\ldots,p_{N-1}$. To do so, we use the condition that the first derivatives must match: these $N-2$ conditions provide exactly the $N-2$ equations needed to solve for the $N-2$ unknowns, the p_i second-derivative values.

To express the a, b, c, and d coefficients in terms of the p second derivative values, then substitute those expressions into the four equations listed above for each spline segment, leads to some unnecessarily complicated expressions. Instead it is convenient to express the equations for the spline segments in a certain canonical form that involves fewer unknown coefficients. If we change variables to $t = (x - x_i)/(x_{i+1} - x_i)$ then the spline can be expressed as:

$$s_i(t) = ty_{i+1} + (1-t)y_i + (x_{i+1} - x_i)^2 \left((t^3 - t)p_{i+1} - ((1-t)^3 - (1-t))p_i\right)/6.$$

Now each spline is defined on the interval $[0, 1]$. This equation is less formidable than it looks because we're mainly interested in the endpoints 0 and 1, and either t or $(1-t)$ is 0 at these points. This representation makes it trivial to check that the spline interpolates and is continuous because $s_{i-1}(1) = s_i(0) = y_i$ for $i = 2,\ldots,N-1$, and it's only slightly more difficult to verify that the second derivative is continuous because $s_i''(1) = s_{i+1}''(0) = p_{i+1}$. These are cubic polynomials which satisfy the requisite conditions at the endpoints, so they are equivalent to the spline segments described above. If we were to substitute for t and find the coefficient of x^3, etc., then we would get the same expressions for the a's, b's, c's, and d's in terms of the x's, y's, and p's as if we were to use the method described in the previous paragraph. But there's no reason to do this, because we've checked that these spline segments satisfy the end conditions, and we can evaluate each at any point in its interval by computing t and using the above formula (once we know the p's).

To solve for the p's we need to set the first derivatives of the spline segments equal at the endpoints. The first derivative (with respect to x) of the above equation is

$$s_i'(t) = z_i + (x_{i+1} - x_i)\left((3t^2 - 1)p_{i+1} + (3(1-t)^2 - 1)p_i\right)/6$$

where $z_i = (y_{i+1} - y_i)/(x_{i+1} - x_i)$. Now, setting $s_{i-1}'(1) = s_i'(0)$ for $i = 2,\ldots,N-1$ yields our system of $N-2$ equations:

$$(x_i - x_{i-1})p_{i-1} + 2(x_{i+1} - x_{i-1})p_i + (x_{i+1} - x_i)p_{i+1} = 6(z_i - z_{i-1}).$$

This system of equations is a simple tridiagonal form that is easily solved with a degenerate version of Gaussian elimination, as we saw in Chapter 37. If we let $u_i = x_{i+1} - x_i$, $d_i = 2(x_{i+1} - x_{i-1})$, and $w_i = 6(z_i - z_{i-1})$, we have, for example, the following simultaneous equations for $N = 7$:

$$\begin{pmatrix} d_2 & u_2 & 0 & 0 & 0 \\ u_2 & d_3 & u_3 & 0 & 0 \\ 0 & u_3 & d_4 & u_4 & 0 \\ 0 & 0 & u_4 & d_5 & u_5 \\ 0 & 0 & 0 & u_5 & d_6 \end{pmatrix} \begin{pmatrix} p_2 \\ p_3 \\ p_4 \\ p_5 \\ p_6 \end{pmatrix} = \begin{pmatrix} w_2 \\ w_3 \\ w_4 \\ w_5 \\ w_6 \end{pmatrix}.$$

In fact, this is a symmetric tridiagonal system, with the diagonal below the main diagonal equal to the diagonal above the main diagonal. It turns out that pivoting on the largest available element is not necessary to get an accurate solution for this system of equations.

The method described in the above paragraph for computing a cubic spline translates very easily into Pascal:

```
procedure makespline;
  var i: integer;
  begin
  readln(N);
  for i:=1 to N do readln(x[i],y[i]);
  for i:=2 to N-1 do d[i]:=2*(x[i+1]-x[i-1]);
  for i:=1 to N-1 do u[i]:=x[i+1]-x[i];
  for i:=2 to N-1 do
    w[i]:=6.0*((y[i+1]-y[i])/u[i]-(y[i]-y[i-1])/u[i-1]);
  p[1]:=0.0; p[N]:=0.0;
  for i:=2 to N-2 do
    begin
    w[i+1]:=w[i+1]-w[i]*u[i]/d[i];
    d[i+1]:=d[i+1]-u[i]*u[i]/d[i]
    end;
  for i:=N-1 downto 2 do
    p[i]:=(w[i]-u[i]*p[i+1])/d[i];
  end;
```

The arrays d and u are the representation of the tridiagonal matrix that is solved using the program in Chapter 37. We use $d[i]$ where $a[i,i]$ is used in that program, $u[i]$ where $a[i+1,i]$ or $a[i,i+1]$ is used, and $z[i]$ where $a[i,N+1]$ is used.

Property 38.1 *A cubic spline on N points can be built in linear time.*

This fact is obvious from the program, which is simply a succession of linear passes through the data. ∎

For an example of the construction of a cubic spline, consider fitting a spline to the five data points

$$(1.0,2.0),\quad (2.0,1.5),\quad (4.0,1.25),\quad (5.0,1.2),\quad (8.0,1.125),\quad (10.0,1.1).$$

(These come from the function $1+1/x$.) The spline parameters are found by solving the system of equations

$$\begin{pmatrix} 6 & 2 & 0 & 0 \\ 2 & 6 & 1 & 0 \\ 0 & 1 & 8 & 3 \\ 0 & 0 & 3 & 10 \end{pmatrix} \begin{pmatrix} p_2 \\ p_3 \\ p_4 \\ p_5 \end{pmatrix} = \begin{pmatrix} 2.250 \\ .450 \\ .150 \\ .075 \end{pmatrix}$$

with the result $p_2 = 0.39541$, $p_3 = -0.06123$, $p_4 = 0.02658$, $p_5 = -0.00047$.

To evaluate the spline for any value of x in the range $[x_1..x_N]$, we simply find the interval $[x_i..x_{i+1}]$ containing x, then compute t and use the formula above for $s_i(x)$ (which, in turn, uses the computed values for p_i and p_{i+1}).

```
function eval(v: real): real;
  var t: real; i: integer;
  function f(x: real): real;
    begin f:=x*x*x-x end;
  begin
  i:=0; repeat i:=i+1 until v<=x[i+1];
  t:=(v-x[i])/u[i];
  eval:=t*y[i+1]+(1-t)*y[i]
        +u[i]*u[i]*(f(t)*p[i+1]+f(1-t)*p[i])/6.0
  end;
```

This program does not check for the error condition when v is not between $x[1]$ and $x[N]$. If the number of spline segments is large (that is, if N is large), some more efficient searching method from Chapter 14 might be used to find the interval containing v.

There are many variations on the idea of curve fitting by piecing together polynomials in a "smooth" way: the computation of splines is a quite well-developed field of study. Other types of splines involve other types of smoothness criteria as well as changes such as relaxing the condition that the spline must exactly touch each data point. Computationally, they involve exactly the same steps to determine the coefficients for each of the spline pieces by solving the system of linear equations derived from imposing constraints on how they are joined.

Method of Least Squares

It frequently arises that, while our data values are not exact, we do have some idea of the form of the function that is to fit the data. The function might depend on some parameters

$$f(x) = f(c_1, c_2, \ldots, c_M, x)$$

and the curve-fitting procedure is to find the choice of parameters that "best" matches the observed values at the given points. If the function were a polynomial (with the parameters being the coefficients) and the values were exact, then this would be interpolation. But now we are considering more general functions and inaccurate data. To simplify the discussion, we'll concentrate on fitting to functions expressed as a linear combination of simpler functions, with the unknown parameters being the coefficients:

$$f(x) = c_1 f_1(x) + c_2 f_2(x) + \cdots + c_M f_M(x).$$

This includes most of the functions we'll be interested in. After studying this case, we'll consider more general functions.

A common way of measuring how well a function fits is the *least-squares criterion*. Here the error is calculated by adding up the squares of the errors at each of the observation points:

$$E = \sum_{1 \le j \le N} (f(x_j) - y_j)^2.$$

This is a very natural measure: the squaring is done to eliminate cancellations among errors with different signs. Obviously, it is most desirable to find the choice of parameters that minimizes E. It turns out that this choice can be computed efficiently: this is the so-called *method of least squares*.

The method follows quite directly from the definition. To simplify the derivation, we'll consider the case $M = 2, N = 3$, but the general method follows directly. Suppose we have three points x_1, x_2, x_3 and corresponding values y_1, y_2, y_3 that are to be fitted to a function of the form $f(x) = c_1 f_1(x) + c_2 f_2(x)$. Our job is to find the choice of the coefficients c_1, c_2 which minimizes the least-squares error

$$\begin{aligned} E = {} & (c_1 f_1(x_1) + c_2 f_2(x_1) - y_1)^2 \\ & + (c_1 f_1(x_2) + c_2 f_2(x_2) - y_2)^2 \\ & + (c_1 f_1(x_3) + c_2 f_2(x_3) - y_3)^2. \end{aligned}$$

To find the choices of c_1 and c_2 which minimize this error, we simply need to set the derivatives dE/dc_1 and dE/dc_2 to zero. For c_1 we have:

$$\begin{aligned} dE/dc_1 = {} & 2(c_1 f_1(x_1) + c_2 f_2(x_1) - y_1) f_1(x_1) \\ & + 2(c_1 f_1(x_2) + c_2 f_2(x_2) - y_2) f_1(x_2) \\ & + 2(c_1 f_1(x_3) + c_2 f_2(x_3) - y_3) f_1(x_3). \end{aligned}$$

Setting the derivative equal to zero leaves an equation which the variables c_1 and c_2 must satisfy ($f_1(x_1)$, etc. are all "constants" with known values):

$$\begin{aligned} c_1\,(f_1(x_1)f_1(x_1) + f_1(x_2)f_1(x_2) + f_1(x_3)f_1(x_3)) \\ +c_2\,(f_2(x_1)f_1(x_1) + f_2(x_2)f_1(x_2) + f_2(x_3)f_1(x_3)) \\ = y_1f_1(x_1) + y_2f_1(x_2) + y_3f_1(x_3). \end{aligned}$$

We get a similar equation when we set the derivative dE/dc_2 to zero. These rather formidable-looking equations can be greatly simplified using vector notation and the "dot product" operation. If we define the vectors $\mathbf{x} = (x_1, x_2, x_3)$ and $\mathbf{y} = (y_1, y_2, y_3)$, then the dot product of $\mathbf{x}$ and $\mathbf{y}$ is the real number defined by

$$\mathbf{x} \cdot \mathbf{y} = x_1y_1 + x_2y_2 + x_3y_3.$$

Now, if we define the vectors $\mathbf{f}_1 = (f_1(x_1), f_1(x_2), f_1(x_3))$ and $\mathbf{f}_2 = (f_2(x_1), f_2(x_2), f_2(x_3))$, then our equations for the coefficients c_1 and c_2 can be very simply expressed:

$$c_1\mathbf{f}_1 \cdot \mathbf{f}_1 + c_2\mathbf{f}_1 \cdot \mathbf{f}_2 = \mathbf{y} \cdot \mathbf{f}_1,$$

$$c_1\mathbf{f}_2 \cdot \mathbf{f}_1 + c_2\mathbf{f}_2 \cdot \mathbf{f}_2 = \mathbf{y} \cdot \mathbf{f}_2.$$

These can be solved with Gaussian elimination to find the desired coefficients.

For example, suppose that we know that the data points

$$(1.0, 2.05) \quad (2.0, 1.53) \quad (4.0, 1.26) \quad (5.0, 1.21) \quad (8.0, 1.13) \quad (10.0, 1.1)$$

should be fit by a function of the form $c_1 + c_2/x$. (These data points are slightly perturbed from the exact values for $1 + 1/x$.) In this case, f_1 is a constant ($\mathbf{f}_1 = (1.0, 1.0, 1.0, 1.0, 1.0, 1.0)$) and $\mathbf{f}_2 = (1.0, 0.5, 0.25, 0.2, 0.125, 0.1)$, so we have to solve the system of equations

$$\begin{pmatrix} 6.000 & 2.175 \\ 2.175 & 1.378 \end{pmatrix} \begin{pmatrix} c_1 \\ c_2 \end{pmatrix} = \begin{pmatrix} 8.280 \\ 3.623 \end{pmatrix}$$

with the result $c_1 = 0.998$ and $c_2 = 1.054$ (both close to one, as expected).

The method outlined above easily generalizes to finding more than two coefficients. To find the constants $c_1, c_2, \ldots, c_M$ in

$$f(x) = c_1f_1(x) + c_2f_2(x) + \cdots + c_Mf_M(x)$$

which minimize the least-squares error for the point and observation vectors

$$\mathbf{x} = (x_1, x_2, \ldots, x_N),$$

$$\mathbf{y} = (y_1, y_2, \ldots, y_N),$$

first compute the function component vectors

$$\mathbf{f}_1 = (f_1(x_1), f_1(x_2), \ldots, f_1(x_N)),$$
$$\mathbf{f}_2 = (f_2(x_1), f_2(x_2), \ldots, f_2(x_N)),$$
$$\vdots$$
$$\mathbf{f}_M = (f_M(x_1), f_M(x_2), \ldots, f_M(x_N)).$$

Then make up an M-by-M linear system of equations $Ac = b$ with

$$a_{ij} = \mathbf{f}_i \cdot \mathbf{f}_j,$$
$$b_j = \mathbf{f}_j \cdot \mathbf{y}.$$

The solution to this system of simultaneous equations yields the required coefficients.

This method is easily implemented by maintaining a two-dimensional array for the **f** vectors, considering **y** as the $(M+1)$st vector. Then an array *a[1..M,1..M+1]* can be filled as follows:

```
for i:=1 to M do
  for j:=1 to M+1 do
    begin
    t:= 0.0;
    for k:=1 to N do t:=t+f[i,k]*f[j,k];
    a[i,j]:=t;
    end;
```

and solved using the Gaussian elimination procedure from Chapter 37.

The method of least squares can be extended to handle nonlinear functions (for example a function such as $f(x) = c_1 e^{-c_2 x} \sin c_3 x$), and it is often used for this type of application. The idea is fundamentally the same; the problem is that the derivatives may not be easy to compute. What is used is an *iterative* method: use some estimate for the coefficients, then use these within the method of least squares to compute the derivatives, thus producing a better estimate for the coefficients. This basic method, which is widely used today, was outlined by Gauss in the 1820s.

□

Exercises

1. Approximate the function $\lg x$ with a degree-four interpolating polynomial at the points 1, 2, 3, 4, and 5. Estimate the quality of the fit by computing the sum of the squares of the errors at 1.5, 2.5, 3.5, and 4.5.
2. Solve the previous problem for the function $\sin x$. Plot the function and the approximation if possible on your computer system.
3. Solve the previous problems using a cubic spline instead of an interpolating polynomial.
4. Approximate the function $\lg x$ with a cubic spline with knots at 2^N for N between 1 and 10. Experiment with different placements of knots in the same range to try to obtain a better fit.
5. What will happen in least-squares data fitting if one of the functions was the function $f_i(x) = 0$ for some i?
6. Use a least-squares curvefitter to find the values of a and b that give the best formula of the form $aN \ln N + bN$ for describing the total number of instructions executed when Quicksort is run on a random file.
7. What values of a, b, c minimize the least-squares error in using the function $f(x) = ax \log x + bx + c$ to approximate the observations $f(1) = 0, f(4) = 13, f(8) = 41$?
8. Excluding the Gaussian elimination phase, how many multiplications are involved in using the method of least squares to find M coefficients based on N observations?
9. Under what circumstances would the matrix arising in least-squares curve fitting be singular?
10. Does the least-squares method work if two different observations are included for the same point?

39

Integration

☐ Computing the integral is a fundamental analytic operation often performed on functions being processed on computers. We want to find the "area under the curve" efficiently and to a reasonable degree of accuracy. In this chapter we examine a number of classical algorithms for solving this basic numerical problem.

First, we'll briefly discuss the situation where an explicit representation of the function is available. In such cases, it may be possible to do *symbolic integration* to transform the representation for a function into a similar representation for the integral. This is appropriate when the functions being processed are in a restricted class of functions whose integrals are available analytically, or in the context of systems that process such representations of functions.

At the other extreme, the function may be defined by a table, so that function values are known for only a few points. In such a case, one can give only an approximate value for the integral, based on assumptions about how the function behaves between the points. The accuracy of the integral is almost wholly dependent on the validity of the assumptions.

The most common situation lies between these extremes: the function to be integrated is represented in such a way that its value at any particular point can be computed. Again, the accuracy of the integral depends on the assumptions about the behavior of the function between whatever points are chosen for evaluation. The goal is to compute a reasonable approximation to the integral of the function, without performing an excessive number of function evaluations. This computation is often called *quadrature* by numerical analysts.

In this chapter we look at several quadrature methods. The methods are elementary—our aim is to gain some experience with such computations as fundamental numerical methods. Many applications actually can benefit from proper application of the elementary techniques we consider, but methods to solve more advanced problems, especially numerical solution of differential equations, are of much more importance in practice.

Symbolic Integration

If full information is available about a function, then it may be worthwhile to use a method that involves manipulating some representation of the function rather than working with numeric values. The goal is to transform a representation of the function into a representation of the integral, in much the same way as indefinite integration is done by hand.

A simple example of this is the integration of polynomials. In Chapter 36 we examined methods for "symbolically" computing sums and products of polynomials, using programs that worked on a particular representation for the polynomials and produced the representation for the answers from the representation for the inputs. The operation of integration (and differentiation) of polynomials can also be done in this way. If a polynomial

$$p(x) = p_0 + p_1 x + p_2 x^2 + \cdots + p_{N-1} x^{N-1}$$

is represented simply by keeping the values of the coefficients in an array p then the integral can be easily computed as follows:

```
for i:=N downto 1 do p[i]:=p[i-1]/i; p[0]:=0;
```

For each term of the polynomial, this program applies the well-known symbolic integration rule $\int_0^x t^{i-1}\,dt = x^i/i$ for $i > 0$. A wider class of functions than just polynomials can be handled by adding more symbolic rules. The addition of composite rules such as *integration by parts,*

$$\int u\,dv = uv - \int v\,du,$$

can greatly expand the set of functions that can be handled. (Integration by parts requires a differentiation capability. Symbolic differentiation is somewhat easier than symbolic integration, since a reasonable set of elementary rules plus the composite *chain rule* will suffice for most common functions.)

The large number of rules available to be applied to a particular function makes symbolic integration a difficult task. Indeed, it has only recently been shown that there is an *algorithm* for this task: a procedure that either returns the integral of any given function or says that the answer cannot be expressed in terms of elementary functions. A description of this algorithm in its full generality would be beyond the scope of this book. However, when the functions being processed are from a small restricted class, symbolic integration can be a powerful tool.

Of course, symbolic techniques have the fundamental limitation that a great many integrals (many of which occur in practice) can't be evaluated symbolically. Next, we'll examine some techniques that have been developed to compute approximations to the values of real integrals.

Simple Quadrature Methods

Perhaps the most obvious way to approximate the value of an integral is the *rectangle method*. Evaluating an integral is the same as computing the area under a curve, and we can estimate the area under a curve by summing the areas of small rectangles that nearly fit under the curve, as diagrammed in Figure 39.1.

To be precise, suppose that we are to compute $\int_a^b f(x)\,dx$, and that the interval $[a,b]$ over which the integral is to be computed is divided into N parts, delimited by the points $x_1, x_2, \ldots, x_{N+1}$. Then we have N rectangles, with the width of the ith rectangle $(1 \le i \le N))$ given by $x_{i+1} - x_i$. For the height of the ith rectangle, we could use $f(x_i)$ or $f(x_{i+1})$, but it would seem that the result would be more accurate if the value of f at the midpoint of the interval $(f((x_i + x_{i+1})/2))$ is used, as in the above diagram. This leads to the quadrature formula

$$r = \sum_{1 \le i \le N} (x_{i+1} - x_i) f\left(\frac{x_i + x_{i+1}}{2}\right)$$

which estimates the value of the integral of $f(x)$ over the interval from $a = x_1$ to $b = x_{N+1}$. In the common case where all the intervals are to be of the same size, say $x_{i+1} - x_i = w$, we have $x_{i+1} + x_i = (2i + 1)w$, so the approximation r to the integral is easily computed.

```
function intrect(a,b: real; N: integer): real;
  var i: integer; w,r: real;
  begin
  r:=0; w:=(b-a)/N;
  for i:=1 to N do r:=r+w*f(a-w/2+i*w);
  intrect :=r;
  end;
```

Of course, as N gets larger, the answer becomes more accurate. Figure 39.2 shows the result of using a smaller interval size for the function shown in Figure 39.1.

Below is a more quantitative example that shows the estimates produced by this function for $\int_1^2 dx/x$ (which we know to be $\ln 2 = 0.6931471805599\ldots$) when invoked with the call *intrect(1.0,2.0,N)* for $N = 10, 100, 1000$:

10	0.6928353604100
100	0.6931440556283
1000	0.6931471493100

When $N = 1000$, our answer is accurate to about seven decimal places. More sophisticated quadrature methods can achieve better accuracy with much less work.

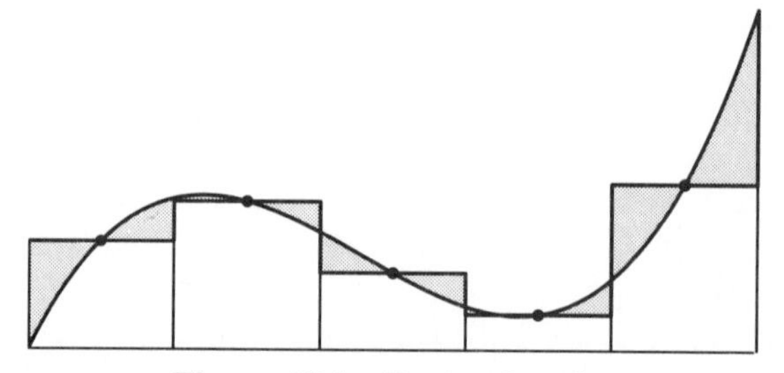

Figure 39.1 Rectangle rule.

It turns out that looking at error estimates for particular methods can often suggest more accurate methods. Consider the analytic expression for the error made in the rectangle method by expanding $f(x)$ in a Taylor series about the midpoint of each interval, integrating, then summing over all intervals. We won't go through the details of this calculation but merely point out that

$$\int_a^b f(x)\,dx = r + w^3 e_3 + w^5 e_5 + \cdots$$

where w is the interval width $((b-a)/N)$ and e_3 depends on the value of the third derivative of f at the interval midpoints, etc. (This is normally a good approximation because most "reasonable" functions have small high-order derivatives, though this is not always true.) For example, if we choose to make $w = .01$ (which would correspond to $N = 200$ in the example above), this formula says the integral computed by the procedure above should be accurate to about six places.

Another way to approximate the integral is to divide the area under the curve into trapezoids, as diagrammed in Figure 39.3. Recall that the area of a trapezoid is one-half the product of the height and the sum of the lengths of the two bases.

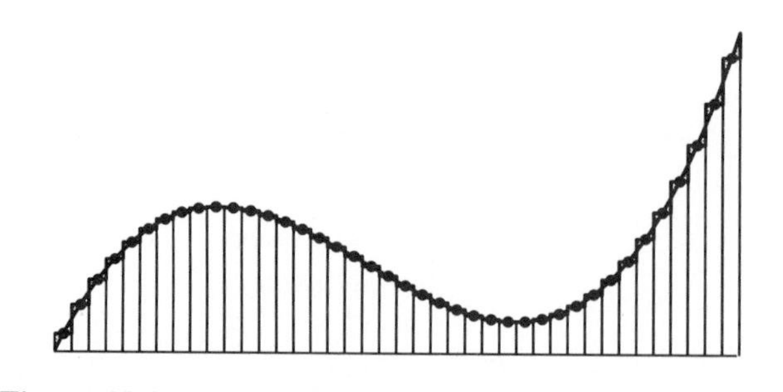

Figure 39.2 Rectangle rule with a smaller interval size.

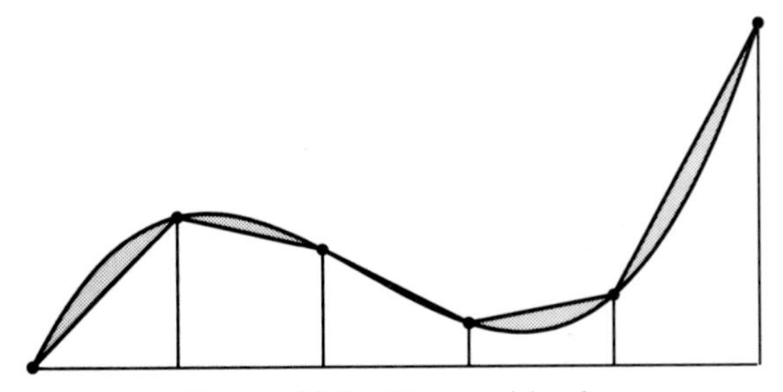

Figure 39.3 Trapezoid rule.

The *trapezoid method* leads to the quadrature formula

$$t = \sum_{1 \le i \le N} (x_{i+1} - x_i) \frac{f(x_i) + f(x_{i+1})}{2}.$$

The following procedure implements the trapezoid method in the common case where all the intervals are the same width:

```
function inttrap(a, b: real; N: integer): real;
  var i: integer; w, t: real;
  begin
  t:=0; w:=(b-a)/N;
  for i:=1 to N do t:=t+w*(f(a+(i-1)*w)+f(a+i*w))/2;
  inttrap:=t;
  end;
```

The error for this method can be derived in a similar way as for the rectangle method. It turns out that

$$\int_a^b f(x)\,dx = t - 2w^3 e_3 - 4w^5 e_5 + \cdots.$$

Thus the rectangle method is twice as accurate as the trapezoid method. This is borne out by our example—this procedure produces the following estimates for $\int_1^2 dx/x$:

10	0.6937714031754
100	0.6931534304818
1000	0.6931472430599

It may seem surprising at first that the rectangle method is more accurate than the trapezoid method. Remember, however, that the rectangles tend to fall partly under the curve, partly over (so that the error within an interval can cancel out), while

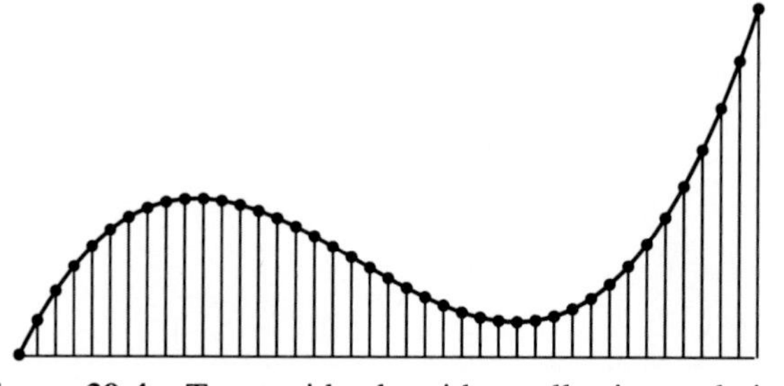

Figure 39.4 Trapezoid rule with smaller interval size.

the trapezoids tend to fall either completely under or completely over the curve. Figure 39.4 shows the trapezoid method with a smaller interval size—it seems to fit the curve exactly, but Figure 39.2 actually gives a better estimate of the area under the curve.

Another perfectly reasonable method is *spline quadrature*: spline interpolation is performed using methods discussed in the previous chapter and then the integral is computed by piecewise application of the trivial symbolic polynomial integration technique described above. This method is actually closely related to the rectangle rule and the trapezoid rule, as we'll see below.

Compound Methods

Examination of the formulas given above for the error of the rectangle and trapezoid methods leads to a simple method that turns out to have much greater accuracy, called *Simpson's method.* The idea is to eliminate the leading term in the error by combining the two methods. Multiplying the formula for the rectangle method by two, adding the formula for the trapezoid method then dividing by three gives the equation

$$\int_a^b f(x)\,dx = \frac{1}{3}(2r + t - 2w^5 e_5 + \cdots).$$

The w^3 term has disappeared, so this formula tells us that we can get a method that is accurate to within w^5 by combining the quadrature formulas in the same way:

$$s = \sum_{1 \le i \le N} \frac{x_{i+1} - x_i}{6} \left(f(x_i) + 4f(\frac{x_i + x_{i+1}}{2}) + f(x_{i+1}) \right).$$

If an interval size of .01 is used for Simpson's rule, then the integral can be computed to about ten-place accuracy. Again, this is borne out in our example. The implementation of Simpson's method is only slightly more complicated than the others (again, we consider the case when the intervals have the same width):

```
function intsimp(a,b: real; N: integer): real;
  var i: integer; w,s: real;
  begin
  s:=0; w:=(b-a)/N;
  for i:=1 to N do
    s:=s+w*(f(a+(i-1)*w)+4*f(a-w/2+i*w)+f(a+i*w))/6;
  intsimp:=s;
  end;
```

This program requires three "function evaluations" (rather than two) in the inner loop, but it produces far more accurate results than do the previous two methods:

10	0.6931473746651
100	0.6931471805795
1000	0.6931471805599

More complicated quadrature methods have been devised that gain accuracy by combining simpler methods with similar errors. The most well-known is *Romberg integration*, which uses two different sets of subintervals for its two "methods."

It turns out that Simpson's method is exactly equivalent to interpolating the data to a piecewise quadratic function, then integrating. It is interesting to note that the four methods we have discussed all can be cast as piecewise interpolation methods: the rectangle rule interpolates to a constant (degree-zero polynomial); the trapezoid rule to a line (degree-one polynomial); Simpson's rule to a quadratic polynomial; and spline quadrature to a cubic polynomial.

Adaptive Quadrature

A major flaw in the methods we have discussed so far is that the errors involved depend not only upon the subinterval size used but also upon the value of the high-order derivatives of the function being integrated. This implies that these methods will not work well at all for certain functions (those with large high-order derivatives). But few functions have large high-order derivatives everywhere. It is reasonable to use small intervals where the derivatives are large and large intervals where the derivatives are small. A method which does this in a systematic way is called an *adaptive quadrature* routine.

The general approach in adaptive quadrature is to use two different quadrature methods for each subinterval, compare the results, and subdivide the interval further if the difference is too great. Of course some care must be exercised, since if two equally bad methods are used, they might agree quite closely on a bad result. One way to avoid this is to ensure that one method always overestimates the result and the other always underestimates the result. Another way to avoid this is to ensure

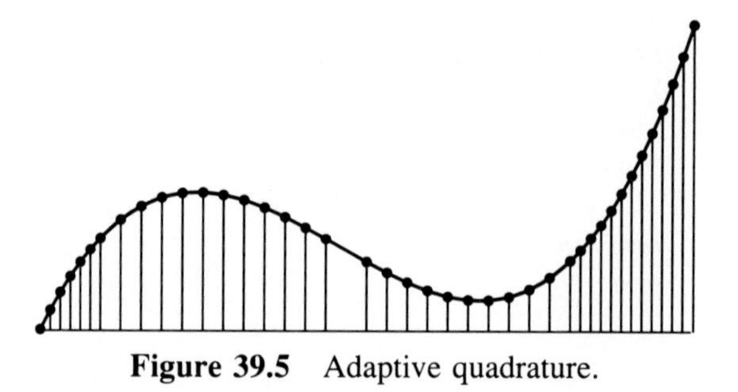

Figure 39.5 Adaptive quadrature.

that one method is more accurate than the other. A method of this latter type is described next.

Significant overhead is involved in recursively subdividing the interval, so it pays to use a good method for estimating the integrals, as in the following implementation:

```
function adapt(a,b: real): real;
  begin
  if abs(intsimp(a,b,10)-intsimp(a,b,5))<tolerance
    then adapt:=intsimp(a,b,10)
    else adapt:=adapt(a, (a+b)/2) + adapt((a+b)/2,b);
  end;
```

Both estimates for the integral are derived from Simpson's method, but one uses twice as many subdivisions as the other. Essentially, this amounts to checking the accuracy of Simpson's method over the interval in question and then subdividing if it is not good enough.

Unlike our other methods, where we decide how much work we want to do and then take whatever accuracy results, in adaptive quadrature we do however much work is necessary to achieve the degree of accuracy decided upon ahead of time. This means that *tolerance* must be chosen carefully, so that the routine doesn't loop indefinitely to achieve an impossibly high tolerance. The number of steps required depends very much on the nature of the function being integrated. A function that fluctuates wildly requires a large number of steps, but such a function would yield a very inaccurate result with the "fixed interval" methods. Figure 39.5 shows the points evaluated when adaptive quadrature (based on the trapezoid rule) is used on the function in Figures 39.1–39.4. Note that the intervals are larger where the function is straight and smooth, smaller where the function bends more quickly.

A smooth function such as our example can be handled with a reasonable number of steps. The following table gives, for various values of t, the value produced and the number of recursive calls required by the above routine to compute $\int_1^2 dx/x$:

0.00001000000	0.6931473746651	1
0.00000010000	0.6931471829695	5
0.00000000100	0.6931471806413	13
0.00000000001	0.6931471805623	33

The above program can be improved in several ways. First, there's certainly no need to call *intsimp(a,b,10)* twice. In fact, the function values for this call can be shared by *intsimp(a,b,5)*. Second, the tolerance bound can be related to the accuracy of the answer more closely if *tolerance* is scaled by the ratio of the size of the current interval to the size of the full interval. Also, a better routine can obviously be developed by using an even better quadrature rule than Simpson's (but it is a basic law of recursion that another *adaptive* routine wouldn't be a good idea). A sophisticated adaptive quadrature routine can provide very accurate results for problems that can't be handled any other way, but careful attention must be paid to the types of functions to be processed.

The "divide-and-conquer" algorithm-design paradigm is thus useful for numerical programs as well. In fact, adaptive methods of this type are very important as solution techniques for advanced numerical problems such as integration in higher dimensions and numerical solution of differential equations.

□

Exercises

1. Write a program to symbolically integrate (and differentiate) polynomials in x and $\ln x$. Use a recursive implementation based on integration by parts.

2. Which quadrature method is likely to produce the best answer for integrating the following functions: $f(x) = 5x$, $f(x) = (3 - x)(4 + x)$, $f(x) = \sin(x)$?

3. Give the result of using each of the four elementary quadrature methods (rectangle, trapezoid, Simpson's, spline) to integrate $y = 1/x$ in the interval $[.1, 10]$.

4. Answer the previous question for the function $y = \sin x$.

5. Discuss what happens when adaptive quadrature is used to integrate the function $y = 1/x$ in the interval $[-1, 2]$.

6. Answer the previous question for the elementary quadrature methods.

7. Give the points of evaluation when adaptive quadrature is used to integrate the function $y = 1/x$ in the interval $[.1, 10]$ with a tolerance of .1.

8. Compare the accuracy of an adaptive quadrature based on Simpson's method to an adaptive quadrature based on the rectangle method for the integral given in the previous problem.

9. Answer the previous question for the function $y = \sin x$.

10. Give a specific example of a function for which adaptive quadrature would be likely to give a drastically more accurate result than the other methods.

SOURCES for Mathematical Algorithms

Much of the material in this section falls within the domain of numerical analysis, and several excellent textbooks are available, for example the book by Conte and de Boor. A book which pays particular attention to computational issues is the 1977 book by Forsythe, Malcomb and Moler. In particular, much of the material discussed here in Chapters 37, 38, and 39 is based on the presentation given in that book. The book by Press *et al.* is also a compendium of useful numerical methods, complete with implementations.

The other major reference for this section is the second volume of D. E. Knuth's comprehensive treatment *The Art of Computer Programming.* Knuth uses the term "seminumerical" to describe algorithms which lie at the interface between numerical and symbolic computation, such as random-number generation and polynomial arithmetic. Among many other topics, Knuth's volume 2 covers in great depth the material given here in Chapters 1, 3, and 4.

The 1975 book by Borodin and Munro is an additional reference for Strassen's matrix multiplication method and general treatment of arithmetic algorithms from a computational-complexity point of view.

Many of the algorithms we've considered (and many others, principally symbolic methods as mentioned in Chapter 39) are embodied in a computer system called MACSYMA, the first of several systems for "symbolic mathematics" that have been developed in recent years. Systems like MACSYMA have become indispensible for mathematicians and scientists for mathematical analysis in a variety of applications.

A. Borodin and I. Munro, *The Computational Complexity of Algebraic and Numerical Problems*, American Elsevier, New York, 1975.

S. Conte and C. de Boor, *Elementary Numerical Analysis*, McGraw-Hill, New York, 1980.

G. E. Forsythe, M. A. Malcomb, and C. B. Moler, *Computer Methods for Mathematical Computations*, Prentice-Hall, Englewood Cliffs, NJ, 1977.

D. E. Knuth, *The Art of Computer Programming. Volume 2: Seminumerical Algorithms*, Addison-Wesley, Reading, MA (second edition), 1981.

W. H. Press, B. P. Flannery, S. A. Teukolsky, and W. T. Vetterling, *Numerical Recipes: The Art of Scientific Computing*, Cambridge University Press, 1986.

MIT Mathlab Group, *MACSYMA Reference Manual*, Laboratory for Computer Science, Massachusetts Institute of Technology, 1977.

Advanced Topics

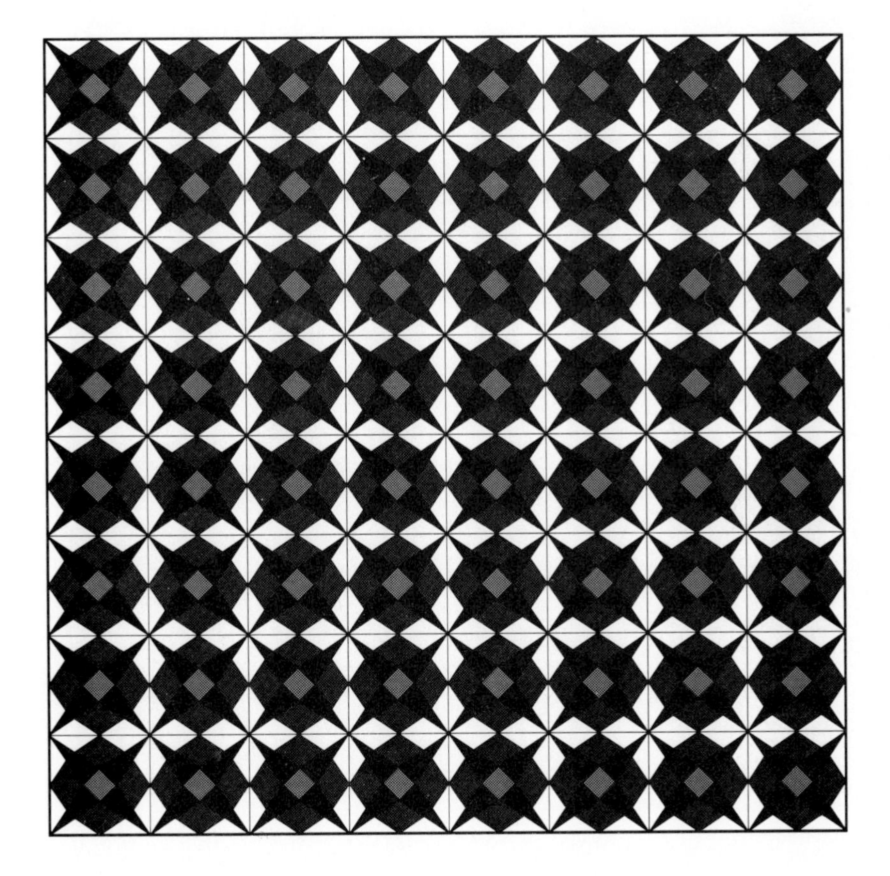

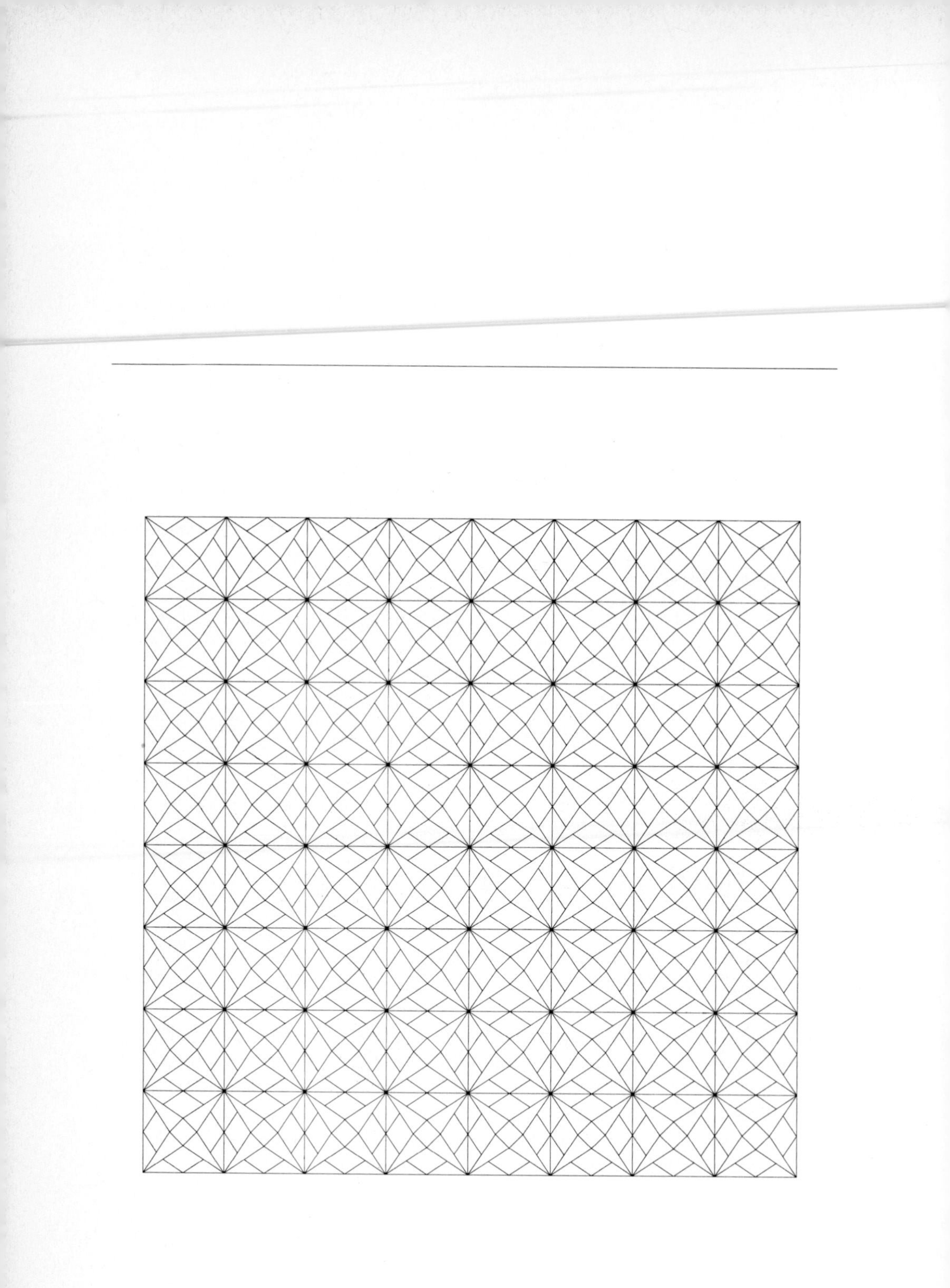

40

Parallel Algorithms

The algorithms we have studied are, for the most part, remarkably robust in their applicability. Most of the methods we've looked at are a decade or more old and have survived many quite radical changes in computer hardware and software. New hardware designs and new software capabilities can certainly have a significant impact on specific algorithms, but good algorithms on old machines are, for the most part, good algorithms on new machines.

One reason for this is that the fundamental design of "conventional" computers has changed little over the years. The design of the vast majority of computing systems is guided by the same underlying principle, one developed by the mathematician J. von Neumann in the early days of modern computing. When we speak of the *von Neumann model of computation*, we refer to a view of computing in which instructions and data are stored in the same memory and a single processor fetches instructions from the memory and executes them (perhaps operating on the data), one by one. Elaborate mechanisms have been developed to make computers cheaper, faster, smaller (physically), and larger (logically), but the architecture of most computer systems can be viewed as variations on the von Neumann theme.

Recently, however, radical changes in the cost of computing components have made it practical to consider radically different types of machines, ones in which a large number of instructions can be executed at each instant in time, or in which the instructions are "wired in" to make special-purpose machines capable of solving only one problem or in which a large number of smaller machines can cooperate to solve the same problem. In short, rather than having a machine execute just one instruction at each time instant, we can think about having a large number of actions being performed simultaneously. In this chapter, we shall consider the potential effect of such ideas on some of the problems and algorithms we have been studying. In particular, we'll consider two approaches to machine architecture that are amenable to the development of parallel algorithms: the *perfect shuffle* and the *systolic array*.

General Approaches

Certain fundamental algorithms are used so frequently and for such large problems that there is always pressure to run them on bigger and faster computers. One result of this has been a series of "supercomputers" which embody the latest technology; they make some concessions to the fundamental von Neumann concept but still are designed to be general-purpose and useful for all programs. The common approach to using such a machine for the type of problem we have been studying is to start with the algorithms that are best on conventional machines and adapt them to the particular features of the new machine. This approach clearly encourages the persistence of old algorithms and old architectures in new machines.

Microprocessors with significant computing capabilities have recently become quite inexpensive. An obvious approach is to try to use a large number of these processors together to solve a large problem. Some algorithms can adapt well to being "distributed" in this way; others simply are not appropriate for this kind of implementation.

The development of inexpensive, relatively powerful processors has also led to the appearance of general-purpose tools for use in designing and building new processors. This in turn has led to increased activity in the development of special-purpose machines for particular problems. If no machine is particularly well-suited to executing some important algorithm, then we can design and build one that is! Appropriate machines can be designed and built for many problems that fit on one (very-large-scale) integrated circuit chip.

A common thread in all of these approaches is *parallelism*: we try to save time by having as many different things as possible happening at any instant. This can lead to chaos if it is not done in an orderly manner. Below, we'll consider two examples that illustrate some techniques for achieving a high degree of parallelism for some specific classes of problems. The idea is to assume that we have not just one but M processors on which our program can run. Thus, if things work out well, we can hope to have our program run M times faster than before.

Several immediate problems are involved in getting M processors to work together to solve the same problem. The most important is that they must communicate in some way: there must be wires interconnecting them and specific mechanisms for sending data back and forth along those wires. Furthermore, there are physical limitations on the type of interconnection allowed. For example, suppose that our "processors" are integrated circuit chips (these can now contain more circuitry than small computers of the past) that have, say, 32 pins to be used for interconnection. Even if we had 1000 such processors, we could connect each to at most 32 others. The choice of how to interconnect the processors is fundamental in parallel computing. Moreover, it's important to remember that this decision must be made ahead of time: a program can change its way of doing things depending on the particular instance of the problem being solved, but a machine generally can't change the way its parts are wired together.

This general view of parallel computation in terms of independent processors with some fixed interconnection pattern applies in each of the three domains described above: a supercomputer has very specific processors and interconnection patterns that are integral to its architecture (and affect many aspects of its performance); interconnected microprocessors involve a relatively small number of powerful processors with simple interconnections; and very-large-scale integrated (VLSI) circuits themselves involve a very large number of simple processors (circuit elements) with complex interconnections.

Many other views of parallel computation have been studied extensively since von Neumann, with renewed interest since inexpensive processors have become available. It would be well beyond the scope of this book to treat all the issues involved. Instead, we'll consider two specific machines that have been proposed for some familiar problems. The machines we consider illustrate the effects of machine architecture on algorithm design and vice versa. A certain symbiosis is at work here: one certainly wouldn't design a new computer without some idea of what it will be used for, and one would like to use the best available computers to execute the most important fundamental algorithms.

Perfect Shuffles

To illustrate some of the issues involved in implementing algorithms as machines instead of programs, we'll look at an interesting method for merging that is suitable for hardware implementation. As we'll see, the same general method can be developed into a design for an "algorithm machine" which incorporates a fundamental interconnection pattern to achieve parallel operation of M processors for solving several problems in addition to merging.

As mentioned above, a fundamental difference between writing a program to solve a problem and designing a machine is that a program can *adapt* its behavior to the particular instance of the problem being solved, while the machine must be "wired" ahead of time always to perform the same sequence of operations. To see the difference, consider the first sorting program we studied, *sort3* in Chapter 8. No matter what three numbers appear in the data, the program always performs the same sequence of three fundamental "compare-exchange" operations. None of the other sorting algorithms we studied have this property. They all perform a sequence of comparisons that depends on the outcome of previous comparisons, and thus presents severe problems for hardware implementation.

Specifically, if we have a piece of hardware with two input wires and two output wires that can compare the two numbers on the input and exchange them if necessary for the output, then we can wire three of these together as shown in Figure 40.1 to produce a sorting machine with three inputs (at the top in the figure) and three outputs (at the bottom). Here, the first box exchanges the C and the B, then the second box exchanges the B and the A, and then the third box exchanges

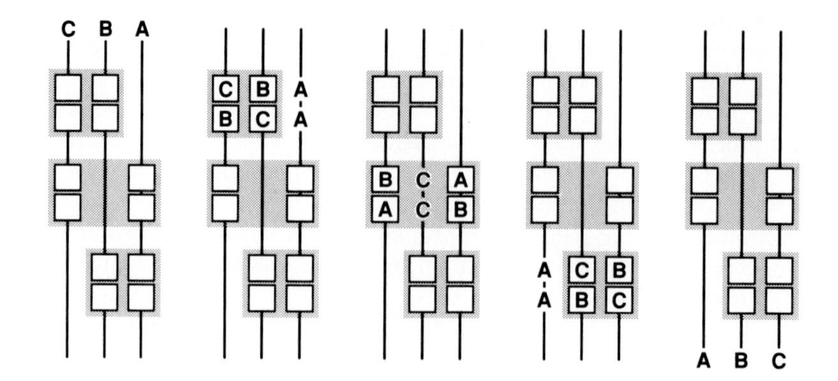

Figure 40.1 Machine that sorts three elements.

the C and the B to produce the sorted result. The machine sorts any permutation of the inputs (just as *sort3* did).

Of course, many details must be worked out before an actual sorting machine based on this scheme can be built. For example, the method of encoding the inputs is left unspecified. One way would be to think of each wire in the diagram above as a "bus" of enough wires to carry the data with one bit per wire; another way is to have the compare-exchangers read their inputs one bit at a time along a single wire (most significant bit first). Also left unspecified is the timing: mechanisms must be included to ensure that no compare-exchanger performs its operation before its input is ready. We clearly won't be able to delve much deeper into such circuit-design questions; we'll concentrate instead on the higher-level issues concerning interconnecting simple processors such as compare-exchangers for solving larger problems.

To begin, we'll consider an algorithm for merging together two sorted files, using a sequence of "compare-exchange" operations that is independent of the particular numbers to be merged and is thus suitable for hardware implementation. Figure 40.2 shows this method in operation on two sorted files of eight keys being merged together into one sorted file.

First we write one file below the other, then we compare those that are vertically adjacent and exchange them if necessary to put the larger one below the smaller one. Next, we split each line in half and interleave the halves, then perform the same compare-exchange operations on the numbers in the second and third lines. (Note that comparisons involving other pairs of lines are not necessary because of the previous sorting.) This leaves both the rows and the columns of the table sorted. This fact is a fundamental property of this method: the reader may wish to check that it is true, though a rigorous proof is a trickier exercise than one might think.

```
A E G G I M N R     A B E E I M N R
A B E E L M P X     A E G G L M P X

        A B E E     A B E E
        I M N R     A E G G
        A E G G     I M N R
        L M P X     L M P X

            A B     A B
            E E     A E
            A E     E E
            G G     G G
            I M     I M
            N R     L M
            L M     N R
            P X     P X

              A   A
              B   A
              A   B
              E   E
              E   E
              E   E
              G   G
              G   G
              I   I
              M   L
              L   M
              M   M
              N   N
              R   P
              P   R
              X   X
```

Figure 40.2 Split-and-interleave merging.

It turns out that this property is preserved in general by the same operation: split each line in half, interleave the halves, and do compare-exchanges between items now vertically adjacent that came from different lines. Each step doubles the number of rows, halves the number of columns, and still keeps the rows and the columns sorted. At the beginning we have 16 columns and one row, then 8 columns and two rows, then 4 columns and 4 rows, then 2 columns and 8 rows, and finally 16 rows and 1 column, which is sorted.

Property 40.1 *Merging two sorted file of N elements can be accomplished in about* $\lg N$ *parallel steps.*

If $N = 2^n$ the method just described obviously takes exactly n steps, each of which requires fewer than $N/2$ independent comparisons. To prove that this method sorts, one need prove that the columns remain sorted: this is left as an exercise, as mentioned above. Other sizes can be handled by adding dummy keys in a straightforward manner. ■

The basic "split each line in half and interleave the halves" operation in the description above is easy to visualize on paper, but how can it be translated into wiring for a machine? There is a surprising and elegant answer to this question which follows directly from writing the tables down in a different way. Rather than writing them down in a two-dimensional fashion, we'll write them down as a simple (one-dimensional) list of numbers, organized in *column-major* order: first put the elements in the first column, then put the elements in the second column, etc. Since compare-exchanges are only done between vertically adjacent items, this means that each stage involves a group of compare-exchange boxes, wired together according to the "split-and-interleave" operation that is necessary to bring items together into the compare-exchange boxes.

This leads to Figure 40.3, which corresponds precisely to the description using tables above, except that the tables are all written in column-major order (including an initial 1-by-16 table with one file, then the other). The reader should be sure to check the correspondence between this diagram and the tables given above. The compare-exchange boxes are drawn explicitly, and lines are drawn showing how elements move in the "split-and-interleave" operation: Surprisingly, in this

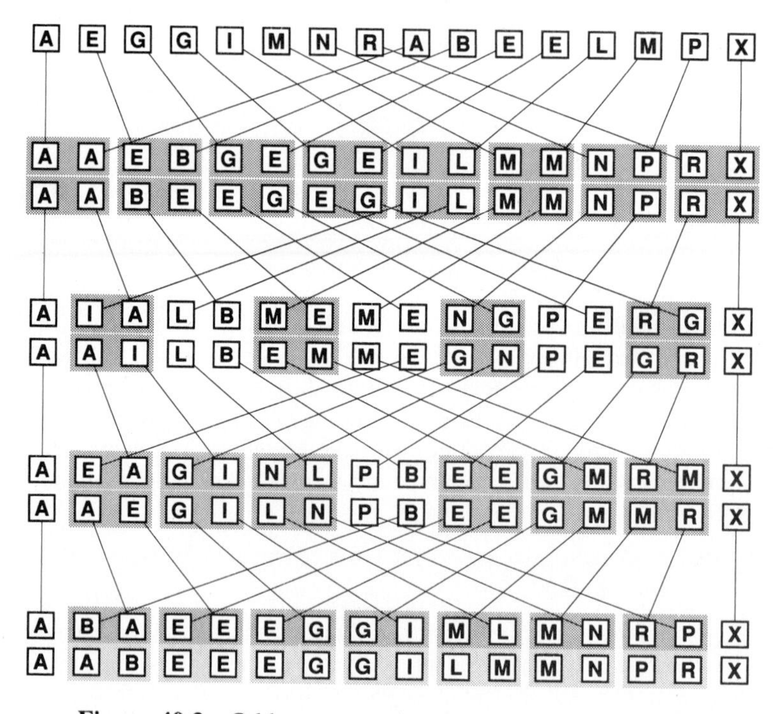

Figure 40.3 Odd-even merging with the perfect shuffle.

representation each "split-and-interleave" operation reduces to precisely the same interconnection pattern. This pattern is called the *perfect shuffle* because the wires are exactly interleaved, in the same way that cards from the two halves are interleaved in an ideal mix of a deck of cards.

This method was named the *odd-even merge* by K. E. Batcher, who invented it in 1968. The essential feature of the method is that all of the compare-exchange operations in each stage can be done in parallel. As Property 40.1 states, it is significant because it clearly demonstrates that two files of N elements can be merged together in $\log N$ parallel steps (the number of rows in the table is halved at every step), using fewer than $N \log N$ compare-exchange boxes. From the description above, this might seem like a straightforward result; actually, the problem of finding such a machine had stumped researchers for quite some time.

Batcher also developed a closely related (but more difficult to understand) merging algorithm, the *bitonic merge*, which leads to the even simpler machine shown in Figure 40.4. This method can be described in terms of the "split-and-interleave" operation on tables exactly as above, except that we begin with the second file in *reverse* sorted order and always do compare-exchanges between

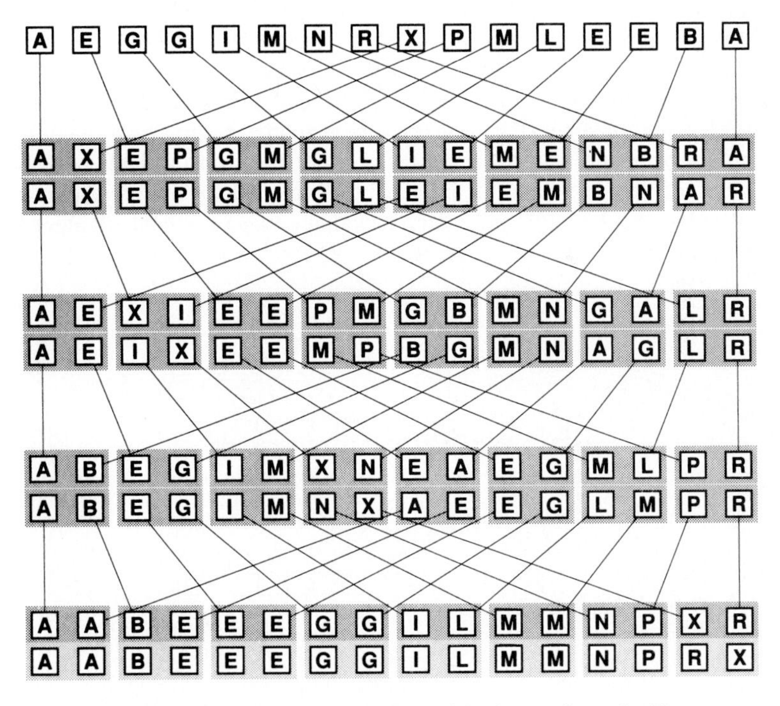

Figure 40.4 Bitonic merging with the perfect shuffle.

vertically adjacent items that came from the *same* lines. We won't go into the proof that this method works: our interest in it is that it removes the annoying feature in the odd-even merge that the compare-exchange boxes in the first stage are shifted one position from those in following stages. As shown in Figure 40.4, each stage of the bitonic merge has exactly the same number of comparators, in exactly the same positions.

Now there is regularity not only in the interconnections but also in the positions of the compare-exchange boxes. There are more compare-exchange boxes than for the odd-even merge, but this is not a problem, since the same number of parallel steps is involved. The importance of this method is that it leads directly to a way to do the merge using only N compare-exchange boxes. The idea is simply to collapse the rows in the table above to just one pair of rows, and thus produce a cycling machine wired together as shown in Figure 40.5. Such a machine can do $\log N$ compare-exchange-shuffle "cycles," one for each of the stages in the figure.

Note carefully that this is not quite "ideal" parallel performance: since we can merge together two files of N elements using one processor in a number of steps proportional to N, we would hope to be able to do the merge in a constant number of steps using N processors. In this case, however, it has been proven that it is impossible to achieve this ideal and that the above machine achieves the best possible parallel performance for merging using compare-exchange boxes.

The perfect-shuffle interconnection pattern is appropriate for a variety of other problems. For example, if a 2^n-by-2^n square matrix is kept in row-major order, then n perfect shuffles will transpose the matrix (convert it to column-major order). More important examples include the fast Fourier transform (which we'll examine in the next chapter), sorting (which can be developed by applying either of the methods above recursively), polynomial evaluation, and a host of others. Each of these problems can be solved using a cycling perfect-shuffle machine with the

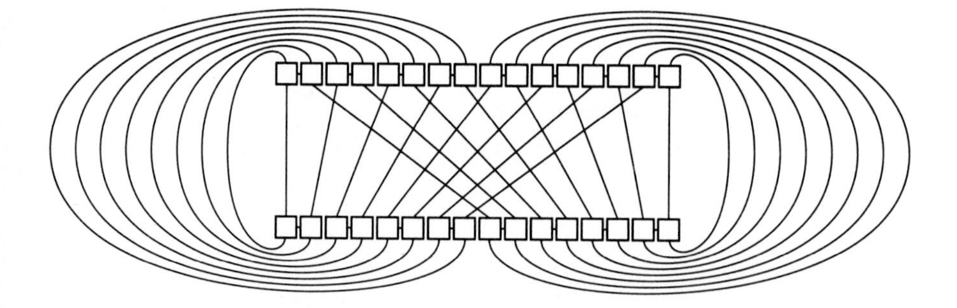

Figure 40.5 A perfect-shuffling machine.

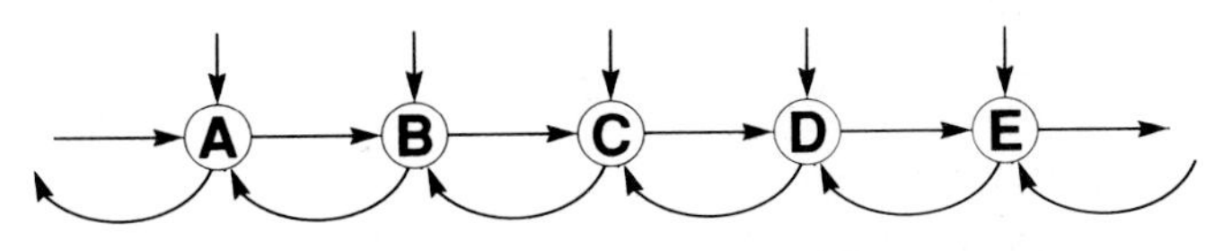

Figure 40.6 A systolic array.

same interconnections as the one diagramed above but with different (somewhat more complicated) processors. Some researchers have even suggested using the the perfect shuffle interconnection for "general-purpose" parallel computers.

Systolic Arrays

One problem with the perfect shuffle is that the wires used for interconnection are long. Furthermore, there are many wire crossings: a shuffle with N wires involves a number of crossings proportional to N^2. These two properties turn out to create difficulties when a perfect shuffle machine is actually constructed: long wires lead to time delays and crossings make the interconnection expensive and inconvenient.

A natural way to avoid both of these problems is to insist that processors be connected only to processors that are physically adjacent. As above, we operate the processors synchronously: at each step, each processor reads inputs from its neighbors, does a computation, and writes outputs to its neighbors. It turns out that this is not necessarily restrictive, and in fact H. T. Kung showed in 1978 that arrays of such processors, which he termed *systolic arrays* (because the way data flows within them is reminiscent of a heartbeat), allow very efficient use of the processors for some fundamental problems.

As a typical application, we'll consider the use of systolic arrays for matrix-vector multiplication. For a particular example, consider the matrix operation

$$\begin{pmatrix} 1 & 3 & -4 \\ 1 & 1 & -2 \\ -1 & -2 & 5 \end{pmatrix} \begin{pmatrix} 1 \\ 5 \\ 2 \end{pmatrix} = \begin{pmatrix} 8 \\ 2 \\ -1 \end{pmatrix}$$

This computation will be carried out on a row of simple processors each of which has three input lines and two output lines, as shown in Figure 40.6. Five processors are used because we'll be presenting the inputs and reading the outputs in a carefully timed manner, as described below.

During each step, each processor reads one input from the *left*, one from the *top*, and one from the *right*; performs a simple computation; and writes one output to the *left* and one output to the *right*. Specifically, the *right* output gets whatever was on the *left* input, and the *left* output gets the result computed by multiplying together the *left* and *top* inputs and adding the *right* input. A crucial characteristic of the processors is that they always perform a dynamic transformation of inputs to outputs; they never have to "remember" computed values. (This is also true of the processors in the perfect shuffle machine.) This is a ground rule imposed by low-level constraints on the hardware design, since the addition of such a "memory" capability can be (relatively) quite expensive.

The paragraph above gives the "program" for the systolic machine; to complete the description of the computation, we also need to describe exactly how the input values are presented. This timing is an essential feature of the systolic machine, in marked contrast to the perfect-shuffle machine, where all the input values are presented at one time and all the output values are available at some later time.

The general plan is to bring in the matrix through the *top* inputs of the processors, reflected about the main diagonal and rotated forty-five degrees, and the vector through the *left* input of processor A, to be passed on to the other processors. Intermediate results are passed from right to left in the array, with the output eventually appearing on the *left* output of processor A. The specific timing for our example is shown in Figure 40.7.

The input vector is presented to the *left* input of processor A at steps 1, 3, and 5 and passed right to the other processors in subsequent steps. The input matrix is presented to the *top* inputs of the processors starting at step 3, skewed so the right-to-left diagonals of the matrix are presented in successive steps. The output vector appears as the *left* output of processor A at steps 6, 8, and 10. (In the diagram, this appears as the *right* input of an imaginary processor to the left of A, which is collecting the answer.)

The actual computation can be traced by following the *right* inputs (*left* outputs) which move from right to left through the array. All computations produce a zero result until step 3; after step 3, processor C has 1 for its *left* input and 1 for its *top* input, so it computes the result 1, which is passed along as processor B's *right* input for step 4. At step 4, processor B has non-zero values for all three of its inputs, and it computes the value 16, to be passed on to processor A for step 5. Meanwhile, processor D computes a value 1 for processor C's use at step 5. Then at step 5, processor A computes the value 8, which is presented as the first output value at step 6; C computes the value 6 for B's use at step 6, and E computes its first nonzero value (-1) for use by D at step 6. The computation of the second output value is completed by B at step 6 and passed through A for output at step 8, and the computation of the third output value is completed by C at step 7 and passed through B and A for output at step 10.

Once the process has been described at a detailed level as above, the method is better understood at a somewhat higher level. The numbers in the middle part of Figure 40.7 (in rectangles) are simply a copy of the input matrix, rotated and reflected as required for presentation to the *top* inputs of the processors. If we check the numbers in the corresponding *left* inputs of the processors receiving the matrix, we find three copies of the input vector, located in exactly the right positions and at the right times for multiplication against the rows of the matrix. Then the *right* outputs of the processors show the intermediate results for each multiplication

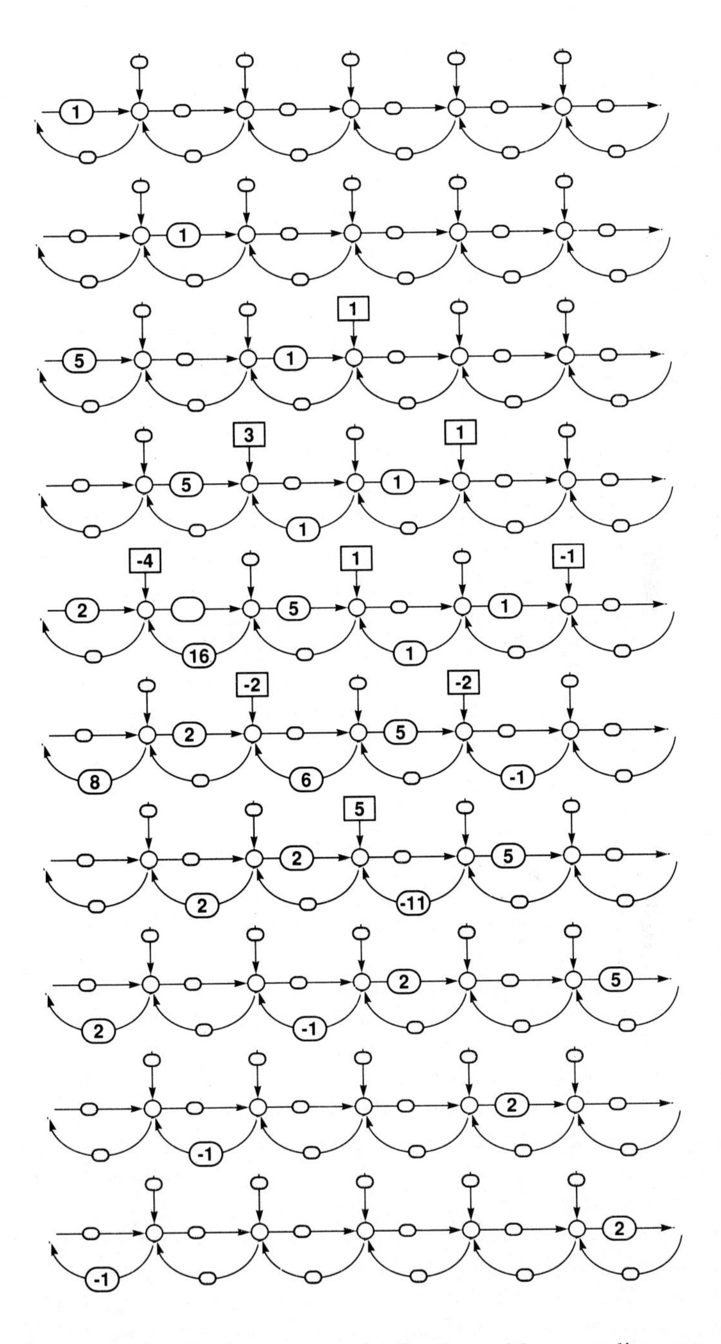

Figure 40.7 Matrix-vector multiplication with a systolic array.

of the input vector and each matrix row. For example, the multiplication of the input vector and the middle matrix row requires the partial computations $1 * 1 = 1$, $1 + 1 * 5 = 6$, and $6 + (-2) * 2 = 2$, and these appear in the entries 1 6 2 (reading diagonally, one step below the middle matrix row 1 1 -2). The systolic machine manages to time things so that each matrix element "meets" the proper input vector entry and the proper partial computation at the processor where it is input, so that it can be incorporated into the partial result.

The method extends in an obvious manner to multiply an N-by-N matrix by an N-by-1 vector using $2N - 1$ processors in $4N - 2$ steps. This comes close to the ideal situation of having every processor perform useful work at every step: a quadratic algorithm is reduced to a linear algorithm using a linear number of processors.

As before, we've only described a general method of parallel computation. Many details in the logical design need to be worked out before such a systolic machine can be constructed. One can appreciate from this example that systolic arrays are at once simple and powerful. The output vector at the edge appears almost as if by magic! However, each individual processor is just performing the simple computation described above: the magic is in the interconnection and the timed presentation of the inputs.

As with perfect shuffle machines, systolic arrays may be used in many different types of problems, including string matching and matrix multiplication among others. Also as with perfect-shuffle machines, some researchers have even suggested using this interconnection pattern for "general-purpose" parallel machines.

Perspective

The study of the perfect shuffle and systolic machines illustrates that hardware design can have a significant effect on algorithm design, clearly suggesting changes that can provide interesting new algorithms and fresh challenges for the algorithm designer.

While this is an interesting and fruitful area for further research, we must conclude with a few sobering notes. First, a great deal of engineering effort is required to translate general schemes for parallel computation such as those sketched above to actual algorithm machines with good performance. For many applications, the resource expenditure required is simply not justified, and a simple "algorithm machine" consisting of a conventional (inexpensive) microprocessor running a conventional algorithm will do quite well. For example, if one has many instances of the same problem to solve and several microprocessors with which to solve them, then ideal parallel performance can be achieved by having each microprocessor (using a conventional algorithm) working on a different instance of the problem, with no interconnection at all. If one has N files to sort and N processors available on which to sort them, why not simply use one processor for each sort, rather than having all N processors labor together on all N sorts?

It is quite difficult to evaluate the effect of various parallel computing strategies on algorithm performance. All the issues discussed in Chapters 6 and 7 must be taken into consideration, with the additional complication that the machine itself becomes a variable. The easiest way to compare machines (and perhaps the most commonly used) can be extremely misleading: running the same program on both machines is exactly the wrong experiment to perform if the underlying algorithm is closely related to the architecture of one of the machines and not the other. The match is bound to do far better than the mismatch. To properly compare machines, one should focus on the *problem* to be solved and look at the best algorithm for that problem on each machine. Does the new architecture allow us to solve a problem that couldn't be solved with an old architecture?

Techniques such as those discussed in this chapter are currently justifiable only for applications with very special time or space requirements. In studying various parallel computation schemes and their effects on the performance of various algorithms, we can look forward to the development of general-purpose parallel computers that will provide improved performance for a wide variety of algorithms.

□

Exercises

1. Outline two possible ways to use parallelism in Quicksort.
2. Prove that the "split-interleave" method preserves the property that the columns are sorted.
3. Write a conventional Pascal program to merge files using Batcher's bitonic method.
4. Write a conventional Pascal program to merge files using Batcher's bitonic method, but without actually doing any shuffles.
5. How many perfect shuffles will bring all the elements in an array of size 2^n back to their original positions?
6. Draw a table like that in Figure 40.7 to illustrate the operation of the systolic matrix-vector multiplier for the following problem:

$$\begin{pmatrix} 2 & 1 & 4 \\ 3 & 0 & 1 \\ 1 & -1 & 3 \end{pmatrix} \begin{pmatrix} 3 \\ 1 \\ -1 \end{pmatrix} = \begin{pmatrix} 3 \\ 8 \\ -1 \end{pmatrix}.$$

7. Write a conventional Pascal program that simulates the operation of the systolic array for multiplying an N-by-N matrix by an N-by-1 vector.
8. Show how to use a systolic array to transpose a matrix.
9. How many processors and how many steps are required for a systolic machine that can multiply an M-by-N matrix by an N-by-1 vector?
10. Give a simple parallel scheme for matrix-vector multiplication using processors which have the capability to "remember" computed values.

41

The Fast Fourier Transform

☐ One of the most widely used arithmetic algorithms is the *fast Fourier transform*, an efficient way to perform a widely used and basic mathematical computation. The Fourier transform is of fundamental importance in mathematical analysis and is the subject of volumes of study. The emergence of an efficient algorithm for this computation was a milestone in the history of computing.

Applications for the Fourier transform are legion. It is the basis for many fundamental manipulations in signal processing, where its use is widespread. Furthermore, as we'll see, it provides a convenient way to improve the efficiency of algorithms for commonplace arithmetic problems. It is beyond the scope of this book to outline the mathematical basis for the Fourier transform or to survey its many applications. Our purpose is to look at the characteristics of a fundamental algorithm for this transform within the context of some of the other algorithms we've been studying.

In particular, we'll examine how to use the algorithm to substantially reduce the time required for polynomial multiplication, a problem we studied in Chapter 36. Only a very few elementary facts from complex analysis are needed to show how the Fourier transform can be used to multiply polynomials, and it is possible to appreciate the fast Fourier transform algorithm without fully understanding the underlying mathematics. This algorithm applies the divide-and-conquer technique in a way similar to other important algorithms we've seen.

Evaluate, Multiply, Interpolate

The general strategy of the improved method for polynomial multiplication we'll be examining takes advantage of the fact that a polynomial of degree $N-1$ is completely determined by its value at N different points. When we multiply two polynomials of degree $N-1$ together, we get a polynomial of degree $2N-2$; if we can find that polynomial's value at $2N-1$ points, then it is completely determined.

But we can find the value of the result at any point simply by evaluating the two polynomials to be multiplied at that point and then multiplying those numbers.

This leads to the following general scheme for multiplying two polynomials of degree $N-1$:

Evaluate the input polynomials at $2N-1$ distinct points.

Multiply the two values obtained at each point.

Interpolate to find the unique result polynomial that has the given value at the given points.

For example, to compute $r(x)=p(x)q(x)$ with $p(x)=1+x+x^2$ and $q(x)=2-x+x^2$, we can evaluate $p(x)$ and $q(x)$ at any five points, say $-2, -1, 0, 1, 2$, to get the values

$$[p(-2), p(-1), p(0), p(1), p(2)] = [3, 1, 1, 3, 7],$$
$$[q(-2), q(-1), q(0), q(1), q(2)] = [8, 4, 2, 2, 4].$$

Multiplying these together term-by-term gives enough values for the product polynomial,

$$[r(-2), r(-1), r(0), r(1), r(2)] = [24, 4, 2, 6, 28],$$

that its coefficients can be found by interpolation. By the Lagrange formula,

$$\begin{aligned} r(x) = {} & 24\frac{x+1}{-2+1}\frac{x-0}{-2-0}\frac{x-1}{-2-1}\frac{x-2}{-2-2} \\ & +4\frac{x+2}{-1+2}\frac{x-0}{-1-0}\frac{x-1}{-1-1}\frac{x-2}{-1-2} \\ & +2\frac{x+2}{0+2}\frac{x+1}{0+1}\frac{x-1}{0-1}\frac{x-2}{0-2} \\ & +6\frac{x+2}{1+2}\frac{x+1}{1+1}\frac{x-0}{1-0}\frac{x-2}{1-2} \\ & +28\frac{x+2}{2+2}\frac{x+1}{2+1}\frac{x-0}{2-0}\frac{x-1}{2-1}, \end{aligned}$$

which simplifies to the result

$$r(x) = 2 + x + 2x^2 + x^4.$$

As described so far, this method is not an attractive algorithm for polynomial multiplication since the best algorithms we have so far for both evaluation (repeated application of Horner's method) and interpolation (Lagrange formula) require N^2 operations. However, there is some hope of finding a better algorithm because the method works for any choice of $2N-1$ distinct points whatsoever, and it is reasonable to expect that evaluation and interpolation will be easier for some sets of points than for others.

Complex Roots of Unity

It turns out that the most convenient points to use for polynomial interpolation and evaluation are complex numbers, in fact, a particular set of complex numbers called the *complex roots of unity.*

A brief review of some facts about complex analysis is necessary. The number $i = \sqrt{-1}$ is an *imaginary* number: though $\sqrt{-1}$ is meaningless as a real number, it is convenient to give it a name, i, and perform algebraic manipulations with it, replacing i^2 with -1 whenever it appears. A *complex number* consists of two parts, real and imaginary; it is usually written as $a + bi$, where a and b are reals. To multiply complex numbers, apply the usual rules, but replace i^2 with -1 whenever it appears. For example,

$$(a + bi)(c + di) = (ac - bd) + (ad + bc)i.$$

Sometimes the real or imaginary part can cancel out when a complex multiplication is performed. For example,

$$(1 - i)(1 - i) = -2i,$$

$$(1 + i)^4 = -4,$$

$$(1 + i)^8 = 16.$$

Scaling this last equation by dividing through by $16 = \sqrt{2}^8$), we find that

$$\left(\frac{1}{\sqrt{2}} + \frac{i}{\sqrt{2}}\right)^8 = 1.$$

In general, there are many complex numbers that evaluate to 1 when raised to a power. These are the so-called complex roots of unity. In fact, it turns out that for each N, there are exactly N complex numbers z with $z^N = 1$. One of these, named w_N, is called the *principal Nth root of unity*; the others are obtained by raising w_N to the kth power, for $k = 0,1,2,\ldots,N - 1$. For example, we can list the eighth roots of unity as follows:

$$w_8^0, w_8^1, w_8^2, w_8^3, w_8^4, w_8^5, w_8^6, w_8^7.$$

The first root, w_N^0, is 1 and the second, w_N^1, is the principal root. Also, for N even, the root $w_N^{N/2}$ is -1 (because $(w_N^{N/2})^2 = 1$). The precise values of the roots are unimportant for the moment. We'll be using only simple properties which can easily be derived from the basic fact that the Nth power of any Nth root of unity must be 1.

Evaluation at the Roots of Unity

The crux of our implementation will be a procedure for evaluating a polynomial of degree $N - 1$ at the Nth roots of unity. That is, this procedure transforms the N coefficients that define the polynomial into the N values resulting from evaluating that polynomial at all of the Nth roots of unity.

This may not seem to be exactly what we want, since for the first step of the polynomial multiplication procedure we need to evaluate polynomials of degree $N - 1$ at $2N - 1$ points. Actually, this is no problem, since we can view a polynomial of degree $N - 1$ as a polynomial of degree $2N - 2$ with $N - 1$ coefficients (those for the terms of largest degree) that are zero.

The algorithm that we'll use to evaluate a polynomial of degree $N - 1$ at N points simultaneously will be based on a simple divide-and-conquer strategy. Rather than dividing the polynomials in the middle (as in the multiplication algorithm in Chapter 4) we'll divide them into two parts by putting alternate terms in each part. This division can easily be expressed in terms of polynomials with half the number of coefficients. For example, for $N = 8$, the rearrangement of terms is as follows:

$$\begin{aligned} p(x) &= p_0 + p_1x + p_2x^2 + p_3x^3 + p_4x^4 + p_5x^5 + p_6x^6 + p_7x^7 \\ &= (p_0 + p_2x^2 + p_4x^4 + p_6x^6) + x(p_1 + p_3x^2 + p_5x^4 + p_7x^6) \\ &\equiv p_e(x^2) + xp_o(x^2). \end{aligned}$$

The Nth roots of unity are convenient for this decomposition because if you square a root of unity, you get another root of unity. In fact, even more is true: for N even, if you square an Nth root of unity, you get an $\frac{1}{2}N$th root of unity (a number which evaluates to 1 when raised to the $\frac{1}{2}N$th power). This is exactly what is needed to make the divide-and-conquer method work. To evaluate a polynomial with N coefficients on N points, we split it into two polynomials with $\frac{1}{2}N$ coefficients. These polynomials need be evaluated only on $\frac{1}{2}N$ points (the $\frac{1}{2}N$th roots of unity) to compute the values needed for the full evaluation.

To see this more clearly, consider the evaluation of a degree-7 polynomial $p(x)$ on the eighth roots of unity

$$W_8 : w_8^0, w_8^1, w_8^2, w_8^3, w_8^4, w_8^5, w_8^6, w_8^7.$$

Since $w_8^4 = -1$, this is the same as the sequence

$$W_8 : w_8^0, w_8^1, w_8^2, w_8^3, -w_8^0, -w_8^1, -w_8^2, -w_8^3.$$

Squaring each term of this sequence gives two copies of the sequence $\{W_4\}$ of the fourth roots of unity:

$$W_8^2 : w_4^0, w_4^1, w_4^2, w_4^3, w_4^0, w_4^1, w_4^2, w_4^3.$$

Now, our equation

$$p(x) = p_e(x^2) + x p_o(x^2)$$

tells us immediately how to evaluate $p(x)$ at the eighth roots of unity from these sequences. First, we evaluate $p_e(x)$ and $p_o(x)$ at the fourth roots of unity. Then we substitute each of the eighth roots of unity for x in the equation above, which requires adding the appropriate p_e value to the product of the appropriate p_o value and the eighth root of unity:

$$\begin{aligned}
p(w_8^0) &= p_e(w_4^0) + w_8^0 p_o(w_4^0),\\
p(w_8^1) &= p_e(w_4^1) + w_8^1 p_o(w_4^1),\\
p(w_8^2) &= p_e(w_4^2) + w_8^2 p_o(w_4^2),\\
p(w_8^3) &= p_e(w_4^3) + w_8^3 p_o(w_4^3),\\
p(w_8^4) &= p_e(w_4^0) - w_8^0 p_o(w_4^0),\\
p(w_8^5) &= p_e(w_4^1) - w_8^1 p_o(w_4^1),\\
p(w_8^6) &= p_e(w_4^2) - w_8^2 p_o(w_4^2),\\
p(w_8^7) &= p_e(w_4^3) - w_8^3 p_o(w_4^3).
\end{aligned}$$

In general, to evaluate $p(x)$ on the Nth roots of unity, we recursively evaluate $p_e(x)$ and $p_o(x)$ on the $\frac{1}{2}N$th roots of unity and perform the N multiplications as above. This works only when N is even, and thus we'll assume from now on that N is a power of two, so that it remains even throughout the recursion. The recursion stops when $N = 2$ and we have $p_0 + p_1 x$ to be evaluated at 1 and -1, with the results $p_0 + p_1$ and $p_0 - p_1$.

Property 41.1 *A polynomial of degree $N - 1$ can be evaluated at the Nth roots of unity with about $N \lg N$ multiplications.*

The number of multiplications used satisfies the fundamental "divide-and-conquer" recurrence $M(N) = 2M(N/2) + N$, which has the solution $M(N) = N \lg N$ (Formula 4 in Chapter 6). This is a substantial improvement over the straightforward N^2 method for interpolation but, of course, it works only at the roots of unity. ■

This gives a method for transforming a polynomial from its representation as N coefficients in the conventional manner to its representation in terms of its values at the roots of unity. This conversion of the polynomial from the first representation to the second is the Fourier transform, and the efficient recursive calculation procedure that we have described is called the "fast" Fourier transform (FFT). (These same techniques apply to functions more general than polynomials. More precisely, then, we're doing the "discrete" Fourier transform.)

Interpolation at the Roots of Unity

Now that we have a fast way to evaluate polynomials at a specific set of points, all we need is a fast way to interpolate polynomials at those same points, and we will have a fast polynomial multiplication method. Surprisingly, it works out that, for the complex roots of unity, running the evaluation program on a particular set of points will do the interpolation! This is a specific instance of a fundamental "inversion" property of the Fourier transform, from which many important mathematical results can be derived.

For our example with $N = 8$, the interpolation problem is to find the polynomial

$$r(x) = r_0 + r_1x + r_2x^2 + r_3x^3 + r_4x^4 + r_5x^5 + r_6x^6 + r_7x^7$$

which has the values

$$r(w_8^0) = s_0, \quad r(w_8^1) = s_1, \quad r(w_8^2) = s_2, \quad r(w_8^3) = s_3,$$
$$r(w_8^4) = s_4, \quad r(w_8^5) = s_5, \quad r(w_8^6) = s_6, \quad r(w_8^7) = s_7.$$

When the points under consideration are the complex roots of unity, it is literally true that the interpolation problem is the "inverse" of the evaluation problem. If we let

$$s(x) = s_0 + s_1x + s_2x^2 + s_3x^3 + s_4x^4 + s_5x^5 + s_6x^6 + s_7x^7$$

then we can get the coefficients

$$r_0, r_1, r_2, r_3, r_4, r_5, r_6, r_7$$

just by evaluating the polynomial $s(x)$ at the inverses of the complex roots of unity

$$W_8^{-1} : w_8^0, w_8^{-1}, w_8^{-2}, w_8^{-3}, w_8^{-4}, w_8^{-5}, w_8^{-6}, w_8^{-7}.$$

But this is the same sequence as the complex roots of unity, but in a different order:

$$W_8^{-1} : w_8^0, w_8^7, w_8^6, w_8^5, w_8^4, w_8^3, w_8^2, w_8^1.$$

In other words, we can use exactly the same routine for interpolation as for evaluation: only a simple rearrangement of the points to be evaluated is required.

The proof of this fact requires some elementary manipulations with finite sums; those unfamiliar with such manipulations may wish to skip to the end of this paragraph. Evaluating $s(x)$ at the inverse of the tth Nth root of unity gives

$$\begin{aligned}
s(w_N^{-t}) &= \sum_{0 \le j < N} s_j (w_N^{-t})^j \\
&= \sum_{0 \le j < N} r(w_N^j)(w_N^{-t})^j \\
&= \sum_{0 \le j < N} \sum_{0 \le i < N} r_i (w_N^j)^i (w_N^{-t})^j \\
&= \sum_{0 \le j < N} \sum_{0 \le i < N} r_i w_N^{j(i-t)} \\
&= \sum_{0 \le i < N} r_i \sum_{0 \le j < N} w_N^{j(i-t)} = N r_t .
\end{aligned}$$

Nearly everything disappears in the last term because the inner sum is trivially N if $i = t$. If $i \neq t$ then it evaluates to

$$\sum_{0 \le j < N} w_N^{j(i-t)} = \frac{w_N^{(i-t)N} - 1}{w_N^{(i-t)} - 1} = 0.$$

Note that an extra scaling factor of N arises. This is the "inversion theorem" for the discrete Fourier transform, which says that the same method will convert a polynomial both ways: between its representation as coefficients and its representation as values at the complex roots of unity.

Property 41.2 *A polynomial of degree $N - 1$ can be interpolated at the Nth roots of unity with about $N \lg N$ multiplications.*

The mathematics above may seem complicated, but the results are quite easy to apply: to interpolate a polynomial on the Nth roots of unity, use the same procedure as for evaluation, using the interpolation values as polynomial coefficients, then rearrange and scale the answers. ■

Implementation

Now we have all the pieces for a divide-and-conquer algorithm to multiply two polynomials using only about $N \lg N$ operations. The general scheme is to:

Evaluate the input polynomials at the $(2N - 1)$st roots of unity.

Multiply the two values obtained at each point.

Interpolate to find the result by evaluating the polynomial defined by the numbers just computed at the $(2N - 1)$st roots of unity.

The description above can be directly translated into a program that uses a procedure that can evaluate a polynomial of degree $N - 1$ at the Nth roots of unity. Unfortunately, all the arithmetic in this algorithm is to be complex arithmetic, and Pascal has no built-in type *complex*. While it is possible to have a user-defined type for the complex numbers, it is then also necessary to define procedures or functions for all the arithmetic operations on the numbers, and this obscures the algorithm unnecessarily. The following implementation assumes a type *complex* for which the obvious arithmetic functions are defined:

```
eval(p, outN, 0);
eval(q, outN, 0);
for i:=0 to outN do r[i]:=p[i]*q[i];
eval(r, outN, 0);
for i:=1 to N do
  begin t:=r[i]; r[i]:=r[outN+1-i]; r[outN+1-i]:=t end;
for i:=0 to outN do r[i]:=r[i]/(outN+1);
```

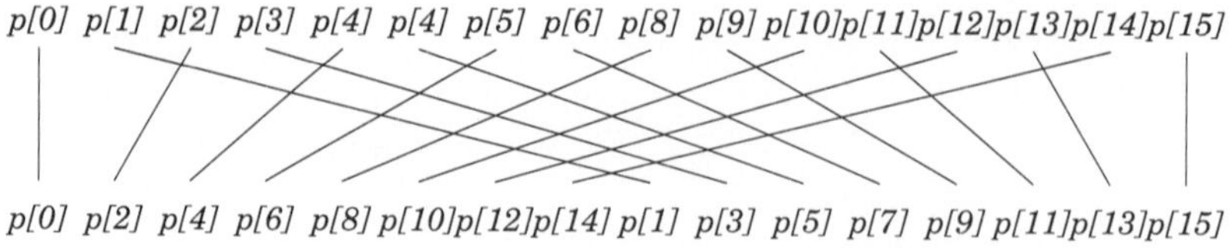

p[0] p[1] p[2] p[3] p[4] p[4] p[5] p[6] p[8] p[9] p[10]p[11]p[12]p[13]p[14]p[15]

p[0] p[2] p[4] p[6] p[8] p[10]p[12]p[14] p[1] p[3] p[5] p[7] p[9] p[11]p[13]p[15]

Figure 41.1 Perfect unshuffle for the FFT.

This program assumes that the global variable *outN* has been set to $2N-1$ and that p, q, and r are arrays indexed from 0 to $2N-1$ that hold complex numbers. The two polynomials to be multiplied, p and q, are of degree $N-1$, and the other coefficients in those arrays are initially set to zero. The procedure *eval* replaces the coefficients of the polynomial given as the first argument by the values obtained when the polynomial is evaluated at the roots of unity. The second argument specifies the degree of the polynomial (one less than the number of coefficients and roots of unity) and the third argument is described below. The above code computes the product of p and q and leaves the result in r.

Now we are left with the implementation of *eval*. As we've seen before, recursive programs involving arrays can be quite cumbersome to implement. It turns out that for this algorithm it is possible to get around the usual problem of storage management by reusing the storage in a clever way. What we would like to do is have a recursive procedure that takes as input a contiguous array of $N+1$ coefficients and returns the $N+1$ values in the same array. But the recursive step involves processing two noncontiguous arrays: the odd and even coefficients. On reflection, the reader will see that the perfect shuffle of the previous chapter is exactly what is needed here. We can get the odd coefficients in a contiguous subarray (the first half) and the even coefficients in a contiguous subarray (the second half) by doing a "perfect unshuffle" of the input, as diagrammed in Figure 41.1 for $N = 15$.

Of course, the actual values of the complex roots of unity are needed to do the implementation. It is well known that

$$w_N^j = \cos\left(\frac{2\pi j}{N+1}\right) + i\sin\left(\frac{2\pi j}{N+1}\right);$$

these values are easily computed using conventional trigonometric functions. In the program below, the array w is assumed to hold the ($outN+1$)st roots of unity.

This leads to the following implementation of the FFT:

```
procedure eval(var p: poly; N,k: integer);
  var i,j: integer;
  begin
  if N=1 then
    begin
    t:=p[k]; p1:=p[k+1];
    p[k]:=t+p1; p[k+1]:=t-p1
    end
  else
    begin
    for i:=0 to N div 2 do
      begin
      j:=k+2*i;
      t[i]:=p[j]; t[i+1+N div 2]:=p[j+1]
      end;
    for i:=0 to N do p[k+i]:=t[i];
    eval(p,N div 2,k);
    eval(p,N div 2,k+1+N div 2);
    j:=(outN+1) div (N+1);
    for i:=0 to N div 2 do
      begin
      t:=w[i*j]*p[k+(N div 2)+1+i];
      t[i]:=p[k+i]+t; t[i+(N div 2)+1]:=p[k+i]-t
      end;
    for i:=0 to N do p[k+i]:=t[i]
    end
  end;
```

This program transforms the polynomial of degree N in place in the subarray $p[k..k+N]$ using the recursive method outlined above. (For simplicity, the code assumes that $N+1$ is a power of two, though this dependence is not hard to remove.) If $N = 1$, then the easy computation to evaluate at 1 and -1 is performed. Otherwise the procedure first shuffles, then recursively calls itself to transform the two halves, and then combines the results of these computations as described above. To get the roots of unity needed, the program selects from the w array at an interval determined by the variable i. For example, if *outN* is 15, the fourth roots of unity are found in $w[0]$, $w[4]$, $w[8]$, and $w[12]$. This eliminates the need to recompute roots of unity each time they are used.

Property 41.3 *Two polynomials of degree N can be multiplied with $2N \lg N + O(N)$ complex multiplications.*

This fact follows directly from Properties 1 and 2. ■

As mentioned at the outset, the scope of applicability of the FFT is far greater than can be indicated here; and the algorithm has been intensively used and studied in a variety of domains. Nevertheless, the fundamental principles of operation in more advanced applications are the same as for the polynomial multiplication problem discussed here. The FFT is a classic example of the application of the "divide-and-conquer" algorithm design paradigm to achieve truly significant computational savings.

□

Exercises

1. How would you improve the simple evaluate-multiply-interpolate algorithm for multiplying together two polynomials $p(x)$ and $q(x)$ with known roots p_0, $p_1,\ldots,p_{N-1}$ and q_0, $q_1,\ldots,q_{N-1}$?
2. Find a set of N real numbers at which a polynomial of degree N can be evaluated using substantially fewer than N^2 operations.
3. Find a set of N real numbers at which a polynomial of degree N can be interpolated using substantially fewer than N^2 operations.
4. What is the value of w_N^M for $M > N$?
5. Is it worthwhile to multiply sparse polynomials using the FFT?
6. The FFT implementation has three calls to *eval*, just as the polynomial multiplication procedure in Chapter 36 has three calls to *mult*. Why is the FFT implementation more efficient?
7. Give a way to multiply two complex numbers together using fewer than four integer multiplication operations.
8. How much storage would be used by the FFT if we didn't circumvent the storage-management problem with the perfect shuffle?
9. Why can't some technique like the perfect shuffle be used to avoid the problems with dynamically declared arrays in the polynomial multiplication procedure of Chapter 36?
10. Write an efficient program to multiply a polynomial of degree N by a polynomial of degree M (not necessarily powers of two).

42

Dynamic Programming

☐ The *divide-and-conquer* principle has guided the design of many of the algorithms we've studied: to solve a large problem, break it up into smaller problems which can be solved independently. In *dynamic programming* this principle is carried to an extreme: when we don't know exactly which smaller problems to solve, we simply solve them all, then store the answers away to be used later in solving larger problems. This approach is widespread in operations research. Here, the term "programming" refers to the process of formulating the constraints of a problem so that the method is applicable. This is an art that we won't pursue in any further detail, except to look at a few examples. (The "programming" we'll be interested in involves writing Pascal programs to find the solutions.)

We have already seen some algorithms that can be cast in a dynamic programming framework. For example, Warshall's algorithm to find the transitive closure of a graph and Floyd's algorithm to find all the shortest paths in a weighted graph (both in Chapter 32) both work by considering the vertices one by one, solving subproblems for the vertex currently under consideration by making use of solutions for all the vertices previously considered.

Two difficulties may arise in any application of dynamic programming. First, it may not always be possible to combine the solutions of smaller problems to form the solution of a larger one. Second, the number of small problems to solve may be unacceptably large. No one has characterized precisely which problems can be effectively solved with dynamic programming; there are many "hard" problems for which it does not seem to be applicable (see Chapters 44 and 45), as well as many "easy" problems for which it is less efficient than standard algorithms.

We'll see several examples in this chapter of problems for which dynamic programming is quite effective. These problems involve looking for the "best" way to do something, and they have the general property that any decision involved in finding the best way to do a small subproblem remains a good one when that subproblem becomes a piece of some larger problem.

Knapsack Problem

A thief robbing a safe finds it filled with N types of items of varying size and value, but has only a small knapsack of capacity M to use to carry the goods. The *knapsack problem* is to find the combination of items which the thief should choose for his knapsack in order to maximize the total value of all the items he takes.

For example, suppose that he has a knapsack of capacity 17 and the safe contains many items of each of the sizes and values shown in Figure 42.1. (As usual, we use single-letter names for the items in the example and integer indices in the programs, with the knowledge that more complicated names can be translated to integers using standard searching techniques.) Then the thief can take five A's (but not six) for a total take of 20, or he can fill up his knapsack with a D and an E for a total take of 24, or he can try many other combinations. But which will maximize his total take?

There are clearly many commercial situations in which a solution to the knapsack problem could be important. For example, a shipping company might wish to know the best way to load a truck or cargo plane with items for shipment. In such applications, other variants to the problem might arise as well: for example, there might be a limited number of each kind of item available. Many such variants can be handled with the same approach that we're about to examine for solving the basic problem stated above.

In a dynamic-programming solution to the knapsack problem, we calculate the best combination for *all* knapsack sizes up to M. It turns out that we can perform this calculation very efficiently by doing things in an appropriate order, as in the following program:

```
for j:=1 to N do
  begin
  for i:=1 to M do
    if i-size[j]>=0 then
      if cost[i]<(cost[i-size[j]]+val[j]) then
        begin
        cost[i]:=cost[i-size[j]]+val[j];
        best[i]:=j
        end;
  end;
```

In this program, *cost*[i] is the highest value that can be achieved with a knapsack of capacity i and *best*[i] is the last item that was added to achieve that maximum (as described below, this is used to recover the contents of the knapsack). First, we calculate the best we can do for all knapsack sizes when only items of type A

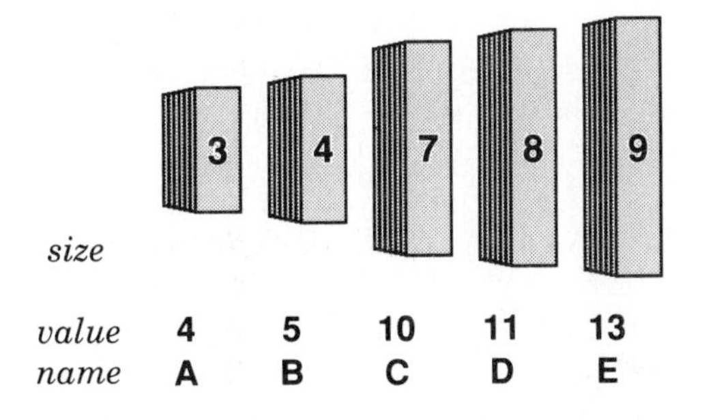

Figure 42.1 Knapsack problem instance.

are taken, then we calculate the best that we can do when only A's and B's are taken, etc. The solution reduces to a simple calculation for *cost*[*i*]. Suppose an item *j* is chosen for the knapsack: then the best value that could be achieved for the total would be *val*[*j*] (for the item) plus *cost*[*i*−*size*[*j*]] (to fill up the rest of the knapsack). If this value exceeds the best value that can be achieved *without* an item *j*, then we update *cost*[*i*] and *best*[*i*]; otherwise we leave them alone. A simple induction proof shows that this strategy solves the problem.

Figure 42.2 traces the computation for our example. The first pair of lines shows the best that can be done (the contents of the *cost* and *best* arrays) with only A's, the second pair of lines shows the best that can be done with only A's and B's, etc.. The highest value that can be achieved with a knapsack of size 17 is 24. In the course of computing this result, we also solved many smaller subproblems. For example, the highest value that can be achieved with a knapsack of size 16 using only A's, B's and C's is 22.

The actual contents of the optimal knapsack can be computed with the aid of the *best* array. By definition, *best*[*M*] is included, and the remaining contents are the same as for the optimal knapsack of size *M*−*size*[*best*[*M*]]. Therefore, *best*[*M*−*size*[*best*[*M*]]] is included, and so forth. For our example, *best*[*17*]=C; then we find another type-C item at size 10, then a type-A item at size 3.

Property 42.1 *The dynamic-programming solution to the knapsack problem takes time proportional to NM.*

This is obvious from inspection of the code. ■

Thus, the knapsack problem is easily solved if M is not large, but the running time can become unacceptable for large capacities. Furthermore, a crucial point that cannot be overlooked is that the method does not work at all if M and the sizes or values are, for example, real numbers instead of integers. This is more than a minor annoyance—it is a fundamental difficulty. No good solution is known for this problem, and we'll see in Chapter 45 that many people believe no good solution exists. To appreciate the difficulty of the problem, the reader might wish

	k	*1*	*2*	*3*	*4*	*5*	*6*	*7*	*8*	*9*	*10*	*11*	*12*	*13*	*14*	*15*	*16*	*17*
j=1																		
	cost[k]	0	0	4	4	4	8	8	8	12	12	12	16	16	16	20	20	20
	best[k]			A	A	A	A	A	A	A	A	A	A	A	A	A	A	A
j=2																		
	cost[k]	0	0	4	5	5	8	9	10	12	13	14	16	17	18	20	21	22
	best[k]			A	B	B	A	B	B	A	B	B	A	B	B	A	B	B
j=3																		
	cost[k]	0	0	4	5	5	8	10	10	12	14	15	16	18	20	20	22	24
	best[k]			A	B	B	A	C	B	A	C	C	A	C	C	A	C	C
j=4																		
	cost[k]	0	0	4	5	5	8	10	11	12	14	15	16	18	20	21	22	24
	best[k]			A	B	B	A	C	D	A	C	C	A	C	C	D	C	C
j=5																		
	cost[k]	0	0	4	5	5	8	10	11	13	14	15	17	18	20	21	23	24
	best[k]			A	B	B	A	C	D	E	C	C	E	C	C	D	E	C

Figure 42.2 Knapsack problem solution.

to try solving the case where the values are all 1, the size of the jth item is $\sqrt{j}$ and M is $N/2$.

But when capacities, sizes and values are all integers, we have the fundamental principle that optimal decisions, once made, need not be changed. Once we know the best way to pack knapsacks of any size with the first j items, we do not need to reexamine those problems, regardless of what the next items are. Any time this general principle can be made to work, dynamic programming is applicable. In this case, the principle works because integer values allow us to make exact, "optimal" decisions.

In this algorithm, only a small amount of information about previous optimal decisions needs to be saved, if M is not large. Different dynamic programming applications have widely different requirements in this regard; we'll see other examples below.

Matrix Chain Product

A classic application of dynamic programming is found in the problem of minimizing the amount of computation needed to multiply a series of matrices of different sizes. Such methods must be considered in any application involving extensive manipulation of matrices.

Suppose that the six matrices

$$\begin{pmatrix} a_{11} & a_{12} \\ a_{21} & a_{22} \\ a_{31} & a_{32} \\ a_{41} & a_{42} \end{pmatrix} \begin{pmatrix} b_{11} & b_{12} & b_{13} \\ b_{21} & b_{22} & b_{23} \end{pmatrix} \begin{pmatrix} c_{11} \\ c_{21} \\ c_{31} \end{pmatrix} \begin{pmatrix} d_{11} & d_{12} \end{pmatrix} \begin{pmatrix} e_{11} & e_{12} \\ e_{21} & e_{22} \end{pmatrix} \begin{pmatrix} f_{11} & f_{12} & f_{13} \\ f_{21} & f_{22} & f_{23} \end{pmatrix}$$

are to be multiplied together. Of course, for the multiplications to be valid, the number of columns in one matrix must be the same as the number of rows in the next. But the total number of scalar multiplications involved depends on the order in which the matrices are multiplied. For example, we could proceed from left to right: multiplying A by B, we get a 4-by-3 matrix after 24 scalar multiplications. Multiplying this result by C gives a 4-by-1 matrix after 12 more scalar multiplications. Multiplying this result by D gives a 4-by-2 matrix after 8 more scalar multiplications. Continuing in this way, we get a 4-by-3 result after a grand total of 84 scalar multiplications. But if we proceed from right to left instead, we get the same 4-by-3 result with only 69 scalar multiplications.

Many other orders are clearly possible. The order of multiplication can be expressed by parenthesization: for example the left-to-right order is the ordering (((((A*B)*C)*D)*E)*F), and the right-to-left order is (A*(B*(C*(D*(E*F))))). Any legal parenthesization will lead to the correct answer, but which one leads to the fewest scalar multiplications?

Very substantial savings can be achieved when large matrices are involved: for example, if matrices B, C, and F in the example above were to each have a dimension of 300 instead of 3, then the left-to-right order requires 6024 scalar multiplications but the right-to-left order uses an astronomical 274,200. (In these calculations we're assuming that the standard method of matrix multiplication is used. Strassen's or some similar method could in principle save some work for large matrices, but the same considerations about the order of multiplications apply. Thus, multiplying a p-by-q matrix by a q-by-r matrix will produce a p-by-r matrix, each entry computed with q multiplications, for a total of pqr multiplications.)

In general, suppose that N matrices are to be multiplied together:

$$M_1 M_2 M_3 \cdots M_N$$

where the matrices satisfy the constraint that M_i has r_i rows and r_{i+1} columns for $1 \leq i < N$. Our task is to find the order of multiplying the matrices that minimizes the total number of multiplications used. Certainly trying all possible orderings is impractical. (The number of orderings is a well-studied combinatorial function called the *Catalan number*: the number of ways to parenthesize N variables is about $4^{N-1}/N\sqrt{\pi N}$.) But it is certainly worthwhile to expend some effort to find a good solution because N is generally quite small compared to the number of multiplications to be done.

As above, the dynamic-programming solution to this problem involves working "bottom up," saving computed answers to small partial problems to avoid recomputation. First, there's only one way to multiply M_1 by M_2, M_2 by M_3, ..., M_{N-1} by M_N; we record those costs. Next, we calculate the best way to multiply successive triples, using all the information computed so far. For example, to find the best way to multiply $M_1M_2M_3$, first we find the cost of computing M_1M_2 from the table we saved and then add the cost of multiplying that result by M_3. This total is compared with the cost of first multiplying M_2M_3, then multiplying by M_1, which can be computed in the same way. The smaller of these is saved, and the same procedure followed for all triples. Next, we calculate the best way to multiply successive groups of four, using all the information gained so far. By continuing in this way we eventually find the best way to multiply together all the matrices.

This leads to the following program:

```
for i:=1 to N do
  for j:=i+1 to N do cost[i,j]:=maxint;
for i:=1 to N do cost[i,i]:=0;
for j:=1 to N-1 do
  for i:=1 to N-j do
    for k:=i+1 to i+j do
      begin
      t:=cost[i,k-1]+cost[k,i+j]+r[i]*r[k]*r[i+j+1];
      if t<cost[i,i+j] then
        begin cost[i,i+j]:=t; best[i,i+j]:=k end;
      end;
```

For $1 \le j \le N-1$, we find the minimum cost of computing

$$M_iM_{i+1} \cdots M_{i+j}$$

by finding, for $1 \le i \le N-j$ and for each k between i and $i+j$, the cost of computing $M_iM_{i+1} \cdots M_{k-1}$ and $M_kM_{k+1} \cdots M_{i+j}$ and then adding the cost of multiplying these results together. Since we always break a group into two smaller groups, the minimum costs for the two groups are merely looked up in a table, not recomputed. In particular, *cost*[*l*,*r*] gives the minimum cost of computing $M_lM_{l+1} \cdots M_r$, the cost of the first group above is *cost*[*i*,*k*−*1*] and the cost of the second group is *cost*[*k*,*i*+*j*]. The cost of the final multiplication is easily determined: $M_iM_{i+1} \cdots M_{k-1}$ is a r_i-by-r_k matrix, and $M_kM_{k+1} \cdots M_{i+j}$ is a r_k-by-r_{i+j+1} matrix, so the cost of multiplying these two is $r_ir_kr_{i+j+1}$. In this way, the program computes *cost*[*i*,*i*+*j*] for $1 \le i \le N-j$ with j increasing from 1 to $N-1$.

	B	C	D	E	F
A	24 [A][B]	14 [A][BC]	22 [ABC][D]	26 [ABC][DE]	36 [ABC][DEF]
B		6 [B][C]	10 [BC][D]	14 [BC][DE]	22 [BC][DEF]
C			6 [C][D]	10 [C][DE]	19 [C][DEF]
D				4 [D][E]	10 [DE][F]
E					12 [E][F]

Figure 42.3 Solution of matrix chain problem.

When we reach $j = N - 1$ (and $i = 1$), then we've found the minimum cost of computing $M_1 M_2 \cdots M_N$, as desired.

As above, we need to keep track of the decisions made in a separate array *best* for later recovery when the actual sequence of multiplications is to be generated. The following program implements this process of extracting the optimal parenthesization from the *cost* and *best* arrays computed by the program above:

```
procedure order(i,j: integer);
  begin
  if i=j then write(name(i)) else
    begin
    write('(');
    order(i, best[i,j]-1); write('*'); order(best[i,j],j);
    write(')')
    end
  end;
```

Figure 42.3 is a table that can be used to trace the progress of these programs for the sample problem given above. It gives the total cost and the optimal "last" multiplication for each subsequence in the list of matrices. For example, the entry in row A and column F says that 36 scalar multiplications are required to multiply

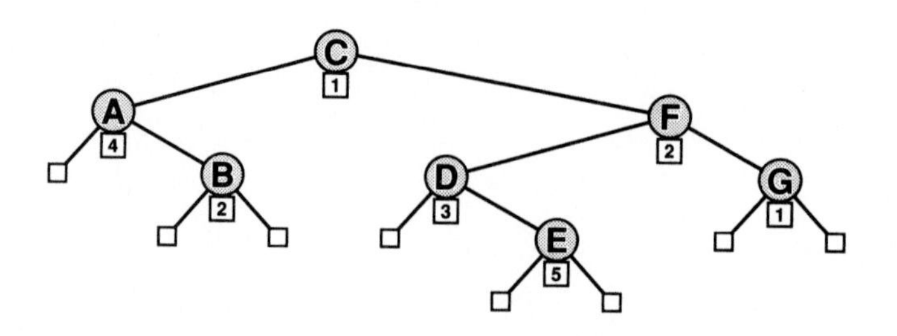

Figure 42.4 A binary search tree with frequencies.

matrices A through F together, and that this can be achieved by multiplying A through C in the optimal way, then multiplying D through F in the optimal way, then multiplying the resulting matrices together. Only D is actually in the *best* array: the full optimal splits are indicated in the diagram for clarity. To find how to multiply A through C in the optimal way, we look in row A and column C, etc. For our example, the parenthesization computed is ((A*(B*C))*((D*E)*F)) which, as mentioned above, requires only 36 scalar multiplications. For the example cited earlier with the dimensions of 3 in B, C and F changed to 300, the same parenthesization is optimal, requiring 2412 scalar multiplications.

Property 42.2 *Dynamic programming solves the matrix chain product problem in time proportional to* N^3 *and space proportional to* N^2.

Again, this follows directly from inspection of the code. The space requirement, in particular, is substantially more than we used for the knapsack problem. But the time and space requirements to find the optimum are likely to be quite negligible compared to the savings achieved. ■

Optimal Binary Search Trees

In many applications of searching, it is known that the search keys may occur with widely varying frequency. For example, a program that checks the spelling of words in English text is likely to look up words like "and" and "the" far more often than words like "dynamic" and "programming." Similarly, a Pascal compiler is likely to see keywords like "**end**" and "**do**" far more often than "**label**" or "**downto**." If binary-tree searching is used, it is clearly advantageous to have the most frequently sought keys near the top of the tree. A dynamic programming algorithm can be used to determine how to arrange the keys in the tree so as to minimize the total cost of searching.

Each node in the binary search tree in Figure 42.4 is labeled with an integer that is assumed to be proportional to its frequency of access. That is, out of every 18 searches in this tree, we expect four to be for A, two to be for B, one to be for

C, etc. Each of the four searches for A requires two node accesses, each of the two searches for B requires three node accesses, and so forth. We can compute a measure of the "cost" of the tree by simply multiplying the frequency for each node by its distance from the root and summing. This is the *weighted internal path length* of the tree. For the tree in Figure 42.4, the weighted internal path length is $4*2+2*3+1*1+3*3+5*4+2*2+1*3=51$. We would like to find the binary search tree for the given keys with the given frequencies that has the smallest internal path length over all such trees.

This problem is similar to the problem of minimizing weighted external path length that we saw in studying Huffman encoding (Chapter 22). In Huffman encoding, however, it was not necessary to maintain the order of the keys; in the binary search tree, we must preserve the property that all nodes to the left of the root have keys which are less, etc. This requirement makes the problem very similar to the matrix-chain-multiplication problem treated above: virtually the same program can be used.

Specifically, assume we are given a set of search keys $K_1 < K_2 < \cdots < K_N$ and associated frequencies $r_0, r_1, \ldots, r_N$, where r_i is the anticipated frequency of reference to key K_i. We want to find the binary search tree that minimizes the sum, over all keys, of these frequencies times the distance of the key from the root (the cost of accessing the associated node).

The dynamic-programming approach to this problem is to compute, for each j increasing from 1 to $N-1$, the best way to build a subtree containing K_i, $K_{i+1}, \ldots, K_{i+j}$ for $1 \le i \le N-j$, as in the following program:

```
for i:=1 to N do
  for j:=i+1 to N+1 do cost[i,j]:=maxint;
for i:=1 to N do cost[i,i]:=f[i];
for i:=1 to N+1 do cost[i,i-1]:=0;
for j:=1 to N-1 do
  for i:=1 to N-j do
    begin
    for k:=i to i+j do
      begin
      t:=cost[i,k-1]+cost[k+1,i+j];
      if t<cost[i,i+j] then
        begin cost[i,i+j]:=t; best[i,i+j]:=k end;
      end;
    t:=0; for k:=i to i+j do t:=t+f[k];
    cost[i,i+j]:=cost[i,i+j]+t;
    end;
```

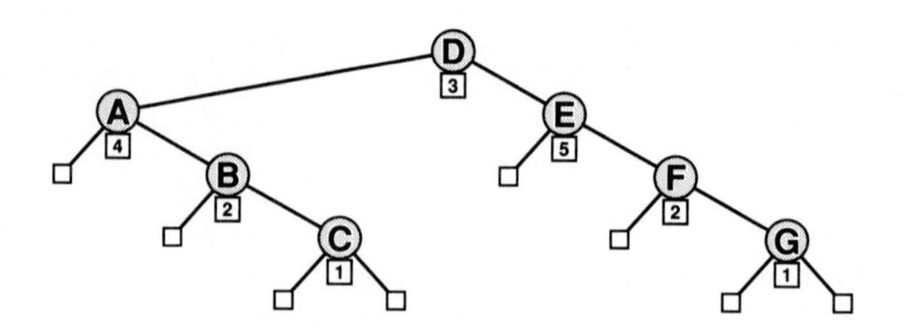

Figure 42.5 Optimal binary search tree.

For each j, the computation is done by trying each node as the root and using precomputed values to determine the best way to do the subtrees. For each k between i and $i+j$, we want to find the optimal tree containing $K_i, K_{i+1}, \ldots, K_{i+j}$ with K_k at the root. This tree is formed by using the optimal tree for $K_i, K_{i+1}, \ldots, K_{k-1}$ as the left subtree and the optimal tree for $K_{k+1}, K_{k+2}, \ldots, K_{i+j}$ as the right subtree. The internal path length of this tree is the sum of the internal path lengths for the two subtrees plus the sum of the frequencies for all the nodes (since each node in the new tree is one step further from the root).

Note that the sum of all the frequencies is added to any cost, and so it is not needed when looking for the minimum. Also, we must have *cost*$[i,i-1]=0$ to cover the possibility that a node could just have one child (there was no analogue to this in the matrix-chain problem).

As before, a short recursive program is required to recover the actual tree from the *best* array computed by the program. The optimal tree computed for the tree in Figure 42.4 is shown in Figure 42.5. The weighted internal path length of this tree is 41.

Property 42.3 *The dynamic-programming method of finding an optimal binary search tree takes time proportional to* N^3 *and space proportional to* N^2.

Again, the algorithm works with a matrix of size N^2 and spends time proportional to N on each entry. It is actually possible in this case to reduce the time requirement to N^2 by exploiting the fact that the optimal position for the root of a tree can't be too far from the optimal position for the root of a slightly smaller tree, so that k need not range over all the values from i to $i+j$ in the program above. ■

Time and Space Requirements

The above examples suggest that dynamic-programming applications can have quite different time and space requirements depending on the amount of information about small subproblems. For the shortest-paths algorithm, no extra space is required; for the knapsack problem, space proportional to the size of the knapsack is needed; for the other problems, N^2 space is needed. For each problem, the time required is a factor of N greater than the space required.

The range of possible applicability of dynamic programming is far larger than covered in the examples. From a dynamic programming point of view, divide-and-conquer recursion can be thought of as a special case in which a minimal amount of information about small cases must be computed and stored, and exhaustive search (which we'll examine in Chapter 44) can be thought of as a special case in which a maximal amount of information about small cases must be computed and stored. Dynamic programming is a natural design technique that appears in many guises to solve problems throughout this range.

□

Exercises

1. In the example given for the knapsack problem, the items are sorted by size. Does the algorithm still work properly if they appear in arbitrary order?
2. Modify the knapsack program to take into account another constraint defined by an array *num[1..N]* which contains the number of available items of each type.
3. What would the knapsack program do if one of the values were negative?
4. True or false: If a matrix chain involves a 1-by-k by k-by-1 multiplication, then there is an optimal solution for which that multiplication is last. Defend your answer.
5. Write a program that actually multiplies together a list of N matrices in an optimal way. Assume that the matrices are stored in a three-dimensional array *matrices[1..Nmax,1..Dmax,1..Dmax]*, where *Dmax* is the maximum dimension, with the ith matrix stored in *matrices[i,1..r[i],1..r[i+1]]*.
6. Draw the optimal binary search tree for the example in the text, but with all the frequencies increased by one.
7. Write the program to construct the optimal binary search tree.
8. Suppose we've computed the optimum binary search tree for some set of keys and frequencies, and say that one frequency is incremented by one. Write a program to compute the new optimum tree.
9. Why not solve the knapsack problem in the same way as the matrix-chain and optimum binary search tree problems: by minimizing, for k from 1 to M, the sum of the best value achievable for a knapsack of size k and the best value achievable for a knapsack of size $M-k$?
10. Extend the program for the shortest-paths problem to include a procedure *paths(i,j: integer)* that fills an array *path* with the shortest path from i to j. This procedure should take time proportional to the length of the path each time it is called, using an auxiliary data structure built up by a modified version of the program given in Chapter 32.

43

Linear Programming

☐ Many practical problems involve complicated interactions between a number of varying quantities. One example of this is the network flow problem discussed in Chapter 33: the flows in the various pipes must obey physical laws over the network. Another example is scheduling various tasks in (say) a manufacturing process in the face of deadlines, priorities, etc. Very often it is possible to develop a precise mathematical formulation which captures the interactions involved and reduces the problem at hand to a more straightforward mathematical problem. This process of deriving a set of mathematical equations whose solution implies the solution of a given practical problem is called *mathematical programming*. Again, as in Chapter 42, the term "programming" here refers to the process of choosing the variables and setting up the equations so that a solution to the equations corresponds to a solution to the problem. In this chapter, we consider a fundamental variety of mathematical programming, *linear programming*, and an efficient algorithm for solving linear programs, the *simplex* method.

Linear programming and the simplex method are of fundamental importance because a wide variety of important problems are amenable to formulation as linear programs and to efficient solution by the simplex method. Better algorithms are known for some specific problems, but few problem-solving techniques are as widely applicable as the process of first formulating the problem as a linear program, then computing the solution using the simplex method. A library routine for the simplex method can be an indispensible tool for attacking complex problems.

Research in linear programming has been extensive, and a full understanding of all the issues involved requires mathematical maturity beyond that assumed for this book. On the other hand, some of the basic ideas are easy to comprehend, and the actual simplex algorithm is not difficult to implement, as we'll see below. As with the fast Fourier transform in Chapter 41, our intent is not to provide a full practical implementation, but rather to learn some of the basic properties of the algorithm and its relationship to other algorithms we've studied.

Linear Programs

Mathematical programs involve a set of *variables* related by a set of mathematical equations (*constraints*) and an *objective function* involving the variables that is to be maximized subject to the constraints. If all of the equations involved are simply linear combinations of the variables, we have the special case that we're considering called *linear programming*.

The following linear program corresponds to the network flow problem in Chapter 33.

Maximize $x_{AB} + x_{AD}$
subject to the constraints

$$\begin{aligned} x_{AB} &\le 6 & x_{CD} &\le 3 \\ x_{AC} &\le 8 & x_{CE} &\le 3 \\ x_{BD} &\le 6 & x_{DF} &\le 8 \\ x_{BE} &\le 3 & x_{EF} &\le 6 \end{aligned}$$

$$\begin{aligned} x_{BD} + x_{BE} &= x_{AB}, \\ x_{CD} + x_{CE} &= x_{AC}, \\ x_{BD} + x_{CD} &= x_{DF}, \\ x_{BE} + x_{CE} &= x_{EF}, \end{aligned}$$

$$x_{AB}, x_{AC}, x_{BD}, x_{BE}, x_{CD}, x_{CE}, x_{DF}, x_{EF} \ge 0.$$

There is one variable in this linear program corresponding to the flow in each of the pipes. These variables satisfy two types of equations: inequalities, corresponding to capacity constraints on the pipes, and equalities, corresponding to flow constraints at every junction. Thus, for example, the inequality $x_{AB} \le 8$ says that pipe AB has capacity 8, and the equation $x_{BD} + x_{BE} = x_{AB}$ says that the outflow must equal the inflow at junction B. Note that all the equalities together give the implicit constraint $x_{AB} + x_{AC} = x_{DF} + x_{EF}$, which says that the inflow must equal the outflow for the whole network. Also, of course, all of the flows must be positive.

This is clearly a mathematical formulation of the network flow problem: a solution to this particular mathematical problem is a solution to the particular instance of the network flow problem. The point of this example is not that linear programming will provide a better algorithm for this particular problem, but rather that linear programming is a quite general technique that can be applied to a variety of problems. For example, if we were to generalize the network flow problem to include costs, say, as well as capacities, the linear programming formulation would not look much different, even though the problem might be significantly more difficult to solve directly.

Not only are linear programs richly expressive, but also an algorithm exists for solving them (the simplex algorithm) that has proven to be quite efficient for many problems arising in practice. For some problems (such as network flow) there may be an algorithm specifically oriented to that problem which can perform better than linear programming/simplex; for other problems (including various extensions of network flow), no better algorithms are known. Even if there is a better algorithm, it may be complicated or difficult to implement, while the procedure of developing a linear program and solving it with a simplex library routine is often quite straightforward. This general-purpose aspect of the method is quite attractive and has led to its widespread use. The danger in relying upon it too heavily is that it may lead to inefficient solutions for some simple problems (for example, many of those studied in this book).

Geometric Interpretation

Linear programs can be cast in a geometric setting. The following linear program is easy to visualize because only two variables are involved:

Maximize $x_1 + x_2$
subject to the constraints

$$\begin{aligned} -x_1 + x_2 &\le 5, \\ x_1 + 4x_2 &\le 45, \\ 2x_1 + x_2 &\le 27, \\ 3x_1 - 4x_2 &\le 24, \\ x_1, x_2 &\ge 0. \end{aligned}$$

This linear program corresponds to the geometric situation diagrammed in Figure 43.1. Each inequality defines a halfplane in which any solution to the linear program must lie. For example, $x_1 \ge 0$ means that any solution must lie to the right of the x_2 axis, and $-x_1 + x_2 \le 5$ means that any solution must lie below and to the right of the line $-x_1 + x_2 = 5$ (which goes through $(0, 5)$ and $(5, 10)$). Any solution to the linear program must satisfy *all* of these constraints, so the region defined by the intersection of all these halfplanes (shaded in the figure) is the set of all possible solutions. To solve the linear program we must find the point within this region which maximizes the objective function.

A region defined by intersecting halfplanes is always convex (we've encountered this fact before, in one of the definitions of the convex hull in Chapter 25). This convex region, called the *simplex*, forms the basis for an algorithm to find the solution to the linear program that maximizes the objective function.

A fundamental property of the simplex that is exploited by the algorithm is that the objective function is maximized at one of the vertices of the simplex; thus

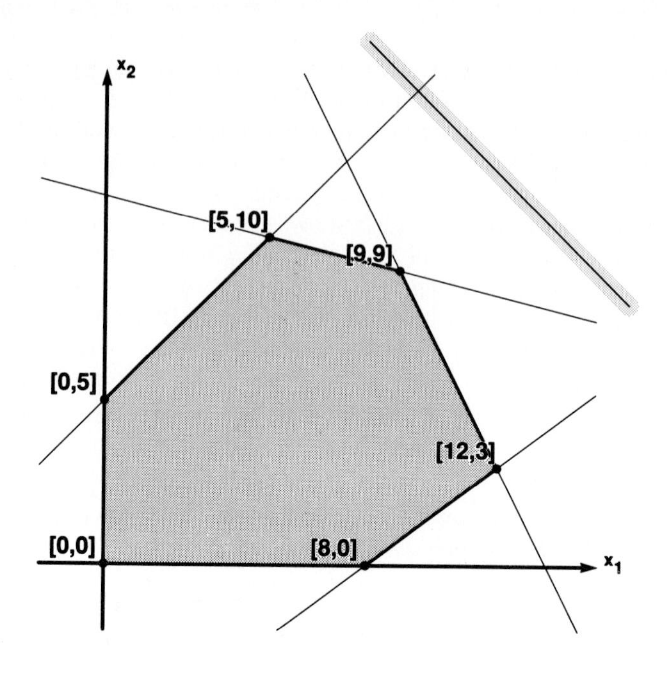

Figure 43.1 A two-dimensional simplex.

only the vertices need be examined, not all the points inside. To see why this is so, consider the shaded bar at the upper right in Figure 43.1, which corresponds to the objective function. The objective function can be thought of as defining a line of known slope (in this case -1) and unknown position. We're interested in the point at which the line hits the simplex as it is moved in from infinity. This point is the solution to the linear program: it satisfies all the inequalities because it is in the simplex, and it maximizes the objective function because no points with larger values were encountered. For our example, the line hits the simplex at (9,9) which maximizes the objective function at 18.

Other objective functions correspond to lines of other slopes, but the maximum always occurs at one of the vertices of the simplex. The algorithm we'll examine below is a systematic way of moving from vertex to vertex in search of the minimum. In two dimensions there's not much choice about what to do, but, as we'll see, the simplex is a much more complicated object when more variables are involved.

One can also appreciate from Figure 43.1 why mathematical programs involving nonlinear functions are so much more difficult to handle. For example, if the objective function is nonlinear, it can be a curve that can strike the simplex along one of its edges, not at a vertex. If the inequalities are also nonlinear, quite complicated geometric shapes which correspond to the simplex can arise.

Geometric intuition makes it clear that various anomalous situations can arise. For example, suppose that we add the inequality $x_1 \geq 13$ to the linear program in the example above. It is quite clear from Figure 43.1 that in this case the intersection of the half-planes is empty. Such a linear program is called *infeasible*: there are no points which satisfy the inequalities, let alone any which maximize the objective function. On the other hand, the inequality $x_1 \leq 13$ is *redundant*: the simplex is entirely contained within its halfplane, so this inequality is not represented in the simplex. Redundant inequalities do not affect the solution at all, but they must be dealt with during the search for the solution.

A more serious problem is that the simplex may be an open (unbounded) region, in which case the solution may not be well-defined. This would be the case in our example if the second and third inequalities were deleted. Even if the simplex is unbounded the solution may be well-defined for some objective functions, but an algorithm to find it might have significant difficulty getting around the unbounded region.

It must be emphasized that, though these problems are quite easy to see when we have two variables and a few inequalities, they are very much less apparent for a general problem with many variables and inequalities. Indeed, detection of these anomalous situations is a significant part of the computational burden of solving linear programs.

The same geometric intuition holds for more variables. In three dimensions the simplex is a convex three-dimensional solid defined by intersecting halfspaces defined by the planes whose equations are given by changing the inequalities to equalities. For example, if we add the inequalities $x_3 \leq 4$ and $x_3 \geq 0$ to the linear program above, the simplex becomes the solid object diagrammed in Figure 43.2.

To make the example more three-dimensional, suppose that we change the objective function to $x_1 + x_2 + x_3$. This defines a plane perpendicular to the line $x1 = x2 = x3$. If we move a plane in from infinity along this line, we hit the simplex at the point $(9, 9, 4)$ which is the solution. (Also shown in Figure 43.2 is a path along the vertices of the simplex from $(0, 0, 0)$ to the solution, for reference in the description of the algorithm below.)

In n dimensions, we intersect halfspaces defined by $(n - 1)$-dimensional hyperplanes to define the n-dimensional simplex, and bring in an $(n - 1)$-dimensional hyperplane from infinity to intersect the simplex at the solution point. As mentioned above, we risk oversimplification by concentrating on intuitive two- and three-dimensional situations, but proofs of the facts above involving convexity, intersecting hyperplanes, etc. involve a facility with linear algebra somewhat beyond the scope of this book. Still, the geometric intuition is valuable, since it can help us understand the fundamental characteristics of the basic method used in practice to solve higher-dimensional problems.

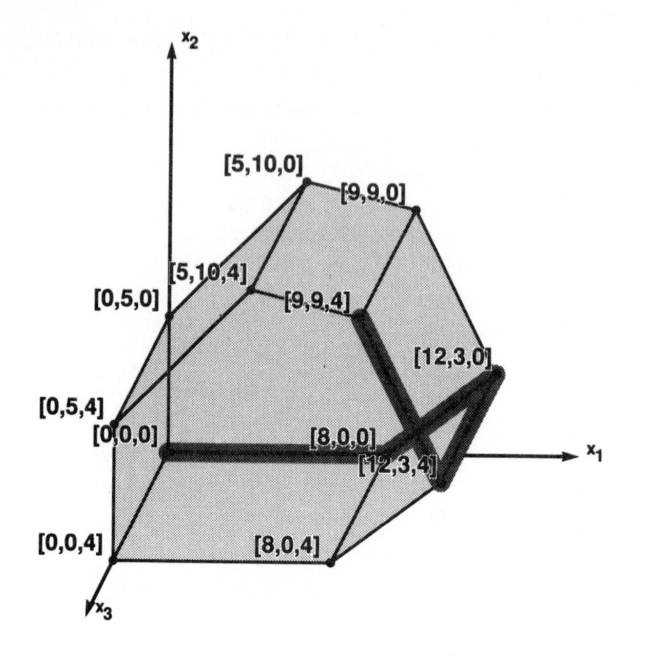

Figure 43.2 A three-dimensional simplex.

The Simplex Method

The *simplex method* is the name commonly used to describe a general approach to solving linear programs by using pivoting, the same fundamental operation used in Gaussian elimination. It turns out that pivoting corresponds in a natural way to the geometric operation of moving from point to point on the simplex in search of the solution. The several algorithms commonly used differ in essential details having to do with the order in which simplex vertices are searched. That is, the well-known "algorithm" for this problem can more precisely be described as a generic method which can be refined in any of several different ways. We've encountered this sort of situation before, for example Gaussian elimination (Chapter 37) or the Ford-Fulkerson algorithm (Chapter 33).

First, it's clear that linear programs can take many different forms. For example, the linear program above for the network flow problem has a mixture of equalities and inequalities, but the geometric examples above use only inequalities. It is convenient to reduce the number of possibilities somewhat by insisting that all linear programs be presented in the same *standard form*, where all the equations are equalities except for an inequality for each variable stating that it is nonnegative. This may seem like a severe restriction, but actually it is not difficult to convert general linear programs to this standard form. The following linear program is the standard form for the three-dimensional example given in Figure 43.2:

Maximize $x_1 + x_2 + x_3$
subject to the constraints

$$\begin{aligned} -x_1 + x_2 + y_1 &= 5 \\ x_1 + 4x_2 + y_2 &= 45 \\ 2x_1 + x_2 + y_3 &= 27 \\ 3x_1 - 4x_2 + y_4 &= 24 \\ x_3 + y_5 &= 4 \\ x_1, x_2, x_3, y_1, y_2, y_3, y_4, y_5 &\geq 0. \end{aligned}$$

Each inequality involving more than one variable is converted into an equality by introducing a new variable. The y's are called *slack* variables because they take up the slack allowed by the inequality. Any inequality involving only one variable can be converted to the standard nonnegative constraint simply by renaming the variable. For example, a constraint such as $x_3 \leq -1$ would be handled by replacing x_3 by $-1 - x_3'$ everywhere that it appears.

This formulation makes obvious the parallels between linear programming and simultaneous equations. We have N equations in M unknown variables, all constrained to be positive. In this case, note that there are N slack variables, one for each equation (since we started out with all inequalities). We assume that $M > N$ which implies that there are many solutions to the equations: the problem is to find the one which maximizes the objective function.

For our example, there is a trivial solution to the equations: take $x_1 = x_2 = x_3 = 0$, then assign appropriate values to the slack variables to satisfy the equalities. This works because in happens that $(0, 0, 0)$ is a point on the simplex in this example. Although this need not be the case in general, in explaining the simplex method we will restrict attention for now to linear programs where it is known to be the case. This is still a quite large class of linear programs: for example, if all the numbers on the right-hand side of the inequalities in the standard form of the linear program are positive and slack variables all have positive coefficients (as in our example) then there is clearly a solution with all the original variables zero. Later we will return to the general case.

Given a solution with $M - N$ variables set to zero, it turns out that we can find another solution with the same property by using a familiar operation, *pivoting*. This is essentially the same operation used in Gaussian elimination: an element $a[p,q]$ is chosen in the matrix of coefficients defined by the equations, then the pth row is multiplied by an appropriate scalar and added to all other rows to make the qth column all zero except for the entry in row q, which is made 1. Consider the following matrix, which represents the linear program given above:

$$\begin{pmatrix} -1.00 & -1.00 & -1.00 & 0.00 & 0.00 & 0.00 & 0.00 & 0.00 & 0.00 \\ -1.00 & 1.00 & 0.00 & 1.00 & 0.00 & 0.00 & 0.00 & 0.00 & 5.00 \\ 1.00 & 4.00 & 0.00 & 0.00 & 1.00 & 0.00 & 0.00 & 0.00 & 45.00 \\ 2.00 & 1.00 & 0.00 & 0.00 & 0.00 & 1.00 & 0.00 & 0.00 & 27.00 \\ 3.00 & -4.00 & 0.00 & 0.00 & 0.00 & 0.00 & 1.00 & 0.00 & 24.00 \\ 0.00 & 0.00 & 1.00 & 0.00 & 0.00 & 0.00 & 0.00 & 1.00 & 4.00 \end{pmatrix}.$$

This $(N + 1)$-by-$(M + 1)$ matrix contains the coefficients of the linear program in standard form, with the $(M + 1)$st column containing the numbers on the right-hand sides of the equations (as in Gaussian elimination), and the 0th row containing the coefficients of the objective function, with the sign reversed. The significance of the 0th row is discussed below; for now we'll treat it just like all of the other rows.

For our example, we'll carry out all computations to two decimal places. Doing this obviously ignores issues such as computational accuracy and accumulated error, which are just as important here as they are in Gaussian elimination.

The variables which correspond to a solution are called the *basis* variables and those which are set to 0 to make the solution are called *non-basis* variables. In the matrix, the columns corresponding to basis variables have exactly one 1 with all other values 0, while non-basis variables correspond to columns with more than one nonzero entry.

Now, suppose that we wish to pivot this matrix for $p = 4$ and $q = 1$. That is, an appropriate multiple of the fourth row is added to each of the other rows to make the first column all zero except for a 1 in row 4. This produces the following result:

$$\begin{pmatrix} 0.00 & -2.33 & -1.00 & 0.00 & 0.00 & 0.00 & 0.33 & 0.00 & 8.00 \\ 0.00 & -0.33 & 0.00 & 1.00 & 0.00 & 0.00 & 0.33 & 0.00 & 13.00 \\ 0.00 & 5.33 & 0.00 & 0.00 & 1.00 & 0.00 & -0.33 & 0.00 & 37.00 \\ 0.00 & 3.67 & 0.00 & 0.00 & 0.00 & 1.00 & -0.67 & 0.00 & 11.00 \\ 1.00 & -1.33 & 0.00 & 0.00 & 0.00 & 0.00 & 0.33 & 0.00 & 8.00 \\ 0.00 & 0.00 & 1.00 & 0.00 & 0.00 & 0.00 & 0.00 & 1.00 & 4.00 \end{pmatrix}.$$

This operation removes the 7th column from the basis and adds the 1st column to the basis. Exactly one basis column is removed because exactly one basis column has a 1 in row p.

By definition, we can get a solution to the linear program by setting all the non-basis variables to zero, then using the trivial solution given in the basis. In the solution corresponding to the above matrix, both x_2 and x_3 are zero because they are non-basis variables and $x_1 = 8$, so the matrix corresponds to the point $(8, 0, 0)$ on the simplex. (We're not interested in the values of the slack variables.) Note that the upper right-hand corner of the matrix (row 0, column $M + 1$) contains the value of the objective function at this point. This is by design, as we shall soon see.

Now suppose that we perform the pivot operation for $p = 3$ and $q = 2$:

$$\begin{pmatrix} 0.00 & 0.00 & -1.00 & 0.00 & 0.00 & 0.64 & -0.09 & 0.00 & 15.00 \\ 0.00 & 0.00 & 0.00 & 1.00 & 0.00 & 0.09 & 0.27 & 0.00 & 14.00 \\ 0.00 & 0.00 & 0.00 & 0.00 & 1.00 & -1.45 & 0.64 & 0.00 & 21.00 \\ 0.00 & 1.00 & 0.00 & 0.00 & 0.00 & 0.27 & -0.18 & 0.00 & 3.00 \\ 1.00 & 0.00 & 0.00 & 0.00 & 0.00 & 0.36 & 0.09 & 0.00 & 12.00 \\ 0.00 & 0.00 & 1.00 & 0.00 & 0.00 & 0.00 & 0.00 & 1.00 & 4.00 \end{pmatrix}.$$

This removes column 6 from the basis and adds column 2. By setting non-basis variables to 0 and solving for basis variables as before, we see that this matrix corresponds to the point $(12, 3, 0)$ on the simplex, for which the objective function has the value 15. Note that the value of the objective function is strictly increasing. Again, this is by design, as we shall soon see.

How do we decide which values of p and q to use for pivoting? This is where row 0 comes in. For each non-basis variable, row 0 contains the amount by which the objective function would increase if that variable were changed from 0 to 1, with the sign reversed. (The sign is reversed so that the standard pivoting operation will maintain row 0, with no changes.) Pivoting using column q amounts to changing the value of the corresponding variable from 0 to some positive value, so we can be sure the objective function will increase if we use any column with a negative entry in row 0.

Now, pivoting on any row with a positive entry for that column will increase the objective function, but we also must make sure that it will result in a matrix corresponding to a point on the simplex. Here the central concern is that one of the entries in column $M + 1$ might become negative. This can be forestalled by finding, among the positive elements in column q (not including row 0), the one that gives the smallest value when divided into the $(M + 1)$st element in the same row. If we take p as the index of the row containing this element and pivot, then we can be sure that the objective function will increase and that none of the entries in column $M + 1$ will become negative; this is enough to ensure that the resulting matrix corresponds to a point on the simplex.

There are two potential difficulties with this procedure for finding the pivot row. First, what if there are no positive entries in the qth column? This is an inconsistent situation: the negative entry in row 0 says that the objective function can be increased, but there is no way to increase it. It turns out that this situation arises if and only if the simplex is unbounded, so the algorithm can terminate and report the problem. A more subtle difficulty arises in the degenerate case when the $(M + 1)$st entry in some row (with a positive entry in column q) is 0. Then this row will be chosen, but the objective function will increase by 0. This is not a difficulty in itself: the difficulty arises when there are two such rows. Certain natural policies for choosing between such rows lead to *cycling*: an infinite sequence of pivots which do not increase the objective function at all. We have been avoiding difficulties such as cycling in our example to make the description of the method clear, but it must be emphasized that such degenerate cases are quite likely to arise in practice. The generality available with linear programming implies that degenerate cases of the general problem will arise in the solution of specific problems.

Several possibilities are available for avoiding cycling. One method is to break ties randomly. This makes cycling extremely unlikely (but not mathematically impossible). Another anti-cycling policy is described below.

In our example, we can pivot again with $q = 3$ (because of the -1 in row 0 and column 3) and $p = 5$ (because 1 is the only positive value in column 3). This gives the following matrix:

$$\begin{pmatrix} 0.00 & 0.00 & 0.00 & 0.00 & 0.00 & 0.64 & -0.09 & 1.00 & 19.00 \\ 0.00 & 0.00 & 0.00 & 1.00 & 0.00 & 0.09 & 0.27 & 0.00 & 14.00 \\ 0.00 & 0.00 & 0.00 & 0.00 & 1.00 & -1.45 & 0.64 & 0.00 & 21.00 \\ 0.00 & 1.00 & 0.00 & 0.00 & 0.00 & 0.27 & -0.18 & 0.00 & 3.00 \\ 1.00 & 0.00 & 0.00 & 0.00 & 0.00 & 0.36 & 0.09 & 0.00 & 12.00 \\ 0.00 & 0.00 & 1.00 & 0.00 & 0.00 & 0.00 & 0.00 & 1.00 & 4.00 \end{pmatrix}$$

This corresponds to the point $(12,3,4)$ on the simplex, for which the value of the objective function is 19.

In general, there might be several negative entries in row 0, and several different strategies for choosing among them have been suggested. We have been proceeding according to one of the most popular methods, the *greatest-increment* method: always choose the column with the smallest value in row 0 (largest in absolute value). This does not necessarily lead to the largest increase in the objective function, since scaling must be done according to the row p chosen. If this column-selection policy is combined with the row-selection policy of using, in case of ties, the row that results in the removal of the column of lowest index from the basis, then cycling cannot happen. (This anticycling policy is due to R. G. Bland.) Another possibility for column selection is actually to calculate the amount by which the objective function would increase for each column, then use the column which gives the largest result. This is called the *steepest-descent* method. Yet another interesting possibility is to choose randomly from among the available columns.

Finally, after one more pivot at $p = 2$ and $q = 7$, we arrive at the solution:

$$\begin{pmatrix} 0.00 & 0.00 & 0.00 & 0.00 & 0.14 & 0.43 & 0.00 & 1.00 & 22.00 \\ 0.00 & 0.00 & 0.00 & 1.00 & -0.43 & 0.71 & 0.00 & 0.00 & 5.00 \\ 0.00 & 0.00 & 0.00 & 0.00 & 1.57 & -2.29 & 1.00 & 0.00 & 33.00 \\ 0.00 & 1.00 & 0.00 & 0.00 & 0.29 & -0.14 & 0.00 & 0.00 & 9.00 \\ 1.00 & 0.00 & 0.00 & 0.00 & -0.14 & 0.57 & 0.00 & 0.00 & 9.00 \\ 0.00 & 0.00 & 1.00 & 0.00 & 0.00 & 0.00 & 0.00 & 1.00 & 4.00 \end{pmatrix}$$

This corresponds to the point $(9,9,4)$ on the simplex, which maximizes the objective function at 22. All the entries in row 0 are nonnegative, so any pivot will only serve to decrease the objective function.

The above example outlines the simplex method for solving linear programs. In summary, if we begin with a matrix of coefficients corresponding to a point on the simplex, we can do a series of pivot steps which move to adjacent points on the simplex, always increasing the objective function, until the maximum is reached.

One fundamental fact not yet noted is crucial to the correct operation of this procedure: once we reach a point at which no single pivot can improve the objective function (a "local" maximum), then we have reached the "global" maximum. This is the basis for the simplex algorithm. As mentioned above, the proof of this (and many other facts which may seem obvious from the geometric interpretation) in general is quite beyond the scope of this book. But the simplex algorithm for the general case operates in essentially the same manner as for the simple problem traced above.

Implementation

The implementation of the simplex method for the case described above is quite straightforward from the description. First, the requisite pivoting procedure uses code similar to our implementation of Gaussian elimination in Chapter 37:

```
procedure pivot(p,q: integer);
  var j,k: integer;
  begin
  for j:=0 to N do
    for k:=M+1 downto 1 do
      if (j<>p) and (k<>q) then
        a[j,k]:=a[j,k]-a[p,k]*a[j,q]/a[p,q];
  for j:=0 to N do if j<>p then a[j,q]:=0;
  for k:=1 to M+1 do if k<>q then a[p,k]:=a[p,k]/a[p,q];
  a[p,q]:=1
  end;
```

This program adds multiples of row p to each row as necessary to make column q all zero except for a 1 in row q, as described above. As in Chapter 37, it is necessary to take care not to change the value of $a[p,q]$ before we've finished using it.

In Gaussian elimination, we processed only rows below p in the matrix during forward elimination and only rows above p during backward substitution using the Gauss-Jordan method. A system of N linear equations in N unknowns could be solved by calling *pivot*(i,i) for i ranging from 1 to N then back down to 1 again.

The simplex algorithm, then, consists simply of finding the values of p and q as described above and calling *pivot*, repeating the process until the optimum is reached or the simplex is determined to be unbounded:

```
repeat
   q:=0; repeat q:=q+1 until (q=M+1) or (a[0,q]<0);
   p:=0; repeat p:=p+1 until (p=N+1) or (a[p,q]>0);
   for i:=p+1 to N do
      if a[i,q]>0 then
         if (a[i,M+1]/a[i,q])<(a[p,M+1]/a[p,q]) then p:=i;
   if (q<M+1) and (p<N+1) then pivot(p,q)
until (q=M+1) or (p=N+1);
```

If the program terminates with $q=M+1$ then an optimal solution has been found: the value achieved for the objective function is in $a[0,M+1]$ and the values for the variables can be recovered from the basis. If the program terminates with $p=N+1$, then an unbounded situation has been detected.

This program ignores the issue of cycle avoidance. To implement Bland's method, it is necessary to keep track of the column that would leave the basis, were a pivot done using row p. This is easily done by setting $outb[p]:=q$ after each pivot. Then the loop to calculate p can be modified to set $p:=i$ also if equality holds in the ratio test and $outb[p] < outb[q]$. Alternatively, a random element could be selected by generating a random integer x and replacing each array reference $a[p,q]$ (or $a[i,q]$) by $a[(p+x)$ **mod** $(N+1),q]$ (or $a[(i+x)$ **mod** $(N+1),q]$). This has the effect of searching through the column q in the same way as before, but starting at a random point instead of the beginning. The same sort of technique could be used to choose a random column (with a negative entry in row 0) to pivot on.

The program and example above treat a simple case that illustrates the principle behind the simplex algorithm but avoids the substantial complications that can arise in actual applications. The main omission is that the program requires the matrix to have a *feasible basis*: a set of rows and columns that can be permuted into the identity matrix. The program starts with the assumptions that there is a solution with the $M - N$ variables appearing in the objective function set to zero and that the N-by-N submatrix involving the slack variables has been "solved" to make that submatrix the identity matrix. This is easy to do for the particular type of linear program we stated (with all inequalities on positive variables), but in general we need to find some point on the simplex. Once we have found one solution, we can make appropriate transformations (mapping that point to the origin) to bring the matrix into the required form, but at the outset we don't even know whether a solution exists.

In fact, it has been shown that *detecting* whether a solution exists is as difficult computationally as finding the optimum solution, given that one exists. Thus it should not be surprising that the technique commonly used to detect the existence of a solution is the simplex algorithm! Specifically, we add another set of artificial variables $s_1, s_2, \ldots, s_N$ and add variable s_i to the ith equation. This is done simply by adding to the matrix N columns filled with the identity matrix. This immediately gives a feasible basis for this new linear program. The trick is to run the above algorithm with the objective function $-s_1 - s_2 - \cdots - s_N$. If there is a solution to the original linear program, then this objective function can be maximized at zero. If the maximum reached is not zero, then the original linear program is infeasible. If the maximum is zero, then the normal situation is that $s_1, s_2, \ldots, s_N$ all become non-basis variables, so we have computed a feasible basis for the original linear program. In degenerate cases, some of the artificial variables may remain in the basis, so it is necessary to do further pivoting to remove them (without changing the cost).

To summarize, a two-phase process is normally used to solve general linear programs. First, we solve a linear program involving the artificial s variables in order to get a point on the simplex for our original problem. Then, we dispose of the s variables and reintroduce our original objective function to proceed from this point to the solution.

Analyzing the running time of the simplex method is extremely complicated, and few results are available. No one knows the "best" pivot selection strategy, because there are no results to tell us how many pivot steps to expect for any reasonable class of problems. It is possible to construct artificial examples for which the running time of the simplex is very large (an exponential function of the number of variables). However, those who have used the algorithm in practical settings are unanimous in testifying to its efficiency in solving actual problems.

The simple version of the simplex algorithm we've considered, while quite useful, is merely part of a general and beautiful mathematical framework providing a complete set of tools that can be used to solve a variety of very important practical problems.

□

Exercises

1. Draw the simplex defined by the inequalities $x_1 \geq 0$, $x_2 \geq 0$, $x_3 \geq 0$, $x_1 + 2x_2 \leq 20$, and $x_1 + x_2 + x_3 \leq 10$.

2. Give the sequence of matrices produced for the example in the text if the pivot column chosen is the largest q for which $a[0,q]$ is negative.

3. Give the sequence of matrices produced for the example in the text for the objective function $x_1 + 5x_2 + x_3$.

4. Describe what happens when the simplex algorithm is run on a matrix with a column of all zeros.

5. Does the simplex algorithm use the same number of steps if the rows of the input matrix are permuted?

6. Give a linear programming formulation of the example for the knapsack problem in Chapter 42.

7. How many pivot steps are required to solve the linear program "Maximize $x_1 + \cdots + x_M$ subject to the constraints $x_1, \ldots, x_M \leq 1$ and $x_1, \ldots, x_M \geq 0$"?

8. Construct a linear program consisting of N inequalities on two variables for which the simplex algorithm requires at least $N/2$ pivots.

9. Give a three-dimensional linear-programming problem that illustrates the difference between the greatest-increment and steepest-descent column-selection methods.

10. Modify the implementation given in the text to write out the coordinates of the optimal solution point.

44

Exhaustive Search

☐ Some problems involve searching through a vast number of potential solutions to find an answer, and simply do not seem to be amenable to solution by efficient algorithms. In this chapter, we'll examine some characteristics of problems of this sort and some techniques that have proven to be useful for solving them.

To begin, we should reorient our thinking somewhat on exactly what constitutes an "efficient" algorithm. For most of the applications that we have discussed, we have become conditioned to thinking that an algorithm must be linear or run in time proportional to something like $N \log N$ or $N^{3/2}$ to be considered efficient. We've generally considered quadratic algorithms bad and cubic algorithms awful. But any computer scientist would be absolutely delighted to know a cubic algorithm for the problems that we'll consider in this and the next chapter. In fact, even an N^{50} algorithm would be pleasing (from a theoretical standpoint) because these problems are believed to require *exponential* time.

Suppose we have an algorithm that takes time proportional to 2^N. If we were to have a computer 1000 times faster than the fastest supercomputer available today, then we could perhaps solve a problem for $N = 50$ in an hour's time (under the most generous assumptions about the simplicity of the algorithm). But in two hour's time we could only do $N = 51$, and even in a year's time we could only get to $N = 59$. And even if a new computer were to be developed with a million times the speed, and we were to have a million such computers available, we couldn't get to $N = 100$ in a year's time. Realistically, we have to settle for N on the order of 25 or 30. A "more efficient" algorithm in this situation may be one that could solve a problem for $N = 100$ with a realistic amount of time and money.

The most famous problem of this type is the *traveling salesman problem*: given a set of N cities, find the shortest route connecting them all, with no city visited twice. This problem arises naturally in a number of important applications, so it has been studied quite extensively. We'll use it as an example in this chapter to examine some fundamental techniques. Many advanced methods have been developed for this problem, but it is still unthinkable to solve an arbitrary instance of the problem for $N = 1000$.

The traveling-salesman problem is difficult because there seems to be no way to avoid having to check the length of a very large number of possible tours. Checking each and every tour is *exhaustive search*: first we'll see how that is done. Then we'll see how to modify that procedure to greatly reduce the number of possibilities checked, by trying to discover incorrect decisions as early as possible in the decision-making process.

As mentioned above, solving a large traveling-salesman problem is impossible in a practical sense, even with the very best techniques known. As we'll see in the next chapter, the same is true of many other important practical problems. But what can be done when such problems arise in practice? Some sort of answer is expected (the traveling salesman has to do something): we can't simply ignore the existence of the problem or state that it's too hard to solve. At the end of this chapter, we'll see examples of some methods developed to cope with practical problems that seem to require exhaustive search. In the next chapter, we'll examine in some detail the reasons why no efficient algorithm is likely to be found for many such problems.

Exhaustive Search in Graphs

If the traveling salesman can travel only between certain pairs of cities (for example, if he is traveling by air), then the problem is directly modeled by a graph: given a weighted (possibly directed) graph, we want to find the shortest simple cycle that connects all the nodes.

This immediately brings to mind another problem that would seem to be easier: given an undirected graph, is there *any* way to connect all the nodes with a simple cycle? That is, starting at some node, can we "visit" all the other nodes and return to the original node, visiting every node in the graph exactly once? This is known as the *Hamilton cycle* problem. In the next chapter, we'll see that it is computationally equivalent to the traveling salesman problem in a strict technical sense.

In Chapters 29 and 31 we saw a number of methods for systematically visiting all the nodes of a graph. For those algorithms, it was possible to arrange the computation so that each node is visited just once, and this led to very efficient algorithms. For the Hamilton-cycle problem, such a solution is not apparent: it seems to be necessary to visit each node many times. For the other problems, we built a tree; when a "dead end" was reached in the search, we could start it up again, working on another part of the tree. For this problem, the tree must have a particular structure (a cycle): if we discover during the search that the tree being built cannot be a cycle, we have to go back and rebuild part of it.

To illustrate some of the issues involved, we'll look at the Hamilton cycle problem for the example graph shown in Figure 44.1. Depth-first search visits

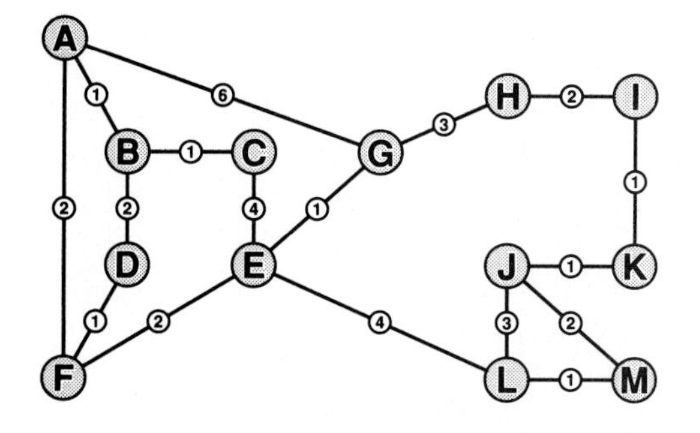

Figure 44.1 An instance of the traveling salesman problem.

the nodes in this graph in the order A B C E F D G H I K J L M (assuming an adjacency-matrix or sorted adjacency-list representation). This is not a simple cycle, and thus to find a Hamilton cycle we have to try another way to visit the nodes. It turns out the we can try all possibilities systematically with a simple modification to the *visit* procedure, as follows:

```
procedure visit(k: integer);
  var t: integer;
  begin
  id:=id+1; val[k]:=id;
  for t:=1 to V do
    if a[k,t] then
      if val[t]=0 then visit(t);
  id:=id-1; val[k]:=0
  end;
```

Rather than marking every node it touches with a nonzero *val* entry, this procedure "cleans up after itself" and leaves *id* and the *val* array exactly as it found them. The only marked nodes are those for which *visit* hasn't completed, and these nodes correspond exactly to a simple path of length *id* in the graph, from the initial node to the one currently being visited. To *visit* a node, we simply visit all unmarked adjacent nodes (marked ones would not correspond to a simple path). The recursive procedure checks all simple paths in the graph that start at the initial node.

Figure 44.2 shows the order in which paths are checked by the above procedure for the example graph in Figure 44.1. Each node in the tree corresponds to a call of *visit*: thus the descendants of each node are adjacent nodes that are unmarked at the time of the call. Each path in the tree from a node to the root corresponds to a simple path in the graph. Thus, the first path checked is A B C E F D. At this

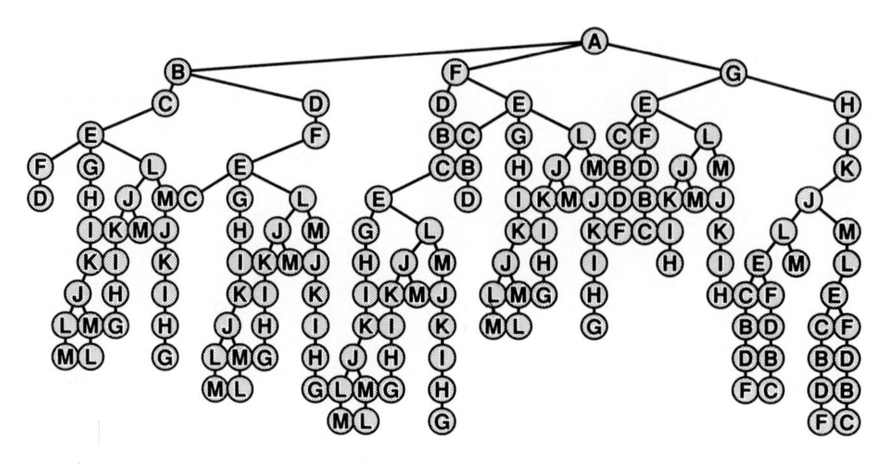

Figure 44.2 Exhaustive search.

point all vertices adjacent to D are marked (have non-zero *val* entries), so *visit* for D unmarks D and returns. Then *visit* for F unmarks F and returns. Then *visit* for E tries G which tries H, etc., eventually leading to the path A B C E G H I K J L M. Note carefully that in depth-first search nodes remain marked after they are visited, but in exhaustive search nodes are visited many times. The "unmarking" of the nodes makes exhaustive search different from depth-first search in an essential way (even though the code is rather similar)—the reader should be sure to understand the distinction.

As mentioned above, *id* is the current length of the path being tried and $val[k]$ is the position of node k on that path. Thus we can make the *visit* procedure above test for the existence of a Hamilton cycle by having it test whether there is an edge from k to 1 when $val[k]=V$. In the example above, there is only one Hamilton cycle, which appears twice in the tree, traversed in both directions (there are three other paths of length V). The program can be made to solve the traveling-salesman problem by keeping track of the length of the current path in the *val* array, then keeping track of the minimum of the lengths of the Hamilton cycles found.

Backtracking

The time taken by the exhaustive search procedure given above is proportional to the number of calls to *visit*, which is the number of nodes in the exhaustive search tree. For large graphs, this will clearly be very large. For example, if the graph is complete (every node connected to every other node), then there are $V!$ simple cycles, one corresponding to each arrangement of the nodes. (This case is studied in more detail below.) Even for the graph of Figure 44.1, it is not easy for a human to find a Hamilton cycle by inspection, so we seek a more efficient way to do it by computer.

Next we'll examine techniques to greatly reduce the number of possibilities tried. All of these techniques involve adding tests to *visit* to discover that recursive calls should not be made for certain nodes. This corresponds to *pruning* the exhaustive search tree: cutting certain branches and deleting everything connected to them.

One important pruning technique is removal of symmetries. The value of this technique is manifested in the above example by the fact that each cycle is traversed in both directions, and thus is found twice. In this case, we can ensure that we find each cycle just once by insisting that three particular nodes appear in a particular order. For example, if we insist that node C appear after node A but before node B, then we don't have to call *visit* for node B unless node C is already on the path. This leads to the drastically smaller tree shown in Figure 44.3.

This technique is not always applicable. Suppose, for example, that we're trying to find a path (not necessarily a cycle) connecting all the vertices. Now the above scheme can't be used, since we can't know in advance whether a path will lead to a cycle or not.

Each time we cut off the search at a node, we avoid searching the entire subtree below that node. For very large trees, this is a very substantial savings. Indeed, the savings is so significant that it is worthwhile to do as much as possible within *visit* to avoid making recursive calls. There are several ways to proceed for our example: one is to notice that some paths might divide the graph in such a way that the unmarked nodes aren't connected, so no cycle could be found. For example, there can't be any simple path starting with ABE in Figure 44.1, since that path

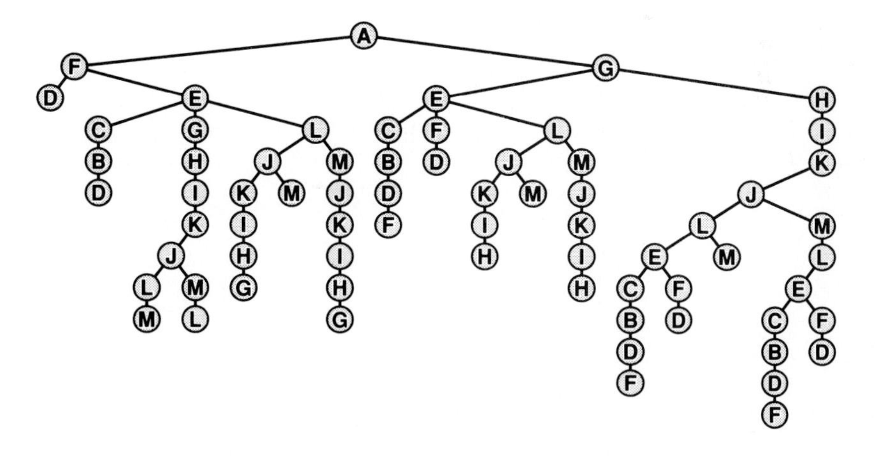

Figure 44.3 Search for cycle with A before C before B.

separates B, C, and D from the rest of the graph. For the cost of one depth-first search to discover this fact, we can avoid 25 recursive calls on *visit* (see Figure 44.3).

Figure 44.4 shows the search tree that results when this rule is applied to the tree of Figure 44.3. Again the tree is drastically smaller: it has only 19 nodes compared to 153 in the full exhaustive search tree (Figure 44.2). It is important to note that the savings achieved for this toy problem is only indicative of the situation for larger problems. A cutoff high in the tree can lead to truly significant savings; missing an obvious cutoff can lead to truly significant waste.

The general procedure described above of solving a problem by systematically generating all possible solutions is called *backtracking*. Whenever partial solutions to a problem can be successively augmented in many ways to produce a complete solution, a recursive implementation like the program above may be appropriate. As above, the process can be described by an exhaustive search tree whose nodes correspond to the partial solutions. Going down in the tree corresponds to progress towards a more complete solution; going up in the tree corresponds to "backtracking" to some previously generated partial solution, from which point it might be worthwhile to proceed forwards again.

For another example, consider the knapsack problem of Chapter 42. As mentioned there, this problem is considerably more difficult when the values need not be integers. For this problem, the partial solutions are clearly some selection of items for the knapsack, and backtracking corresponds to removing an item to try some other combination. Pruning the search tree by removing symmetries is quite effective for this problem, since the order in which objects are put into the knapsack doesn't affect the cost.

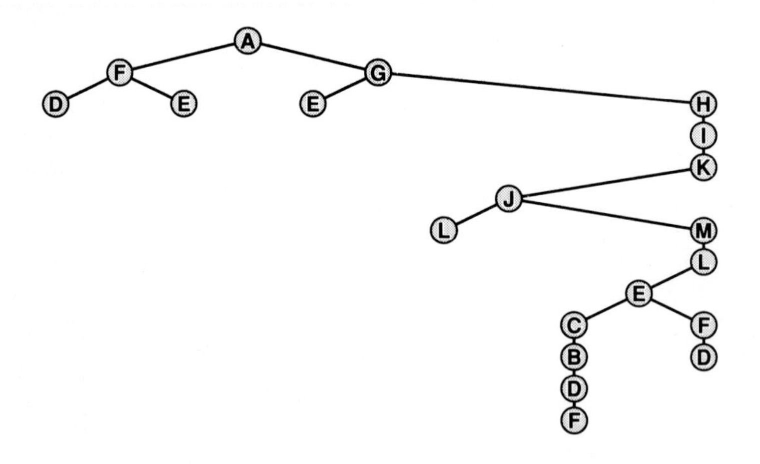

Figure 44.4 Search for a cycle with cutoff when graph is split.

When searching for the *best* path (the traveling-salesman problem), another important pruning technique is available that terminates the search as soon as it is determined that it can't possibly be successful. Suppose a path of cost x through the graph has been found. Then it's fruitless to continue along any path for which the cost so far is greater than x. This can be implemented simply by making no recursive calls in *visit* if the cost of the current partial path is greater than the cost of the best full path found so far. We clearly can't miss the minimum-cost path by adhering to such a policy.

The pruning is more effective if a low-cost path is found early in the search; one way to make this more likely is to *visit* the nodes adjacent to the current node in order of increasing cost. In fact, we can do even better: often, we can compute a bound on the cost of all full paths that begin with a given partial path. For our example, we can get a much better bound on the cost of any full path which starts with the partial path made up of the marked nodes by adding the cost of the minimum spanning tree of the unmarked nodes. (The rest of the path is a spanning tree for the unmarked nodes; its cost will certainly not be lower than the cost of the minimum spanning tree of those nodes.)

This general technique of calculating bounds on partial solutions in order to limit the number of full solutions needing to be examined is sometimes called *branch-and-bound*. Of course, it applies whenever costs are attached to paths (and we are trying to minimize costs). Normally, our aim in such problems is to cut down on a massive number of possibilities in the search for a solution—it is not unusual to apply dozens of heuristic rules like the one described in the previous paragraphs to avoid going down the wrong path.

Backtracking and branch-and-bound are quite widely applicable as general problem-solving techniques. For example, they form the basis for many programs that play games such as chess or checkers. In this case, a partial solution is some legal positioning of all the pieces on the board, and the descendant of a node in the exhaustive search tree is a position that can be the result of some legal move. Ideally, it would be best if a program could exhaustively search through all possibilities and choose a move that leads to a win no matter what the opponent does, but there are normally far too many possibilities to do this, so a backtracking search is typically done with quite sophisticated pruning rules so that only "interesting" positions are examined. Exhaustive search techniques are also used for other applications in artificial intelligence.

In the next chapter we'll see several other problems similar to those we've been studying that can be attacked using these techniques. Solving a particular problem involves the development of sophisticated criteria that can be used to limit the search. We've given only a few examples of the many techniques that have been tried for the traveling salesman problem, and equally sophisticated methods have been developed for other important problems.

However sophisticated the criteria, though, it is generally true that the running time of backtracking algorithms remains exponential. Roughly, if each node in the search tree has α children, on the average, and the length of the solution path is N, then we expect the number of nodes in the tree to be proportional to α^N. Different backtracking rules correspond to reducing the value of α, the number of choices to try at each node. It is worthwhile to expend effort to do this because a reduction in α increases the size of the problem that can be solved. For example, an algorithm which runs in time proportional to 1.1^N can solve a problem perhaps eight times a large as one which runs in time proportional to 2^N. On the other hand, as we have mentioned, neither can do well on very large problems.

Digression: Permutation Generation

It is an interesting computational puzzle to write a program that generates all possible ways of rearranging N distinct items. A simple program for this *permutation generation* problem can be derived directly from the exhaustive search program for graphs that is given above. As noted above, when that program is run on a complete graph, then it must try to visit the vertices of that graph in all possible orders. We get all permutations by keeping the vertices labeled in the order they appear on the search path and printing out all the vertex labels every time we have a path of length V, as in the following program:

```
procedure visit(k: integer);
  var t: integer;
  begin
  id:=id+1; val[k]:=id;
  if id=V then writeperm;
  for t:=1 to V do
    if val[t]=0 then visit(t);
  id:=id-1; val[k]:=0
  end;
```

This program is derived from the procedure above by eliminating all reference to the adjacency matrix (since all edges are present in a complete graph). The procedure *writeperm* simply writes out the entries of the *val* array. This is done each time *id=V*, corresponding to the discovery of a complete path in the graph. (Actually, the program can be improved somewhat by omitting the **for** loop when *id=V*, since at that point is known that all the *val* entries are nonzero.) To print out all permutations of the integers 1 through N, we invoke this procedure with the call *visit(0)* with *id* initialized to -1 and the *val* array initialized to zero. This corresponds to introducing a dummy node to the complete graph, and checking all paths in the graph starting with node 0. When invoked in this way for *N=4*, this procedure produces the following output (printed here in two columns):

```
1 2 3 4        2 3 1 4
1 2 4 3        2 4 1 3
1 3 2 4        3 2 1 4
1 4 2 3        4 2 1 3
1 3 4 2        3 4 1 2
1 4 3 2        4 3 1 2
2 1 3 4        2 3 4 1
2 1 4 3        2 4 3 1
3 1 2 4        3 2 4 1
4 1 2 3        4 2 3 1
3 1 4 2        3 4 2 1
4 1 3 2        4 3 2 1
```

Admittedly, the interpretation of the procedure as generating paths in a complete graph is barely visible. But a direct examination of the procedure reveals that it generates all $N!$ permutations of the integers 1 to N by first generating all $(N - 1)!$ permutations with the 1 in the first position (calling itself recursively to place 2 through N), then generating the $(N - 1)!$ permutations with the 1 in the second position, etc.

Now, it would be unthinkable to use this program even for $N = 16$, because $16! > 2^{50}$. Still, it's an important program to study because it can form the basis of a backtracking program to solve any problem involving reordering a set of elements.

For example, consider the *Euclidean traveling-salesman problem*: given a set of N points in the plane, find the shortest tour that connects them all. Since each ordering of the points corresponds to a legal tour, the above program can be made to exhaustively search for the solution to this problem simply by changing it to keep track of the cost of each tour and the minimum of the costs of the full tours, just as above. Then the same branch-and-bound technique as used above can be applied, as well as various backtracking heuristics specific to the Euclidean problem. (For example, it is easy to prove that the optimal tour cannot cross itself, so the search can be cut off on all partial paths that cross themselves.) Different search heuristics correspond to different ways of ordering the permutations. Such techniques can save an enormous amount of work but always leave an enormous amount of work to be done. It is not at all simple to find an exact solution to the Euclidean traveling salesman problem, even for N as low as 16.

Another reason that permutation generation is of interest is that there are a number of related procedures for generating other combinatorial objects. In some cases, the number of objects generated are not quite so numerous as are permutations, and such procedures can thus be useful in practice for larger N. An example is a procedure to generate all ways of choosing a subset of size k out of a set

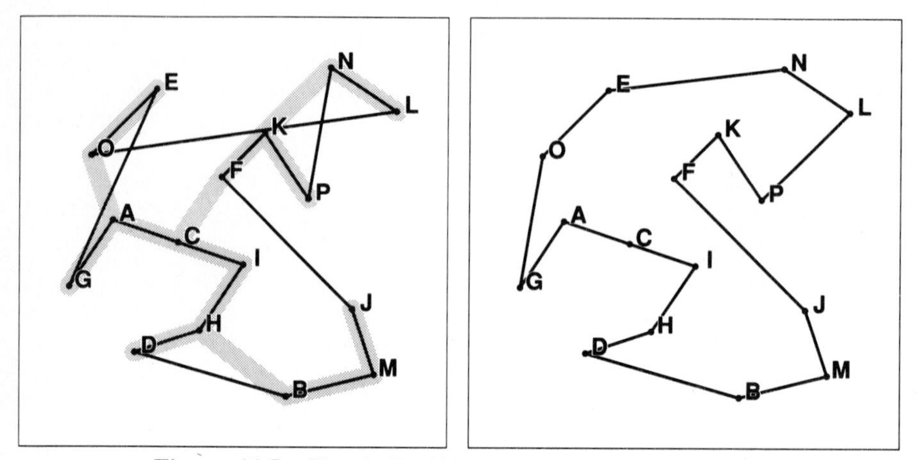

Figure 44.5 Simple Euclidean traveling-salesman tour.

of N items. For large N and small k, the number of ways to do this is roughly proportional to N^k. Such a procedure could be used as the basis for a backtracking program to solve the knapsack problem.

Approximation Algorithms

Since finding the shortest tour seems to require so much computation, it is reasonable to consider whether it might be easier to find a tour that is almost as short as the shortest. If we're willing to relax the restriction that we absolutely must have the shortest possible path, then it turns out that we can deal with problems much larger than is possible with the techniques above.

For example, it's relatively easy to find a tour that is longer by at most a factor of two than the optimal tour. The method is based on simply finding the minimum spanning tree: this not only provides a lower bound on the length of the tour, as mentioned above, but also turns out to provide an *upper bound* on the length of the tour, as follows. Given a minimum spanning tree, produce a tour by visiting the nodes of the minimum spanning tree using the following procedure: to process node x, visit x, then visit each child of x, applying this visiting procedure recursively and returning to node x after each child has been visited, and ending up at node x. This tour traverses every edge in the spanning tree twice, so its cost is twice the cost of the tree. It is not a simple tour, since a node may be visited many times, but it can be converted to a simple tour simply by deleting all but the first occurrence of each node. Deleting an occurrence of a node corresponds to taking a shortcut past that node: certainly it can't increase the cost of the tour. Thus, we have a simple tour which has a cost less than twice that of the minimum spanning tree.

For example, the diagram on the left in Figure 44.5 shows a minimum spanning tree for our set of sample points (computed as described in Chapter 31), along with a corresponding simple tour. This tour is clearly not the optimum, because it self-intersects. For a large random point set, it seems likely that the tour produced in this way will be close to the optimum, though no analysis has been done to support this conclusion.

Another approach that has been tried is to develop techniques to improve an existing tour in the hope that a short tour can be found by applying such improvements repeatedly. For example, in the *Euclidean* traveling-salesman problem, where graph distances are distances between points in the plane, a self-intersecting tour can be improved by removing each intersection as follows. If the line AB intersects the line CD, then a shorter tour is formed by deleting AB and CD and adding AD and CB. Applying this procedure successively will, given any tour, produce a tour that is no longer and which is not self-intersecting. For example, the procedure applied to the tour produced from the minimum spanning tree (on the left in Figure 44.5) gives the shorter tour on the right in Figure 44.5.

In fact, one of the most effective ways of getting approximate solutions to the Euclidean traveling-salesman problem, developed by S. Lin, is to generalize the procedure above to improve tours by switching around three or more edges in an existing tour. Very good results have been obtained by applying such a procedure successively to an initially *random* tour until it no longer leads to an improvement. One might think that it would be better to start with a tour already close to the optimum, but Lin's studies indicate that this may not be the case.

The various approaches to producing approximate solutions to the traveling salesman problem described above are merely indicative of the types of techniques that can be used in order to avoid exhaustive search. The brief descriptions above do not do justice to the many ingenious ideas that have been developed: the formulation and analysis of algorithms of this type is still a quite active area of research in computer science.

One might legitimately question why the traveling-salesman problem and the other problems we've been alluding to *require* exhaustive search. Couldn't a clever algorithm find the minimal tour as easily and quickly as we can find the minimum spanning tree? In the next chapter we'll see why most computer scientists believe that there is no such algorithm and why approximation algorithms of the type discussed in this section must therefore be studied.

□

Exercises

1. Which would you prefer to use, an algorithm requiring N^5 steps or one requiring 2^N steps?
2. Does the "maze" graph in Chapter 29 have a Hamilton cycle?
3. Draw the tree corresponding to Figure 44.4 when you are looking for a Hamilton cycle on the sample graph starting at vertex B instead of vertex A.
4. How long could exhaustive search take to find a Hamilton cycle in a graph in which all nodes are connected to exactly two other nodes? Answer the same question for the case where all nodes are connected to exactly three other nodes.
5. How many calls to *visit* are made (as a function of V) by the permutation generation procedure?
6. Derive a nonrecursive permutation generation procedure from the program given.
7. Write a program which determines whether or not two given adjacency matrices represent the same graph, except with different vertex names.
8. Write a program to solve the knapsack problem of Chapter 42 when the sizes can be real numbers.
9. Write a program to count the number of spanning trees of a set of N points in the plane with no intersecting edges.
10. Solve the Euclidean traveling-salesman problem for our sixteen sample points.

45

NP-Complete Problems

☐ The algorithms we've studied in this book generally are used to solve practical problems and therefore consume reasonable amounts of resources. The practical utility of most of the algorithms is obvious: for many problems we have the luxury of several efficient algorithms to choose from. Unfortunately, as pointed out in the previous chapter, many problems arise in practice which do not admit such efficient solutions. What's worse, for a large class of such problems we can't even tell whether or not an efficient solution might exist.

This state of affairs has been a source of extreme frustration for programmers and algorithm designers, who can't find any efficient algorithm for a wide range of practical problems, and for theoreticians, who have been unable to find any reason why these problems should be difficult. A great deal of research has been done in this area and has led to the development of mechanisms by which new problems can be classified as being "as difficult as" old problems in a particular technical sense. Though much of this work is beyond the scope of this book, the central ideas are not difficult to learn. It is certainly useful when faced with a new problem to have some appreciation for the types of problems for which no one knows any efficient algorithm.

Sometimes the line between "easy" and "hard" problems is a fine one. For example, we saw an efficient algorithm in Chapter 31 for the following problem: "Find the shortest path from vertex x to vertex y in a given weighted graph." But if we ask for the *longest* path (without cycles) from x to y, we have a problem for which no one knows a solution substantially better than checking all possible paths. The fine line is even more striking when we consider similar problems that ask for only "yes-no" answers:

Easy: Is there a path from x to y with weight $\leq M$?
Hard(?): Is there a path from x to y with weight $\geq M$?

Breadth-first search yields to a solution for the first problem in linear time, but all known algorithms for the second problem could take exponential time.

We can be much more precise than "could take exponential time," but that will not be necessary for the present discussion. Generally, it is useful to think of an exponential-time algorithm as one which, for some input of size N, takes time proportional to 2^N (at least). (The substance of the results that we're about to discuss is not changed if 2 is replaced by any number $\alpha > 1$.) This means, for example, that an exponential-time algorithm cannot be guaranteed to work for all problems of size 100 (say) or greater, because no one can wait for an algorithm to take 2^{100} steps, regardless of the speed of the computer. Exponential growth dwarfs technological changes: a supercomputer may be a trillion times faster than an abacus, but neither can come close to solving a problem that requires 2^{100} steps.

Deterministic and Nondeterministic Polynomial-Time Algorithms

The great disparity in performance between "efficient" algorithms of the type we've been studying and brute-force "exponential" algorithms that check each possibility makes it possible to study the interface between them with a simple formal model. In this model, the efficiency of an algorithm is a function of the number of bits used to encode the input, using a "reasonable" encoding scheme. (The precise definition of "reasonable" includes all common methods of encoding things for computers: an example of an unreasonable coding scheme is unary, where M bits are used to represent the number M. Rather, we would expect that the number of bits used to represent the number M should be proportional to $\log M$.) We're interested merely in identifying algorithms guaranteed to run in time proportional to some polynomial in the number of bits of input. Any problem that can be solved by such an algorithm is said to belong to

P: the set of all problems that can be solved by deterministic algorithms in polynomial time.

By *deterministic* we mean that at any time, whatever the algorithm is doing, there is only one thing that it could do next. This very general notion covers the way that programs run on actual computers. Note that the polynomial is not specified at all and that this definition certainly covers the standard algorithms we've studied so far. Sorting belongs to P because (for example) insertion sort runs in time proportional to N^2 (the existence of $N \log N$ sorting algorithms is not relevant in the present context). Also, the time taken by an algorithm obviously depends on the computer used, but it turns out that using a different computer affects the running time by only a polynomial factor (again, assuming reasonable limits), so that this also is not particularly relevant to the present discussion.

Of course, the theoretical results we're discussing are based on a completely specified model of computation within which the general statements we're making here can be proved. Our intent is to examine some of the central ideas, not to develop rigorous definitions and theorem statements. The reader may be assured that any apparent logical flaws are due to the informal nature of the description, not the theory itself.

One "unreasonable" way to extend the power of a computer is to endow it with the power of *nondeterminism*: to assert that when an algorithm is faced with a choice of several options, it has the power to "guess" the right one. For the purposes of the discussion below, we can think of an algorithm for a nondeterministic machine as "guessing" the solution to a problem, then verifying that the solution is correct. In Chapter 20, we saw how nondeterminism can be useful as a tool for algorithm design; here we use it as a theoretical device to help classify problems. We have

NP: the set of all problems that can be solved by nondeterministic algorithms in polynomial time.

Obviously, any problem in P is also in NP. But it seems that there should be many other problems in NP: to show that a problem is in NP, we need only find a polynomial-time algorithm to check that a given solution (the guessed solution) is valid. For example, the "yes-no" version of the longest-path problem is in NP. Another example of a problem in NP is the *satisfiability* problem. Given a logical formula of the form

$$(x_1 + x_3 + x_5) * (x_1 + \overline{x}_2 + x_4) * (\overline{x}_3 + x_4 + x_5) * (x_2 + \overline{x}_3 + x_5)$$

where the x_i's represent Boolean variables (**true** or **false**), "+" represents **or**, "*" represents **and**, and $\overline{x}$ represents **not**, the satisfiability problem is to determine whether or not there exists an assignment of truth values to the variables that makes the formula **true** ("satisfies" it). We'll see below that this particular problem plays a special role in the theory.

Nondeterminism is such a powerful operation that it seems almost absurd to consider it seriously. Why bother considering an imaginary tool that makes difficult problems seem trivial? The answer is that, powerful as nondeterminism may seem, no one has been able to *prove* that it helps for any particular problem! Put another way, no one has been able to find a single problem that can be proven to be in NP but not in P (or even prove that one exists): we do not know whether or not P = NP. This is a quite frustrating situation because many important practical problems belong to NP (they could be solved efficiently on a non-deterministic machine) but may or may not belong to P (we don't know any efficient algorithms for them on a deterministic machine). If we could prove that a problem doesn't belong to P, then we could abandon the search for an efficient solution to it. In the absence of such a proof, there is the possibility that some efficient algorithm has gone undiscovered. In fact, given the current state of our knowledge, there could be some efficient algorithm for *every* problem in NP, which would imply that many efficient algorithms have gone undiscovered. Virtually no one believes that P = NP, and a considerable amount of effort has gone into proving the contrary, but this remains the outstanding open research problem in computer science.

NP-Completeness

Below we'll look at a list of problems that are known to belong to NP but might or might not belong to P. That is, they are easy to solve on a non-deterministic machine but, despite considerable effort, no one has been able to find an efficient algorithm on a conventional machine (or prove that none exists) for any of them. These problems have an additional property that provides convincing evidence that $P \neq NP$: if *any* of the problems can be solved in polynomial time on a deterministic machine, then so can *all* problems in NP (i.e., P = NP). That is, the collective failure of all researchers to find efficient algorithms for all of these problems might be viewed as a collective failure to prove that P = NP. Such problems are said to be *NP-complete*. It turns out that a large number of interesting practical problems have this characteristic.

The primary tool used to prove that problems are NP-complete employs the idea of *polynomial reducibility*. We show that any algorithm to solve a new problem in NP can be used to solve some known NP-complete problem by the following process: transform any instance of the known NP-complete problem to an instance of the new problem, solve the problem using the given algorithm, then transform the solution back to a solution of the NP-complete problem. We saw an example of a similar process in Chapter 34, where we reduced bipartite matching to network flow. By "polynomially reducible," we mean that the transformations can be done in polynomial time: thus the existence of a polynomial-time algorithm for the new problem would imply the existence of a polynomial-time algorithm for the NP-complete problem, and this would (by definition) imply the existence of polynomial-time algorithms for all problems in NP.

The concept of reduction provides a useful mechanism for classifying algorithms. For example, to prove that a problem in NP is NP-complete, we need only show that some known NP-complete problem is polynomially reducible to it: that is, that a polynomial-time algorithm for the new problem can be used to solve the NP-complete problem, and then can, in turn, be used to solve all problems in NP. For an example of reduction, recall the following two problems from Chapter 44:

> TRAVELING SALESMAN: Given a set of cities and distances between all pairs, find a tour of all the cities of distance less than M.
>
> HAMILTON CYCLE: Given a graph, find a simple cycle that includes all the vertices.

Suppose we know the Hamilton-cycle problem to be NP-complete and wish to determine whether or not the traveling-salesman problem is also NP-complete. Any algorithm for solving the traveling-salesman problem can be used to solve the Hamilton-cycle problem, through the following reduction: given an instance of the Hamilton-cycle problem (a graph) construct an instance of the traveling-salesman problem (a set of cities, with distances between all pairs) as follows: for cities for the traveling salesman use the set of vertices in the graph; for distances between each pair of cities use one if there is an edge between the corresponding vertices in the graph, two if there is no edge. Then have the algorithm for the traveling salesman problem find a tour of distance less than or equal to N, the number of vertices in the graph. That tour must correspond precisely to a Hamilton cycle. An efficient algorithm for the traveling salesman problem would also be an efficient algorithm for the Hamilton cycle problem. That is, the Hamilton-cycle problem reduces to the traveling-salesman problem, so the NP-completeness of the Hamilton-cycle problem implies the NP-completeness of the traveling-salesman problem.

The reduction of the Hamilton-cycle problem to the traveling-salesman problem is relatively simple because the problems are so similar. Actually, polynomial-time reductions can be quite complicated indeed and can connect problems which seem to be quite dissimilar. For example, it is possible to reduce the satisfiability problem to the Hamilton cycle problem. Without going into details, let us look at a sketch of the proof.

We wish to show that if we had a polynomial-time solution to the Hamilton cycle problem, then we could get a polynomial-time solution to the satisfiability problem by polynomial reduction. The proof consists of a detailed method of construction showing how, given an instance of the satisfiability problem (a Boolean formula), construct (in polynomial time) an instance of the Hamilton-cycle problem (a graph) with the property that knowing whether the graph has a Hamilton cycle tells us whether the formula is satisfiable. The graph is built from small components (corresponding to the variables) that can be traversed by a simple path in only one of two ways (corresponding to the truth or falsity of the variables). These small components are attached together as specified by the clauses, using more complicated subgraphs that can be traversed by simple paths corresponding to the truth or falsity of the clauses. It is quite a large step from this brief description to the full construction, and development of rigorous detailed proofs of this type is rather a challenge (though experts in the field do get to be rather adept at building such "gadgets" for reductions). Our point here is to illustrate that polynomial reduction can be applied to quite dissimilar problems.

Thus, if we were to have a polynomial-time algorithm for the traveling-salesman problem, then we would also have a polynomial-time algorithm for the Hamilton-cycle problem, which would also give us a polynomial-time algorithm for the satisfiability problem. Each problem that is proven NP-complete provides another potential basis for proving yet another future problem NP-complete. The proof might be as simple as the reduction given above from the Hamilton-cycle problem to the traveling-salesman problem, or as complicated as the transformation sketched above from the satisfiability problem to the Hamilton cycle problem, or somewhere in between. Literally thousands of problems from a wide variety of applications areas have been proven to be NP-complete by transforming one to another in this way.

Cook's Theorem

Reduction uses the NP-completeness of one problem to imply the NP-completeness of another. There is one case, though in which reduction can't be used: how was the *first* problem proven to be NP-complete? This was done by S. A. Cook in 1971. Cook gave a direct proof that satisfiability is NP-complete: that if there is a polynomial time algorithm for satisfiability, then all problems in NP can be solved in polynomial time.

The proof is extremely complicated but the general method can be explained. First, a full mathematical definition of a machine capable of solving any problem in NP is developed. This is a simple model of a general-purpose computer known as a *Turing machine* that can read inputs, perform certain operations, and write outputs. A Turing machine can perform any computation that any other general-purpose computer can, using the same amount of time (to within a polynomial factor), and it has the additional advantage that it can be concisely described mathematically. Endowed with the additional power of nondeterminism, a Turing machine can solve any problem in NP. The next step in the proof is to describe each feature of the machine, including how instructions are executed, in terms of logical formulas such as appear in the satisfiability problem. In this way a correspondence is established between every problem in NP (which can be expressed as a program on the nondeterministic Turing machine) and some instance of satisfiability (the translation of that program into a logical formula). Now, the solution to the satisfiability problem essentially corresponds to a simulation of the machine running the given program on the given input, so it produces a solution to an instance of the given problem. Further details of this proof are well beyond the scope of this book. Fortunately, only one such proof is really necessary: it is much easier to use reduction to prove NP-completeness.

Some NP-Complete Problems

As mentioned above, literally thousands of diverse problems are known to be NP-complete. In this section, we list a few in order to illustrate the wide range of problems that have been studied. Of course, the list begins with *satisfiability* and includes *traveling-salesman* and *Hamilton-cycle*, as well as *longest-path*. The following additional problems are representative:

PARTITION: Given a set of integers, can they be divided into two sets whose sum is equal?

INTEGER LINEAR PROGRAMMING: Given a linear program, is there a solution in integers?

MULTIPROCESSOR SCHEDULING: Given a deadline and a set of tasks of varying length to be performed on two identical processors, can the tasks be arranged so that the deadline is met?

VERTEX COVER: Given a graph and an integer N, is there a set of fewer than N vertices which touches all the edges?

These and many related problems have important natural practical applications, and there has been strong motivation for some time to find good algorithms to solve them. The fact that no good algorithm has been found for any of these problems is surely strong evidence that $P \neq NP$, and most researchers certainly believe this to be the case. (On the other hand, the fact that no one has been able to prove that any of these problems do not belong to P could be construed to comprise a similar body of circumstantial evidence on the other side.) Whether or not P = NP, the practical fact is that we have at present no algorithms guaranteed to solve any of the NP-complete problems efficiently.

As indicated in the previous chapter, several techniques have been developed to cope with this situation, since some sort of solution to these various problems must be found in practice. One approach is to change the problem and find an "approximation" algorithm that finds not the best solution but a solution guaranteed to be close to the best. (Unfortunately, this is sometimes not sufficient to ward off NP-completeness.) Another approach is to rely on "average-time" performance and develop an algorithm that finds the solution in some cases, but doesn't necessarily work in all cases. That is, while it may not be possible to find an algorithm guaranteed to work well on all instances of a problem, it may well be possible to solve efficiently virtually all of the instances that arise in practice. A third approach is to work with "efficient" exponential algorithms, using the backtracking techniques described in the previous chapter. Finally, there is quite a large gap between polynomial and exponential time which is not addressed by the theory. What about an algorithm that runs in time proportional to $N^{\log N}$ or $2^{\sqrt{N}}$?

All the application areas we've studied in this book are touched by NP-completeness: there are NP-complete problems in numerical applications, in sorting and searching, in string processing, in geometry, and in graph processing. The most important practical contribution of the theory of NP-completeness is that it provides a mechanism to discover whether a new problem from any of these diverse areas is "easy" or "hard." If one can find an efficient algorithm to solve a new problem, then there is no difficulty. If not, a proof that the problem is NP-complete at least tells us that developing an efficient algorithm would be a stunning achievement (and suggests that a different approach should perhaps be tried). The scores of efficient algorithms we've examined in this book are testimony that we have learned a great deal about efficient computational methods since Euclid, but the theory of NP-completeness shows that, indeed, we still have a great deal to learn.

□

Exercises

1. Write a program to find the longest simple path from x to y in a given weighted graph.
2. Could there be an algorithm which solves an NP-complete problem in an *average* time of $N \log N$, if P $\neq$ NP? Explain your answer.
3. Give a nondeterministic polynomial-time algorithm for solving the partition problem.
4. Is there an immediate polynomial-time reduction from the traveling-salesman problem on graphs to the Euclidean traveling-salesman problem, or vice versa?
5. What would be the significance of a program that could solve the traveling salesman problem in time proportional to 1.1^N?
6. Is the logical formula given in the text satisfiable?
7. Could one of the "algorithm machines" with full parallelism be used to solve an NP-complete problem in polynomial time, if P $\neq$ NP? Explain your answer.
8. How does the problem "compute the exact value of 2^N" fit into the P–NP classification scheme?
9. Prove that the problem of finding a Hamilton cycle in a *directed* graph is NP-complete, using the NP-completeness of the Hamilton-cycle problem for undirected graphs.
10. Suppose that two problems are known to be NP-complete. Does this imply that there is a polynomial-time reduction from one to the other, if $P \neq NP$?

SOURCES for Advanced Topics

Each of the topics covered in this section is the subject of volumes of reference material. The reader seeking more information should anticipate engaging in serious study; we'll only be able to indicate some basic references here.

The perfect-shuffle machine of Chapter 40 is described in the 1971 paper by Stone, which covers many other applications. One place to look for more information on systolic arrays is the chapter by Kung and Leiserson in Mead and Conway's book on VLSI. The standard text on digital signal processing and the FFT is the book by Oppenheim and Schafer. Further information on dynamic programming (and topics from other chapters) may be found in the book by Hu. Our treatment of linear programming in Chapter 43 is based on the excellent treatment in the book by Papadimitriou and Steiglitz, where all the intuitive arguments are backed up by full mathematical proofs. Further information on exhaustive search techniques may be found in the books by Wells and by Reingold, Nievergelt, and Deo. Finally, the reader interested in more information on NP-completeness may consult the survey article by Lewis and Papadimitriou and the book by Garey and Johnson, which has a full description of various types of NP-completeness and a categorized listing of hundreds of NP-complete problems.

M. R. Garey and D. S. Johnson, *Computers and Intractability: a Guide to the Theory of NP-Completeness*, Freeman, San Francisco, CA, 1979.

T. C. Hu, *Combinatorial Algorithms*, Addison-Wesley, Reading, MA, 1982.

H. R. Lewis and C. H. Papadimitriou, "The efficiency of algorithms," *Scientific American*, 238, 1 (1978).

C. A. Mead and L. C. Conway, *Introduction to VLSI Design*, Addison-Wesley, Reading, MA, 1980.

C. H. Papadimitriou and K. Steiglitz, *Combinatorial Optimization: Algorithms and Complexity*, Prentice-Hall, Englewood Cliffs, NJ, 1982.

E. M. Reingold, J. Nievergelt, and N. Deo, *Combinatorial Algorithms: Theory and Practice*, Prentice-Hall, Englewood Cliffs, NJ, 1982.

A. V. Oppenheim and R. W. Schafer, *Digital Signal Processing*, Prentice-Hall, Englewood Cliffs, NJ, 1975.

H. S. Stone, "Parallel processing with the perfect shuffle," *IEEE Transactions on Computing*, **C-20**, 2 (February, 1971).

M. B. Wells, *Elements of Combinatorial Computing*, Pergamon Press, Oxford, 1971.

Program Index

adapt (adaptive quadrature), 562.
add (polynomial addition), 523.
adjlist (build adjacency structure for a graph), 421.
adjmatrix (build adjacency matrix for a graph), 420.
binarysearch (binary search in a sorted file), 198.
(biconnected components of a graph), 440.
bits (extract bits from an integer), 134, 135, 140, 246, 247, 254, 256, 329.
brutesearch (brute-force string searching), 279.
bstdelete (binary search tree deletion, indirect), 211, 395.
bstinitialize (initialize binary search tree, indirect), 211, 394.
bstinsert (binary search tree insertion, indirect), 211, 394.
bubble (bubble sort), 101.
buildytree (sort for line intersection algorithm), 394.
ccw (test orientation of three points in the plane), 350, 351, 368.
change (change priority in priority queue), 147.
check (distance check for closest-point algorithm), 404.
chisquare (chi-square test for randomness), 517.
construct (construct a priority queue), 146.
(convert infix to postfix), 28.
deletenext (delete from a linked list), 20, 23.
delete (remove an item from a priority queue), 147.
digitalinsert (insertion into a digital search tree), 247.
digitalsearch (search in a digital search tree), 246.
(distribution counting sort), 111, 140.
(double hashing), 239.
downheap (top-down heap repair), 152, 153, 156.
eliminate (forward elimination phase of Gaussian elimination), 539.
euclid (find the gcd of two integers), 8,12.
eval (evaluate a spline), 550.
eval (polynomial evaluation for the FFT), 591.
(evaluate postfix expression), 28.
expression (top-down parsing), 308, 313.
factorial (compute factorial function), 52.
factor (top-down parsing), 309, 314.
fibonacci (compute Fibonacci numbers), 52, 53.
findinit (initialize for equivalence testing), 446, 460.
find (equivalence testing), 444, 446, 460.
get (from a queue), 30, 48, 431.

grahamscan (find the convex hull with the Graham scan), 368.
gridrange (range searching, using grid method), 379.
(hashing with separate chaining), 234.
hashinitialize (create an empty hash table), 237.
hashinsert (insert into a hash table), 237.
heapsort (Heapsort), 153.
initnext (initialization for Knuth-Morris-Pratt string searching), 283.
insertafter (insert into a linked list), 20, 23.
insertion (insertion sort), 100, 105.
insert (insert new item into a priority queue): 147 (unordered list), 154 (heap).
insiderect (test if point is inside rectangle), 375, 379, 384.
inside (test if point is inside polygon), 354.
insitu (permute an array inplace), 106.
intersect (line segment intersection test), 351, 354.
intrect (quadrature, rectangle rule), 557.
intsimp (quadrature, Simpson's rule), 561.
inttrap (quadrature, trapezoid rule), 559.
josephus (simulation with linked lists for the Josephus problem), 22.
kmpsearch (Knuth-Morris-Pratt string searching), 282.
(knapsack problem, dynamic programming solution), 596.
kruskal (compute minimum spanning tree using Kruskal's algorithm), 460.
(least-squares data fitting), 553.
listadd (polynomial addition, linked list), 523.
listbfs (breadth-first search of a graph, adjacency structure), 431.
listdfs (depth-first search of a graph, adjacency structure): 423 (recursive), 428 (nonrecursive).
listinitialize (initialize a linked list), 20, 23.
listinsert (insert after sequential search, linked list implementation), 197.
listpfs (priority-first search of a graph, adjacency structure), 455, 491.
listsearch (sequential search, linked list implementation), 196.
makespline (compute a cubic spline), 549.
match (general regular-expression pattern matching), 301, 315.
matchall (general regular expression pattern matching), 315.
(matrix chain product problem, dynamic programming solution), 600.
(matrix multiplication), 532.
matrixpfs (priority-first search of a graph, adjacency matrix), 466, 491.
(maximum flow in a network), 490–491.
(maximum matching in a bipartite graph), 499.
mergesort (recursive sort based on merging, array), 166.
mergesort (recursive sort based on merging, linked list), 168.
mergesort (nonrecursive sort based on merging, linked list), 170.
(merge two sorted arrays), 164.

merge (merge two sorted linked lists), 165.
merge (two-dimensional merge for closest-point algorithm), 404.
(minimum spanning tree in a graph), 455, 466.
mult (multiplication, avoiding overflow), 513.
mult (polynomial multiplication), 529.
(optimal binary search tree problem, dynamic programming solution), 603.
(parse postfix expression), 41.
patriciainsert (insertion into a Patricia tree), 256.
patriciasearch (search in a Patricia tree), 254.
pivot (pivoting for simplex and gaussian elimination), 617.
pop (from a pushdown stack), 27, 28, 29, 41, 45, 63–65, 122, 301, 428.
pqchange (change priority in a priority queue, indirect), 160, 455.
pqconstruct (build priority queue, indirect), 160, 460.
pqdownheap (priority queue repair, indirect), 160, 327–328.
pqinitialize (initialize priority queue, indirect), 455, 460.
pqempty, (test if priority queue is empty, indirect), 455, 460.
pqinsert (insert into priority queue, indirect), 160, 182, 455.
pqremove (remove from priority queue, indirect), 455, 460.
pqreplace (replace largest item in a priority queue, indirect), 160, 182.
pqupdate (insert or change priority in a priority queue, indirect), 455.
preprocess (build empty 2D tree), 383.
preprocess (build list structure for grid method), 378.
primes (print out prime numbers), 16.
push (onto a pushdown stack), 27, 28, 29, 41, 45, 63–65, 122, 301, 428.
put (onto a queue), 30, 48, 301, 431.
queueempty (test if queue is empty), 30, 48, 431.
queueinitialize (initialize a queue), 30, 431.
quicksort (Quicksort): 116, 118 (recursive); 122 (nonrecursive).
radixexchange (radix exchange sort), 135.
randinit (initialize additive congruential random number generator), 516.
randomint (random integers in a given range): 514 (linear congruential), 516 (additive congruential).
random (linear congruential random number generator), 513.
range (range searching, using a 2D tree), 384.
rbtreeinitialize (create an empty red-black tree), 226.
rbtreeinsert (insert into a red-black tree), 221.
remove (delete largest item from a priority queue): 147 (unordered list), 152 (heap).
replace (replace largest item in a priority queue), 153.
rotate (rotate a link in a binary tree), 225, 226.
rule (draw a ruler), 54, 57.
scan (sweep line algorithm for Manhattan line intersection), 395.
selection (selection sort), 98.

select (find the *k*th smallest), 127 (recursive), 128 (nonrecursive).
seqinsert (insert, after sequential search), 195.
seqsearch (sequential search), 195.
shellsort (Shellsort), 108.
(shortest path tree in a graph), 463, 466.
(shortest paths using Floyd's method), 477.
(simplex algorithm), 618.
sort (two-dimensional mergesort for closest-pair algorithm), 405.
sort3 (sort three elements), 95, 571.
split (split a 4-node in a red-black tree), 221, 226.
(stable marriage matching), 503.
stackempty (test if pushdown stack is empty), 27, 28, 29, 45, 63–65, 122, 428.
stackinit (initialize a pushdown stack), 27, 28, 29, 122, 428.
star (draw a fractal star), 59.
straightradix (straight radix sort), 140.
(strongly connected components of a graph), 482.
substitute (backward substitution phase of Gaussian elimination), 540.
term (top-down parsing), 309, 314.
theta (pseudo-angle computation), 353, 364, 366.
threesort (sort three elements), 95, 571.
(transitive closure using repeated depth-first search), 474.
(transitive closure using Warshall's method), 475.
traverse (tree traversal), 45, 48, 60–65.
treedelete (delete from a binary search tree), 210.
treeinitialize (create empty binary search tree), 204.
treeinsert (insert into a binary search tree), 205.
treeinsert (insertion into a 2D tree), 381.
treeprint (print keys of a binary search tree in order), 206.
treerange (range searching in one dimension), 374.
treesearch (search in a binary search tree), 203.
(tridiagonal system of simultaneous equations solution), 542.
upheap (bottom-up heap repair), 150.
visit (vertex visit for graph and tree searching):
  breadth-first search, adjacency structure, 431.
  depth-first search for articulation points, 431.
  depth-first search for strongly connected components, 431.
  exhaustive search, 623.
  nonrecursive depth-first search, adjacency structure, 428.
  recursive depth-first search, adjacency matrix, 426.
  recursive depth-first search, adjacency structure, 423.
  tree traversal methods, 60–65.
wrap (find the convex hull with package-wrapping), 364.

Index

Abacus, 634.
About, 75.
Abstract data types, 15, 31–33, 146, 154.
Abstract operations, 68.
Access methods, 259–272.
Adaptive quadrature, 561–563.
Additive congruential random number generator, 515–516.
Adjacency list; *see* Adjacency structure.
Adjacency matrix, 418–419.
Adjacency structure, 420–423.
Adleman, L., 338, 343.
Aho, A. V., 89, 343.
Algorithm, 3.
All shortest paths, 476–478.
All-nearest-neighbors problem, 408.
Almost linear, 449.
Analysis of algorithms, 4, 67–79.
Approximate analysis, 75–76.
Approximation algorithms, 630–631.
Arbitrary number, 126, 509.
Archaeology, 377.
Arithmetic expression evaluation (postfix), 26.
Arithmetic, 521–534.
  large integers, 530–532.
  polynomial, 522–530.
Arrays, 15–17, 524.
Articulation point, 439.
Asymptotic analysis, 75–76.
Average case analysis, 67, 74.
AVL trees, 229.
B-trees, 229, 262–265.
Backtracking, 368, 624–527.
Backward substitution, 537–540.
Balanced multiway merging, 178–180.
Balanced trees, 215–230, 374.
Basis variables, 614.
Batcher, K. E., 575.
Bayer, R., 262, 273.
Bentley, J. L., 89, 412.
Betterling, W. T., 565.
Biconnectivity in graphs, 438–441.
Binary search trees, 202–212, 243, 391–396.
  array representation, 211–212.
  indirect representation, 211–212.
  optimal, 602–604.
  standard representation, 202–204.
Binary search, 198–200.
Binary tree search, 202–212.
Binary trees, 37–38.
  complete, 38.
  full, 38, 40.
  properties, 39–40.
Bipartite graphs, 497–499.
Bitonic merging, 575.
Bits, 134, 135, 140, 246, 247, 254, 256, 329.
Bland, R. G., 616.
Borodin, A., 565.
Bottom-up algorithms; *see* Nonrecursive.
Boyer-Moore string searching algorithm, 286–289.
Branch-and-bound, 627.
Breadth-first search.
  priority-first search implementation, 457.
  queue implementation, 430–434.

shortest paths, 462.
Brown, M. R., 189.
Brute-force, 82.
Bubble sort, 100–101, 102, 103.

C, 12.
Caesar cipher, 335–336.
Carroll, L., 35.
Chi-square (χ^2) test, 516–518.
Circular list, 21, 22.
Closest-pair problem, 402–408.
Closest-point problems, 401–411.
Closure, 294, 298.
Clustering, 239, 241.
Combine and conquer, 58, 168.
Comer, D., 273.
Compare-exchange, 95, 571–577.
Compiler, 305, 312–315.
Compiler-compilers, 316.
Complete graph, 418.
Complete tree, 148.
Complex roots of unity, 585–591.
Computational complexity, 71–74.
Concatenation, 294, 297.
Connected components, 437–438.
Connectivity in graphs, 437–449.
Constant factors, 73.
Constant time, 69.
Context-free grammars, 306–308.
Context-sensitive grammars, 308.
Convex hull, 359.
Convex hull algorithms, 359–372.
  Graham scan, 365–371.
  interior elimination, 368–369, 371.
  k-dimensional, 364–365.
  package wrapping, 362–365, 370–371.
Convex polygon, 355, 359–372.
Conway, L. C., 641.
Cook's theorem, 638.
Cook, S. A., 278, 638.
Cross edges, 472–473.
Cryptanalysis, 333–335.
Cryptography, 333–342, 510.
Cryptology, 333–342.
Cryptosystem, 334.
Cryptovariables, 337.
Cubic running time, 70, 71.
Curve fitting, 545–554.
Cycle testing in graphs, 427.
Cycling, 615.

dad (parent link representation of forests), 43, 327–329, 444–448, 455–456.
Dags, 479–481.
Data structures, 3.
  abstract, 15, 31–33, 146, 154.
  adjacency lists, 420–423.
  adjacency matrix, 418–419.
  adjacency structure, 420–423.
  array, 15–17, 524.
  B-tree, 229, 262–265.
  binary search tree, 202–212, 243, 391–396.
  circular list, 21, 22.
  deque, 299–302.
  doubly-linked list, 21.
  heap, 148–161, 180–183, 327–328.
  linked list, 17–22, 523–524.
  priority queue, 145–162, 453–461, 465–468.
  pushdown stack, 25–29, 41, 45, 63–65, 122, 300–301, 428, 457.
  queue, 29–31, 300–301, 431, 457.
  red-black tree, 219–229.
  sorted list, 147, 195.
  string, 277.
  2-3-4 tree, 216, 229, 262.
  top-down 2-3-4 tree, 215–219.
  unordered list, 147, 195.
Data types, 10, 31–33.
Databases, 260, 373.
Decryption, 334–342.

Defensive programming, 86.
Delaunay triangulation, 409–410.
Deletion.
  in binary search trees, 209–211.
  in hash tables, 242.
  in graphs, 422.
Deo, N., 641.
Depth-first search.
  recursive, 423–427, 480–481.
  directed graphs, 472–475.
  nonrecursive, 427–430.
  priority-first search implementation, 457.
Depth-first search forest, 425, 427.
Deque, 299–302.
Deterministic algorithms, 634.
Dictionaries, 193.
Digital search trees, 245–248.
Dijkstra's algorithm (for finding shortest paths), 457.
Dijkstra, E., 457, 465, 506.
Directed graph, 418, 422.
Disk searching, 259.
Distribution counting, 111–113, 140, 143.
Divide-and-conquer. 53–59, 116, 402, 527–530, 532–533, 586–587.
Divide-and-conquer recurrences, 530.
Double buffering, 183.
Double hashing, 239–242.
Doubly-linked lists, 21.
Down edges, 472–473.
Driver programs, 9, 95.
Dummy nodes, 42–43; *see* Head node, *z*.
Duplicate keys; *see* equal keys.
Dynamic programming, 595–606.

Edelsbrunner, H., 412.
Edmonds, J., 490.
Empirical analysis of algorithms, 83–84.
Encryption, 334–342.
Encryption/decryption machines, 337–338.
End-recursion removal, 62, 122.
Equal keys, 194, 200, 247.
Eratosthenes, sieve of, 16, 32.
Escape sequence, 321.
Euclid's algorithm (for finding the gcd), 8–9, 12, 13, 340.
Euclidean minimum spanning tree, 467.
Euclidean traveling salesman problem, 629–631.
Exception dictionary, 242.
Exchange, 95.
Exhaustive search, 621–632.
Exponential running time, 70, 71, 621.
Exponentiation, 339–340, 525–526, 531–532.
Extendible hashing, 265–271.
External nodes, 36, 204, 249.
External path length, 37, 39–40.
External searching, 259–272.
  B-trees, 262–265.
  extendible hashing, 265–271.
  indexed sequential access, 260–262.
External sorting, 94.

Factorial, 51–52.
Factoring, 340.
Fagin, R., 265, 273.
Fast Fourier transform, 576, 583–593.
Fibonacci numbers, 52–53.
FIFO (first in, first out), 30, 49.
File compression, 319–331.
  Huffman decoding, 329–330.
  Huffman encoding, 324–330.
  variable-length encoding, 322–324.
Filters, 96.
Finite-state machine, 284, 285.
  deterministic, 295.
  nondeterministic, 295.
Flannery, B. P., 565.

Floyd's algorithm (for finding shortest paths), 477–478, 595.
Floyd, R., 477.
Ford, L. R., 487.
Forecasting, 184.
Forest, 37, 417, 442.
Forsythe, G. E., 565.
Forward elimination, 537–540.
4-node, 215–216.
Fractals, 59.
Fredman, M. L., 189.
Friedman, J. H., 412.
Fringe vertices, 431.
Fulkerson, D. R., 487.

Garbage collection, 25.
Garey, M. R., 641.
Gauss-Jordan reduction, 540, 542.
Gaussian elimination, 535–544, 553.
gcd; *see* Greatest common divisor, Euclid's method.
General regular-expression pattern matching, 301.
Geometric algorithms, 347–411, 467–468.
  closest-point problems, 401–411.
  convex hull, 359–372.
  elementary, 347–356.
  intersection, 389–399.
  range searching, 373–387.
Gerrymandering, 347.
Golin, M. J., 412.
Gonnet, G. H., 89, 189, 273.
Gosper, R. W., 278.
Goto, 100, 108, 118.
Graham scan, 365–371.
Graham, R. L., 89, 365, 412.
Graph algorithms, 415–508.
  all shortest paths, 476–478.
  breadth-first search, 430–434.
  connected components, 437–438.
  depth-first search, recursive, 480–481.
  Dijkstra's algorithm, 457.
  elementary, 415–436.
  Kruskal's algorithm, 458–460.
  matching, 495–505.
  minimum spanning tree, 452–461, 465–468.
  nonrecursive depth-first search, 427–430.
  Prim's algorithm, 457.
  priority-first search, 453–459.
  recursive depth-first search, 423–427.
  shortest paths, 461–468.
  shortest-path spanning tree, 455, 466, 467.
  stable marriage problem, 499–504.
  strongly connected components, 481–483.
  union-find, 441–449.
Graph, 416.
  adjacency matrix, 418–419.
  adjacency structure, 420–423.
  bipartite, 497–499.
  complete, 418
  directed, 418, 422.
  directed acyclic, 479.
  edge, 416.
  path, 417.
  representation, 418–423, 459–460.
  spanning tree, 417.
  undirected, 418.
  vertex, 416.
  weighted, 418, 422.
Graph isomorphism. 435.
Greatest common divisor, 8.
Greatest increment method, 616.
Grid method, 376–379, 385.
Guibas, L., 273.

Hamilton cycle problem, 622, 633–637.
Hash functions, 232–234.

Hashing, 231–244, 270, 289–290.
advanced methods, 242–243.
clustering, 239, 241.
deletion, 242.
double hashing, 239–242.
linear probing, 236–239.
open addressing, 236–242.
separate chaining, 234–236.
Head node (*head*), 18, 23, 25, 170, 197, 203–205, 226, 254, 256, 381, 383.
Head pointer, 30.
Heap, 148–161, 180–183, 327–328.
condition, 148.
indirect, 159–160.
algorithms, 147, 150–156, 161.
Heapsort, 153–161.
Hoare, C. A. R., 115, 189.
Hoey, D., 389, 412.
Hopcroft, J. E., 89.
Horner's method, 233, 525, 584.
Hu, T. C., 641.
Huffman decoding, 329–330.
Huffman encoding, 324–330.
Huffman frequency tree, 325–326.
Huffman, D. A., 324, 343.
Hybrid searching, 252–253.

Implementation of algorithms, 81–89.
Increment sequence, 108.
Indexed sequential access, 260–262.
Indirect binary search trees, 211–212.
Indirect priority queue, 455.
Indirect sort, 95, 147.
Infeasible linear program, 611.
Infix, 27, 40–42.
Inner loop, 85, 98, 118–119.
Inorder tree traversal; *see* Tree traversal.
Input/output, 11.
Insertion sort, 98–100, 103, 104–105, 125.
Integer linear programming, 638.
Integration, 555–564.
Interior elimination, 368–369, 371.
Internal nodes, 36, 204, 249.
Internal path length, 37, 39–40, 208, 220, 446.
Internal sorting, 94.
Interpolation search, 201–202.
Intersection, 389–399.
general lines, 396–399.
horizontal and vertical lines, 390–396.
two line segments, 349–351.

Jarvis, R. A., 412.
Jensen, K., 7, 89.
Johnson, D. S., 641.
Join operation in priority queues, 147, 161.
Josephus problem, 21–22, 32.

k-dimensional (*k*D) trees, 385–386.
Kahn, D., 343.
Karp, R. M., 279, 343, 490.
Key distribution, 338.
Keys, 94, 193.
Knapsack problem, 596–598, 626.
Knuth, D. E., 89, 189, 273, 278, 343, 506, 512, 515, 565.
Knuth-Morris-Pratt string searching, 280–285.
Konheim, A., 343.
Kruskal's algorithm (for finding the minimum spanning tree), 458–460, 467–468.
Kruskal, J., 458, 506.
Kung, H. T., 577.

Lagrange interpolation formula, 526–527, 584.
Large integers, 530–532.
Lazy deletion, 211.

Leading term, 70.
Leaf pages, 268.
Least-squares data fitting, 551–553.
Left recursion, 311.
Leftmost child, right sibling representation of forests, 44.
Lehmer, D., 511.
Lewis, H. R., 641.
$\lg N$ running time, 71.
LIFO (last in, first out), 30, 49.
Lin S., 631.
Line segment intersection, 349–351.
Line, 348.
Linear congruential random number generator, 511–514.
Linear feedback shift registers, 514–515.
Linear lists, 31–33.
Linear probing, 236–239.
Linear programming, 485, 607–620.
Linear running time, 70, 71.
Linked lists, 17–22, 523–524.
  circular, 21.
  deleting an item, 18–19.
  doubly-linked, 21.
  dummy nodes, 18.
  inserting an item, 18.
List mergesort, 166–168.
Lists; *see* Linked lists.
$\ln N$ running time, 71.
Logarithmic running time, 70.
Longest path, 633.
Lookahead, 309.
Lower bounds, 72.

MACSYMA, 565.
Malcomb, M. A., 565.
Manhattan geometry, 390.
Matching, 495–505.
Matrices, 17.
  inverse, 542.
  tridiagonal, 541–542, 549.
Matrix arithmetic, 532–533.
Matrix chain product problem, 598–602.
Matrix multiplication, 532–533, 577–580, 598–602.
Maxflow-mincut theorem, 489.
Maximum matching, 495–496.
Mazes, 434–435.
McCreight, E. 262, 273.
Mead, C. A., 641.
Median, 126–130.
Median-of-three partitioning, 126.
Mehlhorn, K., 273, 506.
Merge-until-empty, 186–187.
Mergesort, 163–176, 403–406.
Merging, 164–165, 573–577.
Method of least squares, 551–553.
Minimum spanning tree, 452–461, 465–468, 630–631.
  comparison of methods, 467.
  dense graphs, 466.
  sparse graphs, 455.
Mismatched character heuristic, 287–288.
Modulus computation, 531–532.
Moler, C. B., 565.
Morris, J. H., 278, 343.
Morrison, D. R., 253.
Multicommodity flow, 492.
Multidimensional range searching, 385–387.
Multiplication of large integers, 531.
Multiprocessor scheduling, 638.
Multiway merging, 178–180.
Multiway radix searching, 251–253.
Munro, I., 565.

$N \log N$ running time, 70, 71.
NP, 635–640.
NP-complete problems, 346, 633–640, 638–639.
NP-completeness, 636–637.

Nearest-neighbor problem, 401.
Network flow, 485–493.
Network, 487.
Nievergelt, J., 265, 273, 641.
Nondeterminism, 295, 634.
Nonrecursive (implementations of recursive programs).
  Fibonacci numbers, 53.
  depth-first search, 423.
  tree traversal, 60–65.
  mergesort, 170.
  parsing, 41, 311–312.
  quicksort, 122.
  selection, 128.
Nonterminal nodes, 36.
Numerical analysis, 10.

O-notation, 72.
Objective function, 608.
Odd-even merging, 574.
Open addressing, 236–242.
Operations research, 485.
Oppenheim, A. V., 641.
Optimization, 84–85.
Or, 294, 298.
Ordered hashing, 242.
Ottmann, T., 412.

P, 634–640.
Package wrapping, 362–365, 370–371.
Papadimitriou, C. H., 506, 641.
Parallel algorithms, 569–582.
Parallel arrays, 23, 212.
Parent link representation of forests, 43, 327–329, 444–448, 455–456.
Parse trees, 40–42, 306–307.
Parser generator, 316.
Parsing, 305–317.
  bottom-up, 311–312.
  recursive descent, 308–311.
  shift reduce, 312.
  top-down, 308–311.
Partition problem, 638.
Partitioning, 116–118, 126, 135–139.
  median-of-three, 126.
  tree, 124.
Pascal language, 7–13.
Patashnik, O., 89.
Path length.
  internal, 37, 39–40, 208, 220, 446.
  external, 37, 39–40.
  weighted external, 325–327.
  weighted internal, 603.
Patricia, 253–257, 291.
Pattern matching, 293–303.
Perfect shuffle, 575–577, 590–591.
Permutations, 101, 628–629.
Pippenger, N., 265, 273.
Pivoting, 538–540, 614–618.
Planarity, 435.
Point, 348.
Pointer sort, 95, 147, 159–160.
Polish notation, 27.
Polygon, 348.
  convex, 355, 359–372.
  simple closed, 351–353.
  standard representation, 348.
  test if point inside, 353.
Polynomials
  addition, 522–524.
  arithmetic, 522–530.
  evaluation, 524–526, 586–587.
  interpolation, 526–527, 546, 587–589.
  multiplication, 522, 527–530, 583–593.
Polyphase merging, 185–187.
Postfix, 27, 40–42.
Postorder tree traversal; *see* Tree traversal.
Pratt, V. R., 278, 343.
Preference lists, 500–504.
Preorder tree traversal; *see* Tree traversal.

Preparata, F., 412.
Preprocessing, 373–374.
Press, W. H., 565.
Prim's algorithm (for finding the minimum spanning tree), 457.
Prim, R. C., 457, 506.
Prime numbers, 289–290, 339.
Priority queue, 145–162, 453–459, 460–461, 465–468.
Priority-first search, 453–459, 463–464, 490–492.
Profiling, 68.
Program optimization, 84–85, 173–174.
Programming style, 86.
Projection, 376.
Pseudo-random numbers, 509.
Public-key cryptosystems, 338–341.
Pushdown stacks, 25–29, 41, 45, 63–65, 122, 300–301, 428, 457.

Quadratic running time, 70, 71.
Quadrature; *see* Integration.
Quasi-random numbers, 510.
Queues, 29–31, 300–301, 431, 457.
Quicksort, 115–126, 138, 143, 371.

RSA public-key cryptosystem, 338–341, 531–532.
Rabin, M. O., 279, 343.
Rabin-Karp string searching algorithm, 289–291.
Radix exchange sort, 135–139, 141.
Radix searching, 245–258.
  digital search trees, 245–248.
  equal keys, 247.
  multiway radix searching, 251–253.
  Patricia, 253–257.
  tries, 248–253.
Radix sorting, 133–144.
  radix exchange, 135–139, 141.
  straight radix, 139–142.
Random number generators.
  additive congruential, 515–516.
  bad, 513, 518.
  combination, 513, 518.
  linear congruential, 511–514.
Random numbers, 126, 509–520.
Range searching, 373–387.
  application to line intersection, 395, 398.
  grid method, 376–379, 385.
  k-dimensional (*k*D) trees, 385–386.
  multidimensional range searching, 385–387.
  preprocessing, 373–374.
  projection, 376.
  sequential search, 375.
  2-dimensional (2D) trees, 380–386.
Records, 193.
Rectangle method, 557–558.
Recurrence relations, 51–53, 76–78, 120–121, 208, 530.
Recursion removal, 61–65, 85, 122–124.
Recursive programs, 51–66, 174–175.
  call trees, 57, 406.
  closest-pair problem, 405.
  depth-first search, 423.
  left recursion, 311.
  mergesort, 166–168.
  quicksort, 119.
  recursive descent compiler, 312–315.
  recursive descent parsing. 308–311.
  selection, 127.
  tree definitions, 38.
Red-black trees, 219–229.
Reducibility, 636–637.
Regular expression, 294.
Reingold, E. M., 641.
Replacement selection, 180–185.
Representation.
  array for queues, 30.
  array for stacks, 28–29.

binary search trees, 202–204, 211–212.
binary trees, 40–43.
characters, 10.
direct-array for linked lists, 22–23.
directed graphs, 422.
finite state machine, 298.
forests, 43–44.
graphs, 418–423, 459–460.
heaps as arrays, 149.
large integers, 530–532.
linked list for stacks, 26–27.
matrices, 532.
numbers as bits, 134.
polynomials, 522–524.
priority queues, 145–162.
raster font, 321.
sets, 441.
trees as binary trees, 44.
weighted graphs, 422.
Reverse Polish notation, 27.
Rivest, R., 189, 338, 343.
Roberts, E., 89.
Rotation, 223–226.
Run-length encoding, 320–322.

Satisfiability, 635.
Schafer, R. W., 641.
Searching, 193–273.
binary, 198–200.
binary tree, 202–212.
external, 259–272.
hybrid, 252–253.
interpolation, 201–202.
radix, 245–258.
self-organizing, 197.
sequential, 195–197.
Sedgewick, R., 189, 273, 412.
Selection sort, 96–98, 101, 102, 104, 127.
Selection, 126–130.
Self-organizing search, 197.
Sentinel, 100, 108, 118, 137, 150, 164, 165, 166, 195, 291, 349, 363, 366.
Separate chaining, 234–236.
Sequential search, 375.
Sethi, R., 343.
Sets, 441.
Shamir, A., 338, 343.
Shamos, M. I., 389, 412.
Shellsort, 107–111, 116.
Shift-reduce parsing, 312.
Shortest paths, 457, 461–468, 633.
Shortest-path spanning tree.
comparison of methods, 467.
dense graphs, 466.
sparse graphs, 455.
Sieve of Eratosthenes, 16, 32.
Simple closed path, 351–353.
Simplex method, 609–619.
Simpson's method, 560–561.
Singular matrices, 538.
Sink, 487.
Slack variables, 613.
Sleator, D. D., 189.
Sort-merge, 178.
Sorting, 93–189.
bubble, 100–101, 102, 103.
disk, 187.
distribution counting, 111–113.
equal keys, 120.
external, 94, 177–188.
heapsort, 153–161.
indirect, 95, 104–107.
insertion, 98–100, 103, 104–105.
internal, 94.
large records, 104, 105–107.
linear, 142–143.
mergesort, 163–176.
method of choice, 110.
packaging, 95–96.
pointer, 95, 104–107.
priority queues, 145–162.

properties of elementary methods, 102–103.
quicksort, 115–126.
radix exchange, 135–139, 141.
selection, 96–98, 101, 102, 104, 127.
shellsort, 107–111.
stability, 94, 173, 175.
straight radix, 139–142.
tape, 177–187.
Source, 487.
Spanning tree, 417
Sparse matrices, 532, 541.
Sparse polynomials, 524.
Spline, 546–550.
interpolation, 546–550.
quadrature, 560, 561.
Split-and-interleave merging, 573–577.
Splitting, 216–218, 221–226, 264, 265, 266–271.
Stability in sorting, 94, 173, 175.
Stable marriage problem, 499–504.
Stack, 300, 457.
Standish, T. A., 343.
Steepest descent method, 616.
Steiglitz, K., 506, 641.
Stone, H. S., 641.
Storage allocation, 22–25.
Straight radix sort, 139–142.
Strassen's method (for matrix multiplication), 532–533, 543.
String searching, 277–292.
Boyer-Moore algorithm, 286–289.
brute-force, 279–280.
Knuth-Morris-Pratt algorithm, 280–285.
mismatched character heuristic, 287–288.
multiple searches, 291.
Rabin-Karp algorithm, 289–291.
Strings, 277.
Strong, H. R., 265, 273.
Strongly connected components, 481–483.
Supercomputer, 570, 634.
Sweep-line algorithm (for line intersection), 390–396.
Symbol tables, 193.
Symbolic integration, 556.
Syntax, 8.
Systems of simultaneous linear equations, 535–544, 548–549, 553.
Systems programming, 86–87, 183–185, 259–260.
Systolic arrays, 577–580.

Tail pointer, 30.
Tarjan, R. E., 189, 435, 449, 481, 506.
Terminal nodes, 36.
Teukolsky, S. A., 565.
Thompson, K., 343.
3-node, 215–216.
Top-down 2-3-4 trees, 215–219.
Top-down algorithms; *see* Recursive programs.
Topological sorting, 479–481.
Transitive closure, 473–476.
Trapezoid method, 559–560.
Traveling salesman problem, 352, 435, 621, 636.
Tree traversal, 44–49, 374.
inorder, 46, 60–61.
level order, 47.
postorder, 47.
preorder, 45, 62–65.
recursive, 60–65.
removing recursion, 60–65.
symmetric order, 46.
Tree vertices, 431.
Trees, 35–49, 417.
binary, 37–38.
complete, 148.
definitions, 35–39
external path length, 37, 39–40.

height, 37, 40.
internal path length, 37, 39–40, 208, 220, 446.
ordered, 37.
oriented, 39.
parent link representation, 444–448, 455–456.
parse, 40–42.
partitioning, 124.
properties, 38–40.
recursive definitions, 38.
rooted, 39.
traversal; *see* Tree traversal.
weighted external path length, 325–327.
weighted internal path length, 603.
Tries, 248–253, 323–324.
Turing machine, 638.
Two-dimensional (2D) trees, 380–386.
2-node, 216.
2-3 trees, 229.
2-3-4 trees, 216, 229, 262.

Ullman, J. D., 89, 273, 343.
Undirected graph, 418.
Uniform random numbers, 510.
Union-find, 441–449.
Unseen vertices, 431.
Up edges, 472–473.
Upper bounds, 68–69, 72.

Van Leeuwen, J., 506.
Variable-length encoding, 322–324.
Vectors, 17.
Vernam cipher, 336.
Vertex cover, 639.
Vertices.
fringe (in graph searching), 431.
in graphs, 416.
in trees and forests, 36.
tree (in graph searching), 431.
unseen (in graph searching), 431.
Vigenere cipher, 336.
Virtual memory, 187, 271.
Von Neumann, J., 569.
Voronoi diagram, 401, 408–410, 467.
Voronoi dual; *see* Delaunay triangulation.
Vuillemin, J., 189.

Warshall's algorithm (for finding the transitive closure of a graph), 595.
Warshall, S., 475.
Weighted graphs, 418, 422, 451–469, 496.
Weighted external path length, 325–327.
Weighted internal path length, 603.
Wells, M. B., 641.
Wirth, N., 7, 89.
Worst case analysis, 67, 71–74.

z, 18, 23, 25, 27, 41, 45, 48, 60–65, 85, 117, 165, 170, 203–205, 221, 226, 246–247, 254, 256, 381, 383, 520–524.

Epilog

The algorithms in this book have already been used for at least one application: producing the book itself. In large measure, when the text says "the above program generates the figure below," this is literally so. The book was produced on a computer-driven phototypsetting device, and most of the artwork was generated automatically by the programs that appear here. The primary reason for organizing things in this way is that it allows complex artwork to be produced easily; an important side benefit is that it gives some confidence that the programs will work as promised for other applications. This approach was made possible by recent advances in the printing industry, and by judicious use of modern typesetting and systems software.

The book consists of over a thousand computer files, at least one for each figure and each program, and one for the text of each chapter. Typesetting the book involves not only the normal work of positioning the characters in the text, but also running the programs, under control of the figure files, to produce high-level descriptions of the figures that can later be printed. This process is briefly described here.

Programs are individual Pascal files, written so that they can be bound into driver programs for debugging or bound into the text for printing. In the text, a program may be referenced directly for its text, in which case it is run through a formatting filter; or indirectly (through a figure file) for its output, in which case it is executed and its output directed to imaging software that draws a figure. During debugging, the program output was usually simplified, as described below, though sometimes bugs were easiest to see in the figures themselves.

The interface between the programs and the imaging software is a high-level one modeled on the method developed by Marc Brown and the author for an interactive system to provide dynamic views of algorithms in execution for educational and other applications. The algorithms are instrumented to produce "interesting events" at important points during execution that provide information about changes in underlying data structures. Associated with each figure is a program called a "view" that reacts to interesting events and produces descriptions for use by the imaging software. This arrangement allows each algorithm to be used to produce several different figures, since different views can react differently to the same set of interesting events. (In particular, debugging views that trace the progress of an algorithm are simple to build.) The procedure calls in the algorithms that signal interesting events do not appear in the text because they are filtered out in the formatting step.

The imaging package that produces the artwork itself was written specifically for the purpose of producing this book; it again was modeled on many of the visual designs that we developed for our interactive system, but was redone to exploit the high resolution available on the phototypesetting device used to print the book. This package actually resides on the printing device and takes as input rather high-level representations of data structures. Thus the printer arranges characters to form a paragraph at one moment; lines, characters and shading to form a tree, graph or geometric figure at the next. Typically, a figure file consists of the name of a view and a small amount of descriptive information about the size of the picture and the styles of picture elements. A view typically produces direct representations of data structures (permutations are lists of integers, trees are "parent-link arrays," etc.). The imaging software uses all this information to arrange major picture elements and attend to details of drawing.

In the first edition of this book, the figures were pen-and-ink drawings, because at that time it was difficult if not impossible to produce comparable drawings by computer. Now, it is difficult to imagine proceeding without the aid of the computer. Creating these figures with pen and ink would be a daunting task; it would even be difficult to write "by hand" the low-level graphic orders to create the images (recall that the algorithms in the book did most of that work). However, the most important contribution of the computer was not the production of the final images (perhaps that could be done some other way, somehow), but the quick production of interim versions for the *design* of the figures. Most of the figures are the product of a lengthy design cycle including perhaps dozens of versions.

An elusive goal for computer scientists in recent decades has been the development of an "electronic book" that brings the power of the computer to bear in the development of new communications media. On the one hand, this book may be viewed as a step back from interactive computer-based media into a traditional form; on the other hand, it perhaps may be viewed as one small step towards that goal.